FORMAL ENGINEERING DESIGN SYNTHESIS

The development of a new design is often thought of as a fundamentally human, creative act. However, emerging research has demonstrated that aspects of design synthesis can be formalized. First steps in this direction were taken in the early 1960s when systematic techniques were introduced to guide engineers in producing high-quality designs. By the mid-1980s these methods evolved from their informal (guideline-like) origins to more formal (computable) methods. In recent years, highly automated design synthesis techniques have emerged.

This timely book reviews the state of the art in formal design synthesis methods. It also provides an in-depth exploration of several representative projects in formal design synthesis and examines future directions in computational design synthesis research. The chapters are written by internationally renowned experts in engineering and architectural design. Among topics covered are: shape grammars, architectural design, evolutionary techniques, kinematic design, chemical and electronic design, MEMS design, design compilation, function synthesis, and the use of artificial intelligence in synthesis.

Covering topics at the cutting edge of engineering design, *Formal Engineering Design Synthesis* will appeal to engineering, computer science and architecture graduate students, researchers and designers.

Erik K. Antonsson is Executive Officer (Department Chair) and Professor of Mechanical Engineering at the California Institute of Technology, where he organized the Engineering Design Research Laboratory. He was a National Science Foundation Presidential Young Investigator from 1986 to 1992, and he won the 1995 Richard P. Feynman Prize for Excellence in Teaching. His research interests include formal methods for engineering design, design synthesis, representing and manipulating imprecision in preliminary engineering design, rapid assessment of early designs (RAED), structured design synthesis of microelectromechanical systems (MEMS), and digital micropropulsion microthrusters. Dr. Antonsson is a Fellow of the ASME, and he is a member of the IEEE, SME, ACM, ASEE, IFSA, and NAFIPS.

Jonathan Cagan is a Professor of Mechanical Engineering at Carnegie Mellon University, with appointments in the Schools of Design, Biomedical and Health Engineering, and Computer Science. His research, teaching, and consulting are in the area of design theory, methodology, automation, and practice, with an emphasis on the early stages of the design process. Dr. Cagan is a Fellow of the ASME and a member of the Phi Beta Kappa, Tau Beta Pi, and Sigma Xi national honor societies and IDSA, AAAI, SAE, and ASEE professional societies. He is the recipient of the National Science Foundation's NYI Award and the Society of Automotive Engineers Ralph R. Teetor Award for Education.

Formal Engineering Design Synthesis

Edited by

ERIK K. ANTONSSON
California Institute of Technology

JONATHAN CAGAN
Carnegie Mellon University

CAMBRIDGE
UNIVERSITY PRESS

CAMBRIDGE UNIVERSITY PRESS
Cambridge, New York, Melbourne, Madrid, Cape Town, Singapore, São Paulo

Cambridge University Press
The Edinburgh Building, Cambridge CB2 2RU, UK

Published in the United States of America by Cambridge University Press, New York

www.cambridge.org
Information on this title: www.cambridge.org/9780521792479

First published 2001
This digitally printed first paperback version 2005

A catalogue record for this publication is available from the British Library

Library of Congress Cataloguing in Publication data

Formal engineering design synthesis / edited by Erik K. Antonsson,
    Jonathan Cagan.
        p.   cm.
    Includes bibliographical references and index.
    ISBN 0-521-79247-9
    1. Engineering design.   I. Antonsson, Erik K.   II. Cagan, Jonathan, 1961–
    TA174 .F67   2001
    620′.0042 – dc21   2001025164

ISBN-13 978-0-521-79247-9 hardback
ISBN-10 0-521-79247-9 hardback

ISBN-13 978-0-521-01775-6 paperback
ISBN-10 0-521-01775-0 paperback

*"Scientists investigate
that which already is;
Engineers create that
which has never been."*

Albert Einstein

Contents

Contributing Authors

Erik K. Antonsson is Professor and Executive Officer (Dept. Chair) of Mechanical Engineering at the California Institute of Technology, Pasadena, CA 91125-4400, where he founded and supervises the Engineering Design Research Laboratory; he has been a member of the faculty since 1984.

Jonathan Cagan is Professor of Mechanical Engineering at Carnegie Mellon University, Pittsburgh, PA 15213, and Director of the Mechanical Engineering Computational Design Laboratory; he holds appointments in the Schools of Design, Computer Science and Biomedical and Health Engineering.

Scott D. Barnicki is a Research Associate with the Eastman Chemical Company, P. O. Box 1972, Kingsport, TN 37662-5150.

Georges G. E. Gielen is a Professor in the Department of Electrical Engineering at the Katholieke Universiteit Leuven, Kasteelpark Arenbert 10, 3001 Leuven, Belgium, working in analog and mixed-signal integrated circuit design and design automation.

James L. Greer is an Assistant Professor in the Department of Engineering Mechanics at The United States Air Force Academy, 2354 Fairchild Dr., Suite 6H2, Colorado Springs, CO 80840-2944.

Leo Joskowicz is an Associate Professor at the School of Computer Science and Engineering, Ross Building, Room 223, The Hebrew University of Jerusalem Givat Ram, Jerusalem 91904, Israel.

Kenneth Kotovsky is a Professor in the Psychology Department at Carnegie Mellon University, Pittsburgh, PA 15213.

Cin-Young Lee is a Mechanical Engineering Ph.D. candidate in the Engineering Design Research Laboratory at the California Institute of Technology, Pasadena, CA 91125.

Lin Ma is a Mechanical Engineering Ph.D. candidate in the Engineering Design Research Laboratory at the California Institute of Technology, Pasadena, CA 91125.

Lionel March is a Professor in Design and Computation at the University of California, 1200 Dickson, Box 951456, Los Angeles, CA 90095-1456.

J. Michael McCarthy is a Professor in the Department of Mechanical Engineering at the University of California, 4200 Engineering Gateway, Irvine, CA 92697-3975.

William J. Mitchell is a Professor of Architecture and Media Arts and Sciences and the Dean of the School of Architecture and Planning at the Massachusetts Institute of Technology, 77 Massachusetts Ave., Room 7-231, Cambridge, MA 02139.

Panos Y. Papalambros is the Donald C. Graham Professor of Engineering and Professor of Mechanical Engineering at The University of Michigan, 2266 G. G. Brown Lab, Ann Arbor, MI 48109-2125.

Rob A. Rutenbar is Professor of Electrical and Computer Engineering and (by courtesy) Professor of Computer Science at Carnegie Mellon University, Pittsburgh, PA 15213.

Kristina Shea is University Lecturer in the Department of Engineering at the University of Cambridge, Trumpington St., Cambridge CB2 1PZ, U.K.

Jeffrey J. Siirola is a Technology Fellow at the Eastman Chemical Company, P. O. Box 1972, Kingsport, TN 37662-5150.

Herbert A. Simon* is a Professor in the Departments of Computer Science and Psychology at Carnegie Mellon University, Pittsburgh, PA 15213.

Thomas F. Stahovich is an Associate Professor in the Department of Mechanical Engineering at Carnegie Mellon University, Pittsburgh, PA 15213.

George Stiny is a Professor of Design and Computation in the Department of Architecture at the Massachusetts Institute of Technology, 77 Massachusetts Ave., Room 10-431, Cambridge, MA 02139.

Allen C. Ward is an independent consultant on lean development processes at Ward Synthesis, Inc., 15 1/2 East Liberty St., Ann Arbor, MI 48104.

Kristin L. Wood is a Professor of Mechanical Engineering in the Mechanical Systems and Design Division of the Department of Mechanical Engineering at The University of Texas at Austin, ETC 5.140, Austin, TX 78712-1063.

---

* We note with sorrow that Prof. Simon passed away while this books was in press.

Foreword

Lionel March

There are too few authors who reflect on the act of designing. Generally designers, when they publish, like to show off the designs they have made. One time-honored way of instructing in design is by exemplar – the case history approach. Another approach takes so-called 'principles' – usually rules of thumb – which have worked in the past (in most cases). Two centuries ago, the architect Rondelet wrote: "*Théorie est une science qui dirige toutes les opérations de la pratique.*"[1] (Rondelet, 1830) Two millenia ago, the Vitruvian vision was one in which architect and engineer were indistinguishable (Vitruvius, 1914). While today that is generally speaking no longer the case, it gives me special delight to introduce a book on design where the authors hail from both disciplines: authors who, often in lonely circumstances, are contributing to the development of theoretical foundations for the act of designing. Their otherwise separate and isolated efforts are brought together between these covers by two distinguished editors. It is this confluence of ideas that impresses: perhaps, a trickle now, but a potential torrent tomorrow.

There is an apt quotation by Machiavelli that heads the Preface. In the case of theoretical contributions to design, history confirms a persistent resistance to change. Girard Desargues is a good example (Desargues, 1636, 1639; Field and Gray, 1987). He reflected upon the spatial representations that engineers and architects, such as himself, employed when designing. His projective methods enabled him not only to short-cut, but also to add precision to traditional craft techniques. For this he was shown the school door and abandoned to pace the streets posting bills to broadcast his results. Two centuries on, Gaspard Monge was disciplined by his commanding officer for breaking protocol in calculating the *defilade* for a fortress in far too short a time (Aubry, 1954; Monge, 1799a, 1799b; Taton, 1950, 1951). A military secret for some years, Monge's descriptive geometry was to enable many of the advances in mechanical engineering design throughout the nineteenth century.

At the same time, also at L'Ecole Polytechnique, J. N. L. Durand was educating architects in no more than two months of intermittent instruction over two years by adopting a systematic approach to designing: to a knowledge of the architectural elements, to the laws of their combination, and finally to the mechanics of

---

[1] "Theory is a science which directs all the operations of practice."

composition (Durand, 1801, 1891; Hernandez, 1983; Villari, 1990). His books influenced many more than the students in his classes. Franz Reuleaux achieved similar results for the design of machines by defining kinematic elements and a notation for their combination (Reuleaux, 1876). It was he who used the term 'design synthesis' by contrast to the hit-or-miss approach of workshop design. A remarkable demonstration of such design synthesis is Felix Wankel's near exhaustive enumeration of rotary engines (Wankel, 1965). With its colored components and combinatory arrangements, Wankel's achievement reminds me of much more recent work – Lynn Conway and Carver Mead's rules for designing VLSI chips (Mead and Conway, 1980). Not having the $A \rightarrow B$ form for a computational rule, their 'rules' are effectively declarations in which the λ-parameter may be varied. The practical success of this method cannot be denied, yet at the start the methodology came under attack even within the very institution that saw its touch-and-go birthing. Conway, herself, admits that the method is less than optimal, but that the pragmatic consequences of teachability, the demystification of expertise, and rapid design response are worthwhile gains. And so they have proven.

By way of precedents, Conway admires the achievements of Claude Shannon for his Boolean – combinatorial – representation of all possible circuit configurations (Shannon, 1948; Sloane and Wyner, 1993; Weaver and Shannon, 1963) and of Charles Steinmetz at General Electric for devising a symbolic method to simplify and democratize the designing of AC apparatus (Steinmetz, 1920). The later work of Gabriel Kron (Happ, 1973; Kron, 1953), also at General Electric, on large electromagnetic-mechanical systems, employing Poincaré's combinatorial geometry and Clifford algebras to represent and 'tear' a multidimensional system into relatively independent subsystems, continues to be largely ignored by those whose intellectual conservatism finds comfort in the simplifications of linear systems and combinatorial enumerations. Both Steinmetz and Kron made significant contributions to the library of symbolic methods available for the analysis of performance once a form for a design has been delineated; essentially, for tweaking the parameters. But here, surely, is the crux of design synthesis – the novel shaping of candidate schemata to be analyzed in the first place. The analysis of design performance may well shelter under the big tent of 'economic' theorizing, but the shaping of form must weather the climes of creative actions without.

Design synthesis is the primary objective of this book. In its earliest chapters, the book declares a paradigmatic break from symbolic representation and the combinatorial tradition in which design is assumed to be a search for a workable assembly of given parts. At its sharpest, this radical approach says that such parts are themselves an after-thought, a particular decomposition of the design, a consequence of one's perspective looking back over the generative computation. The new radicalism declares that in design there is no necessity for components to be defined a priori, that design is consequentially more than a combinatorial search among rearrangements, and that ambiguity is the essence of the creative act of designing.

Of course, not everyone writing here will agree with so radical a position. But they do all agree that design synthesis is open to formal investigation, and most papers confront one or more of these issues. This is a contentious volume, and so it should be. Formal design synthesis is arguably one of the most exciting and challenging intellectual frontiers ahead of us. Among these authors, you will find a notable advance party of frontiersmen.

REFERENCES

Aubry, P. V. (1954). *Monge, le savant ami de Napoléon Bonaparte, 1746–1818.*

Desargues, G. (1636). *Traité de la section perspective.* (Treatise on the Perspective Section).

Desargues, G. (1639). *Brouillon project d'une atteinte aux événements des rencontres d'un cône ave un plan.* (Proposed Draft of an Attempt to Deal with the Events of the Meeting of a Cone and a Plane.)

Durand, J. N. L. (1801). *Recueil et Parallèle des Edifices de tout genre Anciens et Modernes.* Paris.

Durand, J. N. L. (1819). *Precis des Leçons d'Architecture données à l'Ecole Royale Polytechnique.* Firmin Didot, Paris, 1802–1805, revised 1819.

Field, J. V. and Gray, J. J., Eds. (1987). *The Geometrical Work of Girard Desargues.* Springer-Verlag, New York–Berlin.

Happ, H. H., Ed. (1973). *Gabriel Kron and Systems Theory.* Union College Press, Schenectady, NY.

Hernandez, A. (1983). J. N. L. Durand's architectural theory: A study in the history of rational building design. *Perspecta, 12*:153–9.

Kron, G. (1953). A method to solve very large physical systems in easy stages. *IRE Transactions on Circuit Theory, 2*(1):71–90.

Mead, C. and Conway, L. (1980). *Introduction to VLSI Systems.* Addison-Wesley Publishing Company, Reading, MA.

Monge, G. (1799a). *Feuilles d'analyse appliqueée à la géométrie.* (Analysis Applied to Geometry).

Monge, G. (1799b). *Géométrie descriptive.* (Descriptive Geometry).

Reuleaux, F. (1876). *The Kinematics of Machinery: Outlines of a Theory of Machines.* MacMillian and Co., London. Translated by Alexander B. W. Kennedy.

Rondelet, J.-B. (1830). *Traité su l'are de bâtir.* Chez A. Rondelet, Paris.

Shannon, C. (1948). A mathematical theory of communication. *Bell System Technical Journal, 27*:379–423 and 623–56.

Sloane, N. J. A. and Wyner, A. D., Eds. (1993). *Claude Elwood Shannon: Collected Papers.* IEEE Press, Piscataway, NJ.

Steinmetz, C. P. (1920). *Theory and Calculation of Transient Electric Phenomena and Oscillations.* McGraw-Hill, New York.

Taton, R. (1950). *Gaspard Monge.* Basel.

Taton, R. (1951). *L'Oeuvre scientifique de Monge.*

Villari, S. (1990). *J. N. L. Durand (1760–1834). Art and Science of Architecture.* New York.

Vitruvius, P. (1914). *The Ten Books on Architecture.* Harvard University Press, Cambridge, MA. (First Century, BC), Translated by Morris Hicky Morgan.

Wankel, F. (1965). *Rotary Piston Machines: Classification of Design Principles for Engines, Pumps, and Compressors.* Iliffe, London. Translated and edited by R. F. Ansdale.

Weaver, W. and Shannon, C. (1949, 1963). *The Mathematical Theory of Communication.* University of Illinois Press, Urbana, IL.

Preface

> It must be considered that there is nothing more difficult to carry out, nor more danger-
> ous to conduct, nor more doubtful in its success, than an attempt to introduce changes.
> For the innovator will have for his enemies all those who are well off under the existing
> order of things, and only lukewarm supporters in those who might be better off under
> the new.
>
> – Niccolò Machiavelli,
> *The Prince and The Discourses*, 1513,
> Chapter 6

The idea for this book was initially sparked by a review of prior publications describing the status of engineering design research, particularly, Prof. Yoshikawa's
1985 collection of papers titled *Design and Synthesis*.[1] This review suggested that
a publication presenting the status of design research at the change of the century
would be valuable. Immediately, however, it became clear that the range and scope
of design research are such that no single collection could be thorough and comprehensive without limiting its scope to a portion of the state of the art in engineering
design research. The portion that appears to have the greatest promise, and that has
established its foundations since 1984, is formal engineering design synthesis: the
automatic creation of design configurations.

Over single-malt whiskeys one evening in April 1999, between sessions at George
Stiny's Workshop on Shape Computation, we discussed the idea with enthusiasm.
Thus this edited book project was born. We felt then (and continue to feel now) that
formal design synthesis methods are beginning to be well established, that a sufficient
foundation for research into these methods now exists (perhaps for the first time),
and that productive research in this important area is likely to grow significantly in
the decades ahead.

We felt that we could solicit and include representative work from several areas of
engineering and architecture, surveying leading work in these areas and presenting
an overview of methods being used, as well as illustrating the details of at least one
method in each area. We thought that each invited contributor could also provide an

---

[1] H. Yoshikawa, Ed. *Design and Synthesis*, 1985, North-Holland/Elsevier. (Proceedings of the International Symposium on Design and Synthesis, Tokyo, Japan, July 11–13, 1984.)

outlook of the future for the area from his or her perspective. The collection would serve the present research community and provide a future historical reference to this emerging field.

We then set to perhaps the most critical element of this project, selecting the authors to invite to contribute. We scoured the research literature for work in this area, and had the difficult task of choosing authors to invite. Our choices were guided by several factors. Each author had to be engaged with leading the development of methods for automated synthesis of engineering design configurations, and be able to provide a thoughtful overview of the related work of others. Additionally, we sought contributors who express themselves clearly in writing, and who have a commitment to and thoughtful perspective on the future of formal engineering design synthesis research.

We remain gratified by the enthusiastic responses we received from the contributors, and by their patience with our review and editing of their contributions. The encouragement and thoughtful suggestions of Florence Padgett and Ellen Carlin at Cambridge University Press were the final crucial ingredients. This volume is the happy result of these efforts.

– Erik K. Antonsson and Jonathan Cagan

Introduction

Erik K. Antonsson and Jonathan Cagan

"A science of design not only is possible but is actually emerging at the present time."

Herbert A. Simon, 1916–2001,
The Sciences of the Artificial,
1969, MIT Press, Cambridge

This book is dedicated to describing the most current advances in the state of automated engineering design synthesis. From early references to the subject in *The Sciences of the Artificial* (Simon, 1969), to the significant recent accomplishments of the community, the field is moving into prominence in both ideas and application. This area, as well as the relationship of the words in the title, *Formal Engineering Design Synthesis*, are described below.

Design. Creation of new things is an activity that is crucial to life. Humans and animals build houses to protect their families, as well as develop strategies and tactics for gathering food. Humans have extended this notion beyond the reproduction of known configurations to the creation of novel things. This endeavor is referred to as *design*, and it has been practiced for centuries in many disciplines, both artistic and technical. All of the ancient human cultures have stunning examples of design in the areas of art, architecture, engineering, and war.

Engineering Design. Engineering design is a subset of these design activities. It focuses on the technical aspects of performance of designed systems, rather than the aesthetic. Thus, engineering design concerns itself, for example, with the creation of the structure to support the innovative design of the architect, or with the creation of machines to carry out the novel strategies and tactics of a wartime leader, or with the creation of structures and systems to support and convey people and goods in commerce and daily life.

Engineering Design Synthesis. Engineering design encompasses many activities, including the creative activity of conceiving new devices and systems, as well as the analysis, refinement, and testing of those concepts. Synthesis is the creative step itself: the conception and postulation of possibly new solutions to solve a problem. In most engineering design, this step is performed by creative human minds.

Why *design synthesis*? Isn't this pair of words redundant? No. Design encompasses many activities, creative as well as analytical, and thus the *synthesis* aspects of design are those that create candidate solutions. Synthesis, by itself, can be taken to mean many different things, from the physical combination of atoms and molecules into a new molecule, to the combination of existing parts or elements into a larger structure or system. The two words together (with *design* modifying *synthesis*) means that portion of the design process having to do with the creation of new candidate design concepts. The three words together (*engineering* modifying *design* modifying *synthesis*) means that portion of the engineering design process having to do with the creation of new alternatives and candidate concepts.

Design synthesis is not limited to combinations of existing elements (atoms or points), but intrinsically requires a nonanalytical flexible approach to determining what the elements of a given design are, and the ability to reapportion and reconfigure existing elements and to create new ones.

> [Durand and Guadet] promulgated a view of architectural composition that was essentially particulate and combinatorial.
>
> ...they would develop the design by deploying elements from an established vocabulary of construction elements....
>
> W. Mitchell, Chapter 1

Both William J. Mitchell (in Chapter 1) and George Stiny (in Chapter 2) provide powerful and cogent arguments for why design synthesis cannot be limited to a "Lego-like" composition of existing elements, and perhaps even point to the limitations of the structured approach of VDI 2221 (VDI, 1987) where design requirements are broken down into functional subelements, and then embodiments that provide the subfunctions are combined into the completed design.

Formal Engineering Design Synthesis. *Formal* in this context means computable, structured, and rigorous, not ad hoc. It is the goal of research into formal engineering design synthesis methods not only to develop procedures that can synthesize novel engineering designs but also to understand the fundamental principles by which these methods work, so that this understanding can be applied to new systems and design environments.

> The working knowledge of professionals is almost universally considered intrinsically informal, hence unteachable except by experience. If we express working knowledge formally, in computational terms, we can manipulate it, reflect on it, and transmit it more effectively.
>
> H. Abelson and G. J. Sussman, *Computation: An Introduction to Engineering Design* (1990).
>
> By formal, we mean that the process is founded in a theory, set of theories, or set principles.
>
> K. Wood and J. Greer, Chapter 6
>
> This lets me calculate in the kind of milieu that's expected for creative activity....
>
> G. Stiny, Chapter 2

In the early 1960s, engineering design methodology underwent a renaissance. Methods began to be developed to guide engineers through a process to produce high-quality designs. In the mid-1980s these methods began to evolve from their informal (guideline-like) origins to more formal (i.e., computable) methods. Recently, the foundations of methods to automatically synthesize new designs have begun to be developed.

BACKGROUND

In addition to defining the terms of the title and the focus of the book, a description of *engineering*, its relationship to other technical activities, and the special role of design in engineering are described below.

Engineering. Engineering is the creation of new things. Engineering is distinguished from science, in that the goal of science is the discovery of the behavior and structure of the existing universe; engineering is the creation of new things.

> Science is the study of what *Is*,
> Engineering builds what *Will Be*.

> The scientist merely explores
> that which exists,
> while the engineer creates
> what has never existed before.
>
> T. Von Kármán (ca. 1957)

> The *central* activity of engineering,
> as distinguished from science,
> is the design of new devices, processes and systems.
>
> M. Tribus, *Rational Descriptions, Decisions and Designs* (1969)

As described above, design is the creation of new devices or systems. Although engineering design relies heavily on analysis to establish the behavior or performance of candidate designs, fundamentally (as shown in Figure I.1) the procedure is the reverse of analysis.

Analysis. In analysis, the procedure begins with a physical device or system, and a model (usually computable) is built of the system by decomposing, abstracting, and approximating the system. The objective is to produce a useful description of the behavior of an existing system.

anal·y·sis *n, pl* -y·ses [NL, fr. Gk, fr. *analyein* to break up, fr. *ana-* + *lyein* to loosen – more at LOSE] (1581) **1:** separation of a whole into its component parts **2 a:** the identification or separation of ingredients of a substance **b:** a statement of the constituents of a mixture **3 a:** proof of a mathematical proposition by assuming the result and

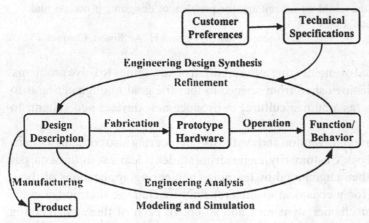

Figure I.1. A flow chart of the engineering design process, distinguishing analysis from synthesis.

deducing a valid statement by a series of reversible steps **b (1):** a branch of mathematics concerned mainly with functions and limits **(2):** CALCULUS 1b **4 a:** an examination of a complex, its elements, and their relations **b:** a statement of such an analysis **5 a:** a method in philosophy of resolving complex expressions into simpler or more basic ones **b :** clarification of an expression by an elucidation of its use in discourse **6:** the use of function words instead of inflectional forms as a characteristic device of a language **7:** PSYCHOANALYSIS

Merriam-Webster's Collegiate Dictionary (2000)

Design. Design (or design synthesis) is the reverse of the analytical process, as shown in Figure I.1. Design begins, not with a description of a device or system, but rather with a description of a desired function or behavior (generally not yet available from an existing system). The objective is to produce a description of a system that will exhibit the desired behavior.

de·sign \di-'zīn*vb* [ME to outline, indicate, mean, fr. M F & M L; M F *designer* to designate, fr. M L *designare*, fr. L, to mark out, fr. *de-* + *signare* to mark—more at SIGN] *vt* (14c) **1:** to create, fashion, execute, or construct according to plan: DEVISE, CONTRIVE **2 a:** to conceive and plan out in the mind <he ~*ed* the perfect crime> **b:** to have as a purpose: INTEND <she ~*ed* to excel in her studies> **c:** to devise for a specific function or end <a book ~*ed* primarily as a college textbook> **3** *archaic*: to indicate with a distinctive mark, sign, or name **4 a:** to make a drawing, pattern, or sketch of **b:** to draw the plans for ~*vi* **1:** to conceive or execute a plan **2:** to draw, lay out, or prepare a design

Merriam-Webster's Collegiate Dictionary (2000)

The goal of science is analytical. That is, the scientific process is to hypothesize a structure, behavior, or mechanism of behavior for an existing system (physical, molecular, chemical, atomic, biological, geological, etc.), and then to conduct experiments to confirm or refute this theory. Although synthesis is clearly an important part of the scientific process (the design of an experiment, the creation of a hypothesis, etc.), in this context, the focus of science is analytical in that its goal is to explore and discover the structure and/or behavior of existing system(s).

They are both the product of human minds trying to solve ... the problem of discovering the nature of the world we inhabit and the problem of designing processes and artifacts that 'improve' that world.

J. Cagan, K. Kotovsky, and H. A. Simon, Chapter 13

At its heart, the goal of engineering is the creation of new things to solve problems. It is fundamentally distinguished from science in that the goal of engineering is to utilize what we know (as a human culture) to produce new devices and systems to solve problems.

Just as science contains invention and synthesis, engineering also contains a significant element of analysis. Customarily, engineering students learn scientific analysis methods, which are then augmented by learning engineering applications of these methods, including (for mechanical engineers) fluid mechanics, thermodynamics, heat transfer, solid mechanics, dynamics, and so on. In each of these engineering disciplines, students are taught how to analyze the behavior of given systems.

With a background in analytical methods, engineering students become ready to apply these analytical methods to solving engineering design or synthesis problems. Here the problem is, essentially, the reverse of analysis problems.

Engineering design synthesis starts with a description of a desired behavior, and the goal is to create a system that will exhibit the desired behavior. Notice that this is exactly the reverse of analysis, where the problem starts with an existing system, and the problem is to establish its behavior. Because engineering design requires synthesis of new solutions, and because it is essentially the reverse of analysis, it depends on methods and approaches that are fundamentally different from either scientific or engineering analysis. The same, of course, is true for synthesis of scientific hypotheses and experiments.

Synthesis. The definition below appears to rely on the notion of combining, or "synthesizing," existing elements into a new configuration; however, synthesis is legitimately a broader notion, one that encompasses the creation of new things in general, and not necessarily only by means of combination of existing elements.

> **syn·the·sis** *n, pl* **-the·ses** [Gk. fr. *syntithenai* to put together, fr. *syn-* + *tithenai* to put, place – more at Do] (1589) **1 a:** the composition or combination of parts or elements so as to form a whole **b:** the production of a substance by the union of chemical elements, groups, or simpler compounds or by the degradation of a complex compound **c:** the combining of often diverse conceptions into a coherent whole; *also:* the complex so formed **2 a:** deductive reasoning **b:** the dialectic combination of thesis and antithesis into a higher stage of truth **3:** the frequent and systematic use of inflected forms as a characteristic device of a language – syn·the·sist *n*
>
> *Merriam-Webster's Collegiate Dictionary* (1994)

The core of formal engineering design synthesis is the act of synthesis itself. In the context of this book the focus is on formal, computable, algorithmic synthesis. Each of the chapters presents formal, repeatable, well-bounded, and well-articulated approaches to design creation. A primary aspect of synthesis is compilation, the act of composing a design from parts. Though none is more explicit than in Allen C. Ward's chapter, all of the work presented in this book addresses synthesis approaches through compilation or other means of building designs. One aspect of synthesis, often limited by compilation yet found in formal approaches to creative design, is *emergence*, the coming into existence of features, behaviors, or geometries not articulated in the original concept or anticipated in the formal representation.

SUMMARY

Success in developing automated design synthesis methods has been achieved in several domains. The domains are narrow (e.g., *digital* very large-scale integration, or VLSI), and other related domains have not achieved similar success (e.g., the automated synthesis of analog or power electronics, or other areas of electrical engineering). This suggests that future design synthesis methodologies are likely to also be focused on similarly narrow domains. However, the existence of operating synthesis methods, coupled with the recent theoretical advances (described in this volume) produce an environment that is ripe for further advances and successes. Research in this area is already being applied to other challenging and exciting fields, including protein design (Dahiyat and Mayo, 1996, 1997; Regan and Wells, 1998; Gordon and Mayo, 1999; Street and Mayo, 1999; Pierce et al., 2000; Rossi et al., 2000; Voigt et al., 2000; Zou and Saven, 2000).

Emerging research has demonstrated that aspects of synthesis can be formalized and the foundations now exist to actively pursue highly automated synthesis techniques. Our goal, as editors of this volume, is to capture a picture of the most advanced state of the art in formal engineering design synthesis. Our motivation is the many recent advances in this field (summarized by the contributing authors) and the promise for further advances.

This collection serves three purposes:

- first, to provide a context by reviewing the state of the art in formal design synthesis methods;
- second, to provide an in-depth exploration of several representative projects in formal design synthesis;
- third, to examine future directions in computational design synthesis research.

This collection covers the following topics: shape grammars and their application in engineering; stochastic optimization; topology optimization; the relationship between invention and discovery; evolutionary techniques; kinematic design; structural, chemical, electronic, and MEMS design; architectural design; design compilation and formal theories of design; the use of artificial intelligence in synthesis; and the representation and use of function for synthesis.

Each of the 13 chapters that follow presents a different topic in this field, reviews its history, presents the latest work, and projects advances that might reasonably be expected to come about in the future. The chapters, the topics they address, their relationship to other chapters, and the author(s) of each chapter are briefly introduced below.

THE CHAPTERS

The 13 chapters are sequenced to move from philosophical issues of shape and shape grammars through applications and methods to philosophical issues of relationships between engineering fields and across seemingly disparate areas of thought.

The book begins with three chapters on shape synthesis based on shape grammars. Shape grammars are a form of production system, but one that supports geometric representation and reasoning, parametric representations, and emergent shape properties. The work on shape grammars originated in the field of architecture, and so the book begins with a chapter by William J. Mitchell that lays out the many issues of synthesis of space and shape that involve humankind. This discussion illustrates the formality and logic of architectural design, but it also argues that such design goes beyond particulate composition. Unfortunately, most computer-aided support for design restricts consideration for the particulate rather than the ambiguous, emergent properties so prevalent in human (here, architectural) design. The chapter argues for the shape grammar representation, introduced in Stiny and Gips (1972) and in Gips and Stiny (1980), as a means to represent, support, and realize the more natural design process.

Chapter 2 by George Stiny discusses in depth the development of and theory behind shape grammars by providing deep insight into issues surrounding ambiguity in design, in general, and in shape, in particular. Certainly architectural design differs from engineering design in its freedom of form creation and more decoupled connection to physical behavior, and its strong connection to the aesthetics. However, many of the ideas, insights, and methods introduced in these first two chapters transfer to the engineering synthesis process.

Chapter 3 by Jonathan Cagan surveys the history and state of the art of the transfer of shape grammars to the engineering community. In addition to showing the emerging success of the approach, the chapter delves into many issues that effect the design of engineering shape grammars as well as potential areas for growth.

Chapter 4 by Panos Y. Papalambros and Kristina Shea focuses on nontraditional developments in the more traditional optimization field as applied to engineering structural synthesis. The chapter nicely explores continuous and discrete structures. The book flows into this application area through many emerging topics in the field, including the use of shape grammars for discrete structural design along with other continuous and discrete approaches to structural topology synthesis. One issue addressed in this chapter, again, is bringing aesthetics into the engineering equation.

Next, Erik K. Antonsson looks further at structural design by means of the synthesis of microelectromechanical systems (MEMS) in Chapter 5. Here, the application of advanced design synthesis methods to microsystems is reviewed, and the relationship of these advances to digital VLSI design synthesis is discussed. Two methods are discussed in detail: an approach based on interconnected combinations of primitive parametrically scalable surface micromachined elements, and a stochastic exploration methodology.

Kristin Wood and James L. Greer then present an exposition on functional representation and its use in automated design systems in Chapter 6. The chapter begins with a much broader discussion of the design process, enabling a unique perspective on the impact of each method surveyed in the chapter. Topics for future work motivate the field to broaden its scope for inclusion in the overall design process.

Thomas F. Stahovich then gives an insightful overview of the field of artificial intelligence (AI), surveying the history of the field as applied to design in Chapter 7. The chapter focuses on techniques for search, knowledge-based systems, machine learning, and qualitative physics, taking the reader through a tutorial, historical review, and view of the future for each area.

Cin-Young Lee, Lin Ma, and Erik K. Antonsson in Chapter 8 look in detail at evolutionary or stochastic design in general, and genetic algorithms in particular, reviewing the state of the art and demonstrating the potential of this approach on a synthesis example.

Chapter 9 by J. Michael McCarthy and Leo Joskowicz focuses on the kinematic synthesis of machines. Mechanism design dates back many hundreds of years. With the advent of computers, their analysis and parameter selection has warranted a significant amount of investigation and success. More recently the field has focused on mechanism synthesis, with many novel ideas emerging into application. This chapter reviews the state of the art in this area and lays out directions for further work.

Although the book begins with the field of architecture, the majority of the discussion in the book centers on the field of mechanical engineering by mechanical engineers. The reason for this focus is twofold: First, the scope of mechanical design is varied, as compared to other types of engineering, leading to a broader array of tools and concepts for use in design; second, as implied, mechanical engineering has many more unanswered issues in automated synthesis. The next two chapters, however, focus on specific nonmechanical engineering domains in which automated synthesis has found great success. In Chapter 10, Scott D. Barnicki and Jeffrey J. Siirola present a survey on chemical process synthesis, namely the transformation of one chemical composition or form of matter into another. The chapter looks at that branch of chemical engineering concerned with the systematic generation of flow sheets for the design, operation, and control of chemical processes.

Georges G. E. Gielen and Rob A. Rutenbar next survey the field of synthesis for analog and mixed-signal integrated circuits in Chapter 11. In the domain of circuits, synthesis came first to digital designs, which enjoy much simpler representations, and models that insulate more of design from the difficult behavior of the underlying semiconductor technology. Recently, shrinking transistor sizes have made it possible to integrate both analog and digital functions on a single chip. However, whereas digital circuits hide the nonlinearities of the underlying fabrication technology, analog circuits harness and exploit these nonlinearities, and use them to connect digital logic to the external world. This chapter explores the motivation for and evolution of analog circuit synthesis, and it contrasts these techniques with more established digital approaches. It concludes with a description of the current state of the art and suggests some emerging directions.

The last two chapters return to a philosophical discussion of design. First, in Chapter 12, Allen C. Ward examines why mechanical engineering is different from other areas of engineering for which synthesis tools are well accepted and used. The chapter explores why compilers for mechanical systems have not worked as well as their electronic counterparts, and what must to be done to change that.

The book ends with a chapter by Jonathan Cagan, Kenneth Kotovsky, and Herbert A. Simon, which explores cognitive models of design invention and scientific discovery, arguing that the cognitive processes used in both activities are the same. The implication of their argument here is that the vast research done on automated scientific discovery may lend insights into techniques for automated design, and that the work discussed in this book may find broader use in the artificial intelligence community. The chapter argues that there is a tight relationship between engineering and science, and a broader outlook by both communities could expand the way automated engineering and science are both accomplished.

REFERENCES

Dahiyat, B. and Mayo, S. (1996). Protein design automation, *Protein Science*, **5**(5):895–903.

Dahiyat, B. and Mayo, S. (1997). *De novo* protein design: Fully automated sequence selection, *Science*, **278**(5335):82–7.

Gips, J. and Stiny, G. (1980). Production systems and grammars: A uniform characterization, *Environment and Planning B*: *Planning and Design*, **7**:399–408.

Gordon, D. and Mayo, S. (1999). Branch-and terminate: A combinatorial optimization algorithm for protein design, *Structure with Folding and Design*, **7**(9):1089–98.

Pierce, N., Spriet, J., Desmet, J., and Mayo, S. (2000). Conformational splitting: A more powerful criterion for dead-end elimination. *Journal of Computational Chemistry*, **21**(11):999–1009.

Regan, L. and Wells, J. (1998). Engineering and design recent adventures in molecular design – Editorial overview. *Current Opinion in Structural Biology*, **8**(4):441–2.

Rossi, A., Maritan, A., and Micheletti, C. (2000). A novel iterative strategy for protein design. *Journal of Chemical Physics*, **112**(4):2050–55.

Simon, H.A. (1969). *The Sciences of the Artificial*. 1st ed., MIT Press, Cambridge, MA.

Stiny, G. and Gips, J. (1972). Shape grammars and the generative specification of painting and sculpture. In *Information Processing 71*, C.V. Freiman (ed.), North-Holland, Amsterdam, pp. 1460–65.

Street, A. and Mayo, S. (1999). Computational protein design. *Structure with Folding and Design*, **7**(5):R105–9.

VDI (1987). Systematic approach to the design of technical systems and products. Technical report, Verein Deutscher Ingenieure Society for Product Development, Design and Marketing: Committee for Systematic Design, D-1000 Berlin, Germany: Copyright, VDI-Verlag GmbH, D-4000 Dusseldorf.

Voigt, C., Gordon, D., and Mayo, S. (2000). Trading accuracy for speed: A quantitative comparison of search algorithms in protein sequence design. *Journal of Molecular Biology*, **299**(3):789–803.

Zou, J. and Saven, J. (2000). Statistical theory of combinatorial libraries of folding proteins: energetic discrimination of a target structure. *Journal of Molecular Biology*, **296**(1): 281–94.

CHAPTER ONE

Vitruvius Redux

Formalized Design Synthesis in Architecture

William J. Mitchell

PARTICULATE COMPOSITION

A quarter of a century ago I published a paper titled "Vitruvius Computatus" (Mitchell, 1975). In it I argued that, if computer systems achieved the capacity to execute nontrivial architectural design tasks, we could "expect the designs which emerge from them to be characterized by particular stylistic traits." I went on to suggest, more specifically, "If computer systems become architects they may be academic classicists, heirs to Durand and Guadet." Sadly, I turned out to be right.

Jean N. L. Durand and Julien Guadet were great French architectural pedagogues, and authors of highly influential texts that were used by generations of students.[1] They provided models to copy, and they promulgated a view of architectural composition that was essentially particulate and combinatorial. Within the tradition that Durand and Guadet established and formalized, designers relied heavily upon abstract ordering devices such as grids and axes, and they would typically begin a project by exploring various combinations of these devices to produce an organizing skeleton of construction lines (Figure 1.1). Guided by this skeleton, they would then consider alternative ways to arrange the major rooms and circulation spaces. Finally, they would develop the design by deploying elements from an established vocabulary of construction elements – columns, entablatures, doors, windows, and so on. As Durand's famous plates illustrated (Figure 1.2), it was a recursive process of top-down substitution.

Thus, learning to design was much like learning Latin prose composition. There was a well-defined vocabulary of discrete construction elements, there were rules of syntax expressed by allowable combinations of grids and axes (the graphic equivalent of sentence schemata found in grammar textbooks), and you could produce acceptable designs by arranging the construction elements according to these rules. Many of Durand's plates were devoted to illustrating alternative "horizontal combinations" (plans) and "vertical combinations" (elevations) that could be produced in

---

[1] See, in particular, Durand's *Précis des Leçons d'Architecture* (Durand, 1819), his *Partie Graphique des Cours d'Architecture* (Durand, 1821), and Guadet's *Éléments et Théories de l'Architecture* (Guadet, 1902).

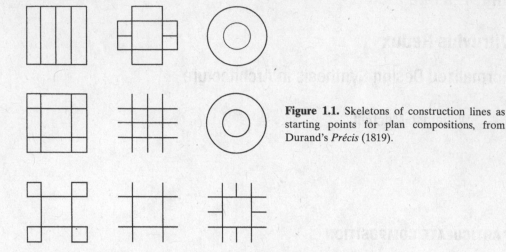

Figure 1.1. Skeletons of construction lines as starting points for plan compositions, from Durand's *Précis* (1819).

this way. In addition to generating a design according to Durand's rules, you could run the process in reverse to parse a completed design into a hierarchy of definite elements and subsystems, just as you could parse a Latin sentence into phrases and parts of speech.

The great figures of early architectural modernism mostly represented themselves publicly as anticlassicists, but they did not forget the intellectual legacy of Durand and Guadet. As Reyner Banham noted, Guadet's *Éléments et Théories de l'Architecture* (1902) "formed the mental climate in which perhaps half the architects of the twentieth century grew up." The compositional strategy that they followed

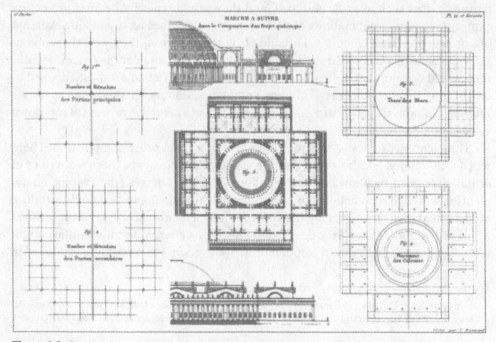

Figure 1.2. Design as a process of top-down substitution, from Durand's *Précis* (1819).

could be characterized as follows:

> The approach is particulate; small structural and functional members (elements of architecture) are assembled to make functional volumes, and these (elements of composition) are assembled to make whole buildings. To do this is to compose in the literal and derivational sense of the word, to put together.
>   R. Banham, *Theory and Design in the First Machine Age* (1967)

The new conditions of industrial mass production clearly strengthened this idea of composing a building from discrete standard parts. The difference, now, was that the drive to achieve economies of scale (rather than the pedagogical strategy of publishing models to copy) motivated standardization and repetition. The extreme response to these conditions was the creation of industrialized component building systems that provided complete, closed "kits of parts" for assembling entire buildings. At a smaller scale, construction toys such as Froebel and Lego blocks reproduced the idea of particulate composition at a smaller scale. Proponents of these systems frequently illustrated their capabilities by producing exhaustive enumerations of possible combinations of elements.

THE BEGINNINGS OF ARCHITECTURAL CAD

Not surprisingly, when the first computer-aided design (CAD) systems appeared, they reified these well-established design traditions.[2] They allowed drawings to be organized in "layers," which reflected both the earlier practice of drafting on overlaid sheets of translucent paper and Durand's notion of successive layers of abstraction. They provided grids, construction lines, and "snap" operations for positioning graphic elements in relation to the established skeleton of a composition. They also relied heavily upon the operations of selecting standard graphic elements, then copying, translating, and rotating them to assemble compositions from repeated instances.

At a deeper level – that of their underlying data structures – these systems also assumed definite, discrete elements. The basic geometric units were points, specified by their Cartesian coordinates. Lines were then described by their bounding points, polygons by their bounding lines, and closed solids by their bounding polygons.[3] More complex graphic constructions were treated as sets of these basic geometric entities, and nongeometric properties (color, mass, material properties, cost information, etc.) might be associated with such groupings. Subshapes were simply subsets, and complete drawings could unambiguously be parsed into hierarchies of subshapes and geometric entities – all the way down to the elementary points. There were many variants on this basic scheme – some of them very clever – but they were all grounded upon the idea of sets of geometric entities encoded as relations on sets of points (Figure 1.3).

Because the usual goal of early CAD systems was efficiency in the production of construction documents, many went a step further and provided higher-level vocabularies of architectural elements such as standard walls, doors, windows,

---

[2] For details of early architectural CAD systems, see Mitchell (1977).

[3] This strategy is grounded in the pioneering work of Steven Coons and Ivan Sutherland. See Stiny, Chapter 2, for a critical discussion of the theoretical foundations.

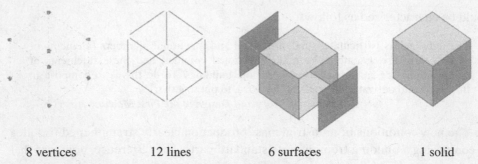

| 8 vertices | 12 lines | 6 surfaces | 1 solid |

Figure 1.3. Selectable geometric entities in a typical CAD system.

columns, desks and chairs, mechanical and electrical equipment items, and so on (Figure 1.4). These became the elements that an architect could quickly select, transform, and assemble in a process that directly (though frequently unknowingly and uncritically) recapitulated Durand. In other words, these systems assumed and enforced a conventionalized architectural ontology. This was reasonable if you were prepared to accept the ontology, but not if you wanted to stretch the boundaries of architectural discourse. You were out of luck if you wanted to explore an architecture of free-form blobs, or of nonrepeating, nonstandard parts, for example.

At an implementation level, however, there were many practical advantages to this approach. It was conceptually simple, and it supported the creation of elegant and efficient graphics pipelines that took advantage of increasing computer resources. It lent itself to some useful elaborations, such as parametric CAD. It also benefited from new approaches to software engineering, such as object-oriented programming.

In contrast, the limitations were obvious enough from the beginning – though, as it turned out, little reflected upon or heeded in what became the headlong post-PC rush to commercialize and market standard computer graphics technology. In "Vitruvius Computatus" I warned:

> Is this continuity of the academic classical tradition of elementary composition in a rather unexpected way merely a historical curiosity of little practical consequence? I suggest not. It indicates that computer systems cannot be regarded as formally neutral tools in the design process. On the contrary, their use can be expected to demand acquiescence to particular formal disciplines. This implies several things. Firstly it becomes clear that the design and development of satisfactory computer-aided design

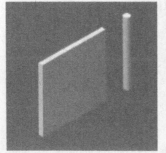 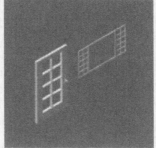 

Figure 1.4. Typical vocabulary elements of an architectural CAD system.

systems is not simply a technical task to be left to systems analysts, engineers, and programmers. It demands a high level of *architectural* skill and capacity for rigorous analysis of built form. Secondly, as architects begin increasingly to work with such systems they will need to devote as much attention to understanding the vocabulary and syntax of form implied by the data structures of those systems as the classical architects of old devoted to study of the orders.

W. Mitchell (1975)

The consequences of implicitly encoding stylistic presuppositions in software eventually showed up in built work. As one prominent and sophisticated designer recently remarked to me, "If you just look at the work of many young designers, you can immediately see what sort of CAD system they are using."

LANGUAGES OF ARCHITECTURAL FORM

Just as Durand conceived of design as a process of recursive substitution of shapes terminated by the insertion of elements from an established architectural lexicon, formal linguistics taught us to understand sentence construction as the recursive substitution of symbols terminated by the insertion of words from a written language's vocabulary. The frequently invoked analogy between classical architectural design and Latin prose composition turns out to be more than superficial!

With the development of the shape grammar formalism, Stiny and Gips (1972) made the idea of shape substitution precise, and they cast it in modern computational form. Whereas formal generative grammars (as the concept is normally understood in linguistics and computer science) generate one-dimensional strings of symbols, shape grammars generate two-dimensional or three-dimensional (or, indeed, *n*-dimensional) arrangements of shapes. They operate by recursively applying shape rewriting rules to a starting shape. A language of designs thus consists of all the shapes that can be derived in this way.

Stiny and Mitchell (1978a, 1978b) first applied shape grammars to nontrivial architectural design by writing a grammar that produced schematic villa plans in the style of the great Italian renaissance architect Andrea Palladio. In some sense, this grammar captured the essence of Palladio's style, and it allowed others to design in that style – thus functioning pedagogically much as Palladio's own *Four Books of Architecture* had been intended to do. Stiny and Mitchell demonstrated the grammar by exhaustively enumerating (at a certain level of abstraction) the plans in the language that it specified. The resulting collection of possibilities included the villa plans that were actually produced by Palladio, together with others that might have been if there had been the time and the opportunity (Figure 1.5).

Subsequently, there have been many other attempts to write grammars that produce particular classes of architectural designs and to enumerate the languages they specify. Numerous of these have been published, over the years, in the journal *Environment and Planning B*. In 1981, for example, Koning and Eizenberg (1981) demonstrated a grammar that produced convincing "prairie houses" in the style of Frank Lloyd Wright (Figure 1.6). A particularly compelling recent example is Duarte's grammar (Duarte, 2000a, 2000b) of house designs in the style of Alvaro

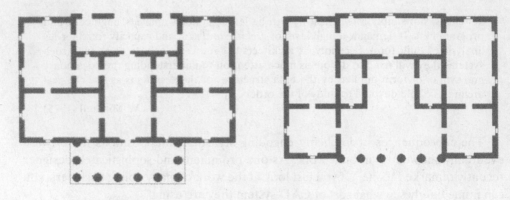

Villa Hollywood   Villa Vine

Figure 1.5. Plans of the Villa Hollywood and the Villa Vine, two of the pseudo-Palladian villas generated by Stiny and Mitchell's grammar.

Siza's Malagueira project (Figure 1.7). Duarte's grammar is presented in the form of a Web site that allows house clients to derive designs that respond to their own particular needs and budgets.

COMBINATORIAL SEARCH

The most obvious way to implement a shape grammar (or some functionally equivalent set of generative rules expressed in a different way) is to begin with the sort of data structure common to CAD systems, as described earlier. Subshapes are thus treated as subsets of geometric elements, and rewriting rules replace one subset with another. This is all a fairly straightforward extension of the computer science tradition dealing with formal languages, automata, production systems, and the like. And it also allows architectural design problems to be assimilated into the intellectual

Figure 1.6. A prairie house in the style of Frank Lloyd Wright, generated by Koning and Eizenberg's grammar.

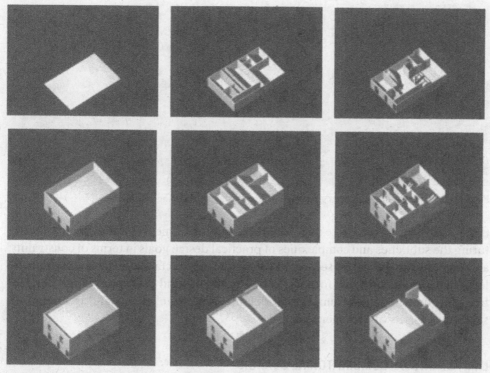

Figure 1.7. Step-by-step derivation, according to Duarte's grammar, of a Malagueira house.

traditions of AI problem-solving (as represented, in particular, by the pioneering work of Newell and Simon) and combinatorial optimization.[4]

Within this framework, design problems can be formulated as problems of searching discrete (or partially discrete) state-spaces to find designs in a language that satisfy specified criteria. (Recall Durand's "horizontal combinations" and "vertical combinations" – variants adapted to different site contexts, functional requirements, and budgets.) These criteria can be expressed, in the usual ways, in the form of constraints and objective functions. The search tasks that result are typically large scale and nastily *np* complete. Heuristic search, simulated annealing, evolutionary strategies, and other such techniques are thus employed to control the search process and produce acceptable results with a reasonable expenditure of computational resources.

This combinatorial approach had its early successes. For example, Mitchell, Steadman and Liggett (1976) produced an efficient and reliable system for designing minimal-cost house plans – much like Siza's Malagueira plans – that satisfied specified area and adjacency requirements for the constituent rooms. More recently, Shea and Cagan (1997, 1999) demonstrated the possibility of deriving unusual but practical and efficient structural designs by means of a shape grammar encoding rules of stability in pin-jointed structures, finite-element analysis of alternatives that are generated for consideration, and the use of simulated annealing to control the search

[4] My early architectural CAD text *Computer-Aided Architectural Design* (Mitchell, 1977) explicitly makes this connection and provides many examples.

Figure 1.8. Unusual dome designs generated by Shea's shape annealing procedure.

(Figure 1.8). However, there were several difficulties. One was the difficulty of capturing the subtleties and complexities of practical design goals in terms of constraints and objective functions. A second was the typical intractability of the search problems that result from such formulations. The third – and most profound – derived from the fundamental assumption of definite, discrete elements and treatment of subshapes as sets of such elements.

LIMITATIONS OF COMBINATORIAL SEARCH

To see the nature of this last difficulty, consider the shape shown in Figure 1.9.[5] It is from master mason Mathias Roriczer's famous pamphlet on the design of cathedral pinnacles (Roriczer, 1486).

Roriczer described its construction, in detail, as follows.

> Would you draw a plan for a pinnacle after the mason's art, by regular geometry? Then heave to and make a square as it is here designated by the letters A, B, C, D. Draw A to B, and B to D, and from D to C, and from C to A, so it may be as in the given figure [Figure 1.10].

> Then you make another square: divide A to B in two equal parts, and place E; likewise, divide BD and there make H; and from D to C and there make an F; similarly from C to A, and there make a G. After that draw a line from E to H, and from H to F, F to G, G to E. An example is in the following figure [Figure 1.11].

> After that you make like the one made above another square: divide FH into two equal parts and there put a K. Similarly on HF place M; also on FG make L, similarly on GE place I. After that draw a line from K to M, from M to L, from L to I, from I to K, as in the following figure [Figure 1.12].
> M. Roriczer, *On the Ordination of Pinnacles* (1486)

It is easy to execute this construction in a standard CAD system, such as AutoCad; the result is a shape represented as 12 lines. Any one of these lines may then be selected and transformed or deleted, to produce design variations of the kind shown in Figure 1.13. (To make this exercise interesting, of course, you have to restrict

[5] See also Stiny, Chapter 2.

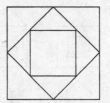

Figure 1.9. Mathias Roriczer's construction of nested squares.

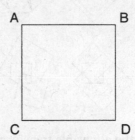

Figure 1.10. The first step in Roriczer's construction procedure.

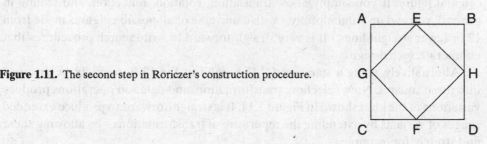

Figure 1.11. The second step in Roriczer's construction procedure.

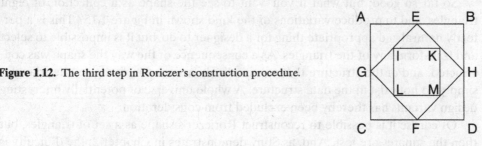

Figure 1.12. The third step in Roriczer's construction procedure.

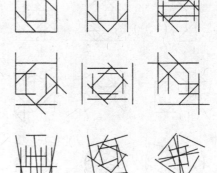

Figure 1.13. Variations on Roriczer's shape produced by selecting and transforming straight lines.

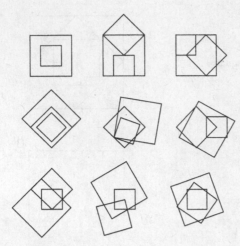

Figure 1.14. Variations on Roriczer's shape produced by selecting and transforming squares.

the allowable transformations such that you preserve some of the properties of the original figure. If you simply allow translation, rotation, reflection, and scaling in general, you end up – unhelpfully – with a universe of all possible figures made from 12 or fewer straight lines.) It is very straightforward to write search procedures that enumerate such designs.

Alternatively, using a standard CAD capability, the 12 lines might be grouped into three squares. Now, selection, transformation, and deletion operations produce variations of the kind shown in Figure 1.14. It is straightforward to produce extended ranges of variants by extending the repertoire of transformations – by allowing shear and stretch, for example.

So far so good; but what if you want to see the shape as a collection of eight triangles, and to produce variations of the kind shown in Figure 1.15? This is a perfectly natural and appropriate thing for a designer to do, but it is impossible to select and transform any of the triangles. As a consequence of the way the shape was constructed, and of the structure that was implicitly assigned in doing so, the triangles simply do not exist in the data structure. A whole universe of potentially interesting design variants has thereby been excluded from consideration.

Of course it is possible to reconstruct Roriczer's shape as a set of triangles, but then the squares are lost. And, as Stiny demonstrates in Chapter 2, the difficulty is

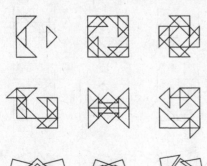

Figure 1.15. Variations on Roriczer's shape produced by selecting and transforming emergent triangles.

far from a trivial annoyance. There are indefinitely many ways to decompose even a very simple shape into subshapes, and any one of these decompositions may be appropriate to a designer's purpose at a particular moment in a design process.

So the classicists and the early modernists got it insidiously wrong. So did the developers of CAD systems, and the authors of design enumeration algorithms, who uncritically accepted the classical idea of particulate composition. Creative "rereading" of shapes, and the subsequent production of variants based upon such rereadings, is an important part of manual design processes. Any computational scheme that prematurely imposes a definite way to parse a composition into parts and subparts will inappropriately constrain a designer's capacity for creative generation of alternatives. It will become a Procrustean bed. The data structures of standard CAD systems, and search strategies that rely upon such data structures, suffer from a very fundamental limitation; they may be useful within their limits, but they do not support a satisfactory general approach to creative design synthesis.

DESIGN DEVELOPMENT AND ASSIGNMENT OF STRUCTURE

Now, however, consider how you might go about *physically* constructing Roriczer's shape – not simply drawing it. Your procedure would obviously depend upon available materials, fabrication techniques, and assembly strategies. If you had some long, rectangular beams, some nails, and a saw, for example, you might create the three-layered structure shown in Figure 1.16. (This is, in fact, a fairly common traditional strategy for framing roofs from heavy beams of limited lengths.) In other words, you would first assign a definite structure – let us call it the "carpenter's structure" – to the shape, and then you would fabricate discrete pieces corresponding to the elements of that structure, and finally you would compose the complete shape by assembling the fabricated elements.

Next, suppose you had a lot of short rods and pin-jointing elements – as in many construction toys. In this case, you might place a jointing element at each node and run rods between, as illustrated in Figure 1.17. Obviously this requires seeing Roriczer's

Figure 1.16. Roriczer's shape developed as a set of heavy roof beams.

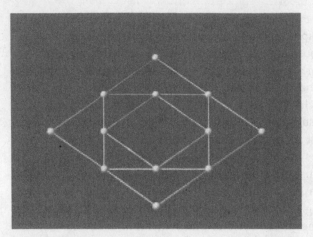

Figure 1.17. Roriczer's shape developed as a pin-jointed space frame.

shape in a different way. And when you see it in this way, you can also quickly see that it might be stretched and folded into the third dimension to produce a beautiful dome (Figure 1.18) – not something that is readily suggested by the carpenter's structure. (However, the carpenter's structure does, of course, suggest *other* interesting things.)

What if you wanted to create the structure from precast concrete components – producing the more complex joints in the factory, and leaving the simpler ones for on-site assembly? You might cast K-shaped and L-shaped elements, as shown in Figure 1.19. Yet another structure – maybe we should call it the "system-builder's structure" – is thereby assigned to Roriczer's shape.

Finally, suppose you had a square piece of plywood and a laser cutter. You might simply cut out four small triangles, to produce the result shown in Figure 1.20. This would particularly have appealed to certain New York painters in the heyday of hard-edge geometric abstraction; let us call it the "minimalist's structure."

Roriczer himself had something else in mind. He thought of the outer square as the plinth of a pinnacle shaft. The second square gave him the size of the shaft, which

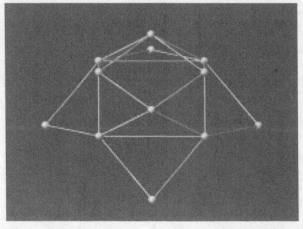

Figure 1.18. The pin-jointed space frame stretched and folded to produce a dome.

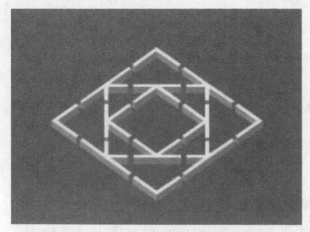

Figure 1.19. Roriczer's shape developed as an assembly of K-shaped and L-shaped precast concrete elements.

worked out to half the area of the plinth. The third square gave the size of the faces of the panels of the shaft. He went on to rotate the second square by 45°, add some additional construction, and produce the plan for a carved stone pinnacle as shown in Figure 1.21.

These examples illustrate a general principle of design invention and development. In the early stages of architectural design processes – when architects freely explore possibilities by producing quick, rough sketches – ambiguity of structure and preservation of the possibility of rereading seem crucial. However, sketches are not buildable. As a design is developed, then, and as an architect begins to think in terms of specific materials, components, and fabrication and assembly possibilities, a building design acquires an increasingly definite structure. An individual sensibility about space, materials, construction, and light comes more prominently into play. Finally, the construction documents describe the design as an assembly of definite

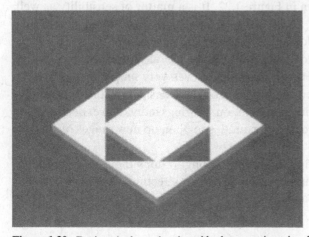

Figure 1.20. Roriczer's shape developed by laser-cutting triangles from a flat sheet.

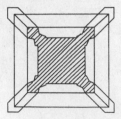

Figure 1.21. Roriczer's own development of the shape as the plan for a cathedral pinnacle.

elements and hierarchically nested subsystems. Ambiguity is a good thing to have at the beginning, but a bad thing to leave there at the end.[6]

FORMAL AND FUNCTIONAL EMERGENCE

Preservation of ambiguity in the early stages of design performs the essential function of allowing a designer to recognize and take advantage of formal and functional *emergence*. For example, if you construct Roriczer's shape by selecting, transforming, and instantiating squares – as an experienced CAD operator might – you never explicitly insert any of the eight visually evident triangles; they simply emerge. Conversely, if you rather perversely construct the shape by selecting, transforming, and instantiating right triangles, then the three squares emerge.

At one level, as was demonstrated in Figure 1.15, recognition and transformation of emergent subshapes supports formal invention. However, there may also be a functional dimension. Recall, for example, that pin-jointed squares are not structurally stable shapes but that pin-jointed triangles are. Thus, recognition of the triangles immediately tells you that you can construct a version of the shape as a pin-jointed frame (as shown, for example, in Figure 1.17 – understood as a truss operating in a vertical plane), that it will then turn out to be structurally stable and statically determinate, and that you will be able to analyze the forces in the members by using elementary techniques of statics.

If you are interested in different functions, you will look for and find different emergent subshapes. Thus, if you read Roriczer's shape as the *parti* for a floor plan, and you know that closed polygons can function as rooms, you can quickly pick out the potential room shapes shown in Figure 1.22. It is a matter of sensibility as well; if you are a postmodernist with a taste for funkiness, then these room shapes may appeal to you, but Durand would have looked straight past them.

Examples could be multiplied indefinitely, but the moral should now be obvious. There are indefinitely many emergent subshapes in even very simple shapes, and the recognition of particular emergent subshapes provides a designer with opportunities to apply knowledge of construction and function. It brings individual experience and sensibility into play, and it provides opportunities to open up new ranges of variants for consideration.

Skilled designers do not simply search for configurations that satisfy predetermined requirements. They watch out for emergent architectural opportunities, they recognize them, and they take advantage of them to achieve unexpected benefits.

---

[6] To put this another way, a design process should begin within the framework of a shape grammar, but it should end up within the framework of a set grammar.

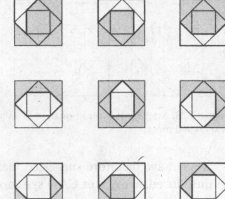

Figure 1.22. Some of the potential room shapes that emerge from Roriczer's shape.

In this way, they frequently achieve unprogrammed goals – desirable features that nobody specified at the outset, and perhaps that were unimaginable before they presented themselves.

Technically, the step-by-step development of a design from abstract and ambiguous shape to precise specification for fabrication and on-site construction work involves recognizing emergent subshapes, reflecting upon their formal, functional, and constructional possibilities, progressively assigning definite structure, and thus progressively fixing functional and constructional properties.

MECHANISMS FOR PROGRESSIVE ASSIGNMENT OF STRUCTURE

Standard architectural drafting techniques employed at successive stages in a typical design process crudely but effectively illustrate this progressive assignment of structure. Early sketches are typically highly ambiguous; they serve as foci for exploration, discussion, and reflection, but they avoid making commitments to details of materiality and construction. At the end of design schematics, single-line representations of walls have become double-line ones, dimensions are precise, and some indication of material and construction choices is given. At the design development stage, specific choices of components and materials are made. And at the construction documents stage, choices of components and materials are precisely and exhaustively documented – often as products chosen from manufacturers' catalogues. At each stage, there are increased opportunities for useful engineering and cost analyses, and for applying search and optimization algorithms.

The usual way to handle the progressive assignment of structure, whether with manual drafting or with a standard CAD system, is to redraw the project at each stage, using the mechanism of superimposed "layers" to transfer critical position and dimension information from one stage to the next. (Drawings are usually made at successively larger scales, as well, so that more detail can be shown.) Figure 1.23, for example, illustrates typical stages in the transformation of a sketched shape into a configuration of walls and rooms (one that might have been produced, for example, by a disciple of Louis Kahn) that is to be constructed in a specific way. This informal and pragmatic strategy works, but it is crude and cumbersome.

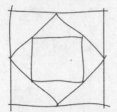 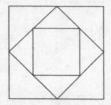 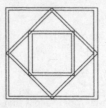 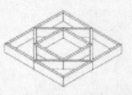

Figure 1.23. Stages in the transformation of a sketched shape into a precise and detailed configuration of walls and rooms.

Many have therefore suggested that digital freehand "sketch" systems should be introduced, in place of CAD systems, at early design stages. It does help to provide a more facile user interface than that of a CAD system, and to allow rough indications of dimensions and geometric relationships rather than precise geometric constructions. However, these provisions deal with only one, rather superficial aspect of ambiguity. Freehand sketching systems miss most of the point unless they provide for structural ambiguity as well as wiggly lines.[7] Sketch recognition systems that interpret wiggly lines as standard shapes and conventional architectural elements just compound the problem. What's really needed is some representational machinery that begins with no commitment to definite parts and structure but that allows such commitment to be made progressively and gracefully as a design process unfolds.

Stiny's theoretical machinery for shape representation and transformation, as discussed in Chapter 2, provides one elegant approach to this task. Within Stiny's system, shapes are transformed by rules that rewrite subshapes. Even seemingly useless rules that rewrite subshapes as identities actually perform the very useful task of assigning structure. Applying a rule expresses a commitment to henceforth "see" a shape in some particular way. Design development may therefore be treated as a process of applying rules that are derived from knowledge of functional, material, fabrication, and on-site assembly possibilities.

EMERGENCE AND CAD/CAM

So far we have focused upon manifestly orderly shapes and compositions made mostly from straight lines, arcs of circles, right angles, simple geometric figures, symmetries, and repeating elements. That is consistent with the traditions of architectural classicism and high modernism, and with the logic of mass production. However, languages of architectural form are now changing; curved-surface modeling and CAD/CAM technologies make it increasingly feasible to fabricate complex curves and curved surfaces, and to construct buildings from nonrepeating elements. Frank Gehry's Bilbao Guggenheim Museum (Figure 1.24) provides a dramatic illustration of this tendency.

---

[7] Wiggliness is much overrated. Long ago, one of my students discovered that he could, by the simple expedient of inserting a loose pen into a plotter, fool professors into believing they were looking at "sensitive" freehand sketches rather than "mechanistic" computer drafting.

Figure 1.24. Frank Gehry's Bilbao Guggenheim Museum, a composition of curved surfaces and nonrepeating elements.

Such radical buildings challenge the traditional architectural ontologies that are built into standard architectural CAD systems, and they render such systems almost useless in the design process. Wall and roof insertion operations are pointless, for example, when there is no clear distinction between walls and roofs. Copy-and-transform operations do not help when there is little repetition.

It might be thought that the use of a commercial curved-surface modeling system, such as that intended for use in product design and computer animation, would solve the problem. (In other words, replace an ontology of boxes with one of blobs.) But in fact, this merely reinstates Durand, yet again, in trendier clothing. The issue of emergence remains – and is perhaps even more crucial.

In the exploration and development of such designs, many complex and subtle emergent shapes may play roles. Generally these shapes do not fall into familiar categories, or have well-known names like "square," "triangle," "arch," or "Doric column." Many begin as abstract shapes made from free-form curves, and are eventually formalized – during design development – as NURBS or the like. Some of them end up being identified as construction elements and they are eventually fabricated by CAD/CAM machinery driven by the three-dimensional CAD model. They were not and could not be elements of some predefined vocabulary, and they were not combined in some Durand-like way to produce the complete composition.

Consider, for example, the metal cladding panels of Gehry's Experience Music Building in Seattle (Figure 1.25). The nonrepeating shapes of the surface panels are not instances of predefined vocabulary elements, but unanticipated shapes that emerged during the design process. Digital descriptions of these shapes were then sent to computer-controlled cutting devices, and the machine-cut shapes were eventually assembled onsite.

Neoclassical and industrial-era architects frequently knew their parts in detail before they began to design. These parts might be displayed in the plates of a textbook, in the pages of a manufacturer's catalogue, or in the menus of a CAD system. They might even be premanufactured. But in CAD/CAM design and construction,

Figure 1.25. Cladding panels of the Experience Music Building, Seattle, under construction.

the parts are established as the design develops, and they are only fully known when design development is complete.

CONCLUSION: BEYOND DURAND

The theoretical underpinning of much architectural CAD and design synthesis software is flawed and limiting. Future progress in formalized design synthesis will depend upon transcending the Procrustean and inadequate representational machinery that has mostly prevailed in computer graphics and CAD for the last quarter century, respecting the fundamental roles of ambiguity and perception in creative design exploration, and accepting that a representation's structure should be the result of progressive design decision-making – not something that is arbitrarily and irrevocably imposed at the outset.

Creative designing is more than just poking around in some predefined combinatorial search tree – no matter how large and complex the tree or how sophisticated the poking. In the most literal sense, what you *see* is what you get.

REFERENCES

Banham, R. (1967). *Theory and Design in the First Machine Age*. 2nd ed., Praeger, New York.
Duarte, J. P. (2000a). "A digital framework for augmenting the architect's creativity." In *Proceedings of the Greenwich 2000: Digital Creativity Symposium*, The University of Greenwich, Greenwich, U.K., 33–42.
Duarte, J. P. (2000b). "The grammar of Siza's houses at Malagueira," *Environment and Planning B: Planning and Design* (to be published).
Durand, J. N. L. (1985). *Précis des Leçons d'Architecture*, Verlag Dr. Alfons Uhl, Nördlingen. (Original work published in 1819).
Durand, J. N. L. (1821). *Leçons d'Architect: Partie Graphique des Cours d'Architecture*, al' École Royale Polytechnique, Paris.
Guadet, J. (1902). *Eléments et Théorie dè l'Architecture*; cours professé a l'École Nationale et Speciale des Beaux-Arts, Librairie 1–32 de la Construction Modern, Paris.
Koning, H. and Eizenberg, J. (1981). "The language of the prairie: Frank Lloyd Wright's prairie houses," *Environment and Planning B: Planning and Design*, **8**(3):295–323.
Mitchell, W. J. (1975). "Vitruvius computatus," in *Models and Systems in Architecture and Building*, D. Hawkes, (ed.), The Construction Press, Hornby, Lancaster, pp. 53–59.

Mitchell, W. J. (1977). *Computer-Aided Architectural Design*, Van Nostrand Reinhold, New York.

Mitchell, W. J., Steadman, P., and Liggett, R. S. (1976). "Synthesis and optimization of small rectangular floor plans," *Environment and Planning B: Planning and Design*, **3**(1): 37–70.

Roriczer, M. (1486). *On the Ordination of Pinnacles*. Translated by J. W. Papworth, 1853; reprinted in E. G. Holt (1957). *A Documentary History of Art*, Vol. 1, Doubleday, New York, pp. 95–101.

Shea, K. and Cagan, J. (1997). "Innovative dome design: applying geodesic patterns with shape annealing," *Artificial Intelligence for Engineering Design, Analysis, and Manufacturing*, **11**:379–394.

Shea, K. and Cagan, J. (1999). "Languages and semantics of grammatical discrete structures," *Artificial Intelligence for Engineering Design Analysis, and Manufacturing*, Special Issue, **13**(4):241–251.

Stiny, G. and Gips, J. (1972). "Shape grammars and the generative specification of painting and sculpture." In *Information Processing 71*, C. V. Freiman (ed.), North-Holland, Amsterdam, pp. 1460–1465.

Stiny, G. and Mitchell, W. J. (1978a). "The Palladian grammar," *Environment and Planning B: Planning and Design*, **5**(1):5–18.

Stiny, G. and Mitchell, W. J. (1978b). "Counting Palladian plans," *Environment and Planning B: Planning and Design*, **5**(2):189–198.

CHAPTER TWO

How to Calculate with Shapes

George Stiny

> An interesting question for a theory of semantic information is whether there is any equivalent for the engineer's concept of noise. For example, if a statement can have more than one interpretation and if one meaning is understood by the hearer and another is intended by the speaker, then there is a kind of semantic noise in the communication even though the physical signals might have been transmitted perfectly.
>
> – George A. Miller

WHAT TO DO ABOUT AMBIGUITY

It's really easy to misunderstand when there's so much noise. Shapes are simply filled with ambiguities. They don't have definite parts. I can divide them anyhow I like anytime I want. The shape

includes two squares

four triangles

and indefinitely many Ks like the ones I have on my computer screen right now.
There are upper case Ks

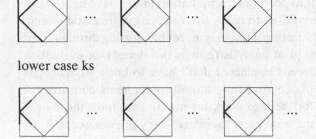

lower case ks

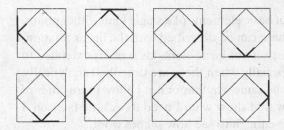

and their rotations and reflections. These are fixed in the symmetry group of the
shape. All of its parts can be flipped and turned in eight ways.

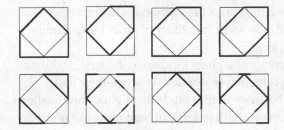

I can divide the shape in myriad myriads of other ways. I can always see something
new. Nothing stops me from changing my mind and doing whatever seems expedient.
These additional parts

scarcely begin to show how easy it is to switch from division to division without rhyme
or reason. (The fantastic number "myriad myriads" is one of Coleridge's poetic
inventions. It turns a noun into an adjective. It's another example of the uncanny
knack of artists and madmen to see triangles and Ks after they draw squares. I do
the trick every time I apply a rule to a shape in a computation. There's no magic in
this. I can show exactly how it works. But maybe I am looking too far ahead. I still
have a question to answer.)

What can I do about ambiguity? It causes misunderstanding, confusion, incoher-
ence, and scandal. Adversaries rarely settle their disputes before they define their
terms, and scientific progress depends on accurate and coherent definitions. But
it's futile trying to remove ambiguity completely with necessary facts, authoritative
standards, or common sense. Ambiguity isn't something to remove. Ambiguity is

something to use. The novelty it brings makes creative design possible. The opportunities go on and on. There's no noise, only the steady hum of invention.

In this chapter, I use rules to calculate with shapes that are ambiguous. This is important for design. I don't want to postpone computation until I have figured out what parts there are and what primitives to use – this happens in heuristic search, evolutionary algorithms, and optimization. Design is more than sorting through combinations of parts that come from prior analysis (how is this done?), or evaluating schemas in which divisions are already in place. I don't have to know what shapes are, or to describe them in terms of definite units – atoms, components, constituents, primitives, simples, and so on – for them to work for me. In fact, units mostly get in the way. How I calculate tells me what parts there are. They're evanescent. They change as rules are tried.

HOW TO MAKE SHAPES WITHOUT DEFINITE PARTS

At the very least, shapes are made up of basic elements of a single kind: either points, lines, planes, or solids. I assume that these can be described with the linear relationships of coordinate geometry. My repertoire of basic elements is easily extended to include curves and exotic surfaces, especially when these are described analytically. However, the results are pretty much the same whether or not I allow more kinds of basic elements, so I'll stay with the few that show what I need to. A little later on, I say why I think straight lines alone are enough to see how shapes work.

Some of the key properties of basic elements are summarized in Table 2.1. Points have no divisions. However, lines, planes, and solids can be cut into discrete pieces – smaller line segments, triangles, and tetrahedrons – so that any two are connected in a series of pieces in which successive ones share a point, an edge, or a face. More generally, every basic element of dimension greater than zero has a distinct basic element of the same dimension embedded in it, and a boundary of other basic elements that have dimension one less.

Shapes are formed when basic elements of a given kind are combined, and their properties follow once the embedding relation is extended to define their parts.

There are conspicuous differences between shapes made up of points and shapes containing lines, planes, or solids. First, shapes with points can be made in just one way if the order in which points are located doesn't matter, and in only finitely many ways even if it does. Distinct shapes result when different points are combined. However, there are indefinitely many ways to make shapes of other kinds. Distinct shapes need not result from different combinations of lines, planes, or solids when these basic elements fuse.

Basic Element	Dimension	Boundary	Content	Embedding
Point	0	none	none	identity
Line	1	two points	length	partial order
Plane	2	three or more lines	area	partial order
Solid	3	four or more planes	volume	partial order

TABLE 2.1.   PROPERTIES OF BASIC ELEMENTS

The shape

can be made with eight lines

as a pair of squares with four sides apiece, with 12 lines

as four triangles with three sides apiece, or with 16 lines as two squares and four triangles; but there's something very odd about this. If the 16 lines exist independently as units that neither fuse nor divide in combination – if they're like points – then the outside square is visually intact when the squares or the triangles are erased. I thought the shape would disappear. And that's not all. The shape looks the same whether the outside square is there or not. There are simply too many lines. Yet they make less than I see. I can find only four upper case Ks, and no lower case ks. Lines are not units, and neither are other basic elements that are not points.

Of course, shapes are ambiguous when they contain points and when they don't. They can be seen in different ways depending on the parts I actually resolve (or as I show below, on how I apply rules to calculate). However, the potential material for seeing is not the same for shapes with points and for shapes made with other basic elements. Points can be grouped to form only a finite number of parts, whereas shapes of other kinds can be divided into any number of parts in any number of ways. This shape

with three points has just eight possible parts: one empty part, and these seven nonempty ones

1	2	3	4	5	6	7

containing one or more of the points used to make the shape. I can show all of the different ways I am able to see the shape – at least as far as parts go – and count

them. But the shape

has indefinitely many parts that need not combine the lines originally used to make it. How the shape is made and how it is divided into parts afterward are independent. What I find in the shape now may alter later. I can always see something new I haven't seen before. There is no end to ambiguity and the novelty it brings once basic elements aren't points.

The contrast between points and basic elements of other kinds that I am beginning to develop leads to alternative models of computation. The first is standard and uncontroversial, but no less speculative for that. The idea simply put is this:

> Computation depends on counting.

Points make counting possible. They provide the units to measure the size or complexity of shapes, and the ultimate constituents of meaning. The second model comes from drawing and looking at shapes. The idea is this:

> Computation depends on seeing.

Now parts of shapes are artifacts of computations not fixed beforehand. There are neither predefined units of measurement nor final constituents of meaning. Units change freely as I calculate. They may be different every time I apply a rule, even the same rule in the same place.

The contrast can be traced to the embedding relation. For points, it is identity, so that each contains only itself. This is what units require. But for the other basic elements, there are indefinitely many of the same kind embedded in each and every one: lines in lines, planes in planes, and solids in solids. This is a key difference between points and basic elements of other kinds. It implies the properties of the algebras I define for shapes in which basic elements fuse and divide, and it lets rules handle ambiguity in computations.

However, the contrast is not categorical. These are reciprocal approaches to computation. Each includes the other in its own way. My algebras of shapes begin with the counting model. It's a special case of the seeing model where the embedding relation and identity coincide. I can deal with the seeing model in the counting one – and approximate it in computer implementations – by using established devices, for example, linear descriptions of basic elements and their boundaries, canonical representations of shapes in terms of maximal elements, and reduction rules. The equivalencies show how complex things made up of units are related to shapes without definite parts.

Whether there are units or not is widely used to distinguish verbal and visual media – discursive and presentational forms of symbolism – as distinct modes of expression. Nonetheless, verbal and visual media may not be as incompatible as this suggests. Shapes made up of points and shapes with lines, planes, or solids belie the idea that there are adverse modes of expression. This makes art and design when they are visual as computational as anything else. The claim is figuratively embedded, if not literally rooted, in the following details.

ALGEBRAS OF SHAPES COME IN A SERIES

Three things go together to define algebras of shapes. First, there are the shapes themselves that are made up of basic elements. Second, there is the part relation on shapes that includes the Boolean operations. Third, there are the Euclidean transformations. The algebras are enumerated up to three dimensions in this series:

$$U_{00} \quad U_{01} \quad U_{02} \quad U_{03}$$
$$U_{11} \quad U_{12} \quad U_{13}$$
$$U_{22} \quad U_{23}$$
$$U_{33}$$

The series is presented as an array to organize the algebras with respect to a pair of numerical indices. (The series goes on in higher dimensions, and not without practical gain. The possibilities are worth thinking about.)

Every shape in an algebra U_{ij} is a finite set of basic elements. These are maximal: any two embedded in a common element are discrete, with boundary elements that are, too. The left index i determines the dimension of the basic elements, and the right index j gives the dimension in which they go together and the dimension of the transformations. The part relation compares shapes two at a time – it holds when the basic elements of one are embedded in basic elements of the other – and the Boolean operations combine shapes. Either sum, product, and difference will do, or equivalently, symmetric difference and product. The transformations are operations on shapes that change them into geometrically similar ones. They distribute over the Boolean operations. The transformations also form a group under composition. Its properties are used to describe the behavior of rules.

A few illustrations help to fix the algebras U_{ij} visually. In fact, this is just the kind of presentation they are meant to support. Once the algebras are defined, what you see is what you get. There is no need to represent shapes in symbols to calculate with them, or for any other reason. Take the shapes in the algebra U_{02}. This one

is an arrangement of eight points in the plane. It is the boundary of the shape

in the algebra U_{12} that contains lines in the plane. The eight longest lines of the shape – the lines someone trained in drafting would choose to draw – are the maximal ones.

TABLE 2.2.   SOME PROPERTIES OF SHAPES

Algebra	Basic Elements	Boundary Shapes	No. of Parts
U_{0j}	points	none	finite
U_{1j}	lines	U_{0j}	infinite
U_{2j}	planes	U_{1j}	infinite
U_{3j}	solids	U_{2j}	infinite

In turn, the shape is the boundary of the shape

in the algebra U_{22}. The four largest triangles in the shape are maximal planes. They are a good example of how this works. They are discrete and their boundary elements are, too, even though they touch. Points are not lines, and they aren't in the boundaries of planes.

For every algebra U_{ij}, i is greater than or equal to zero, and j is greater than or equal to i. Shapes can be manipulated in any dimension at least as big as the dimension of the basic elements combined to make them.

The properties of basic elements are extended to shapes in Table 2.2. The index i is varied to reflect the facts in Table 2.1. The first thing to notice is that an algebra U_{ij} contains shapes that are boundaries of shapes in the algebra U_{i+1j} only if i and j are different. The algebras U_{ii} do not include shapes that are boundaries, and they're the only algebras for which this is so. They also have special properties in terms of embedding and the transformations. These are described a little later on.

The description of parts in Table 2.2 misses some important details. These require the index j. The additional relationships are given in Table 2.3.

BOOLEAN DIVISIONS

The algebras of shapes U_{ij} are usefully classified in terms of their Boolean properties, and in terms of their Euclidean ones. The Boolean classification relies on the variation

TABLE 2.3.   MORE PROPERTIES OF THE PART RELATION

U_{ij}	Every Shape	
	Has Distinct Nonempty Part[a]	Is Distinct Part of Another Shape[b]
$0 = i = j$	no	no
$0 = i < j$	no	yes
$0 < i \leq j$	yes	yes

[a] There is no smallest shape.
[b] There is no largest shape.

of the indices i and j in Table 2.3. The algebras U_{ij} are divided this way:

U_{00}	U_{01}	U_{02}	U_{03}
	U_{11}	U_{12}	U_{13}
		U_{22}	U_{23}
			U_{33}

according to whether or not i and j are zero. There are four regions to examine: the top row and each of its two segments, and the triangular portion of the lower quadrant.

The atomic algebras of shapes are defined when the index i is zero:

U_{00}	U_{01}	U_{02}	U_{03}
	U_{11}	U_{12}	U_{13}
		U_{22}	U_{23}
			U_{33}

Shapes are finite arrangements of points. The atoms – nonempty shapes without distinct nonempty parts – are the shapes with a single point apiece. (Atoms are the same as the units I've been talking about, but with respect to sets of points and the part relation instead of individual points and embedding.) Any product of distinct atoms is the empty shape, and sums of atoms produce all of the other shapes. No two shapes are the same unless their atoms are. A shape is either empty or has an atom as a part. There are two kinds of atomic algebras depending on the value of the index j.

The only Boolean algebra of shapes is U_{00}:

U_{00}	U_{01}	U_{02}	U_{03}
	U_{11}	U_{12}	U_{13}
		U_{22}	U_{23}
			U_{33}

It contains exactly two shapes: the empty shape and an atom made up of a single point. In fact, it is the only finite algebra of shapes, and it is the only complete one in the sense that shapes are defined in all possible sums and products. The algebra is the same as the Boolean algebra for "true" and "false" used to evaluate logical expressions.

The other atomic algebras U_{0j} are defined when j is more than zero:

U_{00}	U_{01}	U_{02}	U_{03}
	U_{11}	U_{12}	U_{13}
		U_{22}	U_{23}
			U_{33}

Each of these algebras contains infinitely many shapes that all have a finite number of points, and a finite number of parts. However, because there are infinitely many points, there is no universal shape that includes all of the others, and so no

complements. This defines a generalized Boolean algebra. It is a relatively comple-
mented, distributive lattice when the operations are sum, product, and difference, or
a Boolean ring with symmetric difference and product. There is a zero in the algebra –
the empty shape – but no universal element, as this would be a universal shape. As
a lattice, the algebra is not complete: every product is always defined, whereas no
infinite sum ever is. Distinct points do not fuse in sums the way basic elements of
other kinds do. The finite subsets of an infinite set, say, the numbers in the counting
sequence 0, 1, 2, 3 . . . , define a comparable algebra.

   This idea comes up again when I consider algebras for decompositions. A decomposition is a finite set of parts (shapes) that add up to make a shape. It gives
the structure of the shape by showing just how it's divided, and how these divisions
interact. The decomposition may have special properties, so that parts are related in
some way. It may be a Boolean algebra on its own, a topology, a hierarchy, or even
something else. For example, suppose that a singleton part contains a single basic
element. These are atoms in the algebras U_{0j} but they do not work this way if i isn't
zero. The set of singleton parts of a shape with points is a decomposition. In fact,
the shape and the decomposition are pretty much alike. However, decompositions
are not defined for shapes with basic elements of any other kind. There are infinitely
many singleton parts.

   Algebras of shapes show what happens when shapes are put together by using
specific operations, but there are complementary accounts that are also useful. Looking at individual shapes in terms of their parts – whether or not these are finite in
number – is another way of describing algebras of shapes with subalgebras. Do not
be alarmed as I move to and fro between these views. It's not unlike seeing shapes
in different ways. For starters, a shape and its parts in an algebra U_{0j} form a complete Boolean algebra that corresponds to the Boolean algebra for a finite set and
its subsets. This is represented neatly in a lattice diagram. For example, the Boolean
algebra for the three points

in $U_{0\,2}$ considered above looks like this:

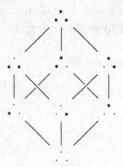

   Pictures like this are compelling, and they make it tempting to consider all shapes,
whether or not they have points, in terms of a finite number of parts combined in
sums and products. It's hard to imagine a better way to describe shapes than to
resolve their parts and to show how these are related. This is what decompositions

do. However, if parts are fixed permanently, then this is a poor way to understand how shapes work in computations. As I describe this below, parts – with and without Boolean complements – vary within computations and from one computation to the next. Parts are decided anew every time I apply a rule. Decompositions – topologies and the like – change dynamically as I go on calculating. I don't know what parts there really are until I stop.

The Boolean algebra for a shape and its parts in an algebra U_{0j} is itself a discrete topology. Every part of the shape is both closed and open and has an empty boundary. Parts are disconnected. (Topological boundaries aren't the boundaries of Table 2.2 that show how shapes in different algebras are related. Now the boundary of a shape is part of it.) This is a formal way of saying what seeing confirms all the time: there is no preferred way to divide the shape into parts. Any division is possible. None is better or worse than any other without ad hoc reasons. These depend on how rules are used to calculate.

To return now to the algebras U_{ij}, the atomless ones are defined when i is more than zero:

$$
\begin{array}{c|ccc}
U_{00} & U_{01} & U_{02} & U_{03} \\
\hline
 & U_{11} & U_{12} & U_{13} \\
 & & U_{22} & U_{23} \\
 & & & U_{33}
\end{array}
$$

Each of these algebras contains infinitely many shapes made up of a finite number of lines, planes, or solids. However, every nonempty shape has infinitely many parts. The empty shape is the zero in the algebra. There is no universal element – a universal shape can't be formed with lines, planes, or solids – and shapes do not have complements. This gives a generalized Boolean algebra, as above for points. But this time, the properties of sums and products are a little more alike: for both, infinite ones may or may not be defined. When infinite sums are formed, basic elements fuse.

In an algebra U_{ij} when i isn't zero, a nonempty shape and its parts form an infinite Boolean algebra, but the algebra is not complete. Infinite sums and products needn't determine shapes. The underlying topology for the shape and its parts is a Stone space. Every part is both closed and open, and the part has an empty boundary. Parts are disconnected. In this, all shapes are exactly the same. They're all ambiguous. There's nothing about shapes by themselves – no matter what basic elements they have – that recommends any division. It's only in computations as a result of what rules do that shapes have definite parts with meaningful interactions, and the possibility of substantial boundaries.

So far, my classification of shapes and their algebras shows that things change significantly once i is bigger than zero. Whether or not the embedding relation implies identity makes a world of difference. However, my algebras are not the only way to think about shapes. There are two alternatives that deserve brief notice.

First, there are the individuals of logic and philosophy. Shapes are like them in important ways. The part relation is crucial for both, but the likeness fades once other details are compared. Individuals form a complete Boolean algebra that has the zero excised. The algebra may have atoms or not, and it may be finite or infinite. Moreover, there are no ancillary operators. Transformations are not defined for individuals. To be honest, though, the nonempty parts of any nonempty shape in an algebra U_{0j} are possible individuals. However, if j isn't zero, then this isn't so for all of the nonempty shapes in the algebra taken together. Individuals are pointless, unless they exercise their right to free assembly finitely.

Second, there are the point sets of solid modeling, but the topology of a shape in an algebra U_{ij} when i isn't zero and the topology of the corresponding point set are strikingly different. The shape is disconnected in the one topology and connected in the other. Among other things, the topology of the point set confuses the boundary of the shape that's not part of it – the boundary of Table 2.2 – and the topological boundary of the shape that is. Points are too small to distinguish boundaries as limits and parts. Point sets are "regular" whenever they're shapes, and Boolean operations are "regularized." This is an artificial way to handle lines, planes, and solids. Of course, shapes made up of points are point sets with topologies in which parts are disconnected. This and my description of individuals confirm what I've already said. The switch from the identity relation to embedding is modest technically, but it has many important consequences for shapes.

EUCLIDEAN EMBEDDINGS

The Euclidean transformations augment the Boolean operations in the algebras of shapes U_{ij} with additional operators. They are defined for basic elements and extend easily to shapes. Any transformation of the empty shape is the empty shape. Further, a transformation of a nonempty shape contains the transformation of each of the basic elements in the shape. This relies on the underlying recursion implicit in Tables 2.1 and 2.2 in which boundaries of basic elements are transformed until new points are defined.

The universe of all shapes, no matter what kind of basic elements are involved, can be described as a set containing certain specific shapes, and other shapes formed by using the sum operation and the transformations. For points and lines, the universe is defined with the empty shape and a shape that contains a single point or a single line. Points are geometrically similar, and so are lines. However, this relationship doesn't hold for planes and solids. As a result, more shapes are needed to start. All shapes with a single triangle or a single tetrahedron do the trick. For a point in zero dimensions, the identity is the only transformation. So there are exactly two shapes: the empty shape and the point itself. In all other cases, there are indefinitely many distinct shapes that come with any number of basic elements. Every nonempty shape is geometrically similar to indefinitely many other shapes.

The algebras of shapes U_{ij} also have a Euclidean classification that refines the Boolean classification given above. Together these classifications form an interlocking taxonomy. The algebras U_{ii} on the diagonal provide the main Euclidean division:

U_{00}	U_{01}	U_{02}	U_{03}
	U_{11}	U_{12}	U_{13}
		U_{22}	U_{23}
			U_{33}

The algebras U_{ii} have an interesting property. A transformation of every shape in each algebra is part of every nonempty shape. This is trivially so in the algebra U_{00}:

U_{00}	U_{01}	U_{02}	U_{03}
	U_{11}	U_{12}	U_{13}
		U_{22}	U_{23}
			U_{33}

for a point in zero dimensions. Both the empty shape and the point are parts of the point. But when i is not zero,

U_{00}	U_{01}	U_{02}	U_{03}
	U_{11}	U_{12}	U_{13}
		U_{22}	U_{23}
			U_{33}

the possibilities multiply. In this case, infinitely many transformations of every shape are parts of every nonempty shape. This is obvious for the empty shape. It is part of every shape under every transformation. But for a nonempty shape, there is a little more to do. The basic elements in the shape are manipulated in their own dimension: lines are in a line, planes are in the plane, and solids are in space. Because these basic elements are separately finite, finite in number, and finitely arranged, they can always be embedded in a single basic element. However, infinitely many transformations of any basic element can be embedded in any other basic element. I can make the one smaller and smaller until it fits in the other and can be moved around. Thus, there are infinitely many ways to make any shape part of any other shape that has at least one basic element. The triangle

in the algebra U_{22} is geometrically similar to parts of a rectangle

and in fact, to parts of itself:

This has important implications for the way rules work in computations. As defined below, a rule applies under a transformation that makes one shape part of another shape. Everything is fine if the rule is given in the algebra U_{00}. It's determinate: if it applies, it does so under a finite number of transformations. And in fact, there are determinate rules in every algebra containing shapes with points. But in the other algebras U_{ii} , a rule is always indeterminate: it applies to every nonempty shape in infinitely many ways as described above. (There are indeterminate rules for points, too, when j is greater than zero.) This eliminates any chance of controlling how the rule is used. It can be applied haphazardly to change shapes anywhere there is an embedded basic element. Indeterminate rules seem to be a burden. They appear to be totally purposeless when they can be applied so freely. Even so, indeterminate rules have some important uses. A few of these are described below.

I like to think that shapes made up of lines in the plane – a few pencil strokes on a scrap of paper – are all that's ever required to study shapes and computations with them. There are reasons for this that are both historical and practical. On the one hand, for example, lines are central in Alberti's famous account of architecture, and on the other hand, the technology of making lines isn't very complicated; but an algebraic reason supersedes all of the others. Shapes containing lines arranged in the plane suit my interests because their algebra U_{12},

U_{00}	U_{01}	U_{02}	U_{03}
	U_{11}	$\boldsymbol{U_{12}}$	U_{13}
		U_{22}	U_{23}
			U_{33}

is the first algebra of shapes in the series U_{ij} in which (1) basic elements have boundaries and the identity relation and embedding are not the same, and (2) there are rules that are determinate and rules that aren't – rules in U_{11} are also indeterminate in U_{12} – when I apply them to shapes to calculate. The algebra U_{12} is representative of all of the algebras where i isn't zero. This lets me show almost everything I want to about these algebras and the shapes they contain in line drawings. Lines on paper are an excellent way to experiment with shapes, and to see how they work in computations.

MORE ALGEBRAS IN NEW SERIES

The algebras U_{ij} can themselves be extended and combined in a host of useful ways to define new algebras of shapes. This provides an open-ended repertoire of shapes and other kinds of expressive devices.

Shapes often come with other things besides basic elements. Labels from a given vocabulary, for example, a, b, c ..., may be associated with basic elements to get shapes like these:

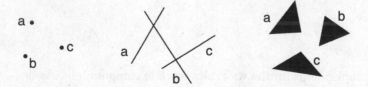

The labels may simply classify basic elements and so parts of shapes, or they may have their own semantics to introduce new kinds of information. In a more ambitious fashion, basic elements may also have properties associated with them that interact as basic elements do when they are combined. I call these *weights*. Weights may go together with basic elements to get shapes like these:

Among other things, weights include different graphical properties such as color, surface texture, and tone, but possibly more abstract things such as sets of labels, numerical values that vary in some way, or combinations of sets and values. (Weights can also be shapes. This may complicate how shapes are classified, but it doesn't compromise the idea that the embedding relation makes a difference in computation.)

Labels and weights let me show how shapes can go together in different ways. I can put a triangle on top of a square to make the shape

in the algebra U_{12}, and then take away the triangle I've added. But the result isn't the square

that might be expected just listening to the words "take away the triangle after adding it to the square." The piece the triangle and square have in common is erased from the top of the square as happens in drawing. This gives the shape

However, I have other options. I can label the lines in the triangle in one way and label the lines in the square in another way, so that the square is still intact after adding and subtracting the triangle. Borrowing a term from computer graphics, the triangle and the square are on *separate layers*. I can do the same thing and more with weights. Either way, this begins to show how labels and weights can be used to modify algebras of shapes. It also anticipates some properties of decompositions I describe in the next section.

When labels and weights are associated with basic elements, additional algebras of shapes are defined. Two new series of algebras are readily formed from the series U_{ij}. The algebras V_{ij} for labeled shapes keep the properties of the algebras U_{ij}, and the algebras W_{ij} for weights may or may not, according to how weights combine. For example, notice that if labels are combined in sets, then the series of algebras W_{ij} includes the series V_{ij}. Of course, it's always possible to extend the algebras U_{ij} in other ways if there's an incentive. There's no reason to be parsimonious – and no elegance is lost – when algebras that fit the bill are so easy to define.

The algebras U_{ij}, V_{ij}, and W_{ij} can also be combined in a variety of ways, for example, in appropriate sums and products (direct products), to obtain new algebras. These algebras typically contain compound shapes with one or more components in which basic elements of various kinds, labels, and weights are mixed and interact, as for example, in the shape

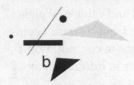

In this way, the algebras U_{ij}, V_{ij}, and W_{ij} and their combinations facilitate the definition of a host of formal devices in a uniform framework. The properties of these devices correspond closely to the properties of traditional media in art and design. Nothing is lost – certainly no ambiguity. This lets me calculate in the kind of milieu that's expected for creative activity, at least as it's normally understood.

The algebras I've been talking about give me the wherewithall to say what I mean by a design. Roughly speaking, designs are used to describe things for making and to show how they work. This may imply a wide range of expressive devices – shapes and the kinds of things in my algebras – in a host of different descriptions that are linked and interact in various ways. Formally then, designs belong to n-ary relations in my algebras. The computational apparatus I'm coming to is meant to define these relations. And in fact, there are already many successful examples of this in architecture and engineering.

SOLIDS, FRACTALS, AND OTHER ZERO-DIMENSIONAL THINGS

I said earlier that decompositions are finite sets of parts that sum to make shapes. They show how shapes are divided into parts for different reasons and how these divisions interact. Of course, decompositions are not just defined for shapes in an algebra U_{ij}. More generally, they're defined for whatever there is in any algebra formed from the algebras in the series U_{ij}. Decompositions have their own algebras. Together with the empty set – it's not a decomposition – they form algebras under the subset relation, the Boolean operations for sets (union, intersection, and so on), and the Euclidean transformations. This makes decompositions zero dimensional. They are the same as shapes in the algebras U_{0j}. The parts they contain behave just like points. Whatever else they may be, the parts in decompositions are units. They're independent in combination and simple beyond analysis.

I like to point out that complex things like fractals and computer models of solids and thought are zero dimensional. Am I obviously mistaken? Solids are clearly three dimensional – I bump into them all of the time – fractal dimensions vary widely, and no one knows for sure about thought. But this misses the point. Fractals and computer models utilize decompositions that are zero dimensional, or other representations such as lists or graphs in which units are likewise given from the start. But what's wrong with this? Computers are meant to manipulate complex things – fractals and the like – that have predefined divisions. They're especially good at counting units – fractal dimensions are defined in this way – and moving them around to go through possible configurations. Yet, computers fail with shapes once they include lines or basic elements of higher dimension. I can describe what happens to shapes as long as I continue to calculate, and measure their complexity as a result of the rules I apply to pick out parts. In this sense, shapes may be just as complex as anything else. There is, however, a telling difference. The complexity of shapes is retrospective. It's an artifact of my computation that can't be defined independently. There aren't any units to count before I calculate. What can I do about this?

I have already shown that units in combination need not correspond with experience. What a computer knows and what I see may be very different, and the computer has zero tolerance for views that aren't its own. The shape

as 16 lines that describe two squares with four sides apiece and four triangles with three sides apiece doesn't behave the way it should. There are too many lines and just too few. For example, some parts are hard to delete. The outside square won't go away, and some parts are impossible to find. The lower case ks aren't there. But even if I'm willing to accept this because the definitions of squares and triangles are exactly what they should be – and for the time being they're more important than ks – there's still a problem. How am I going to find out how the shape is described before I start to use it? Does it include squares or triangles, or is it a combination of both? I can see the shape, but I can't see its decomposition; it's inside the computer. I need to find this hidden structure. Experiments are useful, but they take time and may fail. I'm stuck without help and the occult (personal) knowledge of experts. Shapes aren't like this. What you see is what you get. This is a good reason to use shapes in design. There are times when I don't know what I'm doing and look at shapes to find out. However, there's nothing to gain if I have to ask somebody what shapes are. Can I make decompositions that are closer to the shapes they describe?

It's not hard to guess that my answer is going to be no. There are many reasons for this, but a few are enough to support my conclusion. First, there is the problem of the original analysis. How can I possibly know how to divide a shape into parts that suit my present interests and goals – everything is vague before I start to calculate – and that anticipate whatever I might do later? I can change my mind. Nothing keeps me from seeing triangles and Ks after I draw squares, even if I'm positive squares are all I'll ever need. Contracts aren't hard to break. It's so much easier to see what I want to in an ongoing process than to remember or learn what to do. My immediate perception may take precedence over anything I've decided or know for sure. But there is an option. I don't have to divide the shape into meaningful parts. I can cut it into anonymous units that are small enough to model (approximate) as many new parts as I want. This is a standard practice that does not work. The parts of the shape and their models are too far apart. Shapes containing lines, for example, aren't like shapes that combine points. This is what the algebras U_{0j} and U_{1j} show. Moreover, useful divisions are likely to be infrequent and not very fine. Getting the right ones at the right time is what matters. However, even if units do combine to model the parts I want, they may also multiply unnecessary parts well beyond interest and use to cause unexpected problems. Point sets show this perfectly. I have to deal with the parts I want, and the parts I don't. If I'm clever, this won't stop me; but there is a price to pay. Dividing the shape into smaller and smaller units may exceed my intuitive reach. I can no longer engage the shape directly in terms of what I see. Now there's something definite to know beforehand. Expertise intervenes. I have to think twice about how the shape works as a collection of arbitrary units that go together in an arbitrary way. What happens to novelty and the chance of new experience? Any way you cut it, there's got to be a better way to handle the shape. Decompositions are premature before I calculate.

Decompositions are fine if I remember they're only descriptions. In fact, it's impossible to do without them if I want to talk about shapes, and to say what they're for and to show how they're used. Without decompositions, I could only point at shapes in a vague sort of way. I couldn't write this chapter if there were no decompositions to describe shapes. Still, decompositions are no substitute for shapes in computations. Parts aren't permanent. They may alter erratically every time I use a rule. Is this what it means to calculate? Nothing seems to be fixed in the normal way. The answer is in the algebras U_{ij} , and in the algebras obtained from them.

HOW RULES WORK IN COMPUTATIONS

Most of the things in the algebras I have been describing – I'll call all of these things shapes from now on unless it makes a difference – are useless jumbles. This should come as no surprise to anyone. Most of the numbers in arithmetic are meaningless, too. The shapes that count, such as the numbers I get when I balance my checkbook, are the ones there are when I calculate.

Computations with shapes are defined in different algebras. The way I calculate in each of these algebras depends on the same mechanism. Rules are defined and applied recursively to shapes in terms of the part relation and Boolean sum and difference – or whatever corresponds to the part relation and these operations – and the Euclidean transformations.

Shape rules are defined by ostension: any two shapes whatsoever – empty or not and the same or not – shown one after the other determine a rule. Suppose these shapes are A and B. Then the rule they define is

$$A \rightarrow B$$

The two shapes in the rule are separated by an arrow ($\rightarrow$). The rule has a left side that contains A and a right side that contains B. If A and B are drawn, then registration marks are used to fix the relationship between them. The shape rule

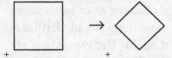

in the algebra U_{12}, for example, turns a square about its center and shrinks it. The relationship is explicit when the mark

in the left side of the rule and the one in the right side register:

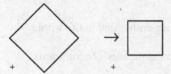

This is a convenient device to show how different shapes line up.

Two rules are the same whenever there is a single transformation that makes the corresponding shapes in both identical. The shape rule

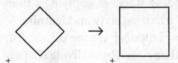

is the same as the one above. But the rule

isn't, even though the shape

is formed when the registration marks in the left and right sides of the rule are made to coincide. The rule is the inverse of the rule defined above. Inverses may or may not be distinct.

The precise details of applying shape rules and saying what they do are straightforward. A rule $A \rightarrow B$ applies to a shape C in two stages.

1. Find a transformation t that makes the shape A part of C. This picks out some part of C that looks like A.
2. Subtract the transformation of A from C, and then add the same transformation of the shape B. This replaces the part of C like A with another part that looks like B.

In the first stage, the rule $A \rightarrow B$ is used to see. It works as an observational device. If A can be embedded in C anyhow at all – no matter what's gone on before – then C has a part like A. In the second stage, the rule changes C in accordance with A and B. Now it's a constructive device. Yet, once something is added, it fuses with what's left of C and may or may not be recognized again. The two stages are intentionally linked by means of the same transformation t. This completes a spatial analogy:

$$t(A) \rightarrow t(B) :: A \rightarrow B$$

The rule $t(A) \rightarrow t(B)$ is the same as the rule $A \rightarrow B$. After transposing, the part of C that's subtracted is to A as the part that's added is to B. But this is as far as the analogy goes. It depends on A and B, and the transformations. There are neither other divisions nor further relationships. The parts of shapes are no less indefinite because they're in rules. The implications of this are elaborated in many ways below.

The formal details of shape rule application shouldn't obscure the main idea behind rules: observation – the ability to divide shapes into definite parts – provides the impetus for meaningful change. This relationship is nothing new, but it's no less important for that. William James already has it at the center of reasoning in *The Principles of Psychology*.

> And the art of the reasoner will consist of two stages:
>   First, *sagacity*, or the ability to discover what part, M, lies embedded in the whole S which is before him;
>   Second, *learning*, or the ability to recall promptly M's consequences, concomitants, or implications.
>
> W. James (1981)

The twin stages in the reasoner's art and the corresponding stages in applying a rule are tellingly alike, if not identical. Every time a rule is tried, sagacity and learning – redescription and inference – work jointly to create a new outcome. It's the sagacity of rules that distinguishes them the most as computational devices. Rules divide shapes anew as they change in an unfolding process in which parts combine and fuse. The root value of observation in this process is reinforced when James goes on to quote John Stuart Mill. For Mill, learning to observe is like learning to invent. The use of shape rules in design elaborates the relationship between observation and invention explicitly.

The way rules are used to calculate is clear in an easy example. The rule

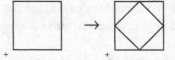

in the algebra U_{12} produces the shapes in this ongoing series

    ...

in computations that begin with the square

In particular, the third shape in the series is produced in two steps in this computation,

and in three steps in this one:

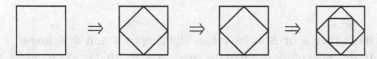

The rule finds squares and inscribes smaller ones. In the first computation, the rule is applied under a different transformation each time it is used to pick out a different square. This is the initial square, or the one the rule has inscribed most recently. But in the following computation, equivalent transformations – in the sense of Table 2.5 – are used in the first and second steps. The same square can be distinguished repeatedly without producing new results. In this case, there are some unexpected implications that I discuss below.

The rule

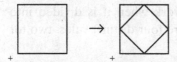

can be described in a nice way. The shape in its right side is the sum of the square in its left side and a transformation of this square. Because there is a transformation that makes the shape in the left side of the rule part of the shape in its right side, the rule can always be applied again. More generally for any shape A, I can define rules in terms of the following scheme:

$$A \rightarrow A + t(A)$$

Rules of this kind are used to produce parts of symmetrical patterns when the transformations t are the generators of a symmetry group. Fractals can also be obtained in approximately the same way, but now according to rules defined in terms of this scheme:

$$A \rightarrow \Sigma t(A)$$

where multiple transformations t of the shape A are added together, and the transformations involve scaling. Fractals may be zero dimensional, but this is scant reason not to define them by using rules in algebras where embedding and identity are not the same. Problems arise in the opposite way, if I try to describe things that are not zero dimensional as if they were.

Of course, there is more to calculating with shapes in design than there is to symmetrical patterns and fractals, even when these are visually striking. In fact, there are more generic schemes to define rules that include both of the schemes above. They have proven very effective in architectural practice and in correlative areas of spatial design.

For any two shapes A and B, I can define rules in terms of this scheme,

$$v \rightarrow A + B$$

and in terms of its inverse,

$$A + B \rightarrow v$$

where the variable v has either A or B as its value. The shapes A and B define a spatial relation – technically, a decomposition of the shape $A + B$ – that works to add shapes and to subtract them in computations. (It's important to stress that neither the relation nor the shapes in it may be preserved in this process.) The schemes can be used together in a general model of computation that includes Turing machines.

The shape

gives some good examples of how these schemes work when it is divided into two squares to define a spatial relation. Then there are four distinct rules: two for addition,

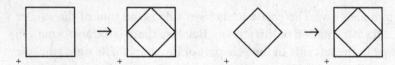

and two for subtraction,

From the initial square

the rules apply to produce the squares in the series described by this segment centered on the initial square,

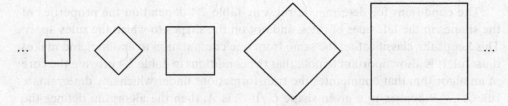

and any finite combination of these squares; for example, the shape

There are a host of wonderful possibilities when multiple spatial relations are used to define rules that apply in concert. Froebel's famous building gifts provide some nice material for design experiments using kindergarten blocks and plane figures in this manner.

CLASSIFYING RULES IN DIFFERENT WAYS

Rules are classified in a variety of important ways. First, it is possible to decide whether they apply determinately or indeterminately. The conditions for determinate rules vary somewhat from algebra to algebra, but they can be framed in terms of a recursive taxonomy of registration marks defined with basic elements. The conditions that make rules determinate in the algebras U_{1j} are specified in Table 2.4 as an example of this. The rules I have defined so far are determinate in the algebra U_{12}, but indeterminate rules are also defined. These include all of the rules in the algebra U_{11}, and more. This rule,

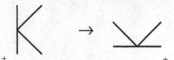

TABLE 2.4. DETERMINATE RULES IN THE ALGEBRAS U_{1j}

Algebra	The Rule $A \rightarrow B$ is Determinate
U_{11}	Never.
U_{12}	Three lines in A do not intersect at a common point. Further, no two of these lines are collinear, and all three are not parallel.
U_{13}	There are two cases: (1) two lines in A are skew; (2) three lines in A do not intersect at a common point, and further, no two of these lines are collinear, and all three are not parallel.

that rotates an upper case K is indeterminate in U_{12}. Its three maximal lines intersect at a common point. Different shapes may be equivalent in terms of their registration marks. For example, the upper case K and the lower case k are the same.

The conditions for determinate rules in Table 2.4 depend on the properties of the shapes in the left sides of rules, and not on the shapes to which the rules apply. This keeps the classification the same from one computation to another, and makes it useful. It is also important to note that the conditions in Table 2.4 provide the core of an algorithm that enumerates the transformations under which any determinate rule $A \rightarrow B$ applies to a given shape C. If C is A, then the algorithm defines the symmetry group of A; but this is merely an aside. More to the point, shape rules can be tried automatically.

The rule

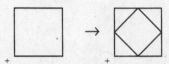

is determinate. The symmetry of the square in its left side, however, lets it apply under eight distinct transformations: four rotations and four reflections. But for each transformation, the results are the same. Why? The reason can be given in terms of Lagrange's famous theorem for subgroups, and it provides another way to classify rules.

Let two transformations be equivalent with respect to a shape if they change it identically. In this example,

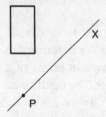

a clockwise rotation of 90° about the point P

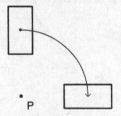

and a reflection across the axis X

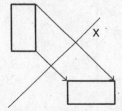

are equivalent relative the rectangle.

If a rule $A \rightarrow B$ has the property that the symmetry group of the shape A is partitioned into q classes with respect to the shape B, then the rule can be used in q distinct ways. Moreover, if the symmetry group of A has n transformations, then q divides n without remainder. The symmetry group of A has a subgroup containing n/q transformations with cosets given by the classes determined by B. Conversely, there is a rule for every subgroup of the symmetry group of A that behaves as the subgroup describes. The distinct uses of the rule $A \rightarrow B$ show the symmetry properties of A, and these serve independently to classify every rule with A in its left side.

This is nicely illustrated with a square in the algebra U_{12} for clockwise rotations, and reflections named by these axes:

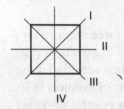

Rules behave in at least six different ways according to the six subgroups of the symmetry group of the square specified in Table 2.5. The symmetry group of the square has other subgroups that are isomorphic to these. One way to see this is in terms of equivalencies among the four axes of the square. There may be 10 subgroups (all of the axes are different), possibly eight (the horizontal axis and the vertical one are the same, and so are the diagonal axes), or the six I have shown (all of the axes are the same).

The rule

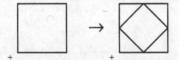

has the symmetry properties of the first rule in Table 2.5, and therefore acts like an identity with only one distinct use. The eight transformations in the symmetry group

TABLE 2.5. CLASSIFICATION OF RULES IN TERMS OF LAGRANGE'S THEOREM

Rule	Subgroup	No. of Cosets
□ → □	0, 90, 180, 270, I, II, III, IV	1
□ → ▱	0, 90, 180, 270	2
□ → ▱	0, 180, I, III	2
□ → ▱	0, 180	4
□ → ▱	0, I	4
□ → □	0	8

of the square in the left side of the rule are equivalent with respect to the shape in its right side.

So far, I have been using transformations to classify rules. This relies on the Euclidean properties of shapes, but it's not the only way to describe rules. I can take a Boolean approach and think about them in terms of how many parts there are in the shapes in their left and right sides. The classification of rules in this way by counting their parts is the crux of the famous Chomsky hierarchy for generative grammars. The hierarchy gives compelling evidence for the idea that computation depends on counting.

Generative grammars produce strings of symbols that are words or well-formed sentences in a language. A rule

$$AqR \rightarrow A'q'L$$

links a pair of strings – in this case, three symbols A, q, and R, and three others A', q', and L – by an arrow. If the strings were shapes, the rule would be like a shape rule. The rule is context free if there's one symbol in its left side, context sensitive if the number of symbols in its left side is never more than the number of symbols in its right side – the above rule is context sensitive – and belongs to a general rewriting system otherwise. Languages vary in complexity according to the rules that define them. The class of languages generated by context free rules is included in the larger class of languages generated by context sensitive rules, and so on. How well does this idea work for shape rules?

There is no problem counting for shapes and other things when they are zero dimensional. Points – or whatever corresponds to them, say, the parts in a decomposition of a shape – are uniquely distinguished in exactly the same way symbols are. But what happens in an algebra when i isn't zero? What can I find in the algebra U_{12}, for example, that corresponds to points? Is there anything I can count on?

The rule

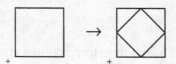

looks as if it should be context free, at least under the unremarkable description

square → square + inscribed square

I have been using all along. There is one square in the left side of the rule, and there are two squares in the right side. In fact, the rule works perfectly in this way in the two computations in the previous section. Recall that the first computation

has two steps, and that the longer computation

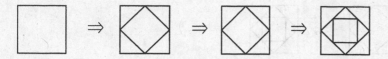

has a redundant step that repeats the preceding shape. Other descriptions of the rule, however, may be more appropriate. What if I want to use it together with another shape rule that finds triangles? The identity

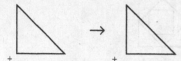

shouldn't change anything. In fact, any rule I add will automatically work as intended with any rule I already have. The part relation lets me recognize triangles – or upper case Ks and lower case ks – after I've combined squares. So it might be necessary to look at the rule

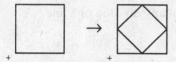

under the alternative description

   square → four triangles

The rule is still context free, but it doesn't work. It applies once to the square

and then can't be applied again to the shape

Now there are four triangles and no squares. I can fix this by dividing triangles and squares into three sides and four sides apiece. After all, this is the way they are normally defined. this gives me a new way of describing the rule

   square with four sides → four triangles with three sides apiece

But if this is the case, then the rule must be context sensitive. Four lines are in the left side of the rule and 12 lines are in the right side. And it is easy to see that the increase in complexity pays off in the computation

where the long sides of four triangles become the four sides of a square. This is proof the rule is context sensitive; but wait, I may be going too fast. There's more to worry about. The rule doesn't work in the longer computation

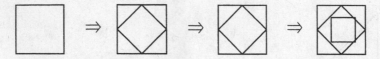

It has to apply to the outside square in the second step. Under the description of the rule I'm using now, each side of the outside square is divided in two. The outside square is four lines at the start, and then eight lines one step later. I can add another rule that changes how squares are described, maybe one with the description

> eight lines → four lines

or a rule that applies recursively to fuse the segments in any division of a line

> two lines → one line

Either way, a general rewriting system is defined. Alternatively, I can represent the rule

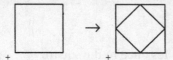

in two ways, so that squares have either four sides or sides described by their eight halves. Notice further that if I have reason to put four triangles in the right side of the rule, then its different versions are of different kinds. What the rule is depends on how it's described. Maybe the rule is context free, maybe it's context sensitive, and maybe it's in a general rewriting system. I have no way of deciding how complex computations are for the shape

in terms of shape rules alone. I have to examine every computation separately and determine how different descriptions of shapes and rules interact. There are too many computations and too many descriptions for this to be practical.

Some classifications aren't worth keeping. They're vacuous. But the Chomsky hierarchy is far from empty. The only problem with it is that the way complexity is measured by counting symbols (units) doesn't work for shapes unless they're zero dimensional. And there's no reason to think that it should. Shape rules depend on seeing. They don't have parts that I can count. These depend on how rules are applied in computations. The description that makes sense for a rule now isn't binding later. What the rule is doing may change as I calculate.

The difference between shape rules and generative grammars is evident in other kinds of computational devices. Their rules are described in alternative ways. But from what I can tell, they're all easy to define in the zero dimensional algebras V_{0j} in which labels are associated with points. This includes devices with or without spatial aspects.

Turing machines are typical. They're defined for symbols on tapes – labeled points evenly spaced on a line – that are modified in accordance with given transitions. For example, the shape rule

$$A \overset{\textstyle .}{} \qquad \overset{\textstyle .}{A'} \overset{\textstyle .}{}$$

$$\overset{\textstyle .}{} \to$$

$$+ \quad \overset{\textstyle .}{} q \qquad + \qquad \overset{\textstyle .}{} q'$$

in the algebra $V_{0\,2}$ corresponds to a machine transition in which the symbol A is read in the state q and replaced by the symbol A'. The state q is changed to the state q', and the tape is moved one unit to the left. I have assumed that the rule applies just under translations. Three points are needed to mimic the transition when all of the transformations are allowed. This shape rule

$$A \overset{\textstyle .}{} \qquad \overset{\textstyle .}{A'} \overset{\textstyle .}{}$$

$$\to$$

$$+ \quad O \overset{\textstyle . \;\; .}{} q \qquad + \quad O \overset{\textstyle . \;\; .}{} q'$$

does the trick in the algebra $V_{0\,2}$. (The previous rule $AqR \to A'q'L$ also matches this transition. However, more than context-sensitive rules are required for generative grammars to simulate Turing machines. Turing machines and general rewriting systems are equivalent.)

There are many other examples. Cellular automata are labeled points in a grid, usually square, that has one or more dimensions, with neighborhood relations specified in rules that are applied in parallel under translations. Lindenmayer systems for plant forms are defined in algebras for labeled points. This is the same for production rules in expert systems, picture languages and pattern languages in architecture and software design, and graph grammars in engineering design and solid modeling. The point is obvious. Computation as normally conceived is zero dimensional. My set grammars for decompositions give added evidence of this. Shape rules are applied in two stages. These are reinterpreted in set grammars in terms of subsets, and set union and difference. The rules in set grammars rely on the identity relation, as rules do in all algebras where i is zero. I can even go on to treat the parts in decompositions as labels, so that set grammars work for shapes containing labeled points in the algebras V_{0j}. Shape rules in algebras where i is greater than zero, however, depend on embedding. This lets me calculate with shapes made up of basic elements that aren't zero dimensional. There's no reason to count points labeled in different ways to calculate. Why not see instead?

I DON'T LIKE RULES, THEY'RE TOO RIGID

It comes as no surprise that the shape

is produced by applying the shape rule

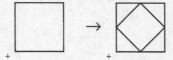

in the algebra $U_{1\,2}$. Squares are inscribed in squares in accordance with a fixed spatial relation between two geometrically similar shapes A and B in the scheme $v \rightarrow A + B$. But what happens if I want to produce the shape

and others like it by inscribing quadrilaterals in quadrilaterals? This is a natural generalization, but one that presents a problem for shape rules. Squares work because they are rigid, whereas relationships between lines and angles can vary arbitrarily in quadrilaterals. Quadrilaterals that are inscribed one in another needn't repeat the same spatial relation. And no two quadrilaterals have to be geometrically similar. I have to define an indefinite number of rules to get the shapes I want with the scheme $v \rightarrow A + B$. I can avoid this embarrassment if I use indeterminate rules – this is an excellent example of how they work to my advantage – but the intuitive idea of inscribing quadrilaterals in quadrilaterals recursively is lost. A much better solution – one that keeps the original scheme $v \rightarrow A + B$ – is to generalize rules.

A shape rule schema

$$x \rightarrow y$$

is a pair of variables x and y that take shapes – or whatever there is in one of my algebras – as values. These are given in an assignment g that satisfies a given predicate. Whenever g is used, a shape rule

$$g(x) \rightarrow g(y)$$

is defined. The rule applies in the usual way. In effect, this allows for shapes and their relations to vary within rules, and it extends the transformations under which rules apply. It provides a nice way to express many intuitive ideas about shapes and how to change them in computations.

My earlier classification of determinate rules (see Table 2.4) implies that there are algorithms to find all of the transformations under which a rule applies to a shape. But are there algorithms that do the same thing for assignments and schemas? What kinds of predicates allow for this, and what kinds of predicates do not? These are good questions that have yet to be fully answered.

What I'm going to say should be obvious by now. Definite descriptions (decompositions) fix the shapes in a rule $g(x) \to g(y)$. The description of $g(x)$, however, may be incompatible with my description of the shape to which the rule applies. This simply cannot be right. It wouldn't be a computation if these descriptions didn't match. But what law says that shapes must be described consistently in order to calculate? The shape $g(x)$ may have indefinitely many descriptions besides the description that defines it. But none of these matter. The part relation is satisfied by shapes – not by descriptions – when I try the rule. My account of how a computation works may use multiple descriptions that jump from here to there erratically. It may even sound crazy or irrational. But whether or not the computation ends favorably with useful results doesn't depend on this.

Here

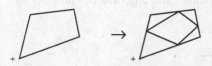

is an example of a shape rule defined in a schema $x \to y$ that produces the shape

and others like it from a given quadrilateral in the algebra U_{12}. (Anytime I wish, I can add another schema, so that I can have any outermost quadrilateral. If this is a square, then I can inscribe squares in squares as above.) An assignment g gives values to the variables x and y according to this predicate:

x is a quadrilateral, and $y = x + z$, where z is a quadrilateral inscribed in x.

The predicate is framed in terms of the scheme $v \to A + B$. It can be elaborated in more detail – for example, the vertices of z are points on the sides of x – but this really isn't necessary to show how shape rule schemas work.

There are indefinitely many predicates equivalent to this one. I can alter the predicate to make y four triangles rather than two quadrilaterals. This gives the same rules, and it confirms what I just said every time I use another rule to replace a quadrilateral. The description of a shape in a rule needn't match the description of the shape to which the rule is applied.

It is also worth noting that the above schema produces the shape

and others like it, because the schema can be applied to the same quadrilateral more than once under different assignments. This is easy to avoid in a number of ways. For example, I can put a notch in the most recent quadrilateral, so that the schema applies just to it. Or I can use a combination of algebras – say, U_{02} and U_{12} – to get a schema in which points distinguish quadrilaterals. A rule defined in such a schema might look like this:

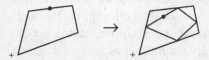

The same idea lets me do a lot of other things as well. For example, it's a cinch to combine two rules that are indeterminate to define a new rule that's not. Further, I can control the symmetry properties of rules to my advantage when I combine them.

Another easy example may be useful. It's one that illustrates once again why indeterminate rules are good to keep around. Because points are geometrically similar and lines are, too, I can define the universe of shapes for the algebras U_{0j} and U_{1j} by using the pair of shape rules

The empty shape is in the left sides of both rules, and a point is in the right side of one and a line is in the right side of the other. However, because planes and solids need not be geometrically similar, I have to use shape rule schemas to obtain the universe of shapes for the algebras U_{2j} and U_{33}. These schemas $x \rightarrow y$ are defined in the same way. The empty shape is always given for the variable x, and either a triangle in U_{2j} or a tetrahedron in U_{33} is assigned to the variable y. A rule in the algebra U_{22} might look like this:

Shape rule schemas have an amazing range of uses in architecture and engineering. Many interesting examples depend on them. In architecture – to enumerate just a few – there are Chinese ice-ray lattice designs, Ancient Greek meander designs, Palladian villas, and original buildings from scratch. In engineering, there are truss and dome designs and MEMS.

PARTS ARE EVANESCENT; THEY CHANGE AS RULES ARE TRIED

I keep saying that the parts that shapes have depend on how rules are used to calculate. How does this work in detail to define decompositions? I'm going to sketch three scenarios, each one included in the next, in which parts are defined and related according to how rules are applied. The scenarios are good for shapes with points as well as lines, and so on. It's sure, though, that they're only examples. There are other ways to talk about rules and what they do to divide shapes into parts.

In the first scenario, only shape rules of a special kind are used to calculate. Every rule

$$A \rightarrow 0$$

erases a shape A, or some transformation of A, by replacing it with the empty shape 0. As a result, a computation has the form

$$C \Rightarrow C - t(A) \Rightarrow \cdots \Rightarrow r.$$

The computation begins with a shape C. Another part of C – in particular, the shape $t(A)$ – is subtracted in each step, when a rule $A \to 0$ is applied under a transformation t. The shape r – the remainder – is left at the conclusion of the computation. Ideally, r is just the empty shape. But whether or not this is so, a Boolean algebra is defined for C. The remainder r when it's not empty and the parts $t(A)$ picked out as rules are tried are the atoms of the algebra. This is an easy way for me to define my vocabulary, or to see how it works in a particular case.

In the next scenario, shape rules are always identities. Every identity

$$A \to A$$

has the same shape A in both its left and right sides. When only identities are applied, a computation has the monotonous form

$$C \Rightarrow C \Rightarrow \cdots \Rightarrow C$$

where as above, C is a given shape. In each step of the computation, another part of C is resolved in accordance with an identity $A \to A$, and nothing else is done. Everything stays exactly the same. Identities are constructively useless. In fact, it's a common practice to ignore them, but this misses their real role as observational devices. The parts $t(A)$ resolved as identities are tried, possibly with certain other parts, and with the empty shape and C define a topology of C when they're combined in sums and products. The first scenario is included in this one if the remainder r is empty, or if r is included with the parts $t(A)$: for each application of a rule $A \to 0$, use the identity $A \to A$. However, a Boolean algebra isn't necessary. Complements are not automatically defined.

In the final scenario, shape rules are not restricted. A rule

$$A \to B$$

applies in a computation

$$C \Rightarrow [C - t(A)] + t(B) \Rightarrow \cdots \Rightarrow D$$

in the usual way. The computation begins with the shape C and ends with the shape D. Whenever a rule $A \to B$ is tried, a transformation t of A is taken away and the same transformation t of B is added back. But this is only the algebraic mechanism for applying rules. What's interesting about the computation is what I can say about it as a continuous process, especially when parts are resolved in surprising ways. In the computation

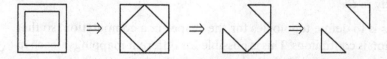

in the algebra $U_{1\,2}$ a pair of squares is turned into a pair of triangles by using the rule

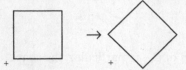

that rotates a square, and the rule

that rotates a triangle. It sounds impossible, but it works. What parts do I need to account for this change, so that there's no break or inconsistency? How are the parts the rules pick out related to one another? When do squares become triangles? And just what does it mean for the computation to be a continuous process?

I'm going to build on a configurational (combinatorial) idea that accords with normal experience to answer these questions. It's the very idea I've been against all along as a way of handling shapes in computations. However, the idea is indispensable when it comes to describing what happens as rules are applied. It's important to keep this foremost in mind. There's a huge difference between describing shapes and calculating with them. The latter doesn't depend on the former. In fact, their relationship is just the opposite. The idea I'm going to use is pretty clear. Things are divided into a finite number of parts. I can recognize these parts, individually and in combination, and I can change them when I like for repair or improvement, or to produce something new. The parts I leave alone stay the same, and they keep their original relationships. Changes are continuous, as long as they correspond in this way to how things are divided into parts.

This idea can be stated in a far more general way with mappings. These describe what rules do to shapes, and relate their topologies (decompositions). The exact details of different mappings may vary – and it's important when they do – but the core idea is pretty much as given above. Suppose a mapping from the parts of a shape C to parts of a shape C' describes a rule $A \rightarrow B$ as it's used to change C into C'. Then this process is continuous if two conditions are met.

1. The part $t(A)$ the rule picks out and replaces is in the topology of C. In terms of the intuitive idea above, the rule recognizes a part that can be changed.
2. For every part x of C, the mapping of the smallest part in the topology of C that includes x is part of the smallest part in the topology of C' that includes the mapping of x. As a result, if two parts are parts of the same parts in the topology of C, then their mappings are parts of the same parts in the topology of C'. The rule changes the shape C to make the shape C', so that the parts of C' and their relationships are consistent with the parts of C. No division in C' implies a division that's not already in C.

The trick to this is to define topologies for the shapes in a computation, so that every rule application is continuous. This is possible for different mappings – each in a host of different ways – working backward after the computation ends. (Of course if I'm impatient, I can redefine topologies after each step, or intermittently.) The mapping

$$h_1(x) = x - t(A)$$

is a good example of this. It preserves every part x of the shape $C - t(A)$ that isn't taken away when a rule $A \rightarrow B$ is used to change a shape C into another shape C'. But mappings may have other properties. If the rule is an identity, then the alternative mapping

$$h_2(x) = x - [t(A) - t(B)]$$

works to include the previous scenario in this one. Now $h_2(x) = x$, and every part of C is left alone. The computation is continuous if the topology of C is the same from step to step. In both cases, parts may not be fixed if there are more rules to try. What I can say about shapes that's definite, at least for parts, may have to wait until I've stopped calculating. Descriptions are retrospective.

The third scenario is also easy to elaborate in useful ways. In particular, the topologies of the shapes in a computation can be filled out when identities are applied to extend the computation without changing its results.

ERASING AND IDENTITY

Some easy illustrations show how these three scenarios work. Suppose I want to use the first scenario to define decompositions for the shape

in terms of the two shape rules

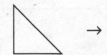   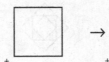

that erase triangles and squares. Five distinct decompositions result when the rules are applied to produce an empty remainder. These are illustrated neatly in this way

by filling in triangles. In the first decomposition, four squares are the atoms in the Boolean algebra that describes the shape. In each of the three succeeding decompositions, four triangles and two squares are picked out as the atoms in Boolean algebras, and in the last decomposition, eight triangles are.

There are two interesting things to notice. First, I can produce the above illustrations in parallel computations – that is, in computations carried out in a direct product of algebras – by recapitulating the use of the rules

in the algebra U_{12}, and building up the corresponding decompositions in a combination of the algebras U_{12} and U_{22}, so that squares contain lines, and triangles are planes. Alternatively, labeled lines or lines with weights fix the layers in this direct product. So a combination of the algebras V_{12} and U_{22} or of W_{12} and U_{22} will do the whole job. The graphics is more concise, but perhaps not as clear. I need two rules whatever I do. These,

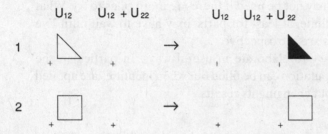

show my original idea. They are applied in the six-step computation

Step	Rule	U_{12}	$U_{12} + U_{22}$
1	2		
2	1		
3, 4, 5	1, 1, 1		
6	2		

to produce the third decomposition above. The rule used in each step of the computation is indicated by number. It's easy to see that the sequence of rule applications
could be different, as long as the same triangles and squares are picked out. In this
scenario and the next one for identities, decompositions do not reflect the order in
which rules are applied. The same rule applications in any sequence have the same
result. However, in the third scenario, decompositions may be sensitive to when rules
are used.

Second, the number $f(n)$ of different decompositions for any shape in the series

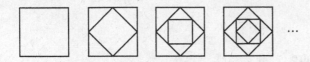

can be given in terms of its numerical position n. The first shape in the series is
described in one way – it's a square – and the second shape can be described in two
ways – it's either two squares or four triangles. Moreover, the number of possible
descriptions for each of the succeeding shapes – so long as I want the remainder to
be empty – is the corresponding term in the series of Fibonacci numbers defined by
$f(n) = f(n-1) + f(n-2)$. It's easy to see exactly how this works in the following
way:

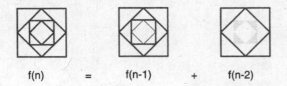

$$f(n) \qquad = \qquad f(n-1) \qquad + \qquad f(n-2)$$

If a rule resolves the inmost square in the shape $f(n)$, then there are $f(n-1)$ ways
to describe the remaining part of the shape; this part is one shape back in the series.
Alternatively, if a rule resolves the four inmost triangles in the shape $f(n)$, then there
are $f(n-2)$ descriptions; the remaining part is now two shapes back in the series.
So the total number of descriptions is $f(n) = f(n-1) + f(n-2)$. I can prove this
for any square and any four triangles that include the square. Still, the visual proof
is immediate enough to make the formal proof superfluous. Seeing is believing.

Now suppose I recast the rules

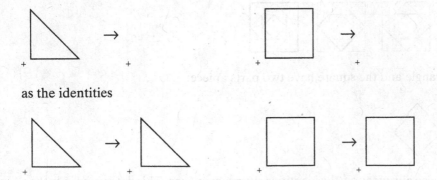

as the identities

and then use them to define decompositions according to the second scenario. Once

again, consider the shape

If I apply an identity to pick out this triangle

and an identity to resolve this square

then the following topology (decomposition)

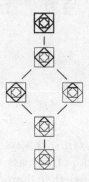

is defined. In this case, I have shown the topology as a lattice. It is easy to see that the right angle of the triangle and a corner of the square are distinguished, but other parts of the triangle and the square go unnoticed. If I recognize complements in addition to the parts I resolve when I use the identities, then a Boolean algebra is defined that has these four atoms:

Now the triangle and the square have two parts apiece:

Of course, I can always apply the identities everywhere I can. This determines another

Boolean algebra with the 24 atoms

There are a couple of things to notice. Triangles come in three kinds

with three parts or four, and squares come in three kinds

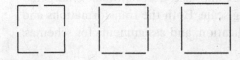

with four parts or eight. Moreover, the different decompositions of the shape may be regarded as subalgebras of the Boolean algebra, including the above five decompositions obtained in the first scenario. However, there's no guarantee that things will always work this way. A Boolean algebra need not be defined for every shape when identities apply wherever they can.

COMPUTATIONS ARE CONTINUOUS

I'm going to try another shape rule to show how the third scenario works. The rule

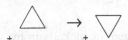

rotates an equilateral triangle about its center, so that this point is permanently fixed. The rule is applied repeatedly to define the computation

in the algebra U_{12}. The final shape in the computation is a rotation of the initial shape about its center. The transformation is surprising because the centers of the triangles in the initial shape change position as it's rotated:

The rule doesn't move the centers of triangles, but they move anyhow. What kind of paradox is this?

The answer is something to see. The rule can be applied to the fourth shape in the computation in two ways. In one way, the rule picks out the three triangles that correspond to the three triangles in the initial shape. But none of these triangles is resolved in the other way. Instead, the rule divides the fourth shape into two triangles – the large outside triangle and the small inside one – that have sides formed from sides of the three triangles that come from the ones in the initial shape. The rule rotates the two triangles in turn to get the fifth and sixth shapes. Now the rule can be applied in alternative ways to the sixth shape in the computation. Either the rule resolves both of the triangles that correspond to the ones in the fourth shape, or the three triangles – one at each corner of the sixth shape – that have sides formed from segments of sides of the triangles in this pair. The rule rotates the three corner triangles one at a time to complete the computation. (This way of using the rule almost certainly subverts another well-respected classification. Does the rule apply locally or globally? The rule generates shapes, and also detects their "emergent" properties. There's more to this than changing scale. Both the transformations and the part relation in the first stage of rule application, and assignments for schemas, are crucial.)

The nine shapes in the computation

are all made up of triangles. Their numbers summarize the action of the rule as it is applied from step to step. Consider the following three series:

$$
\begin{array}{ccccccccc}
3 & 3 & 3 & \boxed{5} & 2 & \boxed{5} & 3 & 3 & 3 \\
3 & 3 & 3 & \boxed{3} & 2 & \boxed{2} & 3 & 3 & 3 \\
3 & 3 & 3 & \boxed{2} & 2 & \boxed{3} & 3 & 3 & 3
\end{array}
$$

The first series shows the maximum number of triangles (these can be picked out by using an identity) in each of the shapes. The next series gives the number of triangles in a shape after the rule has been applied, and the last series gives the number of triangles in a shape as the rule is being applied. In both of these cases, the number of triangles depends on what the rule does either by rotating an existing triangle or seeing a new one to rotate, and then counting in terms of the triangle and its complement. The resulting inconsistencies in the fourth and sixth shapes tell the story. Counting goes awry: two triangles just can't be three, and neither number is five. (It's blind luck that two and three make five. In many cases, there's no easy relationship like this.) Nevertheless, the computation is a continuous process if appropriate topologies are given for the nine shapes.

The topologies for the shapes in the computation are defined as Boolean algebras with the atoms in Table 2.6. This is feasible by using the above mapping $h_1(x) = x - t(A)$. First a topology is given for the final shape. Any finite topology will do. I've used the trivial topology – the final shape is an atom – because the rule doesn't divide it. There's not much for me to say. The final shape is vague without parts or purpose, but ready for more calculation. Once the topology of the final shape is

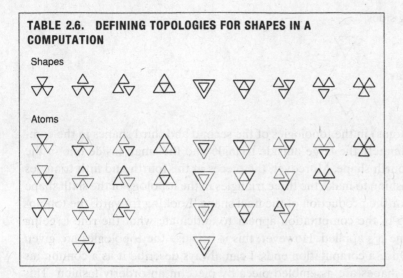

TABLE 2.6. DEFINING TOPOLOGIES FOR SHAPES IN A COMPUTATION

decided, the topologies of the preceding shapes are defined in reverse order to keep the application of the rule continuous. Each topology contains the triangle resolved by the rule – the topmost atom in Table 2.6 – and its complement. I can do this in different ways, with or without complements.

It's instructive to keep a record of the divisions in the triangle in the right side of the rule

formed with respect to these topologies. The triangle is cut in alternative ways in the eight steps of the computation:

The parts defined in the triangle combine sequentially to build up the different triangles that the rule picks out in its subsequent applications. The pieces

from the first three triangles combine in this way,

and the remaining sides

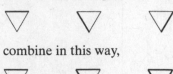

combine in this way,

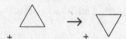

to define parts (atoms) in the topologies of the second and third shapes in the computation, and ultimately the large outside triangle and the small inside one in the topology of the fourth shape. Moreover, the pieces in the fourth and fifth triangles combine in like fashion to make the three triangles in the topology of the sixth shape that are needed for the production of the final shape. Looking forward, the topologies of the shapes in the computation appear to anticipate what the rule is going to do the next time it's applied. However, this is because the topologies are given retrospectively. Once a computation ends, I can always describe it as a continuous process in which shapes are assembled piece by piece in an orderly fashion. This makes a good story and a good explanation.

Every time the rule

is tried, its right side is divided with respect to a different topology. This shows in yet another way that the shapes in a rule have alternative descriptions that change as it's used to calculate. But again, the rule doesn't apply in terms of any fixed description. How the rule is described as it's being used is just an artifact of the computation that may help to explain what's going on.

Before I continue, it's worth noting that the computation

is not isolated. There are many others like it. And in fact, they can be defined around each of the shapes in this infinite array of nested polygons

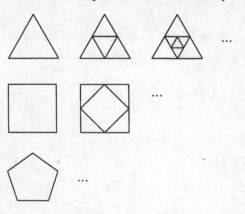

by using only a single shape rule schema $x \rightarrow y$, where x is a triangle or a regular polygon, and y is an appropriate transformation of x. The number of ways I can apply the schema to a shape in the array without remainder corresponds to the Fibonacci number of its column. This is described above for the series of squares inscribed in squares in the second row. Shapes are filled with ambiguity, and there are always rules to take advantage of it in one way or another.

Decompositions are defined for a purpose. At the very least, they're a nice way to talk about computations, and one way to explain what's going on. But there's a lot more. The decompositions that come from computations may describe shapes and provide the wherewithal to evaluate them when they are used in designs. Questions of function and use, manufacture and assembly, or repair and disposal are the kinds of things that can be addressed. This is often possible simply with rules that erase shapes or with identities. And many other things may also be involved, including any of the topological considerations – boundaries, complements, and so on – mentioned above. Moreover, stylistic analysis and the exploration of stylistic change rely on decompositions. Now they can be used to define new rules in terms of the scheme $v \rightarrow A + B$ and its inverse. This provides some innovative ways to calculate with shapes. Already, there are clear applications in architecture and product design.

Most of what I've been saying about rules and computations with shapes is anecdotal. I've been presenting decompositions as case studies. This is no surprise. The way ambiguity is used in a particular computation may have little to do with the way it's used in another. Ambiguity just won't go away. It's there whenever I apply shape rules to calculate. However, it's not handled in the same way all of the time. This is why ambiguity is so interesting, and why it's so valuable. The trick is to use ambiguity without limiting it by trying to generalize. There's no chance of this, as long as rules can vary as widely as shapes themselves. Shape rules fit the bill.

DESIGN WITHOUT THOUGHT

What is design? I used to agree wholeheartedly with Franz Reuleaux:

> ... Invention, in those cases especially where it succeeds, is Thought ...
> F. Reuleaux, *The Kinematics of Machinery* (1963)

But I've changed my mind. This is a dubious metaphor, even with Reuleaux's optimistic qualification, if shapes play a role in design. The kind of reasoning shapes allow – the kind of reasoning William James describes as the ability to handle ambiguity and the novelty it brings – implies something more than the combinatorial, configurational, deductive, zero-dimensional variety of thought Reuleaux takes for granted. Yet, this is clearly computation. Reuleaux's variety of thought is evident in expert systems and other computer models. Whatever these use – artificial intelligence and heuristic search, evolutionary algorithms and optimization, formal logic, or common sense – they're invariably going to be zero dimensional. (It appears this is an empirical fact. Nothing keeps me from using heuristics or optimizing when I calculate with shapes. However, if evolution requires definite units, then evolutionary algorithms are limited to points. This bodes ill for more than design. Life itself may hang in the balance.) So what is design? Right now, I would say perception rather than thought. This is a provocative way to emphasize an important aspect of design. The creative activity I have in mind isn't thought. Design depends on new ways of looking at things. This is exactly what shapes and computations with shapes are all about.

If I have tried to show anything in this chapter, it's that shapes with points aren't like shapes with lines, planes, and solids. Going from zero-dimensional shapes to higher-dimensional ones – from shapes where there are permanent units that I can count to shapes where unforeseen parts may pop in and out of existence whenever I look – alters almost everything. Identity and embedding simply aren't the same, and counting may only discourage seeing. These are philosophical claims as much as algebraic or geometrical ones. I can always see things in new ways when I calculate with shapes made up of lines, planes, and solids. Shapes with points – and like devices whether decompositions or representations in computer models – limit what I can see and how much I can change my mind. Knowing how many points there are doesn't make up for the loss. The last thing I want to do is count.

Is all this freedom really necessary? What do others say about design? The consensus among those who have thought about it is that design practice is flexible and open ended. Donald Schon provides a good sense of this in his flattering account of the reflective practitioner. Schon puts "reframing" and "back talk" at the center of ideal practice in design and in other professions. This is the ability to interact with your work in the same unstructured way that you argue about something new in vague and shifting terms that haven't been defined. (When it's finally time to define your terms, there isn't very much to say.) It's the ability to reconfigure what you're doing before and after you act, to respond freely as you see things in different ways, and to try new ideas in an ongoing process whenever you like. This is what shape rules are for. Nothing prevents me from seeing triangles and Ks after drawing squares, and then acting as if that's what was there all along. There's no reason to assume I have to know anything definite before I see. When shape rules are used in computations, they meld redescription and inference. I can handle ambiguity and the flow of shifting interests and competing goals all at once. Multiple perspectives interact in intricate ways in the flux of practice. This provides the creative means to deal with new situations. Shape rules and the algebras in which they are defined promise a useful account of this kind of process.

BACKGROUND

James Gips and I published the original idea for shape rules in 1972. Today, the literature on shape rules and the way they are used to calculate is much too large to cite item by item. Most of the relevant material has appeared in the journal *Environment and Planning B: Planning and Design*. Papers date from my first one in 1976, – "Two exercises in formal composition," – in which I introduced the scheme $v \rightarrow A + B$ to define shape rules by using spatial relations. These were enumerated from scratch, and fixed in decompositions formed by applying identities. The first use of shape rules to generate designs – Chinese ice-ray lattices – in a given style was described in my next paper the following year. With respect to the themes covered in this chapter, I would recommend papers by the authors in this list: alphabetically, Chris Earl (shape boundaries and topologies of shapes), Ulrich Flemming (architectural styles), Terry Knight (shape rules defined from scratch, stylistic analysis and stylistic change, and more), Ramesh Krishnamurti (computer implementations), Lionel March (miracles and what they mean), and myself. There are many other authors, though, and a lot more of equal interest.

The material from William James is in *The Principles of Psychology* (1981, pp. 957–958) in the chapter on reasoning. The section is aptly headed "In reasoning, we pick out essential qualities." For James, the essential qualities of things may "vary from man to man and from hour to hour" (p. 961). This is the same for shapes. Their parts are redefined every time a rule is tried. It strikes me sometimes that James is especially good at describing what happens when rules are used to calculate with shapes. This may be self-indulgent, but I can't resist quoting him at length at least once. The following passage resonates perfectly with one side of what I've been saying about counting and seeing. (It doesn't deal with the details of calculating that establish equivalences between complex things with numerable units and shapes without definite parts.)

> The relation of numbers to experience is just like that of 'kinds' in logic. So long as an experience will keep its kind we can handle it by logic. So long as it will keep its number we can deal with it by arithmetic. *Sensibly*, however, things are constantly changing their numbers, just as they are changing their kinds. They are forever breaking apart and fusing. Compounds and their elements are never numerically identical, for the elements are sensibly many and the compounds sensibly one. Unless our arithmetic is to remain without application to life, we must somehow *make* more numerical continuity than we spontaneously find. Accordingly Lavoisier discovers his weight-units which remain the same in compounds and elements, though volume-units and quality-units all have changed. A great discovery! And modern science outdoes it by denying that compounds exist at all. There is no such thing as 'water' for 'science'; that is only a handy name for H_2 and O when they have got into the position H-O-H, and then affect our senses in a novel way. The modern theories of atoms, of heat, and of gases are, in fact, only intensely artificial devices for gaining that constancy in the numbers of things which sensible experience will not show. "Sensible things are not the things for me," says Science, "because in their changes they will not keep their numbers the same. Sensible qualities are not the qualities for me, because they can with difficulty be numbered at all. These hypothetic atoms, however, are the things, these hypothetic masses and velocities are the qualities for me; they will stay numbered all the time."
>
> By such elaborate inventions, and at such a cost to the imagination, do men succeed in making for themselves a world in which real things shall be coerced *per fas aut nefas* under arithmetical law (p. 1250).
>
> W. James (1981)

The cost to the imagination is twofold: there's the effort to think up "elaborate inventions," and then there are the limitations these put on what I can make of my future experience. I'm free to use my imagination anyhow I like when I calculate with shapes.

The quotation from Franz Reuleaux is in *The Kinematics of Machinery* (1963, p. 20). Donald Schon tells the story of the reflective practitioner every chance he gets. I like his account in *Educating the Reflective Practitioner* (1987) the most.

George Miller's speculative comments at the beginning of this chapter are in his paper "Information Theory in Psychology" in the omnibus volume *The Study of Information: Interdisciplinary Messages*, edited by Fritz Machlup and Una Mansfield (1983, pp. 495–496). Miller's extension of the engineer's idea of noise to cover ambiguity proves the value of the latter in creative work. How else could he invent this unlikely formula:

ambiguity = noise

The equivalence, however, is within the ambit of common sense: ambiguity is just an annoyance that can always be removed. But shapes imply that both of the clauses in this statement are false. What do others have to say in Machlup and Mansfield's book? Allen Newell (pp. 222–223) is unconcerned that units are used to describe things in computer models of thought. This is a "nonissue" in computer science, at least for artificial intelligence. And Douglas Hofstadter (p. 270) thinks visual puzzles – IQ test problems of the kind invented by Alfred Binet and Lewis Terman – probe the mechanisms of intelligence. These are visual analogies, and may include those defined in shape rules and schemas. But according to Hofstadter, all visual puzzles aren't alike. Some invented by other, less talented people – do they have lower IQs? – have the unfortunate property of being filled with ambiguity. Hofstadter wants to keep visual puzzles as experimental tools, and he hopes to save them in the obvious way by removing any ambiguity. How is he going to do this, so that they're still visual and useful?

What happens to rules and schemas? Despite their bickering in Machlup and Mansfield, both Newell and Hofstadter seem to be in agreement with Francis Galton. He had the idea of measuring mental ability before Binet and Terman. Galton never tired of saying, "Wherever you can, count." This becomes a habit that's hard to break. But why start? You can count me out. There's more to computation than counting. And there's more to design with shapes. Seeing makes the difference. Shapes hold the creative choices ambiguity brings, so that decompositions, definitions, and the evanescent segments and shards of analysis are now afterthoughts of computation. Seeing makes it worthwhile to calculate with shapes.

ACKNOWLEDGMENT

Mine Ozkar did a wonderful job on the figures. She also made sure I said something about everything I'd drawn.

REFERENCES

Environment and Planning B: Planning and Design (1976–2000). Vols. 3–27.

James, W. (1981). *The Principles of Psychology*. Harvard University Press, Cambridge, MA.

Machlup, F. and Mansfield, U., Eds. (1983). *The Study of Information: Interdisciplinary Messages*. Wiley, New York.

Reuleaux, F. (1963). *The Kinematics of Machinery*. Dover, New York.

Schon, D. A. (1987). *Educating the Reflective Practitioner*. Jossey-Bass, San Francisco.

CHAPTER THREE

Engineering Shape Grammars

Where We Have Been and Where We Are Going

Jonathan Cagan

INTRODUCTION

Shape grammars (Stiny, 1980; see also Chapter 2 in this book), originally presented in the architecture literature, have successfully been used to generate a variety of architectural designs including irregular Chinese lattice designs (Stiny, 1977), villas in the style of Palladio (Stiny and Mitchell, 1978), Mughul gardens (Stiny and Mitchell, 1980), prairie houses in the style of Frank Lloyd Wright (Koning and Eizenberg, 1981), Greek meander patterns (Knight, 1986), suburban Queen Anne houses (Flemming, 1987), and windows in the style of Frank Lloyd Wright (Rollo, 1995). The derivation of the grammars and the designs generated by their use have revealed systematic logic in the generation of classes of architectural designs. For example, Koning and Eizenberg's Frank Lloyd Wright prairie house grammar demonstrated that, through the application of a relatively simple set of predefined rules, an infinite number of designs illustrative of Frank Lloyd Wright's own prairie houses can be created. One could question whether Wright consciously or unconsciously used such a set of rules to generate his houses; or one could not care but rather just marvel at the power of a simple rule set, based in the geometry of shape, to generate such rich and representative designs.

We can next ask, if the logic of architecture can be revealed and capitalized upon to represent and generate creative designs within known styles, can the same be done in engineering? In other words, is there a logical structure for the synthesis of engineering design and, if so, can it be modeled and used to create new engineering artifacts? This chapter attempts to address these questions. The simple answers are yes: there is a logical structure for synthesis within engineering, rooted in the interaction between function and form, and styles that emerge as a result of this interaction, to drive realizable design details. Through shape grammars, classes of known artifacts can be recreated, but also through shape grammars, and often the same grammar, unique and novel solutions can be created. Further, within engineering design, spaces of artifacts can be explored through directed search algorithms to drive designs toward characteristics that best meet any given set of design objectives and constraints.

The remainder of this chapter will review shape grammars and the current state of the art in application to engineering design. Next we will explore the issues surrounding the design of an engineering shape grammar and the issues that remain to be solved as we look toward future theoretical and application contributions using shape grammars.

SHAPE GRAMMARS

A shape grammar is a form of production system (see Stiny and Gips, 1980). What differentiates a shape grammar production system from a traditional (first predicate) rule-based system are (1) geometry, (2) parameterization, and (3) emergence (Agarwal and Cagan, 2000). A shape grammar (Stiny, 1980, 1991) derives designs in the language that it specifies by successive application of shape transformation rules to some evolving shape, starting with an initial shape (I). In particular, given a finite set of shapes (S) and a finite set of labels (L), a finite set of shape rules (R) of the form $\alpha \rightarrow \beta$ transform a labeled shape α in $(S, L)^+$ into a labeled shape β in $(S, L)^0$, where $(S, L)^+$ is the set of all labeled shapes made up of shapes in the set S and symbols in the set L and $(S, L)^0$ is the set that contains in addition to all of the labeled shapes in the set $(S, L)^+$ the empty labeled shape $\langle S_\phi, \phi \rangle$. Parametric shape grammars are an extension of shape grammars in which shape rules are defined by filling in the open terms in a general schema. An assignment g that gives specific values to all the variables in α and β determines a shape rule $g(\alpha) \rightarrow g(\beta)$, which can then be applied on a labeled shape in the usual way to generate a new labeled shape.

The shape grammar formalism is defined in a variety of algebras built up in terms of a basic hierarchy (Stiny, 1992). The initial algebras $U_{ij}, 0 \leq i \leq j$, in the hierarchy are given in this table:

$$
\begin{array}{cccc}
U_{0\,0} & U_{0\,1} & U_{0\,2} & U_{0\,3} \\
 & U_{1\,1} & U_{1\,2} & U_{1\,3} \\
 & & U_{2\,2} & U_{2\,3} \\
 & & & U_{3\,3}
\end{array}
$$

Each algebra U_{ij} in the table contains shapes made up of finitely many basic elements of dimension i that are combined in an underlying space of dimension $j \geq i$. For example, in the algebras $U_{0\,2}, U_{1\,2}$, and $U_{2\,2}$, shapes are defined in a plane with points, lines, and planes, respectively. Most shape grammars have been written as a Cartesian product of $U_{0\,2}$ and $U_{1\,2}$, using points and lines that act in planes, though more recent work discussed in this chapter presents $U_{2\,2}$ grammars. Algebras of shapes can also be augmented by labels ($V_{ij}, 0 \leq i \leq j$) or weights ($W_{ij}, 0 \leq i \leq j$) to obtain new algebras in which the shape operations have been redefined to reflect different possible labels and weights on different entities (Stiny, 1992).

Semantics of shape grammars can be considered descriptions of the function, purpose, or use of the shapes generated by a shape grammar. Stiny (1981) defines such descriptions by a description function $h: L_G \rightarrow D$, which maps the language L_G described by grammar G onto descriptions D. These descriptions form the theoretical basis to combine function and form so important in engineering design, because the semantics of the grammar enable the form to be viewed from a given functional perspective.

WHERE WE HAVE BEEN

EARLY WORK

To date there has been quite limited exploration into the application of shape grammars in engineering design. Within engineering, and in particular the mechanical, civil, and electromechanical disciplines, function and form and their relation are critical for understanding the completeness of any design. Initial explorations into the use of grammars in engineering design focused more on function than on shape. Early work centered on discrete function grammars through the use of labels and symbols alone. Mitchell (1991) presented function grammars as shape grammars that are limited to the generation of both realizable and functional designs, and he illustrated this point with an example of a function grammar for the design of primitive huts. Fenves and Baker (1987) presented a function grammar for the conceptual design of structures, using architectural and structural critics to guide the design configuration. In a similar fashion, Rinderle (1991) presented an attribute grammar, a string grammar with a set of attributes that describe parametric and behavioral properties that are attached to every symbol, for the configuration of discrete planar structures. Finger and Rinderle (1989) described a bond graph grammar for the form and function configuration of mechanical systems. In all of these grammars, function was the focus, and any use of form came in a symbolic fashion based on the functional properties.

Early work on shape-oriented grammars can be found in Spillers (1985), where heuristics based on a grammar approach were used to introduce new structural members into a truss structure. Here, as with Rinderle (1991) above, trusses could be generated, but only in a user-directed fashion to generate limited form types. Fitzhorn (1990) and Longenecker and Fitzhorn (1991) presented shape grammars specifying the languages of both manifold and nonmanifold constructive solid geometry and boundary representations (i.e., realizable solids).

All of these efforts provided an early view of how to model the relationship between function and form within a grammar formalism for engineering application. Two aspects of these works missing and critical for the advancement of shape grammars in engineering fields were (1) a detailed application illustrating the potential power of shape grammar representations, and (2) a means to seek out useful and preferred designs within a design space generated by a shape grammar. These two aspects were critical because until a shape grammar could model a detailed engineering process and be used to generate significant design solutions, it was unproven that shape grammars were practical, even on a research scale, in engineering domains. Also, pure generate and test does not in itself effectively solve engineering problems; rather engineers desire solutions that best meet often competing objectives within tight constraints. If such a solution were to fall out of the language of a shape grammar directly and consistently, then the grammar would have to be narrowly defined and obvious, thus prompting the question of its value in solving real engineering problems. Instead, desirable solutions had to be obtained from a more general grammar in a directed manner. Two somewhat concurrent efforts focused on these two aspects of engineering shape grammars as now discussed.

THE LATHE GRAMMAR

Brown, McMahon, and Sims-Williams (1994) developed a grammar that specifies the language of shapes manufacturable on a simplified axisymmetric lathe. The grammar represents drilling, external turning, boring, and external and internal gripping, and it provides a formal computable specification of the capabilities of the lathe, represented by ordered sets of simple two-dimensional parametric shapes. Labels attached to the shapes, together with their associated attribute values, indicate the type, position, and status of the machine tools and the grips. Each grammar rule represents a transition between a combined state and machine and a new state. Individual rules are represented as rewrite rules on sets of parametric attributed labeled shapes by using an extension of Stiny's attributed set grammar formalism (Stiny, 1982).

There are over 120 rules in the grammar, divided into six sets – one each for external gripping, internal gripping, drilling, turning, boring, and retracting the tool. The majority of the rules do not modify the shape, but examine the workpiece and collect data (using attribute values) on the state of the shape and the machine. These data then act as constraints on subsequent rule applications. At any given time, the state of the workpiece is represented by the current shape, and the state of the machine is represented by the attribute values and positions of the shape labels. The legal transitions between states of the workpiece and lathe are defined by the grammar rules. The changing values of the attributes, and the propagation of the labels over the shape as the shape changes, model the physical process that creates the finished object. The shapes in the language of the grammar then represent the parts that can be machined from the initial workpiece by using the lathe. An example rule from Brown's grammar is found in Figure 3.1, and an example of its application is found in Figure 3.2. Figure 3.3 illustrates a part generated by the grammar. This grammar illustrated the ability of shape grammars to model detailed realistic engineering systems.

SHAPE ANNEALING

In focusing on the second problem, that of directed search, Cagan and Mitchell (1993) introduced the shape annealing method, which combines a grammatical

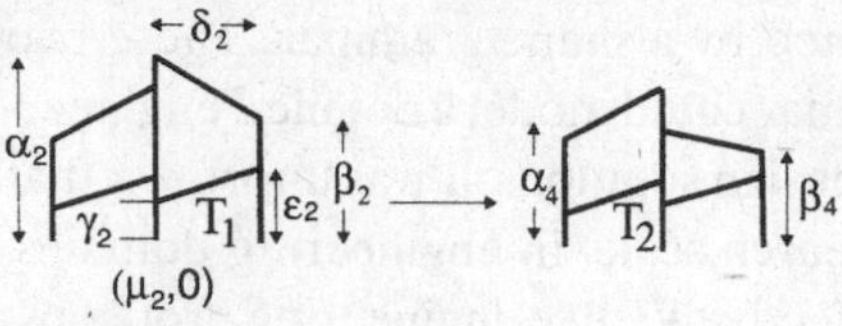

$C_L: \delta_2 <= T_1.l, \ \alpha_2, \beta_2 < \eta$
$\quad \gamma_2 < T_1.h + \delta_2 * T_1.m <= \alpha_2$
$\quad \varepsilon_2 < T_1.h <= \beta_2$

$C_R: \alpha_4 = T_1.h + \delta_2 * T_1.m, \ \beta_4 = T_1.h$
$\quad T_2.h = \alpha_4, \ T_2.l = T_1.l - \delta_2$

Figure 3.1. Example shape rule from Brown et al. (1994) (taken from Brown and Cagan, 1997).

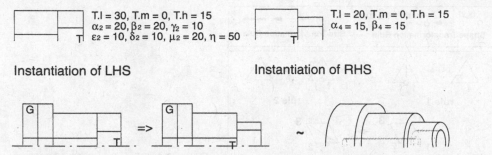

T.l = 30, T.m = 0, T.h = 15
$\alpha_2 = 20$, $\beta_2 = 20$, $\gamma_2 = 10$
$\varepsilon_2 = 10$, $\delta_2 = 10$, $\mu_2 = 20$, $\eta = 50$

T.l = 20, T.m = 0, T.h = 15
$\alpha_4 = 15$, $\beta_4 = 15$

Instantiation of LHS   Instantiation of RHS

Figure 3.2. Application of rule from Figure 3.1 (Brown et al., 1994) (taken from Brown and Cagan, 1997).

formalism to define the language of design alternatives and stochastic optimization to search this language for optimally directed designs. The need for a stochastic search technique was emphasized based on the discrete and multimodal characteristics of the objective function evaluations of the spaces generated by shape grammars. Though any stochastic method was usable, we chose the simulated annealing method (Kirkpatrick, Gelatt, and Vecchi, 1983). The basic idea is that a rule satisfying the left-hand side is (randomly) selected and applied; the resulting design is then evaluated and compared with the previous state. If the new design is an improvement, it is chosen as the current design and the process continues from that design state. If, however, the new design is evaluated to be worse than the previous state, then there is still a probability that the new design is chosen as the current state; this probability is based on the metropolis algorithm (Metropolis et al., 1953) and starts out high and decreases to zero as the algorithm proceeds. The 1993 paper by Cagan and Mitchell described a simple "vanilla" annealing schedule; more sophisticated, dynamic schedules were introduced in more recent work as described below (see, e.g., Shea, Cagan, and Fenves, 1997).

Until the shape annealing technique was introduced, the basic assumption in shape grammars was that the right rule was chosen at the right time; the question of how the rule was chosen was never addressed. The motivation of the shape annealing technique was to do this automatically, driven by the objective function. The significance of this becomes clear when focusing on automated engineering synthesis; useful and desirable engineering systems are generally guided toward preferred instantiations through directed or optimizing search.

The primary application of the shape annealing technique has been in the area of structural design. The design representation used in the shape annealing method is based on an analogy to a network (Cagan and Mitchell, 1994). A network is a means of representing design objects and the coupled relations between these objects, where design objects and their relations can have both functional and spatial attributes. It

Figure 3.3. Standard part generated by Brown's lathe grammar (Brown et al., 1994) (taken from Brown and Cagan, 1997).

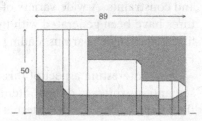

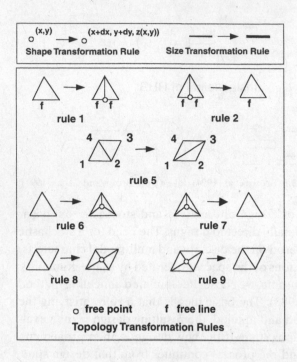

Figure 3.4. Space truss grammar (Shea and Cagan, 1997).

is these capabilities that make it amenable to structural synthesis: a network applied to discrete structures represents structural shape, where the design objects are joints and the relations between joints are the flow of force through structural members. A network, and thus a structure, can be evaluated and optimized based on global and local design goals of design object and relation attributes and acceptable levels of system flow, that is, behavioral constraints. The structural generation and optimization using shape grammars and shape annealing was originally introduced by Reddy and Cagan (1995a, 1995b). However, it was the work of Shea and Cagan (Shea, 1997; Shea et al., 1997; Shea and Cagan, 1997, 1999a, 1999b) that brought the technique to maturity as a consistent, robust method for structural synthesis. In this method, a simple shape grammar that represents a language of structures (mostly truss structures) is used; an example grammar, shown in Figure 3.4, represents a space truss grammar used to generate geodesic-like domes (Shea and Cagan, 1997). During shape annealing, a finite-element analysis is completed at every iteration to provide evaluation of the design objectives and constraints such as minimum weight and surface area while not failing by yield stress and buckling conditions. Dynamic annealing schedules and dynamic probability-based rule selection techniques are one contribution toward making the technique robust and efficient (see Shea, 1997, for details). Further, the method supports any articulable objective functions and constraints. A wide variety of both traditional and very interesting novel structures have been generated with the shape annealing technique. Example geodesic-like domes under various loading conditions and design objectives can be found in Figure 3.5.

One interesting aspect of this work is that syntax defines legal designs whereas semantics expresses design intent and supports evaluation. Function is implied in the grammar, but it is the external finite-element analysis that provides performance

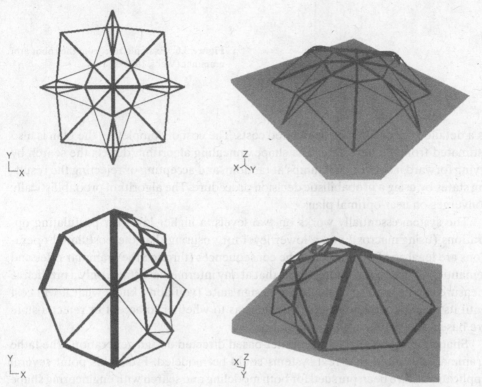

Figure 3.5. Symmetric and nonsymmetric geodesic-like domes generated by shape annealing (Shea and Cagan, 1997).

evaluation. Directed search through the simulated annealing algorithm seeks out desirable designs based on given objectives. The dome study (Shea and Cagan, 1997) and a roof study (Shea and Cagan, 1999a) best illustrate the varied design characteristics (called "design essays") that result from differing design objectives. The grammar itself, then, restricts the design space to geometrically feasible (though not necessarily functional) designs; the optimization drives solutions to be functional and high performing. This separation is important in engineering shape grammars; almost any specific problem will have a different set of objectives and constraints, so for a grammar to be useful beyond a single application it must be general, relying on the search strategy to filter and direct the generation of the solution.

Another significant application of shape annealing comes from the merging of Brown and coworkers' manufacturing grammar with shape annealing, resulting in a method to create optimal manufacturing process plans (Brown and Cagan, 1997). The underlying process model is developed by using a grammar of shape as described above. The formal semantics of the grammar interpret shape operations as manufacturing information, and the grammar and semantics are intended to provide a complete and correct specification of the domain. Planning is then a search process within the space defined by the grammar, using the semantics to guide the search. At any given stage, the approach can generate a possible operation by inspecting the current workpiece and the target part. Proposed operations are then translated into a sequence of grammar rules, which are applied to the current shape, and interpreted

Figure 3.6. Example rule from the robot arm grammar (Wells, 1994).

as a detailed operation with associated costs. The cost of completing the plan is also estimated from the new state. The shape annealing algorithm directs the search by trying forward moves or backjumps at random, and accepting or rejecting the resulting states by using a probabilistic decision procedure. The algorithm probabilistically converges on near-optimal plans.

The system essentially works on two levels: a higher level for postulating operations (using macros), and a lower level for ensuring that the postulated operations are legal and for generating the consequences (through the grammar rules and semantics). What is also interesting is that at any intermediate stage, only a predictive measure can be used to evaluate the design state (you don't know what it will cost until its plan is completed); thus decisions as to whether to accept or reject a state are based on a heuristic cost formulation.

Shape annealing enabled computer-based directed design generation. The lathe grammar illustrated that real systems could be modeled. From this point several applications have been pursued for both modeling and search with engineering shape grammars, as next discussed.

THE ROBOT ARM GRAMMAR

Wells and Antonsson (Wells, 1994) presented an engineering shape grammar (that they term an *engineering grammar*) to generate configurations of modular reconfigurable robot arms. The grammar generates all nonisomorphic assembly configurations while simultaneously calculating kinematic properties of the arms. The grammar models the properties of dyads and connects them together by means of arms. Within the grammar is modeled knowledge about how dyads behave; thus function is modeled through constraints embodied within the grammar that restrict the possible configurations that can be generated. An example rule can be found in Figure 3.6, and example configurations generated by the grammar can be found in Figure 3.7.

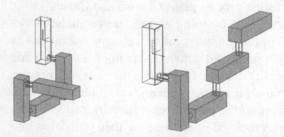

Figure 3.7. Example robot arms generated from Well's grammar (Wells, 1994).

MOVING ON: A COMMENT ON FUNCTION AND FORM
BY MEANS OF ABSTRACTION GRAMMARS

While focusing on engineering design problems, the relationship between the form of the device being designed and the function of the device is, as mentioned, critical. Within the structures work discussed above, the grammar itself was simple, modeling fundamental stability relationships while relying on external evaluators to interpret the function and behavior of the design. Within the robot arm grammar, function is directly mapped to the rules that constrain configurations. As machines and other more intricate products are considered, function must be addressed in a more fundamental manner. Representing function and its relationship to form, however, is a challenging problem. Early work on function described above, and placed within the scope of design descriptions as laid out by Stiny (1981), begins to address issues of discretely separated relationships between function and form.

Schmidt and Cagan pursued function grammars as applied to machine design within a directed search framework (Schmidt, 1995; Schmidt and Cagan, 1992, 1995, 1997, 1998). Originally conceived of as a shape grammar formalism application within shape annealing (Schmidt and Cagan, 1992) in which machines would evolve out of a stochastic annealing process with machine components represented by shapes, it was quickly recognized that the intricacies of functional descriptions within a directed search framework were in and of themselves a difficult problem that had to be addressed. A hierarchical *abstraction grammar* was laid out to discretize a continuum of function to form into useful levels for product function and form representation. One implementation uses three levels of abstraction: energy, kinematic mode plus energy, and component level. The component level represents form, but not, in practice, shape; for atomic components this may be sufficient, but the assumption of a decoupling between function and shape is strong here. Because metrics do not exist on the abstract level, but only on the form levels, and yet design conceptualization occurs on the function levels, directed search occurs along the hierarchy through a recursive simulated annealing algorithm. Both string and graph grammars were developed, generating concepts for optimally performing drills and carts. More recently, Schmidt extended the graph grammar work to the design of planar mechanisms (Schmidt, Shetty, and Chase, 1998), in which interesting kinematic structures of epicyclic gear trains are generated from the grammar. Still, the coupling that can occur between function and form has to be addressed. See Chapter 6 in this book for more discussion on functional representation and synthesis.

CONSUMER PRODUCT DESIGN: THE COFFEE MAKER GRAMMAR

The next direction for shape grammar development has been in the area of product development, with consumer products as the initial focus. With the consideration of products again comes the consideration of function as well as shape. The design of consumer products (such as coffee makers, telephones, toasters, and flashlights) often tends to be driven by a basic functional decomposition, but the products themselves are differentiated by form. Thus function and related form can be treated through discrete subsystems. This is true of the coffee maker shape grammar introduced in

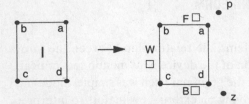

Figure 3.8. An initial shape rule from the coffee maker grammar that chooses a one-heater unit (Agarwal and Cagan, 1998).

Agarwal and Cagan (1998), the first to focus on a class of products. The grammar consists of 100 parametric rules, grouped within functional subsystems of the coffee maker, namely the filter, water storage container, and base, and each rule targets either the function or the form aspect of the coffee maker; not all 100 rules are used, but each functional unit must be designed. The grammar is a $V_{0\,2} \times V_{1\,2}$ grammar operating on three views (top, side, front), resulting in a complete three-dimensional representation of the product.

The initial shape for the coffee maker, that of the coffee pot region, is shown on the left-hand side of Figure 3.8. Points labeled a, b, c, and d are parametric points of the vertical cross section that allow the design of coffee makers of different sizes. The first set of rules, called initial shape rules (rules 1–5), distinguish between the two main classes of coffee makers, those with one heating element and those with two, and break apart the basic form into three regions (for the filter, base, and water storage units). Labels F, B, and W are used to distinguish among the three regions. A square label is associated with each of the three regions for a one-heater design, and a triangle label for a two-heater design. The rule shown in Figure 3.8, for example, chooses a one-heater unit. The points labeled by p and z indicate the rightmost bounds on the design. The points are constrained to have the same rightmost (or $x-$) coordinate.

The next step (rules 6–26) is to design the details of the filter unit, first specifying functionality like sliding versus rotating filters, cone versus flat filters, and whether or not there is a drip stop in the filter. Then the form of the filter is designed to meet the functional specifications; see, for example, Figure 3.9(a). Next the details of the base design are specified (rules 27–37), enabling a polygonal or oval–circular base design; see Figure 3.9(b). Included in the base is the specification of the appropriate heater.

The process next turns to the water storage unit, the main differentiator among different form designs. Initially the cross-section details are specified (rules 38–42), including the placement of the water tubing (Figure 3.10). Rules 43–100 then enable the form design of the water storage unit and its integration with the filter and base units. Here four planes are specified (top and bottom of the filter and top and bottom of the base), and at each plane the shape of the top view section is specified; the final three-dimensional form is then blended between the four planes. Almost any cross-sectional shape can be specified by selecting any number of circles or squares in sequence and then sweeping each in a straight line or circular arc, possibly changing the size or aspects of the circles and square through the sweep. Each shape is then blended together. Figure 3.11 shows the top view of such a sequence, including its merging with the circular filter unit.

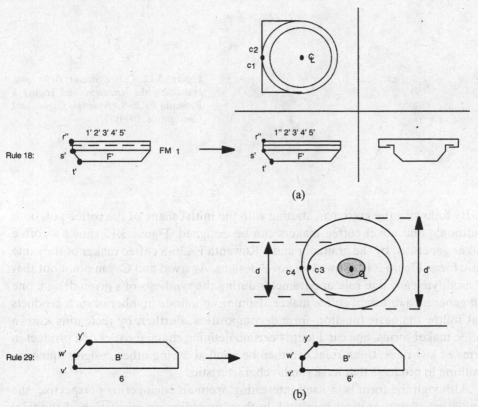

(a)

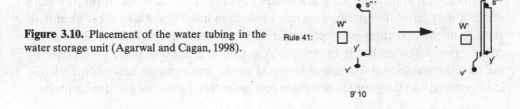

(b)

Figure 3.9. Example rules for the (a) filter and (b) base design from the coffee maker grammar; different views are shown (Agarwal and Cagan, 1998).

Figure 3.10. Placement of the water tubing in the water storage unit (Agarwal and Cagan, 1998).

Figure 3.11. Generation and blending of the top view of a water storage unit (Agarwal and Cagan, 1997).

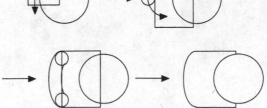

Figure 3.12. Coffee maker (left) generated by the grammar and (right) a Rowenta FK26-S (Agarwal, Cagan, and Constantine, 1999).

By following the grammar, starting with the initial shape of the coffee pot, both traditional and novel coffee makers can be designed. Figure 3.12 shows a coffee maker generated by the grammar and a Rowenta FK26-S coffee maker of the same basic form; Figure 3.13 shows two novel designs. Agarwal and Cagan point out that by modifying any one rule or parameter during the synthesis of a given design, one can generate a different coffee maker, defining an infinite number of such products that follow the basic function–form decomposition. Further, by recreating known coffee maker forms, one can identify certain defining characteristics of a product in terms of key rules; these rules can then be applied during other design sequences, resulting in products that have similar characteristics.

Although the form is in itself interesting, from an engineering perspective, the evaluation of any product is critical, both in providing an evaluation of the final design and also in providing feedback to aid in choosing from various rules during the generation process. Agarwal, Cagan, and Constantine (1999) argue that using performance metrics along with a grammar-based generative system will create a powerful feedback mechanism for the designer during the design generation process. To explore the ability of shape grammars to support the analysis of the resulting designs, they have incorporated costing evaluation into the coffee maker grammar. As any of the 100 rules is selected and its parameter values defined, the current cost of the coffee maker is evaluated. A manufacturing cost structure is defined for injection-molded parts, metal stamped parts, and product assembly, and it is incorporated into the grammar to obtain cost estimates during the generation process.

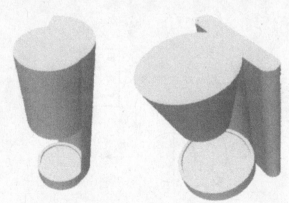

Figure 3.13. Two novel coffee makers generated by the grammar (Agarwal and Cagan, 1998).

The manufacturing structure is incorporated into the grammar by relating the part geometry and rule semantics to the various manufacturing processes. The cost of the coffee maker generated by the grammar shown in Figure 3.12 is estimated as $7.32 for the cost of manufactured and purchased parts and the cost of assembly; the same coffee maker is estimated as costing within 3% of this estimate within the literature.

A MEMS GRAMMAR

This next shape grammar requires a deeper exploration into the coupling that can exist between function and form. Agarwal, Cagan, and Stiny have developed a shape grammar to generate microelectromechanical systems (MEMS) resonators (Agarwal and Cagan, 1999; Agarwal, Cagan, and Stiny, 2000). One aspect of the design of MEMS devices that makes the problem difficult is the strong form–function coupling; a small change in the topology of a design may result in significant performance changes. The grammar is designed to first satisfy the minimum required functionality of the device (by including at least one actuator and one spring) and then modify the device to obtain the desired specifications. The grammar is a parametric labeled U_{22} grammar augmented by weights, that is, a $U_{22} \times$ {weights, labels} grammar, to differentiate and bring precedence to various types of components. Note that a two-dimensional representation is sufficient because MEMS resonators are 2-1/2 dimensional devices. The four main elements of a resonator are the central mass, actuators, anchors, and springs; each of these is assigned a different weight. The original grammar was a nonweighted labeled U_{12} grammar of 50 rules. The weighted labeled U_{22} grammar consists of only 15 rules to accomplish the same design characteristics.

Example rules are found in Figure 3.14, illustrating the addition of a comb drive actuator, the addition of a spring, and the modification of a spring. Figure 3.15 shows a scanning electron microscopy (SEM) image of an actual MEMS device, a similar MEMS device generated by the grammar, and a novel MEMS device also generated by the grammar.

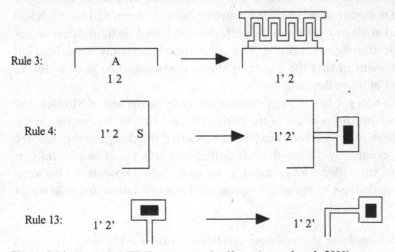

Figure 3.14. Example MEMS grammar rules (from Agarwal et al., 2000).

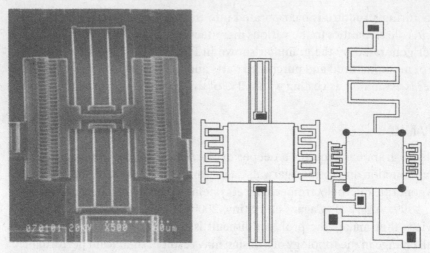

Figure 3.15. SEM image of a MEMS device, a similar device generated by the MEMS grammar, and a novel MEMS device generated by the grammar (from Agarwal et al., 2000).

The idea is not to just create a grammar, but rather a generative system – a designer assist tool – thus there is a need to align the grammar with a directed search algorithm such as shape annealing. Regardless of the final technique employed, the project emphasizes that in engineering design some form of directed generative design in conjunction with the shape grammar is what will lead to an effective design tool.

DESIGNING ARTIFICIAL HEARTS

McCormack and Cagan at Carnegie Mellon, and Antaki at the University of Pittsburgh's McGowan Center for Artificial Organs (McCormack et al., 1999) have developed a novel shape grammar[1] for the design of an artificial heart. The group at the McGowan Center have invented a new class of turbine-based artificial hearts. This grammar models the class of hearts and all variations, an infinite number of them. The grammar demonstrates a strong coupling between form and function and requires a detailed analysis to evaluate its performance, based on fluid dynamic, rotodynamic, and electromagnetic criteria. The application is certainly appealing, but it is also very pertinent in that the grammar shows potential to be used in future design generations of these devices.

The grammar is a $V_{0\,2} \times V_{1\,2} \times V_{2\,2}$ grammar currently composed of 51 rules. The grammar does not control the order of the design of the different subsystems of the heart, although there is precise termination. Labels drive the design and guarantee that all parts are eventually designed, even though any subsystem can be fully or partially defined in any order. The grammar is set up to define boundaries between subsystems that are defined by variables represented through labels; the four major

[1] The grammar was developed with additional input from Bradley Paden from University of California at Santa Barbara.

subsystems are the design of bearings, impellers, motors, and stators. Figure 3.16 illustrates two artificial hearts generated from the grammar. The first, Figure 3.16(b), is a re-creation of one of the current designs from the University of Pittsburgh, Figure 3.16(a), whereas the second, Figure 3.16(c), is a novel design generated by the grammar; in these figures the half-plane of a rotationally symmetric design is shown.

In addition to the shape grammar, engineering analyses can be associated with the rules to perform approximate evaluations of the devices generated. The vision is to first create a tool to assist designers in conceptualizing, visualizing, and gaining immediate feedback on new designs. Next the optimal selection of parameters and shape rules can be explored to help identify superior concepts and instantiations from the infinite set of devices modeled by the grammar.

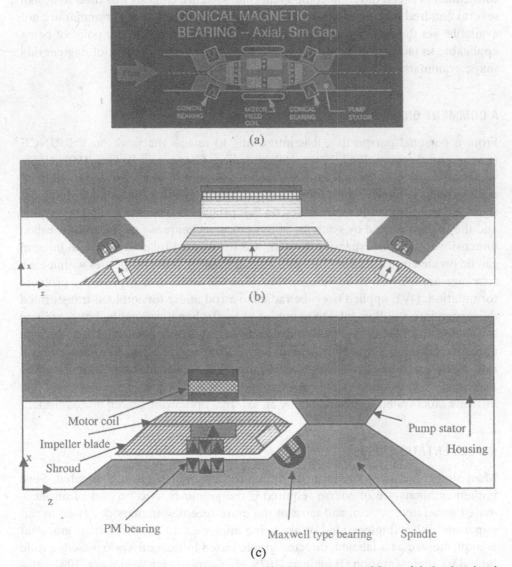

Figure 3.16. Artificial hearts generated from the artificial heart grammar: (a) actual design (reprinted with permission of J. Antaki), (b) similar design from the grammar, and (c) new design concept.

AIRPLANE TUBE ROUTING GRAMMAR

There is one example of a shape grammar being used in industry today. Heisserman led an effort at Boeing that has developed a shape grammar to route systems tubing through an airplane by using their Genesis generative design system (Heisserman, 1991, 1994; Heisserman and Callahan, 1996). The design representation in Genesis includes geometric models (boundary representation solid models), parts and assemblies, part classifications, part interfaces, and functional schematics. Genesis is used to interactively generate complete CAD geometry of aircraft tubing, including the associated fittings, clamps, and mounting brackets. The rules for this application include the manufacturing constraints for the geometry of the tubes, constructive rules for tube routing, fittings types and compatibilities, materials, and clearance constraints between different components and systems. Genesis was used to design several hundred tube assemblies on the 767-400ER. Details of the grammar are not available, yet the recognition that this technology has moved to the point of being applicable to industry is important to motivate future development of engineering shape grammars.

A COMMENT ON 1stPRINCE

From a personal perspective, it is interesting to review the work on 1stPRINCE (Cagan and Agogino, 1987, 1991a, 1991b; Aelion, Cagan, and Powers, 1991, 1992), which introduced mathematical expansion techniques to modify a design state by transforming geometries with the goal of creating innovative, better performing designs. All transformations were based on first principle formulations and reasoning, and they were directed by symbolic optimization techniques. One technique, called dimensional variable expansion (DVE), was formulated by noting that an integral can be divided into a series of integrals over subranges; if the properties within each subrange are allowed to be independent, then the effect is to expand the problem formulation. DVE applied over the radius to a rod under torsion load transformed the design into a hollow tube to minimize weight; when it was applied over volume to a mixed flow reactor, a series of mixed flow reactors were generated that, through inductive techniques, transformed into a plug flow reactor to maximize conversion rate. Note that those expansion techniques could be viewed as shape grammar rules, associated with detailed analytical expressions, driven by a highly constrained and intricate label system, and directed by an external optimizing search mechanism.

IMPLEMENTATION ISSUES

Most shape grammars found in the literature have not been implemented. Such implementations are of course required if the grammar is to be part of an automated generative system, and most of the more recent grammars discussed in this paper are either implemented or are being implemented. Our $V_{1\,2}$ truss grammar is implemented as a labeled, directed graph, based loosely on a winged-edge solid boundary representation (Baumgart, 1975; Heisserman and Woodbury, 1994). $U_{2\,2}$ grammars carry the difficulty of reasoning and manipulating planes directly (and not just the line and points that define them). One approach used in the MEMS grammar was to formalize the grammar as a $U_{2\,2}$ grammar with associated labels and weights yet implement the system through an equivalent $V_{1\,2}$ grammar; although not a pure one-to-one mapping of rules, there is a clear mapping of one to several between the two grammars. The artificial heart $V_{2\,2}$ grammar was implemented as a $V_{2\,2}$ grammar directly. The coffee maker grammar has been implemented in Java and, at this time can be accessed via the author's research web page.

All of these grammars have been driven by labels rather than emergent properties; except for the MEMS grammar, all of the them are set grammars, a subset of shape grammars. So long as emergence will not be required within an implementation, there is no need to implement the grammar using maximal lines (see Krishnamurti, 1980, 1981) or other means to detect emergent shapes; however, as emergence becomes more relevant, such fundamental work will become more pertinent. The difficulty in writing a grammar interpreter for engineering application is that shapes are not simple orthogonal lines or blocks; but tend to be more complex. McCormack and Cagan (2000) have introduced a parametric grammar interpreter able to recognize parametric emergent shapes by introducing a decomposition of shape through a hierarchy based on engineering properties. The result is that any $U_{1\,2}$ or $V_{1\,2}$ grammar (in the current implementation, though the theory extends to other algebras) can be quickly implemented. Further, the interpreter enables grammars to be written that utilize emergent shapes; such a feature is important for creative design. This interpreter should open up the scope of realizable engineering shape grammars. For example, we are using the interpreter to develop and implement a shape grammar for General Motors for the design of vehicle panels that uses emergent properties of shapes.

CREATING AN ENGINEERING SHAPE GRAMMAR

The current state of engineering shape grammars has been reviewed. Recent progress has demonstrated their use in very real settings, hopefully motivating others to think about applications. Below the issues that need to be considered in creating a shape grammar and the research issues that still must be addressed are considered.

DESIGNING A NEW GRAMMAR: WHAT ARE THE ISSUES?

There are eight fundamental issues that need to be discussed in relation to designing engineering shape grammars:

Simple Grammar Versus Knowledge Intensive Grammar. A simple grammar is one like Shea and Cagan's truss grammar, quite basic in that only a half-dozen rules are able to describe all legal geometries. A knowledge-intensive grammar, on the other side, is one like Agarwal and Cagan's coffee maker or McCormack et al.'s artificial heart grammar; these grammars act as a sophisticated expert system, one rich in geometric reasoning where creativity can be supported (as discussed below). The knowledge-intensive grammars generate feasible, functional designs, while the simple, more naïve grammars generate topologically valid but not necessarily feasible or functional solutions – a truss can be generated which protrudes through an obstacle and is overstressed. The knowledge-intensive grammars can be directed through search to generate designs of certain characteristics, but the simple grammars *require* such directed search.

Generation Versus Search. In the generation mode, use of a grammar produces designs that are feasible but are not likely to best meet certain design criteria. In the search mode, the grammar is used to search the design space for designs of certain characteristics and performance; in particular, directed search techniques such as shape annealing seek out optimal solutions among the many feasible designs within the language of the grammar.

These two issues, that of simple versus knowledge intensive and generate versus search, are interconnected as seen in Figure 3.17, which plots the amount of directedness versus the level of knowledge. The structural shape annealing method, combining a simple grammar with directed search, appears in the lower right. The robot arm grammar, because of its relative simplicity, lies in the lower left. The coffee maker grammar, in its current form a knowledge-intensive grammar that can generate designs without search, appears in the upper left. If the artificial heart grammar were to be used within an optimization framework, it would move from the upper left to the upper right. The same happened to the lathe grammar when merged with shape annealing. Knowledge tends to improve the efficiency of the use of a grammar but also restricts the amount of exploration done and novelty within the designs generated. The MEMS grammar, if used within an optimizing search strategy, would likely show a balance between how much knowledge is put into the grammar and the amount of pure exploration that can occur.

Independence Versus Coupling of Form and Function. The coffee maker grammar clearly was designed based on the decomposition (a.k.a. independence) of the functions of the different units. The MEMS grammar, though still identifying discrete function, is dependent on the coupling of the form and function, with the addition and modification of springs profoundly effecting the performance of the device; this effect is due to the continuum nature of the spring structures. The need for coupling over independence comes not from any notion of a more or less pure engineering representation, but rather from the need of the particular class of designs. That said, there is still much research to be done in modeling the potentially tight coupling between the function and form (and resulting behavior) of engineering systems within shape grammars.

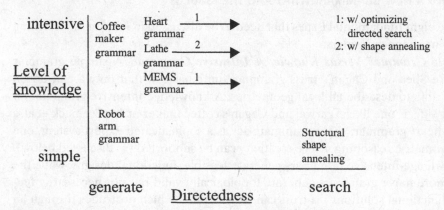

Figure 3.17. Connection between knowledge level and search level of grammars.

Symbolic–Atomic Grammar Versus Emergent Grammar. To date, emergence has not occurred in the pure sense in implementations of an engineering grammar. Some of the early truss shape annealing work showed trapezoids in the final design generated only from triangles; however, this came from the conscious decision to remove a truss member. For configuration problems of atomic components, emergence can happen only through chunking of components into subsystems; however, pure form emergence could result from models of continua or from nonsymbolic representations (e.g., maximal lines). Current work on applications of shape grammars in industry are indicating the benefit of supporting such emergence.

What makes these emergent forms interesting from an engineering perspective is really the implied function within that form; in engineering design, form often is designed to follow function. Thus the consideration of functional emergence becomes even more important. Chunked components effectively represent functional emergence such as a gear on top of a rack being grouped into a rack and pinion. Less obvious functional emergence might occur if a rib is used for structural support but also improves heat transfer capabilities. It may be that functional emergence stems from changing perspectives on the same representation.

Fixed Versus Parametric Grammar. Generally speaking, parametric grammars have more flexibility that is appropriate for most engineering applications. All of the shape grammars since Brown's lathe grammar have been parametric. The parametric nature of shape grammars enables the grammar to concisely represent large (sometimes infinite) variations within a class of designs. It is that characteristic that makes them most attractive for application to intricate engineering domains. However, fixed grammars may have use in engineering design when combinatoric enumeration of symbolic–atomic components is desired (such as with the robot arm grammar).

Type of Algebra that Will Best Model a Given Application. At a minimum, lines (a $U_{1\,2}$ grammar) will be necessary to represent any two-dimensional grammar, and thus any interesting engineering system. A $U_{2\,2}$ grammar may have more concise and powerful expressive properties if significant transformation occurs in the plane; however, with the $U_{2\,2}$ grammar comes complexities in implementation. Because the third dimension is important to many engineering artifacts, $U_{i\,3}$ grammars will become important as the maturity of engineering shape grammars continue to develop. To date we have been able to ignore the third dimension directly by considering symmetry (e.g., the heart grammar) or working only in planes with simple abstractions into the third dimension (e.g., the coffee maker grammar). Note that label ($V_{i\,j}$) and weight ($W_{i\,j}$) grammars each play an important role as well; the issues discussed in this chapter for shape ($U_{i\,j}$) grammars relate as well to label and weight grammars.

Including Evaluation with the Grammar. Evaluation is time intensive, and the effort to associate the proper level of evaluation with the grammar rules is high. It may be that from an education perspective, or when exploring the form space, evaluation is not necessary. However, engineering systems are driven by performance, and without an evaluation system design generation will occur in a vacuum. Thus evaluation of engineering designs is, in practice, necessary. The issue of when to evaluate, however, is dependent on the application and intended use of the grammar. If one wishes only to explore the design space and then observe behavior and performance, then postgeneration evaluation is acceptable. Further, if the cost of generation and evaluation is small then it may be acceptable to generate complete designs and then evaluate them. However, the power of shape grammars can be better illustrated if evaluation is provided during the generation process itself to provide feedback to the design engine (human or computer). The problem becomes *how* to evaluate intermediate designs. With the coffee maker grammar, intermediate evaluation was straightforward because manufacturing cost was directly dependent on the current geometry and context. With the lathe grammar, intermediate evaluation was much more difficult because the cost of the manufacturing plan was dependent on all of the steps toward creating the part; there an approximation was used as an intermediate evaluation.

Routine Versus Creative Design. Within bounds, both can be supported by shape grammars. Clearly shape grammars, as a form of production systems, can support routine design. A question (and potential criticism) of shape grammars is whether they support creative solutions. Brown and Cagan (1996), described the space for creativity within a shape grammar formalism as *bounded creativity*. They argue that, for a given design problem, there might exist a body of standard techniques, taking a designer through a series of actions, specifying the order in which subsolutions should be specified, and prescribing methods for achieving certain objectives, all specifying a routine design problem. Creativity, in their definition, occurs when a designer diverges from these standard procedures in the process of solving the problem. They contend that grammatical design is compatible with this definition, and, in particular, with bounded creativity. Realistically, a grammar does restrict the world of possible designs that can be generated. However, it also supports more focused designs within a world of legal configurations, and exploration beyond the few standard design instantiations often considered; see Figure 3.13 for novel coffee makers, or Figure 3.15(c) for a novel MEMS device. If leaps of technologies or new fundamental ways to look at a class of problems are desired, then a given shape grammar is unlikely to be useful and is unlikely to support that level of creativity. If, however, within an open but reasonably well-defined space novel configurations are desired, then a shape grammar is an ideal choice of representation. The idea of bounded creativity is that the grammar helps the designer move out of the traditional way of looking at the design problem.

WHAT SHOULD A GRAMMAR MODEL?

Next we explore what a grammar should model and how to determine whether it meets the needs specified. The issue becomes how expressive is the language of the grammar and how large is the design space it models. According to the discussion on bounded creativity, the larger the design space the more likely we are to discover novel designs; the smaller the design space, the more efficiently we can explore it. Figure 3.18 shows the space of all topologically valid designs ($\mathcal{S}$) and the space of all feasible designs ($\mathcal{F} : \mathcal{F} \subseteq \mathcal{S}$), where feasibility here is defined based on the satisfaction of design constraints. How much of these spaces should the grammar model? The Language A (L_A), which generates designs D_A, illustrates a grammar that describes a

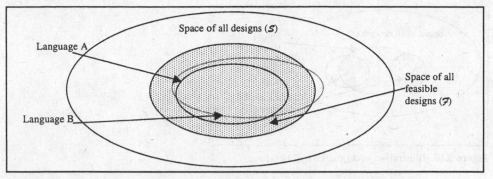

Figure 3.18. Mapping of the feasibility of the space generated by various types of grammars.

proportion of the space of feasible designs ($D_A : \forall D_A \in \mathcal{F}$); all designs in the language are feasible. The dotted Language B (L_B), which generates designs D_B, illustrates a grammar that describes a portion of the feasible designs but also some designs that are not feasible ($D_B : \exists D_B \in \mathcal{F}, \exists D_B \notin \mathcal{F}$). If one wishes to model all feasible designs, then A would be the desirable grammar where the size of A is as large as possible *within* the feasible space. However, assume that a grammar within the feasible space covers only a small portion of that space, whereas a different grammar includes a large portion of the feasible space but also a few designs that are not feasible. In this case, which is better?

An argument could be made that the shape grammar should model only feasible designs – in this case Language A would be preferred. However, for automated design generation, if the cost of generating designs is low and the cost of evaluating those designs is also low, then one may wish a more complete exploration of the space with an external evaluator pruning out the infeasible, as well as suboptimal, designs – in which case Language B would be preferred. As the cost of searching the space or evaluating designs increases, the desire for a more compact language (and thus grammar) is desired. Along with this argument is an understanding of the characteristics of the resulting design space and where the preferred solutions (e.g., optima) lie; if the grammar can concisely pinpoint an optimal solution then the compact language would be desirable – why search the space if the one solution desired is known? If the optima, or good solutions, or at least those solutions that one wishes to explore, are distributed throughout the feasible space, then a more comprehensive grammar would be desired, possibly even if several infeasible designs were modeled.

The ideal grammar would be comprehensive yet model only feasible designs. However, we require the ability to explore the design space and seek out novel solutions. In engineering applications, evaluation is critical, and thus a good evaluator will help one search the space. If the only way to search a larger feasible space is to include some infeasible designs, then the best solution is to allow infeasible designs in the grammar and let the evaluator filter them out. All of the knowledge-intensive grammars that have been created have modeled only feasible designs based on the set of constraints imposed (those that dealt with function rather than performance), yet all of them have represented very large design spaces in which novel designs have been found. With simpler grammars as used in structural shape annealing, the

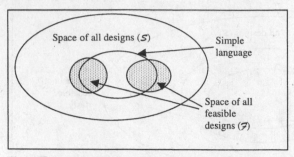

Figure 3.19. Illustration of disjointed feasible space.

grammars modeled only legal truss configurations (topologies), but they did allow for designs that were structurally infeasible (too large a stress in an element) or penetrated a geometric obstacle; those "infeasibilities" were easily ascertained from the evaluator and penalized until they became feasible; the annealer allowed for inequality constraint violations and then pushed the design into the feasible region as the annealing progressed with the idea that moving through an infeasible design could lead to a better feasible solution. This is seen in Figure 3.19, where feasible designs are formed by disjoint clusters. A simple grammar might include both clusters but also the infeasible designs in between (e.g., with the truss grammar the infeasible designs may be mapped by an obstacle). Here an external evaluator and directed search mechanism (shape annealing with finite-element analysis) would push the designs from one feasible cluster through the infeasible region into the other feasible cluster. Without the external evaluator and search strategy the grammar would have had limited use, but with the evaluator and directed search method the combination of the grammar and optimizer became a powerful design tool.

HOW TO CREATE AN ENGINEERING SHAPE GRAMMAR

With an understanding of each of the issues laid out above, we can now focus on the design of the grammar itself. Only directions and insights are given here, as each grammar will be dependent on its context of knowledge and use based on the results of the above analyses. Both the creation of the grammar and the evaluation of the generated designs are briefly addressed.

Creating the Grammar. The creation of each of our grammars began with observation. The truss grammar used basic definitions of truss kinematics. The coffee maker grammar began by reverse engineering several coffee makers and viewing their form and function. The MEMS grammar was developed by reading through the MEMS resonator literature, understanding what types of designs had been generated to date, and what the purpose of a resonator is; this last point led to extended capabilities of the grammar to generate designs beyond those under current consideration.

Once an understanding of functionality, functional decomposition, and basic form requirements are understood, the next focus is on defining form and function subsystems. The preference is for a consistency between the decompositions of form and function, but this may not be possible depending on the application. This function and form decomposition of the coffee makers into the central coffee pot, the filter unit, the base unit, and the water storage unit made the breakdown straightforward. The grammar begins with the coffee pot and builds the other units off of it; integration occurs by parameter instantiation and the merging the different units together. The artificial heart grammar required an understanding of overall integration as well as the intended function of each individual subsystem; the bearings, impellers, motors, and stators are each discrete subsystems, but their performance is tightly coupled as is highlighted through its evaluation. The MEMS grammar builds from the central mass and requires minimal functionality prior to more detailed design options.

For the knowledge-intensive grammars, labels typically control the generation process. Modifying, adding, and removing labels directs the design engine to access different parts of the grammar based on what parts of the device have been designed. For example, the form details of the water storage unit of the coffee maker can only be designed once the parameter values of the filter unit have been defined; this is necessary in order to guarantee that the water storage unit will properly blend with the core unit. Labels also direct the generation of each subsystem within the heart grammar. Although the grammar is designed so that the design engine can jump back and forth between the different subsystems, because each part does not have to be completed at one time, labels become the critical mechanism to guarantee that only those parts for which there is enough information can be designed (i.e., only those rules with sufficient left-hand-side information can be applied). As general parametric shape interpreters evolve, the dependence on labels may give way to direct properties of the shape geometry.

In designing a grammar a certain number of known designs should be removed from the observation process to act as a control group. The grammar can be judged to be effective if it recreates not only the observed designs but also those in the control group; of course, it has to generate new designs as well. In the coffee maker grammar we took apart three coffee makers, but we re-created six (four of which were not included in the three that were reverse engineered).

Design Evaluation. In designing an engineering grammar, various levels of evaluation may be necessary to provide feedback to the design engine that uses the grammar. As discussed, this evaluation could occur after completion of the design as a check on performance, or during the generation process itself. For evaluation to be included during the generation process, criteria that can be evaluated must be associated with individual grammar rules. When the evaluation is dependent only on the current geometry and context, the appropriate analyses can be performed at any point in the design generation process. Each perturbation to the design caused by any rule must be directly linked to a set of analyses or simulations.

If the intermediate stages are less directly associated with the current geometry state, then the issue of how to evaluate the state of a design based on the application of a rule becomes a much more challenging problem. Often an intermediate evaluation is impossible to know accurately because it is dependent on downstream design details. For example, a configuration problem must have the complete configuration to evaluate overall performance or cost. This was the case with an attempt to use shape annealing to perform component placement (Szykman and Cagan, 1993). Although it was able to perform simple component packing in three dimensions, as more realistic product layout problems were considered, the ability to evaluate the quality of a layout required all of the components to be present. In this case an approximation has to be used. The creation of the approximation may in itself be a difficult task, requiring insight into the domain of application and engineering judgment as to what details can be sacrificed to deliver an appropriate approximation. In the case of the component placement problem, the solution was to move away from shape grammars and focus more on optimizing perturbation methods by using move sets (e.g., Szykman and Cagan, 1995). For the lathe process planning with shape annealing, a reasonable approximation was created based on a detailed description of the operation of the lathe, including material to be removed, the different tools used, the speed of rotation of a workpiece, and the rate at which a tool cuts material.

CONCLUDING REMARKS: WHERE WE ARE GOING

This chapter has discussed the state of the art in engineering shape grammars, recognizing that there has been limited work on applying the technology to engineering domains. By examining the evolution of work in this field, we hope the reader recognizes that the time has come; significant applications can be created for many classes of artifacts and are beginning to be used in practice. The last part of this exploration is designed to help the reader consider the motivation for developing grammars and the remaining research issues that will further advance the field.

In general why should researchers, teachers, and design practitioners care about shape grammars? There are four basic reasons.

The first is representation and control. A shape grammar is a logical way to concisely model, generate, and search large design spaces. For many applications it is an ideal way to do so. It also supports analysis that is critical to the design of engineering systems, completing the synthesize–evaluate loop.

The second is design to order. As industry moves toward one-of-a-kind design and manufacture within a class of products, the means to rapidly model and generate that class of products for any given set of parameters will be needed. Shape grammars have demonstrated that ability, as with the coffee maker grammar. Other applications require exploration into many alternatives within a class to find the best configuration for a given design situation. This is the intended use of the artificial heart shape grammar by the designers at the McGowan Center at the University of Pittsburgh. It is the repeatable nature of the design of classes of artifacts that makes shape grammars an appropriate representation for these types of design activity.

Another aspect of consumer products that makes shape grammars interesting and potentially important is in modeling corporate identity or brand. Once grammars for classes of products are modeled, certain features that can be identified within a set of products can be mapped into a set of rules from the grammar; whenever that set of rules is invoked, those features are included in the generated design. Those features can include basic forms. This concept is useful in modeling and identifying corporate identity. Companies often attempt to have a consistent style across their product lines. By identifying grammar rules that lead to features that are associated with the corporate style and requiring those rules and associated parameter values to be applied to all designs generated, the grammar can be used to generate products all of which map to the corporate identity. This idea was discussed in the coffee maker grammar paper (Agarwal and Cagan, 1998), in which Black & Decker, Braun, Krups, and Proctor Silex coffee makers were all generated and differentiated based on the rules included in their design.

The third reason is <u>logic of engineering</u>. Fundamentally there is a logical structure to the synthesis of engineering design. Much like the logic of architecture that has been articulated through shape grammars, shape grammars have the ability to capture and illuminate the logic of designing classes of engineering artifacts. Some of the theoretical and practical issues that shape grammar designers face have been discussed in this paper. Shape grammars support the representation, generation, analysis, and search of the design space. Do shape grammars model human design thinking? It may be that shape grammars provide a theory of engineering design. This question is open, and only future exploration will begin to answer it.

The fourth reason is <u>education</u>. Modeling a class of designs and the accompanying logic of engineering may better position educators to teach the design process to students and practitioners. When the methodological framework of a grammatical representation is used, a design process can be articulated in a very clear, usable way. For example, shape grammars are compatible with methods such as reverse engineering and other observation techniques; it is through observation and patterns that can be identified through reverse engineering that shape grammars for product classes are themselves created. Thus the result of a reverse engineering process can very well be a formal shape grammar that models the space of products of interest. Such exercises would enable students and practitioners to dissect a design process, model it, and then use the results to explore alternatives.

Finally, this chapter concludes with a summary of what the difficult issues are that still must be solved to make the application of shape grammars transparent to the user and creator.

1. There must be a better understanding and modeling of the coupling between function and form.
2. There must be a better ability to demonstrate emergence, be it form or functional; like formal theories for form emergence, is there a similar mathematical formulation for the theory of functional emergence?
3. Grammars must be able to be adaptable to changes (such as through the introduction of new technologies) that will allow for the dynamic expansion of the grammar space.
4. There must be the ability to rapidly implement grammars and have the flexibility to add new rules, most likely through the development of robust grammar interpreters.
5. The ability must exist to couple the implementation of the grammar with simulators used to evaluate their incremental performance.
6. Means must be constantly improved to perform optimizing directed search on a design language.
7. There must be the ability to explore many applications; they are out there and require engineers with an understanding of shape grammars to seek them out and model them!

ACKNOWLEDGMENTS

The author is grateful to Manish Agarwal, Matthew Campbell, Jay McCormack, and George Stiny for their discussions on these ideas and for comments on this manuscript. This manuscript was originally prepared for the National Science Foundation Workshop on Shape Grammars, held at the Massachusetts Institute of Technology in 1999.

REFERENCES

Aelion, V., Cagan, J., and Powers, G. (1991). "Inducing optimally directed innovative designs from chemical engineering first principles," *Computers and Chemical Engineering*, **15**(9):619–627.

Aelion, V., Cagan, J., and Powers, G. (1992). "Input variable expansion – an algorithmic design generation technique," *Research in Engineering Design*, **4**:101–113.

Agarwal, M. and Cagan, J. (1997). "Shape grammars and their languages – a methodology for product design and product representation." In *Proceedings of the 1997 ASME Design Engineering Technical Conferences and Computers in Engineering Conference: Design Theory and Methodology Conference*, ASME, New York, DETC97/DTM-3867.

Agarwal, M. and Cagan, J. (1998). "A blend of different tastes: the language of coffee makers," *Environment and Planning B: Planning and Design*, **25**(2):205–226.

Agarwal, M. and Cagan, J. (1999). "Systematic form and function design of MEMS resonators using shape grammars." In *ICED'99*, Technische Universität München.

Agarwal, M. and Cagan, J. (2000). "On the use of shape grammars as expert systems for geometry based engineering design," *Artificial Intelligence in Engineering Design, Analysis, and Manufacturing*, **14**:431–439.

Agarwal, M., Cagan, J., and Constantine, K. (1999). "Influencing generative design through continuous evaluation: associating costs with the coffee maker shape grammar," *Artificial Intelligence in Engineering Design, Analysis, and Manufacturing*, Special Issue, **13**:253–275.

Agarwal, M., Cagan, J., and Stiny, G. (2000). "A micro language: generating MEMS resonators using a coupled form-function shape grammar," *Environment and Planning B*, **27**:615–626.

Baumgart, B. G. (1975). "A polyhedron representation for computer vision," *AFIPS Conference Proceedings*, **44**:589–596.

Brown, K. N. and Cagan, J. (1996). "Grammatical design and bounded creativity," EDRC Report 24-124-96, Engineering Design Research Center, Carnegie Mellon University, Pittsburgh, PA.

Brown, K. N. and Cagan, J. (1997). "Optimized process planning by generative simulated annealing," *Artificial Intelligence in Engineering Design, Analysis, and Manufacturing*, **11**:219–235.

Brown, K. N., McMahon, C. A., and Sims-Williams, J. H. (1994). "A formal language for the design of manufacturable objects." In *Formal Design Methods for CAD (B-18)*, J. S. Gero and E. Tyugu (eds.), North-Holland, Amsterdam, pp. 135–155.

Cagan, J. and Agogino, A. M. (1987). "Innovative design of mechanical structures from first principles," *Artificial Intelligence in Engineering Design, Analysis, and Manufacturing*, **1**(3):169–189.

Cagan, J. and Agogino, A. M. (1991a). "Inducing constraint activity in innovative design," *Artificial Intelligence in Engineering Design, Analysis, and Manufacturing*, **5**(1):47–61.

Cagan, J. and Agogino, A. M. (1991b). "Dimensional variable expansion – a formal approach to innovative design," *Research in Engineering Design*, **3**:75–85.

Cagan, J. and Mitchell, W. J. (1993). "Optimally directed shape generation by shape annealing," *Environment and Planning B*, **20**:5–12.

Cagan, J. and Mitchell, W. J. (1994). "A grammatical approach to network flow synthesis." In *Formal Design Methods for CAD (B-18)*, J. S. Gero and E. Tyugu (eds.), North-Holland, Amsterdam, pp. 173–189.

Fenves, S. and Baker, N. (1987). "Spatial and functional representation language for structural design." In *Expert Systems in Computer-Aided Design*, Elsevier, New York.

Finger, S. and Rinderele, J. R. (1989). "A transformational approach to mechanical design using a bond graph grammar." In *Proceedings of the First ASME Design Theory and Methodology Conference*, ASME, New York.

Fitzhorn, P. (1990). "Formal graph languages of shape," *Artificial Intelligence for Engineering Design, Analysis, and Manufacturing*, **4**(3):151–163.

Flemming, U. (1987). "More than the sum of the parts: the grammar of Queen Anne houses," *Environment and Planning B*, **14**:323–350.

Heisserman, J. (1991). "Generative geometric design and boundary solid grammars," Ph.D. Dissertation, Carnegie Mellon University, Pittsburgh, PA.

Heisserman, J. (1994). "Generative geometric design," *IEEE Computer Graphics and Applications*, **14**(2):37–45.

Heisserman, J. and Callahan, S. (1996). "Interactive grammatical design." Presented at AI in Design '96, Workshop Notes on Grammatical Design, Stanford, CA, June 24–27.

Heisserman, J. and Woodbury, R. (1994). "Geometric design with boundary solid grammars." In *Formal Design Methods for CAD (B-18)*, J. S. Gero and E. Tyugu (eds.), North-Holland, Amsterdam, pp. 85–105.

Knight, T. W. (1986). "Transformations of the meander motif on greek geometric pottery," *Design Computing*, **1**:29–67.

Kirkpatrick, S., Gelatt, Jr., C. D., and Vecchi, M. P. (1983). "Optimization by simulated annealing," *Science*, **220**(4598):671–679.

Koning, H. and Eizenberg, J. (1981). "The language of the prairie: Frank Lloyd Wright's prairie houses," *Environment and Planning B: Planning and Design*, **8**:295–323.

Krishnamurti, R. (1980). "The arithmetic of shapes," *Environment and Planning B: Planning and Design*, **7**:463–484.

Krishnamurti, R. (1981). "The construction of shapes," *Environment and Planning B: Planning and Design*, **8**:5–40.

Longenecker, S. N. and Fitzhorn, P. A. (1991). "A shape grammar for non-manifold modeling," *Research in Engineering Design*, **2**:159–170.

McCormack, J. and Cagan, J. (2000). "Enabling the use of shape grammars: shape grammar interpretation through general shape recognition." In *Proceedings of the 2000 ASME Design Engineering Technical Conferences: Design Theory and Methodology Conference*, ASME, New York, DETC2000/DTM-14555.

McCormack, J., Cagan, J., and Antaki, J. (1999). "A shape grammar for the streamliner artificial heart," (working paper).

Metropolis, N., Rosenbluth, A., Rosenbluth, M., Teller, A., and Teller, E. (1953). "Equation of state calculations by fast computing machines," *Journal of Chemical Physics*, **21**:1087–1092.

Mitchell, W. J. (1991). "Functional grammars: an introduction." In *Computer Aided Design in Architecture '91: Reality and Virtual Reality*, G. Goldman and M. Zdepski (eds.), New York: New Jersey Institute of Technology, pp. 167–176.

Reddy, G. and Cagan, J. (1995a). "Optimally directed truss topology generation using shape annealing," *ASME Journal of Mechanical Design*, **117**(1):206–209.

Reddy, G. and Cagan, J. (1995b). "An improved shape annealing algorithm for truss topology generation," *ASME Journal of Mechanical Design*, **117**(2A): 315–321.

Rinderle, J. (1991). "Grammatical approaches to engineering design, Part II: melding configuration and parametric design using attribute grammars," *Research in Engineering Design*, **2**:137–146.

Rollo, J. (1995), "Triangle and T-square: the windows of Frank Lloyd Wright," *Environment and Planning B: Planning and Design*, **22**:75–92.

Schmidt, L. C. (1995). "An implementation using grammars of an abstraction-based model of mechanical design for design optimization and design space characterization," Ph.D. Dissertation, Carnegie Mellon University, Pittsburgh, PA.

Schmidt, L. and Cagan, J. (1992). "A recursive shape annealing approach to machine design." Preprints of *The Second International Round-Table Conference on Computational Models of Creative Design*, Heron Island, Queensland, December 7–11, pp. 145–171.

Schmidt, L. C. and Cagan, J. (1995). "Recursive annealing: a computational model for machine design," *Research in Engineering Design*, **7**:102–125.

Schmidt, L. C. and Cagan, J. (1997). "GGREADA: a graph grammar-based machine design algorithm," *Research in Engineering Design*, **9**(4):195–213.

Schmidt, L. C. and Cagan, J. (1998). "Optimal configuration design: an integrated approach using grammars," *ASME Journal of Mechanical Design*, **120**(1):2–9.

Schmidt, L. C. Shetty, H., and Chase, S. C. (1998). "A graph grammar approach for structure synthesis of mechanisms." In *Proceedings of the 1998 ASME Design Engineering Technical Conference*, ASME, New York, DETC98/DTM-5668.

Shea, K. (1997). "Essays of discrete structures: purposeful design of grammatical structures by directed stochastic search," Ph.D. Dissertation, Carnegie Mellon University, Pittsburgh, PA.

Shea, K. and Cagan, J. (1997). "Innovative dome design: applying geodesic patterns with shape annealing," *Artificial Intelligence in Engineering Design, Analysis, and Manufacturing*, **11**:379–394.

Shea, K. and Cagan, J. (1999a). "The design of novel roof trusses with shape annealing: assessing the ability of a computational method in aiding structural designers with varying design intent," *Design Studies*, **20**:3–23.

Shea, K. and Cagan, J. (1999b). "Languages and semantics of grammatical discrete structures," *Artificial Intelligence in Engineering Design, Analysis, and Manufacturing*, Special Issue, **13**:241–251.

Shea, K., Cagan, J., and Fenves, S. J. (1997). "A shape annealing approach to optimal truss design with dynamic grouping of members," *ASME Journal of Mechanical Design*, **119**(3): 388–394.

Spillers, W. (1985). "Shape Optimization of Structures." In *Design Optimization*, Academic Press Inc., Orlando, pp. 41–70.

Stiny, G. (1977). "Ice-ray: a note on the generation of Chinese lattice designs," *Environment and Planning B: Planning and Design*, **4**:89–98.

Stiny, G. (1980). "Introduction to shape and shape grammars," *Environment and Planning B: Planning and Design*, **7**:343–351.

Stiny, G. (1981). "A note on the description of designs," *Environment and Planning B: Planning and Design*, **8**:257–267.

Stiny, G. (1982). "Spatial relations and grammars," *Environment and Planning B: Planning and Design*, **9**:113–114.

Stiny, G. (1991). "The algebras of design," *Research in Engineering Design*, **2**:171–181

Stiny, G. (1992). "Weights," *Environment and Planning B: Planning and Design*, **19**, 413–430.

Stiny, G. and Gips, J. (1980). "Production systems and grammars: a uniform characterization," *Environment and Planning B: Planning and Design*, **7**:399–408.

Stiny, G. and Mitchell, W. J. (1978). "The Palladian grammar," *Environment and Planning B: Planning and Design*, **5**:5–18.

Stiny, G. and Mitchell, W. J. (1980). "The grammar of paradise: on the generation of mughul gardens," *Environment and Planning B: Planning and Design*, **7**:209–226.

Szykman, S. and Cagan, J. (1993). "Automated generation of optimally directed three dimensional component layouts." In *Advances in Design Automation*, ASME, New York, **65**(1):527–537.

Szykman, S. and Cagan, J. (1995). "A simulated annealing-based approach to three-dimensional component packing," *ASME Journal of Mechanical Design*, **117**(2A):308–314.

Wells, A. B. (1994). "Grammars for engineering design," Ph.D. Dissertation, California Institute of Technology, Pasadena, CA.

CHAPTER FOUR

Creating Structural Configurations

Panos Y. Papalambros and Kristina Shea

INTRODUCTION

Objects with a physical existence in space must embody a structure that provides support for the mechanical function of the object, namely, dealing with mechanical loads, forces, and moments. In biological objects, structure usually is classified as endoskeletal or exoskeletal, see Figure 4.1. Endoskeletal structures provide an "internal skeleton" or a frame to which the rest of the object is essentially attached, like the bodies of humans and other vertebrates. Exoskeletal structures provide an "external skeleton" or shell within which the rest of the object is contained, like the body of a crab and other crustaceans. Inorganic object structures can be classified the same way. Early automobile designs used a frame-chassis structure to which everything was attached. More recent designs use a single (unibody or monocoque) main shell structure, whereas most recently a return to endoskeletal structures is being explored, combining metal frames with plastic or composite panels. The choice of structure clearly depends on many factors, including material use and operating environments.

Designing the structure of an artifact is probably the earliest engineering design task undertaken – building a tool, a shelter, or a bridge. For several reasons, structural design remains an important and challenging task. One challenge is designing a structure that meets not only behavioral criteria (load support and transfer) but also other requirements, for example, accessibility, manufacturability, and aesthetics. A second challenge is proper material use, recognizing that material properties often vary or are not precisely known. A third challenge is the dramatic variety of solutions a design problem can have in terms of the *connectivity* or *topology* of the structure – what we will henceforth call the *structural configuration*.

In this chapter we will explore how structural configuration design has come of age as a rigorous design synthesis capability based on a good mix of engineering science, mathematics, and the requisite practical heuristics. In several cases, the progress achieved during the past 15 years has turned a once-intractable problem into a relatively routine one. Our goal here is not to provide a rigorous, scholarly review. Rather, we will try to show how a few good ideas, slowly maturing over a couple of decades, provided a breakthrough design synthesis ability. The Resources

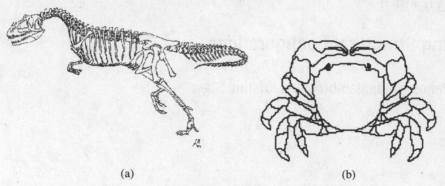

(a)   (b)

Figure 4.1. Endoskeletal and exoskeletal structures: (a) allosaurus, (b) crab.

section at the end of this chapter provides some attributions and credits for the work described throughout the chapter, so referencing will be sparse.

Inevitably, the authors' personal work and experiences color the exposition. In a bittersweet irony, circa 1986, the first author wrote:

> Drastic changes in the design topology cannot be dealt with mathematically in any reasonably useful way that the authors are aware of. Current developments in artificial intelligence are a potential promise of results in that direction.
> P. Papalambros and D. J. Wilde, *Principles of Optimal Design* (1988)

By the time the book was published in 1988, this statement was becoming rapidly obsolete. In this sense our present discussion should serve as an urge for strengthening our faith and perseverance in addressing truly difficult problems.

THE STRUCTURAL CONFIGURATION PROBLEM

The structural configuration problem is difficult because it addresses what is typically associated with creativity and experience, rather than with analysis: determine what the structure should "look like" in order to meet its design intent. Treating design intent in a quantitative fashion can be difficult because it often includes a wide range of both measurable and subjective goals. To illustrate the point, consider the now classic "bracket design" problem, assigned to students in a senior design class (Papalambros, 1988; Papalambros and Chirehdast, 1990).

A rocker arm or bracket device is to be designed and built to transmit a force at right angles, magnifying it by a factor of 3; see Figure 4.2. The device must pivot about a 0.5-in. (1.27 cm) steel rod and carry a maximum load of ±500 lb (226.5 kg). The forces are applied to the device by steel pins through holes located as shown in

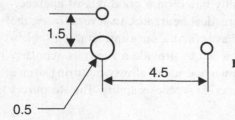

Figure 4.2. The bracket problem.

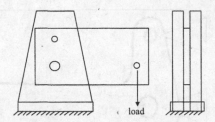

Figure 4.3. Testing rig schematic for bracket problem.

Figure 4.2. The precise sizes of pins and holes are to be supplied by the lab technician. The tools allowable for use in the student machine shop are band saw, drill press, coping saw, and file. The device must be constructed from a single $3.0 \times 6.0 \times 0.25$ in. ($\sim 7.6 \times 15.2 \times 0.63$ cm) rolled aluminum blank (the exact material specifications to be provided by the lab technician). Material may be removed but not added. The device must carry the loads without appreciable yielding of the material or buckling. The design load already includes a factor of safety, so the device is to be designed right up to the indicated failure level.

The device will be tested in a machine that measures deflection and force. A plot of deflection versus force will show a deviation from a straight line when yielding or buckling begins. The point of "significant deviation" is the failure point. The schematic for the test rig is given in Figure 4.3. Design criteria are the ability to carry the design load without failure, minimum weight, and ease of manufacture. A panel of judges will determine the winning design by awarding points based on a formula that weighs the above criteria.

In the 1988 experiment at the University of Michigan, the student teams were given 10 days to develop and build a solution with limited-time access to the machine shop and test lab. Four designs produced are shown in Figure 4.4, with weights A = 35.1 g, B = 37 g, C = 70.8 g, and D = 60.4 g. The winner was Design A. Design B took advantage from the actual test rig that allowed almost no load transfer between the pivot and the upper pin. If no load transfer is assumed, then the problem is a cantilever with one fixed end. Design C has a rather pleasant form but it is heavy. Design D is the simplest to make but it is also heavy. The students worked primarily intuitively, using only some basic finite-element analysis and cardboard prototype experimentation.

Looking at this example, we can observe a few things. If we ignore the manufacturing constraints, the configuration problem can be formulated as follows (Figure 4.5). Given the rectangular design domain and the boundary conditions at the indicated points, find the lightest structure that can carry the loads across the three points

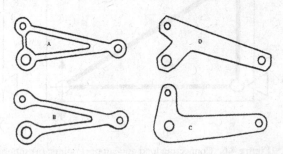

Figure 4.4. Alternative bracket designs developed by student teams.

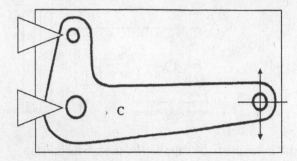

Figure 4.5. Placing a structure in the design domain.

without failure. Here by "boundary conditions" we mean collectively all points of interaction of the design domain with its environment, namely, both load points and support points. One can start with distributing material in the entire domain and then removing all material that is not needed. For example, Design C from Figure 4.4 was intuitively derived by using this approach by placing load-carrying emphasis on the lower support. The question, of course, is how much material can be safely removed and from where; that is, how to determine the exact topology and shape of such a design.

Alternatively, one can look at the connectivity implied by the boundary conditions, that is, examine all possible load paths between loads and supports and pick the most efficient combination. In the bracket problem, three points can be connected only by the triangular combination shown in Figure 4.6(a), which corresponds to Design A. Because the test rig takes care of load transfer between the supports, the connectivity shown in Figure 4.6(b) will suffice, resulting in Design B. This simple type of connectivity assumes straight line connections, namely, a classical truss design.

However, does this simple connectivity result in the most efficient design? In a more general problem where support locations and loads change, the elements of the truss may become relatively large in order to support the load, and self-weight may be comparable in magnitude to the external load. A more elaborate connectivity should then be examined, perhaps by the addition of intermediate joints at "strategic" locations. One possibility is shown in Figure 4.7, where three new joints were added. The problem now is how to determine the best number and locations of these intermediate joints.

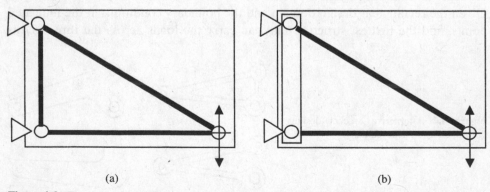

(a)   (b)

Figure 4.6. Connecting load and support points: (a) supports unconnected and (b) supports connected.

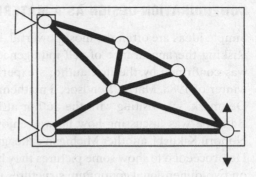

Figure 4.7. Adding intermediate joints to improve load transfer.

Thus far we have used the bracket example to illustrate two approaches for posing the configuration problem. The example also illustrates that two classes of structural configurations may be broadly considered as problem solutions. Continuous topologies, such as those in Figure 4.5 or Design C, are associated with the continuum mechanics viewpoint. Discrete topologies, such as in Figure 4.6 or Designs A and B, can be viewed as decisions about discrete elements: whether or not to use a particular connection member (or node) at a particular location.

In the remainder of this chapter we will see how these simple ideas have led to some very powerful design tools. Before we proceed, there is one remaining idea to bring forth. The entire exposition above approaches the design process as a process for making decisions. For a given problem, alternative designs that meet the design requirements are generated, and one is selected based on some comparison criteria. If we can find the proper mathematical representation, we can pose the following design optimization problem:

minimize

$$f(\mathbf{x})$$

subject to

$$\mathbf{h}(\mathbf{x}) = \mathbf{0},$$
$$\mathbf{g}(\mathbf{x}) \leq \mathbf{0}. \tag{1}$$

Here we assume that we can identify a vector of design variables $\mathbf{x}$ that fully describe all possible design alternatives. The vector functions $\mathbf{h}$ and $\mathbf{g}$ are expressed in terms of these variables, and the equalities and inequalities represent constraints that an acceptable or feasible design must satisfy – written in the standard form of Equation (1). Finally, the objective function f provides the comparison criteria to make the selection of the best design. Mathematically, the problem is a finite-dimensional nonlinear programming problem if $\mathbf{x}$ is a finite vector of real variables, and an infinite-dimensional variational problem if $\mathbf{x}$ is itself a function of another variable, say $\mathbf{x}(s)$, and f is a functional, usually an integral.

The structural configuration problem can be posed and solved as a design optimization problem. The challenges are how to formulate the design decision problem properly, and how easily that problem can be solved once formulated. We will now explore how these questions can be answered in a design context. The mathematics can be quite involved and are cumbersome in our present context. The references provided at the end of the chapter provide ample detail and mathematical rigor.

CONFIGURATION DESIGN AS A MATERIAL DISTRIBUTION PROBLEM

Simple ideas are often the most powerful, but they do not usually start as "simple." Risking the appearance of self-indulgence, we will convey here how this maxim was confirmed by the first author's experience with the problem at hand. In the winter of 1988, Martin Bendsøe, a mathematician from the Technical University of Denmark, was visiting with the author at his home base in Ann Arbor, Michigan. Bendsøe was discussing how he was enjoying the sabbatical visit in Denmark of Noboru Kikuchi, another Michigan colleague and next-door neighbor of the author. He proceeded to show some pictures they had generated, motivated by earlier work on two-dimensional continuum structures using "smear out" techniques.

Those fuzzy-looking pictures essentially represented a way of distributing a fixed amount of material inside a domain so that the boundary conditions could be met by the stiffest structure. Ironically, this was the same semester that the author had given the bracket problem to "warm up" his senior design student teams before they undertook their semester-long project. The author then presented the bracket problem and asked Bendsøe if he could solve it mathematically. Computational resources at the time, at programming and hardware levels, were limited, so they settled for a simpler problem that had symmetric loading. A few days later Bendsøe sent the solutions shown in Figure 4.8. The different pictures corresponded to different amounts of available material. Later that year Bendsøe and Kikuchi published a now-classic article (Bendsøe and Kikuchi, 1988) that ushered an era of rapid development in structural topology design.

The problem was cracked because a fundamentally different problem representation was introduced along with the mathematical finesse required for posing and solving it in a computationally tractable manner. The particular method used by Bendsøe and Kikuchi was based on the mathematical formalism of homogenization. Since then other similar approaches have been successfully introduced, but the breakthrough idea was still a very simple one: look at the structural configuration problem not as a problem of determining sizes and shapes but as one in which material must be distributed inside some given space.

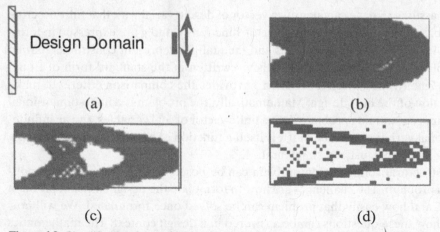

Figure 4.8. Simplified bracket problem and early homogenization solutions: (a) problem definition and (b) optimal material distribution. Progressively less material is made available for distribution in images (b) to (d).

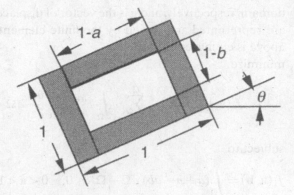

Figure 4.9. A unit cell of the micro-structure.

CONTINUOUS STRUCTURES

The main idea in topology optimization with homogenization is to compute an optimal material distribution on a given design domain, under given loads and boundary conditions. The distributed material is constructed by periodic microstructures. A typical microstructure model consists of square unit cells with rectangular holes; see Figure 4.9. Sizes a and b as well as the orientation angle θ of *each* hole are treated as design variables. Here $a = b = 1$ represents solid material, and $a = b = 0$ represents a void. The average elastic tensor $\mathbf{E}_{ijkl}^G$ of the anisotropic material determined by homogenization is a function of a, b, and θ.

The topology optimization model is defined as

> minimize   mean compliance
> subject to equilibrium equations,   (2)
>   volume constraint,
>   and bound constraints on the design variables.

Mean compliance is a global measure of structural stiffness, so the formulation above creates a trade-off between stiffness and material use. The solution will give the most rigid structure for a given material resource. The problem is equivalent to minimizing the weight for a given stiffness (or stress) level, but the model in Equation (2) is mathematically and computationally more tractable. The problem can be also seen as one with two objectives, compliance and volume. Varying the fixed bound value on the volume constraint and solving the resulting optimization problems will generate a Pareto trade-off curve between the two objectives, because the volume constraint is always active.

In the two-dimensional case the model is
minimize

$$f_0(\mathbf{a}, \mathbf{b}, \theta) = \sum_{i=1}^{2} \int_{\Omega} f_i u_i \, d\Omega + \sum_{i=1}^{2} \int_{\Gamma_\mathrm{T}} t_i u_i \, d\Gamma$$

subject to equilibrium equations

$$f_1(\mathbf{a}, \mathbf{b}) = \int_{\Omega} (a + b - ab) \, d\Omega - \Omega_S \leq 0, \quad 0 < \mathbf{a} < 1, 0 < \mathbf{b} < 1, \text{ and} -\pi/2 < \theta < \pi/2,$$

$$(3)$$

where $\mathbf{a}$, $\mathbf{b}$, and θ are design variables, $\mathbf{f}$ and $\mathbf{t}$ are applied body forces in the design domain Ω and tractions on a portion of the specified boundary Γ_T of the design domain, respectively, and $\mathbf{u}$ is the vector of displacements. The equilibrium equations are represented and solved by the finite-element method equations, so the model above is equivalent to

minimize

$$f_0(\mathbf{a}, \mathbf{b}, \theta) = \frac{1}{2} \sum_{(i,j,k,l=1)}^{2} \int_{\Omega} E_{ijkl}^{G} \frac{\partial u_k}{\partial z_l} \frac{\partial u_l}{\partial z_j} d\Omega$$

subject to

$$f_1(\mathbf{a}, \mathbf{b}) = \int_{\Omega} (a + b - ab)\, d\Omega - \Omega_S \leq 0, \quad 0 < \mathbf{a} < 1, 0 < \mathbf{b} < 1, \text{ and } -\pi/2 < \theta < \pi/2,$$

$$(4)$$

where $\mathbf{z}$ is the vector of the position coordinates. This optimization model has only one constraint other than the simple bound ones.

Using this formulation, a simple two-dimensional (plane elasticity) example is shown in Figures 4.10 and 4.11 for an eyebolt. Two alternative designs are produced by using different "designable" domains for the bolt head. The model for the simpler design is shown in Figure 4.11. The initial image of the topology is processed so that a realistic design can be derived from it. In general these homogenization images offer the basic topology concept but do not provide a complete design. The human designer can interpret and modify these images. Alternatively, automated techniques from image processing and computational geometry can be brought to bear, as illustrated in Figure 4.12.

These configuration design concepts have been extended to complex shell and three-dimensional problems. An example is shown in Figure 4.13.

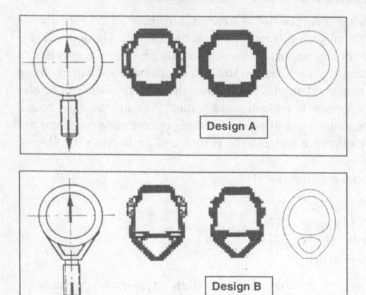

Figure 4.10. Eyebolt designs A and B (Chirehdast et al., 1992).

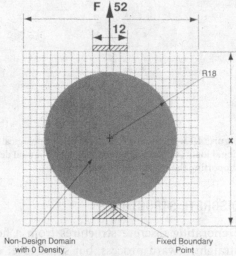

Figure 4.11. Problem formulation for eyebolt Design A.

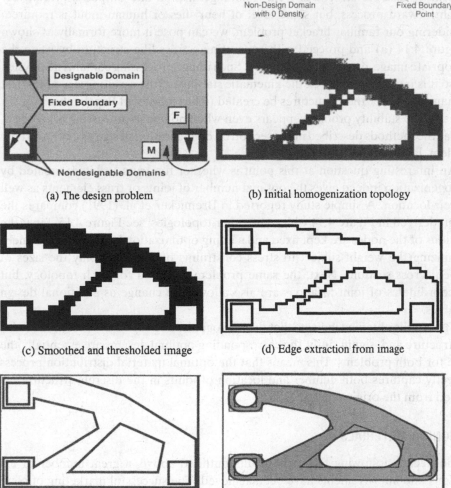

(a) The design problem

(b) Initial homogenization topology

(c) Smoothed and thresholded image

(d) Edge extraction from image

(e) Polygon approximation

(f) Configuration and shape for casting fabrication

Figure 4.12. Image processing and interpretation of homogenization results (Chirehdast, 1992): (a) the design problem; (b) initial homogenization topology; (c) smoothed and thresholded image; (d) edge extraction from image; (e) polygon approximation; (f) configuration and shape for casting lubrication.

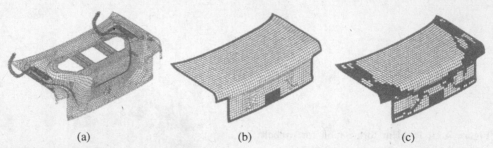

(a)   (b)   (c)

Figure 4.13. Generation of a complex topological object – an automobile trunk lid subjected to torsional load: (a) baseline inner panel lid, (b) initial design domain, and (c) computed topology of stiffener supporting the panel skin (Johanson, 1996).

DISCRETE STRUCTURES

Generating discrete structures from a homogenization-like output is a relatively straightforward process, but some level of heuristics or human input is required. Considering our familiar bracket problem, we can pose it more formally as shown in Figure 4.14 (a) and proceed with generating a truss-like structure by using the appropriate image-processing techniques. Such image processing contains no physics, and so it is important to check the kinematic stability of the resulting structures, lest mechanisms rather than structures be created (Chirehdast and Papalambros, 1992). Actually, the stability problem appears even when no image processing is involved, such as the methods described in the section on configuration design as a connectivity problem, later in this chapter.

An interesting question at this point is whether the truss structure created by homogenization indeed gives the optimal number of joints or truss elements as well as their location. A simple study reported in Bremicker et al. (1991) compares the design derived in Figure 4.14 with some other topologies; see Figure 4.15. First, the locations of the nodes are kept fixed and a sizing optimization is conducted, namely, minimizing the weight subject to stress constraints by changing only the sizes of element cross sections. Next, the same problem is solved for each topology, but the coordinates of joint locations are also allowed to change as additional design variables.

The results obtained indicate that the homogenization topology leads to the lightest structure and, significantly, the corresponding optimal designs are essentially the same for both problems. This means that the optimal material distribution process correctly captures both number and location of joints in the discrete structure extracted from the original image.

DESIGNING MATERIALS

In the decade following the early results outlined above, a great number of extensions and improvements have been achieved. The successful marketing of commercial software by several vendors has resulted in an expanding use of the above techniques in industry worldwide. This use is a good indication that these structural configuration problems can be now considered routine, at least within a "high tech" context in the automotive, aerospace, microelectronics, and biomedical industries.

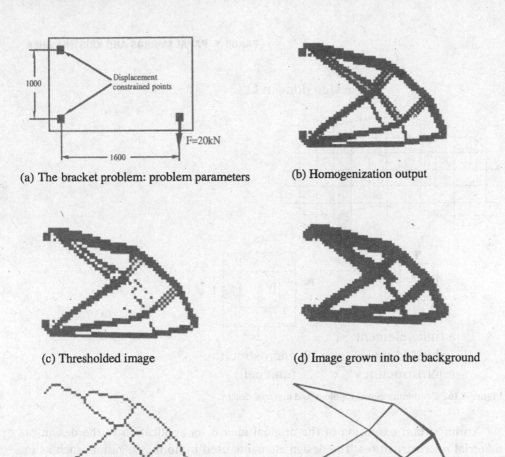

(a) The bracket problem: problem parameters

(b) Homogenization output

(c) Thresholded image

(d) Image grown into the background

(e) Image skeleton

(f) Final truss

Figure 4.14. Converting a homogenization output to a discrete structure (Chirehdast et al., 1993): (a) the bracket problem: problem parameters; (b) homogenization output; (c) thresholed image; (d) image grown into the background; (e) image skeleton; (f) final truss.

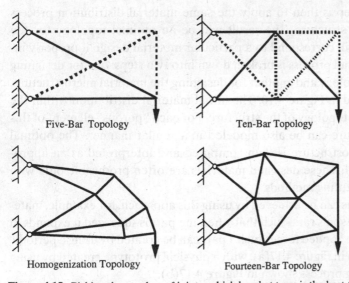

Five-Bar Topology

Ten-Bar Topology

Homogenization Topology

Fourteen-Bar Topology

Figure 4.15. Picking the number of joints: which bracket truss is the best (Bremicker et al., 1991)?

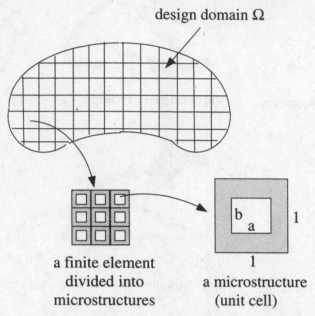

Figure 4.16. Combining global topology and material design.

An important extension of the original idea is its application to the design of material microstructures. The design elements used in homogenization, such as in Figure 4.9, allow us to represent how much material is placed within a pixel or voxel (three-dimensional pixel) in space. The material distribution output is really a collection of elements with holes, with the size of the hole corresponding to an overall material density. A homogeneous, isotropic material is extracted by a filtering method that converts all elements to either full or empty of material. However, the true output is a perforated material that can have realizable structural properties without filtering, if a composite material structure is considered.

The next obvious step is then to apply the same material distribution process to *each* element of the structure. The result will be an optimal topology for the microstructure, which will correspond to a particular material design if properly interpreted. Thus the overall process is broken down into two steps: one for designing the global structural topology and another for designing the material microstructure. The process is illustrated in Figure 4.16. The initial material distribution within the design domain gives the topology of the structure. For each "porous" element of this structure, a microstructure can be also modeled in a similar manner. The optimal "topology" of this microstructure is then computed and interpreted as an appropriate designed material. These designed materials are often producible only with nontraditional manufacturing methods.

Some unusual designs can be achieved by using this approach, for example, materials with a negative Poisson's ratio such that when the part is squeezed it expands in the same direction as the applied force. Such a part can be created by using a periodic microstructure as shown in Figure 4.17(a), with a physical prototype created by using a layered manufacturing process shown in Figure 4.17(b).

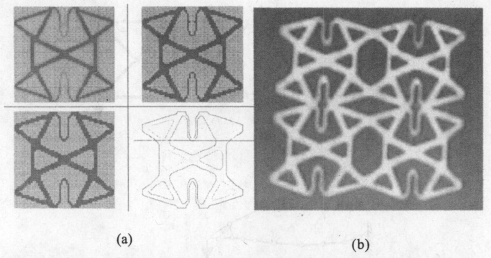

(a)   (b)

Figure 4.17. Material with a negative Poisson's ratio: (a) design generated by homogenization, (b) manufactured part (Marsan, 1999).

Mentioning a porous material above brings to mind some of the most efficient structures found anywhere, the bones of vertebrates. Indeed, mechanics models of bone-like materials have been developed and used to create optimal porous structures that are akin to organic ones, like the layered manufactured part in Figure 4.18. As novel manufacturing processes continue to evolve, our ability to design and produce biological-like artifacts is becoming a growing reality.

Other exciting problem areas involve the design of compliant mechanisms, flexible structures, and electromagnetic design problems. We will briefly discuss these topics in the last section of this chapter.

CONFIGURATION DESIGN AS A CONNECTIVITY PROBLEM

Whereas we can generate optimal discrete structures by interpreting material distribution solutions, as shown in the section on discrete structures above, we can also treat the structural configuration problem as one of directly determining the optimal

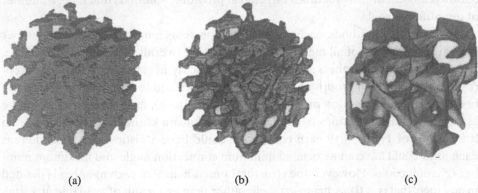

(a)   (b)   (c)

Figure 4.18. A biological (bonelike) structure: (a) three-dimensional image; (b) solid model; (c) close-up of solid model structure (Marsan, 1999).

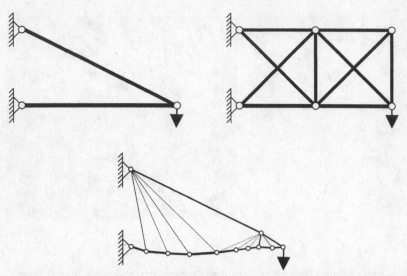

Figure 4.19. Possible discrete brackets.

connectivity of the structure. If we consider the bracket design problem again, this correlates to the case in which students are given bars or sections to cut and join rather than a sheet of material to sculpt. The problem still has a finite amount of material that can be used, but the question remains as to how to determine the best number of joints and connections to transfer a given load to the support locations. Some conventional and novel possibilities are shown in Figure 4.19.

Structural configuration formulated as a connectivity problem has applications in the design of civil structures, such as roof trusses, towers, and bridges, as well as aerospace structures, for example, lens supports. In addition to requirements for withstanding multiple loading conditions, typically these applications involve numerous constraints, often arising from structural codes and known construction techniques, as well as multiple performance goals that can include subjective goals related to elegance and politics. Because of the complexity of considerations, in practice structures are often overdesigned, sometimes unknowingly, resulting in unnecessary costs. Although it would be difficult to formulate the complete problem scenario, these situations can benefit from methods targeted at providing solutions that meet a number of quantifiable goals.

With discrete methods, each connection represents a unique structural member and the connectivity of all members determines the overall structural performance. Because structural members are represented explicitly, in contrast to continuous representations in which groups of elements were combined to create parametric parts (as discussed in the section on discrete structures above), design parameters can be associated with connections to explicitly model domain knowledge. For instance, in the designs of Figure 4.19 each connection could have a distinct material whereas each joint could have an associated minimum connection angle and maximum number of connections. However, the structural functionality of each member is defined in advance, that is, a truss, beam, or shell, rather than as a result of available material.

So far, discrete methods have not directly addressed the determination of the appropriate structural functionality for a particular design scenario. Work mostly

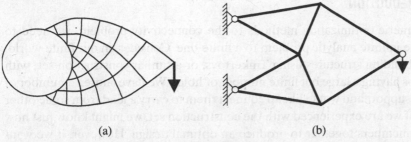

Figure 4.20. True versus practical optimality: (a) adapted from Michell (1904); (b) adapted from Save and Prager (1985).

has focused on the optimal layout of truss structures with a few studies of frames and grillages, which are grids of beam elements.

EARLY ANALYTIC WORK

The first investigation of the discrete structural configuration problem considered the problem of determining the optimal connectivity from an infinite number of connections (Michell, 1904). Using Maxwell's Theorem (Maxwell, 1864), Michell developed analytic methods for the optimal configuration of planar truss structures subject to a single load given a uniform tensile and compressive stress constraint; see Figure 4.20(a). These general shapes have been widely used as reference points for absolute minima. Analytic methods continued to develop though, because Michell's continuum structures are fairly impractical as they are only optimal with respect to consideration of an infinite number of joints and connections and do not take into account local and global stability. Building on Michell's basic idea, Save and Prager (1985) took an engineering approach to derive optimality criteria for a number of structural types, including trusses, frames, grillages, and shells.

If we remove one of the impracticalities of Michell structures by accounting for the reality that adding joints to a structure not only increases the overall mass of the structure but also increases construction costs, the optimal configuration changes, as shown in Figure 4.20(b). From this formulation, relations were derived between the optimal number of joints, the load of the structure, the distance between the supports, and the cost of a single joint (Save and Prager, 1985). Prager's optimality criteria for trusses do not consider the practical limitation of stability. This is treated later but, as with Michell's layouts, Prager's results continue to serve as a reference for benchmarking current methods.

The ideas developed by analytic methods in the early years continue to affect the development of numeric methods. For example, analytic work has shown that optimal truss layouts are nonunique (Prager, 1974), providing motivation for the use of stochastic methods capable of generating multiple equal-quality solutions from a single starting point. Much work continues in the area of analytic methods. However, whereas mathematicians focus on methods capable of providing exact analytic solutions to simple, idealized problems, engineers look for numeric methods that provide approximate solutions to real, complex applications (Rozvany, 1995).

TOPOLOGY REDUCTION

To apply numeric optimization methods to the connectivity problem, we need to transform the infinite analytic problem to a finite one. Consider, in our finite world, that we are building structures using Tinkertoys, or a similar construction set, with fictitious hubs having a large but finite number of holes. We have chosen a number of hubs we can support and would like to connect them to carry a load from some other point. Now, if we are experienced with the construction set, we might know just how to place the members together to produce an optimal design. However, if we want to explore designs outside of our experience, we might construct a structure using all of our hubs and connect the available pieces between them. We can then make decisions as to which pieces are unnecessary for our situation and can be removed. This process can be described as *topology reduction*.

The formulation of the connectivity problem as topology reduction was first put forward in the mid-1960s by research scientists at IBM who used a *ground structure* to define admissible structures that could then be optimized (Dorn, Gomory, and Greenberg, 1964). A ground structure is formed by defining a grid of joints, which includes both the support and external load locations, and creating connections between every joint. For example, defining J joints gives a maximum connectivity of $J(J - 1)/2$ possible connections and their associated cross-sectional area design variables; see Figure 4.21(a). However, although describing the maximum connectivity leads to an unbiased solution, expertise and intuition can help define a more limited ground structure; see Figure 4.21(b). Most often the number of connections described by the ground structure far exceeds the resulting optimal design.

Once a ground structure has been defined and a vector **x** of cross-sectional areas specified as design variables, numerical optimization can be used to solve for the set of areas that yields a least-weight design. Building on the general optimization model given previously, for a chosen material with mass density, ρ, and maximum allowable stress, σ_{allow}, in both tension and compression the optimization problem can be stated as:

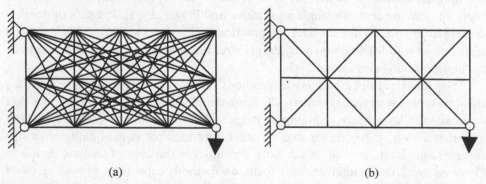

         (a)   (b)

Figure 4.21. Representing the connectivity problem as a ground structure: (a) maximum connectivity, (b) reduced connectivity.

minimize

$$f(x) = \sum_{i=1}^{N} \rho x_i l_i$$

subject to

$$\sigma_i(x) \leq \sigma_{\text{allow}}, \tag{5}$$

where N is the total number of connections in the ground structure. For this simple model, linear programming techniques are commonly used. If in the optimal set of cross-sectional areas one variable has reduced to either zero or a specified minimum, the corresponding connection can be removed from the structure. However, as stability was not modeled in the optimization problem, similar to the material distribution formulation, the optimal topology is often unstable. An interesting finding is that for a given ground structure there exists a family of optimal topologies, which most often only contain a limited number of stable designs (Kirsch, 1989).

In the years following the ground structure formulation of the connectivity problem, development largely focused on extending the optimization models and associated solution algorithms to include more complexity. Widely used approaches include optimality criteria methods, gradient-based methods, and more recently heuristic search. For example, the discretized continuum-type optimality criteria method by Zhou and Rozvany (1992, 1993) has been shown effective for determining optimal layouts considering combinations of displacement, stress, Euler buckling, and global buckling constraints as well as self-weight and multiple load conditions. Moving even further toward practical design constraints, the method has since been extended to provide solutions that are optimal in relation to German design codes (Libermann et al., 1994).

Selecting the fineness of joints and connections defined in the initial ground structure is a function of both the problem and the computational resources. To aid in the construction of ground structures, da Silva Smith (1998) has developed tools for generating planar and three-dimensional ground structures. The resulting designs are dependent on the density of distribution of the joints, which was a recognized limitation of the original research but necessary at the time to create a problem that could be solved. This bias has since been shown to increase for larger-scale structures (Bendsøe, 1995). One means of alleviating a certain amount of bias is to extend the optimization problem to include determination of the location of each joint in the ground structure defined as two (or three) geometric design variables. This increases the number of design variables for a planar grid with J joints from $J(J-1)/2$ cross-sectional areas to $J(J+3)/2$ cross-sectional areas and planar locations.

An approach to including shape variables is to formulate the optimization problem hierarchically so that first the optimal sizes of a ground structure are found, thus reducing the ground structure, and then the optimal joint locations are determined, perhaps using gradient-based methods such as in Bendsøe, Ben-Tal, and Zowe (1994). However, as the joint locations and connection sizes are coupled, it is advantageous to optimize with joint positions and size variables simultaneously (e.g., Rozvany and Zhou, 1991). By the inclusion of geometric variables, it has been shown that the removal of zero-length members arising from joints that have coalesced can result in singular global optima (Kirsch, 1996).

For a given starting point, most optimization methods determine one optimal configuration. But, we know that families of optimal topologies exist even for the simplest of problem formulations (Kirsch, 1989). This has motivated the use of stochastic methods for the optimization of ground structures because the starting point does not bias them. Additionally, stochastic methods easily accommodate the removal and addition of connections throughout optimization. Using genetic algorithms in combination with ground structures where design variables include connection existence in addition to connection size, Hajela, Lee, and Lin (1993) illustrate that families of optimal topologies can be generated. This property of the problem is not unique to the connectivity representation, because similar findings have resulted from combining genetic algorithms with continuous material representations (Chapman and Jakiela, 1994). Genetic algorithms with variable string lengths have also been used in the design of larger-scale structures, for example, microwave towers (Rajeev and Krishnamoorthy, 1997). An inherent difficulty with applying genetic algorithms, though, to the connectivity problem is mapping the representation into strings. A method that addresses this difficulty uses genetic programming (Soh and Yang, 2000), which has a tree-like representation, and shows potential for a more constructional approach. Although stochastic methods do not ensure optimal solutions, the "optimally directed" solutions generated are practical for most design applications.

TOPOLOGY CONSTRUCTION METHODS

A different way to reduce the infinite analytic problem to a finite solvable one can be illustrated again by using Tinkertoys. Rather than using all the hubs and pieces at once and then removing unnecessary connections, we could start with a small number of hubs and pieces that fully connect the load to the support locations. We then proceed to create our design by adding, and possibly removing, pieces and hubs in a prescribed manner until we are satisfied that the design is the best alternative to meet the intended purpose. This approach to the connectivity problem can be called *topology construction*.

It is widely argued that the simultaneous consideration of topology, shape, and size transformations throughout optimization will yield better results than shape and sizing alone. However, as Kirsch (1989) pointed out over 10 years ago and is still true today, creating automatic methods that introduce new connections and joints during the optimization process is a challenge. This is true for a few reasons. For one, we must have a scheme to introduce connections and joints in a systematic and meaningful manner. This results in a dynamic problem representation as the number of design variables changes throughout the optimization. The dynamic nature of the transformation problem presents difficulties to most numeric optimization methods stemming from discontinuities in the objective function. However, using a dynamic representation alleviates the bias associated with ground structure granularity.

Let's first consider the process for transforming our initial structure into an optimal one. Starting from the initial minimal design described previously, we can add

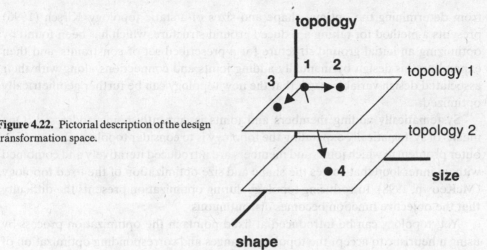

Figure 4.22. Pictorial description of the design transformation space.

joints and connections in possibly "key" locations. This process can be seen as a series of design transformations. Including the connectivity or "topology" as a design variable in addition to the shape and cross-sectional size variables considered in the previous section makes the space of possible designs, at the transformation level, three-dimensional. A pictorial description of this three-dimensional space is shown in Figure 4.22. The axes represent possible design states and are not indexed so as not to imply any sort of ordering of designs. If a given topology, say topology 1, at design state 1, is transformed by changing the cross-sectional area of one or more connections, the design moves in the topology 1 plane to state 2. Likewise, if design state 1 is transformed by moving one or more joint locations, the design moves along the topology 1 plane to state 3. But, if the topology of design state 1 is transformed to a new topology 2, the new design, state 4, exists on a new plane of potential shape and size transformations.

The first heuristic approach for the discrete topology problem was presented by Spillers (1975), who wanted to move away from defining a static ground structure by creating heuristics for introducing new connections based on known truss building techniques (Figure 4.23). The drawback to this method, especially considering computational power at the time, was that it depended on exhaustive topological search from every design state and a complete shape optimization for each change in topology. This corresponds to jumping to all possible topology planes from a given state, finding the best design within that plane, and then selecting the best design comparing among planes. This tactic, although thorough, requires much computation.

Topology construction is not easily incorporated into most optimization techniques. One remedy to this problem is to consider topology construction separately

Figure 4.23. Early truss construction rules (adapted from Spillers, 1975).

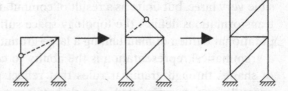

from determining the optimal shape and sizes of a static topology. Kirsch (1996) presents a method for taking a reduced ground structure, which has been found by optimizing an initial ground structure for a prescribed set of constraints, and then expanding this design by manually adding joints and connections along with their associated design variables such that the new topology can be further geometrically optimized.

Systematically adding members and joints is currently being addressed. One means of systematically expanding the topology is to consider topology change as an outer problem in which joints and members are introduced iteratively and combined with an inner loop that solves the shape and size optimization of the fixed topology (McKeown, 1998). Introducing topology during optimization presents the difficulty that the objective function becomes discontinuous.

Yet, topology can be introduced at fixed points in the optimization process by using a heuristic to accept the topology changes and corresponding optimization of shape and size, if the result is a lower objective value (Bojczuk and Mroz, 1999). Whereas the previous approaches simply added joints and connections to the topology uniformly throughout the current design, this method proposes the use of different modes of topology introduction dependent on connectivity and behavior.

STRUCTURAL SHAPE ANNEALING

To this point we have presented approaches to the connectivity problem that treat the problem as either the reduction of a large initial topology or the expansion of smaller topologies through manual or automatic application of heuristics. Adding to the topology construction methods, structural shape annealing is an approach that takes a broader view of the design issues involved in the connectivity problem while providing for automatic expansion and reduction of topology during optimization. This approach, developed by the second author in collaboration with Jonathan Cagan at Carnegie Mellon Unversity (Shea and Cagan, 1998); stems from the idea that optimal connectivity can develop during the optimization through a series of combined size, shape, and topology design transformations. The simplest topology transformations have been defined from observations in conventional truss design. For instance, for planar truss layout, Maxwell's rule states that for every joint added two connections must be added so that the determinacy of the structure is not affected (Figure 4.24). This rule also applies in reverse.

Applying the created set of design transformations recursively then defines a language of structural shapes; an example truss generation is shown in Figure 4.24. Although this language is in theory infinite, creating a computational method requires placing a limit on the number of generated connections. Even with this upper bound, parametric instantiations of each possible topology makes the space of designs possible very large, but finite as a result of computational precision. This compact set of transformations defines the topology space sufficiently without requiring the computational burden of maintaining a large ground structure. An additional benefit of a grammatical representation is the ability to constrain geometric transformations of shapes through grammar rules that reflect practical design considerations, for example, imposing a minimum angle between connections.

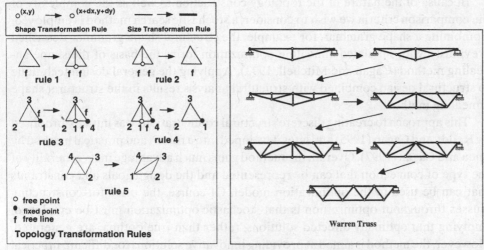

Figure 4.24. Constructing truss structures by using a shape grammar (Shea, Cagan, and Fenves, 1997).

We must now define our comparison criteria to assess the relative merits of designs in the structural language. A motivation for this method and the resulting optimization model has been from both structural and architectural design perspectives. All previous methods have focused on generating structural solutions in which minimizing structural mass was the most important design criteria. However, in many cases, mass is only one of many important design criteria. We can then define the comparison criteria of structural forms to include metrics for structural efficiency, economy, utility, and elegance (Shea and Cagan, 1999b).

The multicriteria problem is formulated in the usual manner, with a scalar objective function that is the weighted sum of all design criteria metrics. Constraints can be included as design criteria using a constraint violation metric that reduces to zero for a feasible design. The problem can then be stated as follows:

minimize objective cost + constraint cost, where

$$objective\ cost = \sum_{i=1}^{l}(objective\ weight_i \times objective\ value_i),$$

$$constraint\ cost = \sum_{j=1}^{m}(constraint\ weight_j \times constraint\ violation_j),$$

(6)

l is the number of objectives, and
m is the number of constraints.

Weighting factors for both the objectives and constraints can be set to determine the relative trade-off among terms and obtain an optimum in a Pareto sense. This cost function formulation can also support multiple independent loading conditions. For each independent loading condition, the structure is analyzed for violations of the behavioral limits (stress, buckling, and displacement) and the dominant violation for all loading conditions is used when the constraint violation is calculated for each connection. (Shea et al., 1997; Shea and Smith, 1999).

Because of the nature of the topology construction as well as the complexity of the comparison criteria we wish to consider, a stochastic search method is employed. Combining a shape grammar, for example, the structural shape grammar described previously, with simulated annealing optimization forms the basis of the shape annealing method (Cagan and Mitchell, 1993). Applying the general design technique to structural design combined with structural analysis results in the structural shape annealing method.

This approach to creating discrete structural configurations was initially proposed by Reddy and Cagan (1995) and later developed into a robust and practical method by Shea and Cagan (1997). Overall, the method gains much advantage in its generality of the type of connection that can be represented and the design goals and constraints that can be used in the optimization model. Of course, the price for constructing trusses throughout optimization is that stochastic optimization must be employed, implying that optimally directed solutions, rather than true optima, are generated. However, the method is aimed at providing innovative solutions for difficult, practical applications. In these cases, solutions not constrained to a limited definition of the design problem may prove advantageous.

Because of the combination of a stochastic optimization technique and topology transformation, structural shape annealing often results in unique comparisons to previous published results. For a standard arch problem (Dorn et al., 1964) considering only a stress constraint, the method was shown to generate solutions just lighter (5%) than those previously generated. These designs could not have been generated by the previous ground structures and were found to have slight asymmetries. As the method does not require a symmetric structure and will not necessarily converge to a global optimum design, slightly asymmetric designs may result with only small increases in mass.

We can also use the method to explore a standard structural design situation of designing planar roof trusses (Shea and Cagan, 1999a). This study explored the relation between design context and structural form. Given an initial design with two fixed supports and three load points, a reflection of the problem shown in Figure 4.25, two lightweight structures were generated; see Figures 4.26(a) and 4.26(b). However, because these structures proved too deep as a result of building regulations, a constraint was placed on depth, producing the heavier but more conventional structure shown in Figure 4.26(c). So, the addition of only one constraint resulted in the generation of the optimally directed bowed Warren truss. It is only through the consideration of the unconventional that we can discover what drives a conventional structural form evolved over 100 years.

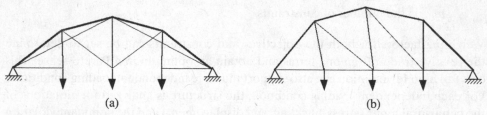

(a)   (b)

Figure 4.25. Symmetric and asymmetric truss structures (Shea and Cagan, 1998): (a) mass = 74.05 kg and (b) mass = 73.95 kg.

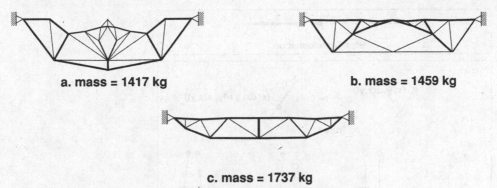

a. mass = 1417 kg   **b. mass = 1459 kg**

c. mass = 1737 kg

Figure 4.26. What drives conventional structures? Masses are equal to (a) 1417 kg; (b) 1459 kg; and (c) 1737 kg (Shea and Cagan, 1999a).

Structural designers are generally very capable in creating planar structures but often have difficulty with free-form three-dimensional structures. To provide some computational guidance, we can extend the planar grammar to model the implicit function of a single-layer space truss and create a three-dimensional structural form by projecting planar structures onto a defined surface. This extension makes a strong shift from the high importance placed on generating least-weight structures to focus on exploring expanded models for directing generation of a structural form. For a specified set of support locations and point loads as before, designs are generated for various performance goals. If we consider only the least-weight design, the structure in Figure 4.27(a) results. Extending the model of performance to incorporate geometric design goals stemming from the definition of a hemisphere, maximum volume or enclosure space, and minimum surface area, combined with maximizing uniformity, results in the design shown in Figure 4.27(b).

Redesign problems are an area where topology construction techniques may have an advantage over topology reduction. The design of transmission towers has evolved over many years to meet an extensive number of spatial constraints related to electrical requirements and current construction techniques as well as numerous load cases. We can use structural shape annealing for redesign rather than innovative design, by taking the current existing design as a starting point and defining a set of design transformations that encodes the problem limitations. For example,

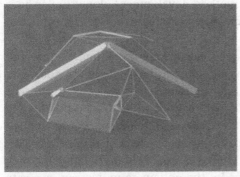
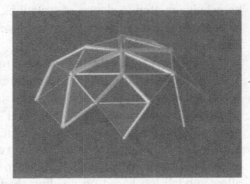

Figure 4.27. The influence of geometric design objectives on three-dimensional structural form (Shea and Cagan, 1997): (a) minimum mass and (b) pseudo-geodesic domes.

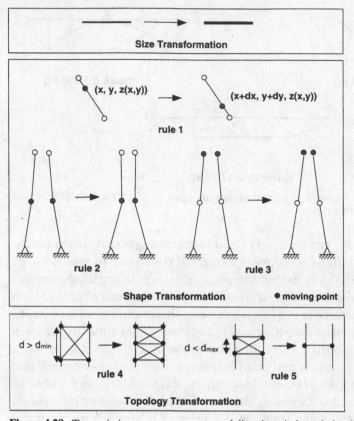

Figure 4.28. Transmission tower grammar: modeling domain knowledge (Shea and Smith, 1999).

we may wish to only consider certain topologies based on standard construction techniques and a corresponding structural construction grammar, Figure 4.28. When generation was begun with the design shown in Figure 4.29(a), which has 105 joints and 322 connections, and restricted changes were allowed to the size, shape, and topology, the reduced topology design in Figure 4.29(b) was generated, which has 89 joints and 262 connections and meets all problem constraints. A maximum of 600 connections were allowed at any one point in the optimization, creating a maximum of 215 shape and size variables (Shea and Smith, 1999). As this method uses knowledge related to classes of structures, method development along similar lines to structures such as communications towers and supporting structures for roofs and tents is possible.

We have expanded the optimization model and structural grammar to incorporate design goals further than structural mass related to both geometric goals and construction capabilities. But, will structurally motivated goals always result in a structure that we consider to "look right"? Although it can be argued that a truly optimal solution should always look right, it is also possible to incorporate explicit knowledge in the topology modification rules that reflect aesthetic proportions. For example, if we begin with an initial design composed from golden triangles and only allow subdivision based on dividing shapes by the golden ratio, we can generate structures that adhere to this proportional model (Figure 4.30).

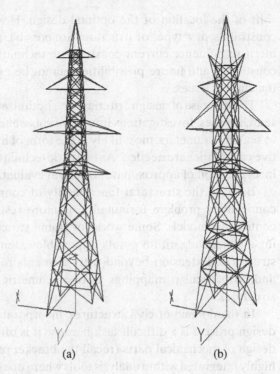

Figure 4.29. Transmission tower redesign (Shea and Smith, 1999): (a) current and (b) reduced topology designs.

USE IN PRACTICE

Although continuous representation methods have recently been integrated within commercial analysis tools, discrete methods are still used primarily as research tools. There is difficulty in creating robust mathematical models that are capable of reflecting practical goals and constraints. As applications reside mainly in the domain of civil structures, discrete methods must incorporate the current design considerations required to make the optimization results practical. For instance, Kirsch (1996) and others have shown that not taking into account local buckling results in a major

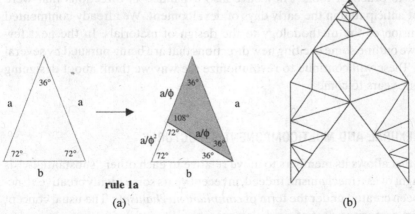

Figure 4.30. Optimal aesthetic configuration (Shea and Cagan, 1999). The (a) new division rule and (b) penrose-like truss configuration are shown.

shift of the location of the optimal design. However, through relaxation of some constraints, new types of structures are possible that could be seen to have sufficient merit to influence current construction techniques. The trade-off between current constraints and future possibilities cannot be explored without creating models of the current issues.

The increase of design criteria in the optimization model can lead to very complex search spaces. Investigations into more robust heuristic methods with less sensitivity to search parameters, most likely in the form of hybrid methods, and true multiobjective capabilities are needed. As heuristic techniques are effective but highly iterative, incorporation of approximate structural evaluation must be investigated.

Because the structural functionality of connecting elements is predefined, the connectivity problem formulation is more restricted in applications than that of continuous models. Some work for frame structures (Grierson and Pak, 1993) using a genetic algorithm points to possible extensions to frame, cable net, and shell structures. Extension beyond truss elements requires appropriate representations and, in particular, mappings among geometric models, analysis models, and real structures.

In the domain of civil structures, incorporation of topology optimization in the design process is a difficult task because it is often considered an art. The topology design of mechanical parts (recall the bracket problem) in the recent past has been highly integrated within analysis tools where one designer often performs both design and analysis tasks. However, discrete structural design is more often than not two disjoint tasks. One hopes that a more integrated design approach will emerge in traditional structural engineering. In cases in which the resulting structures are quite different from conventional norms, constructional methods may prove advantageous because they have the capability of being used in both manual and semiautomated manners to lend credibility to the resulting design.

AN EXPANDING DESIGN HORIZON

The early topology design concepts and the attendant computational methods have now evolved to practical tools. They have also expanded in directions that were probably not anticipated in the early days of development. We already commented on the extension of the methodology to the design of materials. In the next few paragraphs we outline some exciting new directions that are being pursued by several researchers. These will continue to revolutionize the way we think about designing artifacts in the years to come.

MOVING, FLEXURAL AND MULTICOMPONENT STRUCTURES

A structure that allows its members to move relative to each other "substantially" is usually thought of as a mechanism. Indeed, in recent years some highly creative structures have been created under the term of *compliant mechanisms*. The usual concept of mechanism comprises a set of rigid elements joined together. Any compliance in the mechanism is concentrated at the joints or hinges.

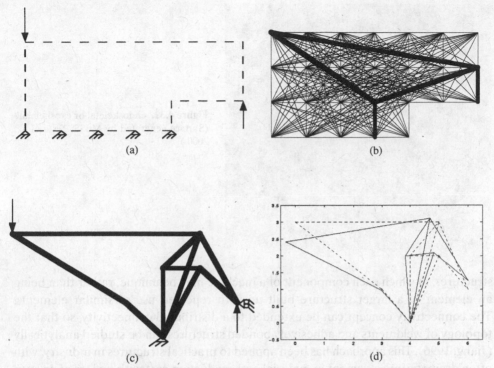

Figure 4.31. Compliant mechanisms: (a) design domain; (b)ground structure; (c) final topology; (d) range of motion (Frecker, 1997).

Another idea often observed in nature is to distribute compliance along members of a structure in a controlled way, so that a relatively small displacement input results in a substantially larger output. The observed relative motion can be quite large and so the designed artifact acts as a mechanism, yet it is a single piece with no joints and so it can be viewed as a structure.

An example of a compliant mechanism is shown in Figure 4.31. The concept has interesting and practical applications to the design of microelectromechanical systems (MEMS), biomechanical systems, and "no assembly" mechanisms that are made out of a single, say, injection-molded part. An extension of this idea is to think of structures as altogether flexible, because "the stiffest structures are not rigid structures but the minimum flexible structures" (Nishiwaki, 1998). Designing flexibility as part of an integrated approach to structural design can have some exciting implications. Apart from compliant mechanisms, controlled flexibility or rigidity can be part of the overall functionality of an artifact. The body of an automobile can be designed so it is actually part of the suspension (Nishiwaki, 1998). Combining this with the compliant mechanism idea, future suspension designs may have only one or just a few parts. Controlled flexibility can enhance the primary function of an artifact, such as flextensional piezoelectric actuators (Nishiwaki, 1998).

Multicomponent and multimaterial structural topology problems come from the idea that real structures are made out of separate components that must be connected. The topology of each component and the connectivity among them interact and must be studied jointly (Johanson, 1996). This is particularly true for mechanical

Figure 4.32. endoskeletal or exoskeletal? (Surface generated by MoSS; Testa et al., 2000.)

structures, in which each component of a machine may be unique, rather than being an element of a larger structure built up with repeated use of similar elements. The connectivity concept can be extended to a distributed connectivity, so that the topology of weldments and adhesively bonded structures can be studied analytically (Jiang, 1996). This approach has been applied to practical structures in industry, with often dramatic improvement in material and production cost with no loss of or even increased functionality (e.g., Gea et al., 1997).

The subjects above are developing into disciplines in their own right, and they are mentioned here in passing simply to indicate how pervasive and inspiring the early concepts on topology design have become.

HYBRID ENDOSKELETAL – EXOSKELETAL STRUCTURES

We discussed continuous and discrete representations independently, but is there scope for a hybrid approach? Although most natural structures are either endoskeletal or exoskeletal, for inorganic structures mutations are a promising possibility. Should the frame support the surface and should the surface enclose the frame (Figure 4.31)? In the example shown, the structure is endoskeletal–exoskeletal, where the frame actually is carrying the load with the surface woven through it. However, current methods are moving toward integration of continuous and discrete representations that would expand the possibilities for structural configuration design. Automotive body designs are likely to evolve into hybrid configurations as the mainstream design of preference.

INTELLIGENT ARTIFACTS

Continuing advances in structural measurement, sensor technology, and control motivate increased desire for artifacts that adapt to changes in their operating boundary conditions – their environment. Designing the artifact configuration and its associated controller must be done concurrently to achieve system optimality. Combining the above technologies with structural design has led to the development of intelligent structures, that is, structures that can adapt their geometry and topology in response to a changing environment.

For structural configuration design, this complicates the problem because the design criteria have now changed. Consider a simple example of multiple load conditions. The best solution is often a compromise such that the static structure meets each load condition independently or in defined combinations. For at least one load case, a member must be at its limit, but it is often the situation that for some loading conditions a member is underutilized. What if, though, the structure can actively configure itself in response to varying loads? Now the configuration problem becomes a compromise between material use and capacity for adapting. The representations and methods discussed here show promise for extensions to the design of these new "live" artifacts.

BEYOND STRUCTURAL PROBLEMS

The theory behind structural design relies mostly on the assumption of an elastic field as the underlying mathematical model for the behavior of the artifact. One could easily see that the same approach could be used in designs of artifacts analyzed with different field models, for example, magnetic and electromagnetic fields. Work in these areas has recently commenced and it is quite promising. One example is recent work on structural optimization in magnetic fields (Yoo, 1999). The homogenization method is extended to problems involving frequency-response optimization of structures excited by magnetic forces, and maximization of the magnetic energy of a structure.

Another example not involving a structure per se is the design of electromagnetic patch antennas used on aircraft and other applications where conformity to the carrying surface is desired; see Figure 4.33 (Li, Volakis, and Papalambros, 2000). Metallic patches are placed on a dielectric substrate to form a single layer. Multiple layers can be combined to produce the desired performance. The configuration of the patch itself as well as that of the substrate have profound influence on antenna performance. Although the technology is highly advanced, rigorous design synthesis

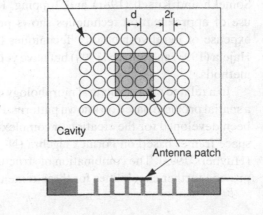

Figure 4.33. Patch antenna configuration problem.

methods are still relatively undeveloped. This is an area in which topology problems can be addressed also with genetic algorithms, as long as simplified physics models are used.

Structures are a good starting point for investigating configurations of physical objects. Many of the ideas related to design configuration problem representations and associated decision criteria are being extended to domains governed by different physics. The future appears to hold much excitement and promise for design configuration research.

RESOURCES

The literature on topology as a material distribution problem is now vast. Much of the work that commenced in the later part of the 1980s is presented in Bensøe's review book (Bendsøe, 1995), which includes work by other authors. The ideas described on the formulation of the topology problem and its extensions as presented here, including many of the figures used, are influenced primarily by work conducted in the course of several doctoral dissertations at the University of Michigan. Some of those that have a stronger design flavor are the dissertations by Suzuki (1991), Chirehdast (1992), Cheng (1992), Gea (1993), Ananthasuresh (1994), Johanson (1996), Jiang (1996), Fonseca (1996), Min (1997), Frecker (1997), Silva (1998), Nishiwaki (1998), and Marsan (1999).

Topology design in the context of an integrated design synthesis approach is described in Chiredast et al. (1992, 1993). A website maintained by S. Kota has interesting information on recent research and applications of compliant mechanisms (www.engin.umich.edu /labs/csdl). A lot of the current work couples topology design with computed tomography scanning, spatial image processing, and layered manufacturing techniques for a variety of applications, ranging from automotive design and nondestructive evaluation to biomedical implants. See, for example, the websites by Quint Corporation (www.quint.co.jp/NewWeb_eng/eng/) and S. Hollister (www.personal.umich.edu/~scottho/scaffoldengin).

There is much literature on configuration as a connectivity problem. For in-depth discussions of optimal truss layout, see Hemp (1973), Cox (1965), and Rozvany (1996, 1997). A concise review of analytic methods is provided in Kirsch (1989). For further methods that focus on near-optimal rather than optimal solutions, see Somekh and Kirsch (1981) and Topping, Khan, and de Barros Leite (1993). The use of approximation techniques shows promise for reducing the computational expense of iterative methods. Techniques have been developed by Szewczyk and Hajela (1993) and Kirsch (1999) but have yet to be used within topology optimization methods.

In a related area, structural morphology considers the construction of topology as a spatial problem based on known patterns. Programs for structural morphology have been developed for the creation of complex geometric patterns, often used for large space frames based on Formex algebra (Nooshin, 1984), and polyhedra CORELLI (Huybers, 1993). The combination of structural morphology and optimization could provide further capabilities for three-dimensional topology optimization.

REFERENCES

Ananthasuresh, G. K. (1994). "A new design paradigm for micro-electro-mechanical systems & investigations on the compliant mechanism synthesis," Ph.D. Dissertation, University of Michigan, Ann Arbor.

Bendsøe, M. P. (1995). *Optimization of Structural Topology, Shape, and Material*. Springer-Verlag, Berlin.

Bendsøe, M. P., Ben-Tal A., and Zowe, J. (1994). "Optimization methods for truss geometry and topology design," *Structural Optimization*, **7**:141–159.

Bendsøe, M. P. and Kikuchi, N. (1988). "Generating optimal topologies in structural design using a homogenization method," *Computational Methods in Applied Mechanics and Engineering*, **71**:197–224.

Bojczuk, D. and Mroz, Z. (1999). "Optimal topology and configuration design of trusses with stress and buckling constraints," *Structural Optimization*, **17**:25–35.

Bremicker, M., Chirehdast, M., Kikuchi, M., and Papalambros, P. Y. (1991). "Integrated topology and shape optimization in structural design," *Mechanics of Structures and Machines International Journal*, **19**(4):551–587.

Cagan, J. and Mitchell, W. J. (1993). "Optimally directed shape generation by shape annealing," *Environment and Planning B: Planning and Design*, **20**:5–12.

Chapman, C. D. and Jakiela, M. J. (1994). "Genetic algorithm-based structural topology design with compliance and manufacturability Considerations," *Advances in Design Automation*, ASME, New York, DE69-2, pp. 309–322.

Cheng, C. H. (1992). "Modeling of the thermal elasto-plastic behavior for composite materials using the homogenization method," Ph.D. Dissertation, University of Michigan, Ann Arbor.

Chirehdast, M. (1992). "An integrated optimization environment for structural configuration design," Ph.D. Dissertation, University of Michigan, Ann Arbor.

Chirehdast, M., Linder, B., Yang, J., and Papalambros, P. Y. (1993). "Concurrent engineering in optimal structural design." In *Concurrent Engineering: Automation, Tools, and Techniques*, A. Kusiak (ed.), Wiley, New York, pp. 75–109.

Chirehdast, M., Gea, H. C., Kikuchi, N., and Papalambros, P. Y. (1992). "Structural configuration examples of an integrated optimal design process." In *Advances in Design Automation – 1992*, D. Hoelzel (ed.), ASME, New York, Vol. 1, pp. 11–20; also *ASME Journal of Mechanical Design*, **116**(4):997–1004.

Chirehdast, M. and Papalambros, P. Y. (1992). "A note on automated detection of mobility of skeletal structures," *Computers and Structures*, **45**(1):197–207.

Cox, H. L. (1965). *The Design of Structures of Least Weight*. Pergamon, Oxford.

da Silva Smith, O. (1998). "Generation of ground structures for 2D and 3D design domains," *Advances in Engineering Software*, **27**:167–178.

Dorn, W. S., Gomory, R. E., and Greenberg, H. J. (1964). "Automatic design of optimal structures," *Journal de Mécanique*, **3**(1):25–52.

Fonseca, J. S. O. (1996). "Design of microstructures of periodic composite materials," Ph.D. Dissertation, University of Michigan, Ann Arbor.

Frecker, M. (1997). "Optimal design of compliant mechanisms," Ph.D. Dissertation, University of Michigan, Ann Arbor.

Gea, H. C. (1993). "Structural optimization for static and dynamic responses using an integrated system approach," Ph.D. Dissertation, University of Michigan, Ann Arbor.

Gea, H. C., Chickermane, H., Yang, R. J., and Chung, C. H. (1997). "Optimal fastener pattern design considering bearing loads." In *ASME Design Engineering Technical Conference*, ASME, New York, Paper No. DAC-3770.

Grierson, D. E. and Pak, W. H. (1993). "Optimal sizing, geometrical and topological design using a genetic algorithm," *Structural Optimization*, **6**:151–159.

Hajela, P., Lee, E., and Lin, C.-Y. (1993). "Genetic algorithms in structural topology optimization." In *Topology Design of Structures*, NATO ASI Series, Kluwer Academic, Dordrecht, The Netherlands, pp. 117–133.

Hemp, W. S. (1973). *Optimum Structures*. Claredon Press, Oxford.

Huybers, P. (1993). "Super-elliptic geometry as a design tool for the optimization of dome structures." In *Computer Aided Optimum Design of Structures III: Optimization of Structural Systems and Applications*, pp. 387–399, Hernandez, S. and C. A. Brebbia (eds.), Computational Mechanics/WIT Press, Southampton/Billerica, MA.

Jiang, T. (1996). "Topology optimization of structural systems using convex approximation methods," Ph.D. Dissertation, University of Michigan, Ann Arbor.

Johanson, R. P. (1996). "Topology optimization of multicomponent structural systems." Ph.D. Dissertation, University of Michigan, Ann Arbor.

Kirsch, U. (1999). "Efficient, accurate reanalysis for structural optimization," *AIAA Journal*, **37**(12):1663–1669.

Kirsch, U. (1989). "Optimal topologies of structures," *Applied Mechanics Review*, **42**(8): 223–238.

Kirsch, U. (1996). "Integration of reduction and expansion processes in layout optimization," *Structural Optimization*, **11**:13–18.

Liebermann, S., Gerdes, D., Birker, T., Rozvany, G. I. N., and Zhou, M. (1994). "Topology design of tubular structures with a variety of design constraints." In *Tubular Structures VI*, Holgate and Wong (eds.), Balkema, Rotterdam, pp. 423–429.

Li, Z., Volakis J. L., and Papalambros, P. Y. (2000). "Performance enhancement of bandgap printed antennas using finite element method and size/topology optimization methods." In *IEEE Millennium Conference On Antennas and Propagation*, Davos.

Marsan, A. L. (1999). "Solid model construction from 3D images," Ph.D. Dissertation, University of Michigan, Ann Arbor.

Maxwell, J. C. (1864). "On the calculation of the equilibrium and stiffness of frames," *Philosophical Magazine*, **27**:294. Reprinted in *The Scientific Papers of James Clerk Maxwell*, W. D. Niven (ed.), New York, Dover Publications, Volume 1, pp. 598–604.

McKeown, J. J. (1998). "Growing optimal pin-jointed frames," *Structural Optimization*, **15**: 92–100.

Michell, A. G. M. (1904). "The limits of economy of material in frame-structures," *Philosophical Magazine*, **8**(47):589–597.

Min, S. J. (1997). "Optimum structural topology design for multi-objective, stability, and trasient problems using the homogenization design method," Ph.D. Dissertation, University of Michigan, Ann Arbor.

Nishiwaki, S. (1998). "Optimum structural topology design considering flexibility," Ph.D. Dissertation, University of Michigan, Ann Arbor.

Nooshin, H. (1984). "Formex configuration processing in structural engineering." Elsevier Applied Science Publishers, London (obtainable from Chapman & Hall Publishers).

Papalambros, P. (1988). "Interdisciplinary experiments in design research and education." Presented at the 1988 International Conference on Engineering Design, Budapest.

Papalambros, P. Y. and Chirehdast, M. (1990). "An integrated environment for structural configuration design," *Journal of Engineering Design*, **1**(1):73–96.

Papalambros P. Y. and Wilde, D. J. (1988). *Principles of Optimal Design: Modeling and Computation*. 1st ed., Cambridge University Press, New York.

Prager, W. (1974). *Introduction to Structural Optimization*. Springer, Vienna.

Rajeev S. and Krishnamoorthy, C. S. (1997). "Genetic algorithms-based methodologies for design optimization of trusses," *Journal of Structural Engineering*, **123**(3):350–358.

Reddy, G. and Cagan, J. (1995). "An improved shape annealing algorithm for truss topology generation," *ASME Journal of Mechanical Design*, **117**(2A):315–321.

Rozvany, G. I. N. (1997), "On the validity of Prager's example of nonunique Michell structures," *Structural Optimization*, **13**:191–194.

Rozvany, G. I. N. (1996). "Some shortcomings in Michell's truss theory," *Structural Optimization*, **12**:244–250.

Rozvany, G. I. N. (1995). "What is meaningful in topology design? An engineer's viewpoint." In *Advances in Structural Optimization*, Kluwer Academic, Boston, MA.

Rozvany, G. I. N., and Zhou, M. (1991). "A new direction in cross-section and layout optimization: the COC algorithm," presented at Computer Aided Optimum Design of Structures 91, Optimization of Structural Systems and Industrial Applications, Cambridge, MA, June 25–27, pp. 39–50.

Save, M. and Prager, W. (1985). *Structural Optimization: Volume 1: Optimality Criteria*, Plenum, New York.

Save, M. and Prager, W. (1990). *Structural Optimization: Volume 2: Mathematical Programming*, Plenum, New York.

Shea, K. and Cagan, J. (1999a). "The design of novel roof trusses with shape annealing: Assessing the ability of a computational method in aiding structural designers with varying design Intent," *Design Studies*, 20(1):3–23.

Shea, K. and Cagan, J. (1999b). "Languages and semantics of grammatical discrete structures," *Artificial Intelligence for Engineering Design, Analysis, and Manufacturing*, Special Issue, 13(4):241–251.

Shea, K. and Cagan, J. (1998). "Topology design of truss structures by shape annealing," in *Proceedings of DETC98: 1998 ASME Design Engineering Technical Conferences*, ASME, New York, DETC98/DAC-5624, pp. 1–11.

Shea, K. and Cagan, J. (1997). "Innovative dome design: applying geodesic patterns with shape annealing," *Artificial Intelligence for Engineering Design, Analysis, and Manufacturing*, 11:379–394.

Shea, K., Cagan, J., and Fenves, S. J. (1997). "A shape annealing approach to optimal truss design with dynamic grouping of members," *ASME Journal of Mechanical Design*, 119(3):388–394.

Shea, K. and Smith, I. (1999). "Applying shape annealing to full-scale transmission tower redesign." In *Proceedings of DETC99: 999 ASME Design Engineering Technical Conferences*, ASME, New York, DETC99/DAC-8681, pp. 1–9.

Silva, E. C. N. (1998). "Design of piezocomposite materials and piezoelectric tranducers using topology optimization," Ph.D. Dissertation, University of Michigan, Ann Arbor.

Soh, C. K. and Yang, Y. (2000). "Genetic programming-based approach for structural optimization," *Journal of Computing in Civil Engineering*, 15(1):31–37.

Somekh, E. and Kirsch, U. (1981). "Interactive optimal design of truss structures," *Computer-Aided Design*, 13(5):253–259.

Spillers, W. R. (1975). *Iterative Structural Design*, North-Holland Monographs In Design Theory, North-Holland/American Elsevier, New York.

Suzuki, K. (1991). "Shape and layout optimization using homogenization method," Ph.D. Dissertation, University of Michigan, Ann Arbor.

Szewczyk, Z. and Hajela, P. (1993). "Neural network approximations in a simulated annealing based optimal structural design," *Structural Optimization*, 5:159–165.

Testa, P., O'Reilly, U., Kangas, M., and Kilian, A. (2000). "MoSS: morphogenetic surface structure: a software tool for design exploration." In *Proceedings of Greenwich 2000: Digital Creativity Symposium*, pp. 71–80.

Topping, B. H. V., Khan, A. I., and de Barros Leite, J. P. (1993). "Topological design of truss structures using simulated annealing." In *Neural Networks and Combinatorial Optimization in Civil and Structural Engineering*; Civil-Comp Press, Edinburgh, U.K., pp. 151–165.

Yoo, J. (1999). "Structural optimization in magnetic fields using the homogenization design method," Ph.D. Dissertation, University of Michigan, Ann Arbor.

Zhou, M. and Rozvany, G. I. N. (1992). "DCOC: an optimality criteria method for large systems: Part I: Theory," *Structural Optimization*, 5:12–25.

Zhou, M. and Rozvany, G. I. N. (1993). "DCOC: an optimality criteria method for large systems: Part II: Algorithm," *Structural Optimization*, 6:250–262.

CHAPTER FIVE

Microsystem Design Synthesis

Erik K. Antonsson

It is typical of many kinds of design problems that the inner system consists of components whose fundamental laws of behavior are well known. The difficulty of the design problem often resides in predicting how an assemblage of such components will behave.

– H.A. Simon, *The Sciences of the Artificial* (1969)

INTRODUCTION

Electromechanical devices and systems, constructed at length scales of micrometers to millimeters and fabricated with patterned deposition and removal processes borrowed from microelectronics, were first seriously proposed in Petersen (1982). These devices, now called microelectromechanical systems (MEMS) or microsystems, have been the object of intense research and development since that time,[1] and significant advances in the fabrication and understanding of these devices and systems have been made. A wide variety of devices and systems have been built with a length scale of tens of micrometers (Fan, Tai, and Muller, 1998; Bryzek, Petersen, and McCulley, 1994; Gabriel, 1995; Trimmer, 1997), including

- sliders and revolute joints (Fan et al., 1988),
- gears (Mehregany et al., 1987; Dopper et al., 1997),
- hinges (Pister et al., 1992),
- electrostatic actuators (Tang, Nguyen, and Howe, 1989; Tang et al., 1990; Horsley et al., 1999),
- motors (Trimmer and Gabriel, 1987; Tai and Muller, 1989; Suzuki and Tangigawa, 1991),
- valves (Huff et al., 1990; Jerman, 1990; Emmer et al., 1992; Bosch et al., 1993;

---

[1] Several scholarly journals chronicle these developments, including the ASME/IEEE *Journal of Microelectromechanical Systems*, the Institute of Physics *Journal of Micromechanics and Microengineering*, the *Journal of Modeling and Simulation of Microsystems*, the Japanese journal *Sensors and Materials*, and *Sensors and Actuators*. Several conferences are regularly organized for the exchange of research and development results in this area, including the IEEE International Conference on Micro Electro Mechanical Systems (IEEE–MEMS), the Solid-State Sensor and Actuator Workshop (held at Hilton Head, SC.), the International Conference on Solid-State Sensors and Actuators (Transducers), and the International Conference on Modeling and Simulation of Microsystems (MSM).

Robertson and Wise, 1998; Chakraborty et al., 2000; Hamn et al., 2000; Papavasiliou, Liepmann, and Pisano, 2000; Rich and Wise, 2000),
- capacitive sensors (Schmidt et al., 1987),
- accelerometers (Roylance and Angell, 1979; Angell, Terry, and Barth, 1983; Starr, 1990; Ciarlo, 1992; Yee et al., 2000),
- pumps (Saif, Alaca, and Sehitoglu, 1999; Deshmukh, Liepmann, and Pisano, 2000; Gong et al., 2000; Li et al., 2000),
- gyroscopes and inertial rate sensors (Vandemeer, Kranz, and Fedder, 1998; Ayazi et al., 2000; Jiang, 2000),
- pressure sensors (Borky and Wise, 1979),
- rocket thrusters and pyrotechnics (Marcuccio, Genovese, and Andrenucci, 1998; Rossi et al., 1998a; Rossi, Esteve, and Mingues, 1999; Lewis et al., 2000),
- heaters and thermal actuators (Lin and Pisano, 1994; Rossi, Scheid, and Esteve, 1997; Rossi, Temple-Boyer, and Esteve, 1998b; Robertson, 2000),
- thermal sensors (Kurabayashi and Goodson, 1998; Ju, Kurabayashi, and Goodson, 1999; Zhang et al., 2000), and
- fuel cells (Savinell and Litt, 2000; Tompsett et al., 2000).

These systems incorporate truly mixed technology, integrating combinations of digital and analog electronics, mechanical structures, electromagnetic actuators, and fluidic chambers.

T. Mukherjee and G.K. Fedder, *Proceedings of the 34th Design Automation Conference* (1997)

These devices range from proof-of-concept laboratory specimens to commercially available off-the-shelf products incorporated into millions of consumer products, such as accelerometers for collision detection in passenger automobiles to initiate airbag deployment.[2]

This chapter addresses the issues relating to the design and automatic synthesis of designs of microsystems. Currently, much of the design synthesis work in this area is performed manually, with computational assistance to evaluate the performance of proposed designs. Because of the strong relationship between microsystems and other areas with highly automated design synthesis, microsystems are a promising area for development of such methods, and significant research results have been obtained by several groups.

This chapter briefly reviews the current status of microsystem design and the latest research results and literature, presents an example of automated microsystem design synthesis, and concludes with a discussion of the future needs and trends in this area.

MICROSYSTEM FABRICATION METHODS

The primary method of fabrication of microsystems is the controlled deposition of one or more layers of material on a substrate (usually a silicon wafer), followed by the removal of selected regions of the deposited layer in a desired pattern.

---

[2] For example, Analog Devices ADXL 150/250 accelerometers (Analog Devices, 2000), and Motorola MMAS40G10D accelerometers (Verma, Baskett, and Loggins, 1998).

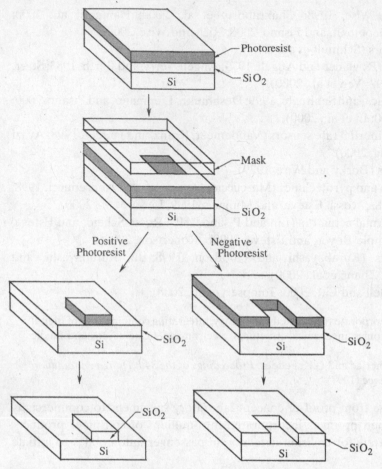

Figure 5.1. Photolithography process. (adapted from May et al., 2000).

The process of deposition, patterning, and removal may be repeated several times. A common method for patterning is optical lithography, adapted from the semiconductor electronics industry, illustrated in Figure 5.1. A mask containing the desired pattern is constructed. A typical mask is a glass photographic plate, with opaque and transparent regions defining the desired pattern. As illustrated in Figure 5.2, this mask is held above the substrate and light is focused through the mask and onto a layer of photosensitive polymer (photoresist), which has been applied on top of the layers on the substrate. Photoresist is applied to the surface of the wafer as a liquid, and the substrate is spun to produce a thin layer of uniform thickness. Exposing the photoresist to ultraviolet light alters its rate of removal in the presence of a developing chemical, and by this process the pattern on the mask can be transferred to the photoresist on the top of the wafer. Photoresist has the additional property that, after it is baked, it is resistant to chemicals that etch the layer(s) underneath, so that when the wafer is exposed to an etchant, the pattern in the photoresist is etched into the wafer.

Photolithography can be applied to the wafer substrate itself, a process called *bulk micromachining* and illustrated in Figure 5.3 (Bassous, 1978; Bean, 1978; Petersen, 1982; Bryzek et al., 1994), or to one or more layers deposited on the substrate in a

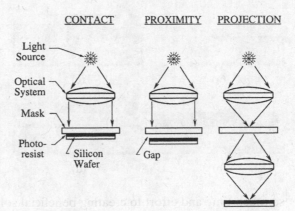

Figure 5.2. Photolithography illumination. (adapted from May et al., 2000).

process called *surface micromachining* (Howe and Muller, 1983; Wolf and Tauber, 1986; Jaeger, 1988; Mehregany, Gabriel, and Trimmer, 1988; Runyan and Bean, 1990; Farooqui and Evans, 1991; Suzuki and Tangigawa, 1991; Wise and Najafi, 1991; Madou, 1997). In surface micromachining, as illustrated in Figure 5.4, one or more layers can be deposited and patterned, with additional layers then deposited and patterned on top. After patterning of the topmost layers, one or more intermediate layers can be removed with an etchant. These intermediate layers are referred to as *sacrificial layers*. Dry plasma etching, or reactive ion etching, can be performed on surface layers, or the substrate itself (Abe, Sonobe, and Enomoto, 1973; Steinbruchel, Lehmann, and Frick, 1985; Flamm, Donnelly, and Ibbotson, 1990; Ayon, 1998).

ATTRIBUTES OF MICROSYSTEMS

Microsystems as a technical discipline appears to share many of the beneficial attributes with other areas that have highly developed and automated design methodologies. These attributes include the following: the fabrication uses methods similar to those used for digital very large-scale integration (VLSI); the systems are assembled from many instances of a small number of components, largely planar ($2^1/_2$ dimensional) geometry; and the mechanical coupling between components is limited (e.g., arrays of sensing elements or actuators).

Additionally, microsystems are a new and still-developing research area, and as a result there is the potential to be able to develop novel solutions to interesting and important problems through the use of automated synthesis methods (whereas in more highly developed areas, e.g., mechanical design, human culture has applied

Figure 5.3. Bulk micromachining cross-section (Tang, 1997, Figure 1): (a) deposit and pattern nitride mask; (b) KoH etch, partial; (c) KoH etch, complete; (d) remove nitride mask. © 1997 ACM. Figure reprinted with permission.

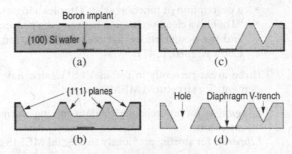

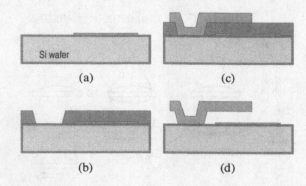

Figure 5.4. Surface micromachining cross section (Tang, 1997, Figure 2): (a) deposit and pattern bottom layer on passivated wafer; (b) deposit and pattern sacrificial oxide; (c) deposit and pattern structural layer; (d) remove sacrificial oxide. © 1997 ACM. Figure reprinted with permission.

so much time and effort to creating beneficial solutions that it is difficult to surpass these well-established solutions with an automated system).

> No rapid design process is available today for MEMS. Only one or two CAD iterations involving simple functional simulations are usually attempted during prototype design. As a result, fabrication replaces simulation in the iterative loop. This is very expensive, since fabricated prototypes often do not meet performance specifications and, sometimes, are not even functional. Full verification of designs requires months of effort, and design optimization is not realistic in all but the simplest of cases.
>
> G.K. Fedder, S. Iyer, and T. Mukherjee, *9th International Conference on Solid State Sensors and Actuators* (1997)

Although automated synthesis of microsystem designs was not contemplated, a list of challenges for microsystem design were articulated in the report of the National Science Foundation (NSF) Workshop on Microelectromechanical Systems Research, held in July, 1987 (NSF, 1988). An excerpt of this report, pertaining to microsystem design, is reproduced in Appendix A.

The potential for advanced design synthesis methods in the microsystems area was identified in the findings and recommendations of the NSF Workshop on Structured Methods for MEMS Design, held at Caltech in November, 1995 (Antonsson, 1996). Among the detailed conclusions from the workshop, the participants observed strong parallels between MEMS and VLSI.

> Several elements appear to contribute to the successes in developing structured design methods for VLSI:
>
> - a small (but growing) number of functional elements,
> - a largely planar topology,
> - a largely rectangular (Manhattan) geometry,
> - the independence of form and function,
> - conservative design rules can eliminate complicating effects,
> - a description of function exists (Boolean logic),
> - "There is a clean separation between the processing done during wafer fabrication and the design effort that creates the pattern to be implemented." (Mead and Conway, 1980, p. 47)

Three areas currently in use in VLSI design have also been identified as common elements for structured MEMS design:

Languages for interchange of data among designers and between designers and fabricators,

Libraries for storing previously successful MEMS device designs for reuse, and

Simulation of desired function and of the fabrication processes.

Each of these has played a crucial role in the development of design methodologies for VLSI, and building on those prior VLSI developments will form the basis for structured design methods for MEMS.

The summary of this report is reproduced in Appendix B.

DIGITAL VLSI DESIGN SYNTHESIS

The early state of automated design techniques in microsystems is similar to the state of solid-state semiconductor design in the early 1960s. At that time, digital electronics design was the exclusive province of highly experienced experts, who manually developed mask layouts. With the work of Carver Mead and Lynn Conway (Mead and Conway, 1980) and others, a methodology employing conservative design rules was introduced, greatly expanding the range of engineers who could perform digital design, and greatly reducing the number of prototypes required before a working device was produced. With time, additional methods have been introduced to automate the creation and verification of mask layouts for digital VLSI, including libraries of increasingly complex reusable elements (Hunt and Rowson, 1996).

As with most automated design environments (including software), the designs produced are of high quality, but they may be somewhat suboptimal compared with designs produced by a highly experienced domain expert. However, when the savings in design time and cost are included in the comparison, the automated methodology has won out, with increasing benefit realized as the complexity of the system increases.[3]

The structured approach to digital VLSI design has enabled devices to be designed that far surpass the complexity of systems that could be designed manually, for example, contemporary microprocessors.

STRUCTURED DESIGN OF MICROSYSTEMS

Today most microsystem design work remains the province of highly experienced domain experts, and it is performed without the benefit of automated synthesis methods or techniques. The development of a new devices and systems requires many prototypes and considerable manual effort.

This is not to say that all work related to microsystem design remains manual. Considerable progress has been made in the development of tools for representation,

---

[3] The slightly reduced performance of designs produced by automated procedures has been a historical barrier to acceptance. The history of software assemblers and compilers provides a particularly instructive example, in which the early machine programmers felt that automated creation of software that didn't take full advantage of every machine cycle or bit of storage was unacceptable (Backus et al., 1967; Rosen, 1967; Sammet, 1969; Brooks, 1995). With time, the savings in time and effort to produce working code (by use of software assemblers and compilers) made clear the advantages of utilizing these techniques, despite the slight performance penalty in the resulting executable program.

analysis, and verification of microsystem designs, including:

- finite-element analysis (FEA) of electric charge distribution and forces (Trimmer and Gabriel, 1987; Senturia et al., 1992; Jaecklin et al., 1993; Sun et al., 1993; Garcia and Sniegowski, 1995; Lee et al., 1995; Ollier, Labeye, and Revol, 1995; Uenishi, Tsugai, and Mehregany, 1995; Shi, Ramesh, and Mukherjee, 1996; Toshiyoshi and Fujita, 1996; Gao et al., 1997; Horenstein et al., 1997, 1999; Phillips and White, 1997; Sakakibara et al., 1997; Zappe et al., 1997; Fan et al., 1999; Hameyer and Belmans, 1999; Horsley et al., 1999; Saif et al., 1999; Ye and Mukherjee, 1999; Ono, Sim, and Esashi, 2000);
- fluid damping (Lim, Varadan, and Varadan, 1997; Nakano, Maeda, and Yamanaka, 1997; Hirai et al., 1998);
- oscillation and resonance (Senturia, Aluru, and White, 1997; Hung and Senturia, 1999; Gabbay, Mehner, and Senturia, 2000; Mehner, Gabbay, and Senturia, 2000);
- thermal effects (Flik, Choi, and Goodson, 1992; Goodson et al., 1995; Chen et al., 1997; Goodson and Asheghi, 1997; Ju et al., 1997, 1999; Ju and Goodson, 1997, 1998, 1999; Kurabayashi and Goodson, 1998; Chui et al., 1999; Goodson and Ju, 1999; Kurabayashi et al., 1999);
- and others (Koppleman, 1989; Crary and Zhang, 1990; Maseeh, Harris, and Senturia, 1990; Pourahmadi and Twerdok, 1990; Zhang, Crary, and Wise, 1990; Buser and de Rooij, 1991; Wise, 1991; Wise and Najafi, 1991; Buser, Crary, and Juma, 1992; DeLapierre, 1992; Senturia and Schmidt, 1992; Schwarzenbach et al., 1993; Kota et al., 1994; Senturia, 1996, 1998a, 1998b; Asaumi, Iriye, and Sato, 1997; Bart, 1997; Elliott, Allan, and Walton, 1997; Lien et al., 1997; Maseeh, 1997; Rao, 1997; Sato et al., 1997; Tabata, 1997; van Suchtelen et al., 1997; van Veenedaal et al., 1997; Gilbert, 1998; Ljung and Bachtold, 1998).

Several authors have presented lists of needs for microsystem design methods, including the need for the integration of MEMS design tools and FEA packages for modeling complex systems (Maseeh et al., 1990; Senturia et al., 1992, 1997; Senturia, 1996, 1998a, 1998b, 2000; Tang, 1997).

Many commercial software products for MEMS design are available (Karem et al., 1997; Maseeh, 1997; Athavale, Li, and Przekwas, 1998; Gilbert, 1998; Ljung and Bachtold, 1998; Krishnamoorthy et al., 2000; Lee et al., 2000; Ljung, Bachtold, and Spasojevic, 2000; Pindera et al., 2000; Przekwas, Turowski, and Chen, 2000; Spasojevic, Ljung, and Bachtold, 2000; Tan et al., 2000). These packages enable a designer to describe a desired two-dimensional layout, extrude it into a three-dimensional shape, create a mesh, and perform finite-element analyses including the distribution of stress, static charge, heat flow, dynamics, and fluid flow.

However, in nearly all microsystem development, the creation of the mask layouts that are ultimately used to produce the device is performed manually, with the help of computer-based drawing tools. As the complexity of microsystems grows, the need for structured design methodologies is also growing.

In addition to the analysis work mentioned above, a steady effort has been undertaken in the microsystems research community to develop simulations of microsystem *fabrication* processes, in an effort to reduce the number of prototypes needed.

Considerable work has been conducted and reported on semiconductor fabrication simulation (Bean, 1978; Runyan and Bean, 1990). One of the earliest reported efforts to extend fabrication simulation into the area of micromechanical systems is the Wulff–Jaccodine method (Jaccodine, 1962; DeLapierre, 1992; Fruhauf et al., 1993). This method uses plane waves that propagate outward at a rate given by the etch rate diagram. At each point on the initial surface, a tangent plane is moved outward a distance equal to the appropriate rate multiplied by the time. At sharp corners there is a geometrical test to determine if new planes appear. The final shape is the envelope of all these planes. An analysis tool called ASEP, which can predict the output shape based on traveling planes, is presented in Buser and de Rooij (1991) and in Thurgate (1991).

The slowness method (Sequin, 1992) uses the inverse of the rate (the slowness) to calculate the trajectories of points or lines in the shape. The trajectory of a corner is given by a vector relation involving the slowness vectors of the two lines that form the corner. This relation states that the trajectory of the corner lies along the normal to the difference of the two-line slowness vectors. The trajectories of the corners are then used to determine when lines disappear and the procedure is iterated to find the shape at any time. Sequin has successfully used the slowness method to model changing shapes (Sequin, 1992).

A cellular automata (CA) approach is introduced in Than and Buttgenbach (1994), Hubbard and Antonsson (1997), and Zhu and Liu (1998). This simulation approach divides a volume of material to be fabricated into small cubic cells. Each simulation cell contains a set of rules prescribing the conditions and rate of removal of a portion of the cell.

The E-shape approach is introduced in Hubbard and Antonsson (1994). The mask layout and etched shapes are modeled as polygons, and therefore the model of the etching process must properly account for the behavior of the sides of the polygons (planes, or tangents in two dimensions) and the corners. The corners are modeled as polygonal approximations to circular curves with an infinitesimal radius. Thus each corner contains a very large number of nearly coincident planes or tangents. The E-shape model focuses on the evolution of these planes as the etching takes place, with particular attention paid to the corners (because all planes or tangents in the entire polygonal shape appear at the corners). An illustration of this geometric approach to etch modeling is shown in Figure 5.5.

A hybrid approach (SEGS) is introduced in Hubbard and Antonsson (1996) and in Li, Hubbard, and Antonsson (1998). The model represents the shapes as a large number of small segments or facets (as in the CA method), but they also retain geometrical information as in the E-shape method. This is analogous to finite-element analysis, which customarily uses elements that conform to the local shape of the object being modeled. The basic approach is to start with the polygonal boundary of a vector method, and then subdivide each straight line segment into many smaller segments. This implements a spatially accurate discretization of the shape, but with many fewer elements than a CA method. Additionally, a rectangular grid of cells is superimposed on top of the line segments to identify and locate global intersections. Because there are now a large number of primitives (segments plus geometry), the method is robust and nonlinear effects can be modeled. Because the primitives

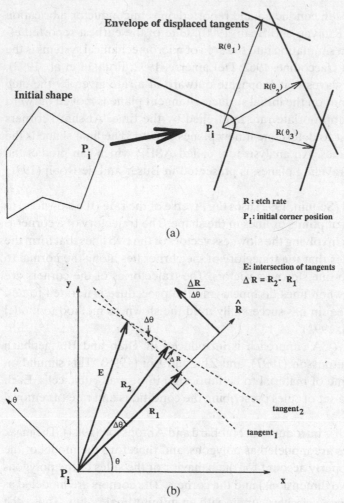

Envelope of displaced tangents

$R(\theta_1)$

$R(\theta_2)$

$R(\theta_3)$

Initial shape

P_i

P_i

R : etch rate

P_i : initial corner position

(a)

E: intersection of tangents

$\Delta R = R_2 - R_1$

$\frac{\Delta R}{\Delta \theta}$

$\hat{R}$

y

$\Delta \theta$

ΔR

$\hat{\theta}$

E

R_2

R_1

tangent$_2$

$\Delta \theta$

tangent$_1$

θ

P_i x

(b)

Figure 5.5. Two-dimensional E-shape etch simulation approach: (a) corner evolving to an envelope of tangents: (b) Intersection of two adjacent tangents near a corner (Hubbard and Antonsson, 1994).

also include geometry information, many fewer primitives are needed to accurately model a shape. This greatly decreases the computational cost. Two illustrations of this hybrid approach to etch modeling are shown in Figures 5.6 and 5.7.

A methodology to determine the fabrication process from a description of a set of layers in surface micromachining is introduced in Gogoi, Yeun, and Mastrangelo (1994) and in Hasanuzzaman and Mastrangelo (1995, 1996). The level-sets methodology to compute moving wave fronts is presented in Sethian (1999). This approach has been applied to three-dimensional microsystem fabrication simulation (Adalsteinsson and Sethian, 1995a, 1995b, 1996, 1997; Sethian, 1996; Sethian and Adalsteinsson, 1997). This research has resulted in significant capabilities in geometrically accurate simulation of the fabrication of microsystems.

MICROSYSTEMS DESIGN SYNTHESIS RESEARCH

Several research groups have initiated projects focusing on developing design synthesis methodologies for microsystems.

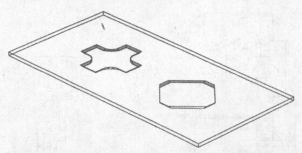

Figure 5.6. Three-dimensional SEGS etch simulation results: holes (Hubbard and Antonsson, 1996).

Tamal Mukherjee and Gary Fedder at Carnegie Mellon University have introduced an approach to microsystem design synthesis based on interconnected combinations of primitive parametrically scalable surface micromachined elements (masses, springs, electrostatic gaps, and anchors); see Mukherjee and Fedder (1997); Fedder et al. (1997); Iyer (1998); Mukherjee, Iyer, and Fedder (1998); Vandemeer (1998); Zhou (1998); Mukherjee, Zhou, and Fedder (1999); Fedder and Jing (1999); and Fedder (1999).

> The initial library focuses on surface micromachined suspended MEMS technology
> from which inertial sensors and other mechanical mechanisms can be designed.
> T. Mukherjee and G. K. Fedder, *34th Design Automation Conference* (1997)

This approach for folded flexure micromechanical resonators begins with a schematic diagram, similar to an electrical circuit diagram. An example microsystem schematic from Mukherjee and Fedder (1997) is shown in Figure 5.8.

A procedure then optimizes the parameters describing each element to best meet the performance specifications desired for the device, based on simulations of behavioral models of the primitive elements. Once the parameters of the elements are

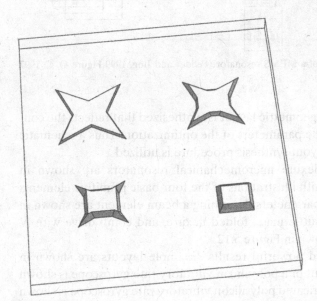

Figure 5.7. Three-dimensional SEGS etch simulation results: compensated pegs (Hubbard and Antonsson, 1996).

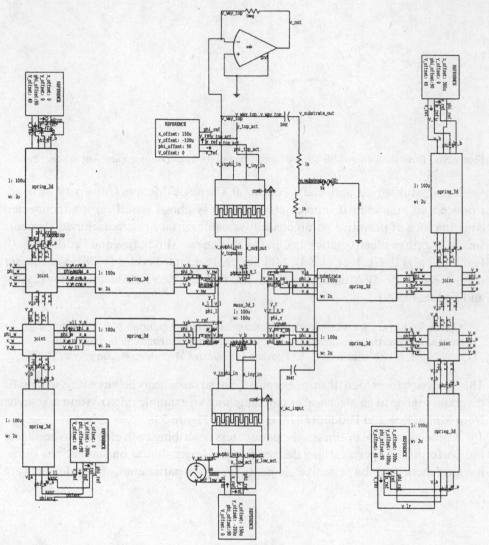

Figure 5.8. Schematic representation of a MEMS resonator (Fedder and Jing, 1999 Figure 4). © 1997 IEEE. Figure reprinted with permission.

determined, a two-dimensional geometric layout is synthesized that reflects the connectivity of the schematic and the parameters of the optimization. Thus a schematic to primitives to simulation to layout synthesis procedure is utilized.

Two examples of folded flexure micromechanical resonators are shown in Figures 5.9 and 5.10(a), along with illustrations of the four basic primitive elements in Figure 5.10(b). The lumped parameters describing a beam element are shown in Figure 5.11, and those for a shuttle mass, folded flexure, and comb drive with N movable "rotor" fingers are shown in Figure 5.12.

This approach has produced powerful results. Example layouts are shown in Figure 5.13. A synthesized layout of a polysilicon vibratory rate gyroscope is shown in Figure 5.14. An SEM of a fabricated polysilicon vibratory rate gyroscope is shown in Figure 5.15.

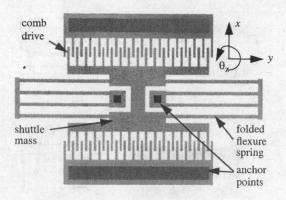

Figure 5.9. Layout of the lateral folded flexure, comb drive microresonator. The black areas are the places where the 2-mm polysilicon structure is anchored to the bottom layer. The rest of the structure is suspended 2 mm above the bottom layer (Iyer, 1998, Figure 1). © S. Iyer. Figure reprinted with permission.

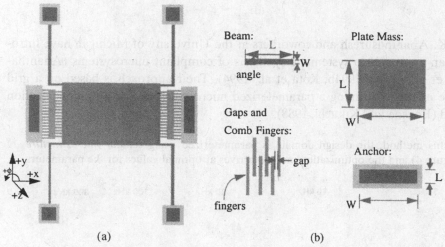

(a)   (b)

Figure 5.10. (a) MEMS design (capacitive accelerometer); (b) hierarchical set of MEMS components (Vandemeer, 1998, Figure 1). © J. Vandemeer. Figure reprinted with permission.

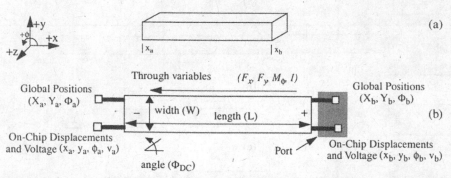

Figure 5.11. Beam component: (a) physical model; (b) schematic representation, showing global position nodes and on-chip displacement nodes (Vandemeer, 1998, Figure 5). © J. Vandemeer. Figure reprinted with permission.

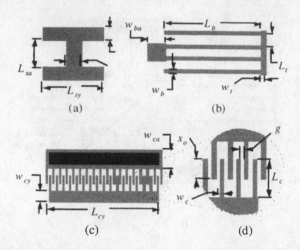

Figure 5.12. Dimensions of the microresonator elements: (a) shuttle mass; (b) folded flexure; (c) comb drive with N movable rotor fingers; (d) close-up view of comb fingers (Iyer, 1998, Figure 3). © S. Iyer. Figure reprinted with permission.

G. K. Ananthasuresh and coworkers at the University of Michigan have introduced an approach to systematic synthesis of compliant microsystems (Ananthasuresh et al., 1994a, 1994b; Kota et al., 1994). Their approach is based on a grid of finite elements forming a parameterized microstructure and a homogenization method (Bensøe and Kikuchi, 1988).

In this method, the design domain is parameterized using *cellular microstructure* (Figure 4) and the optimization method arrives at optimal values for the parameters

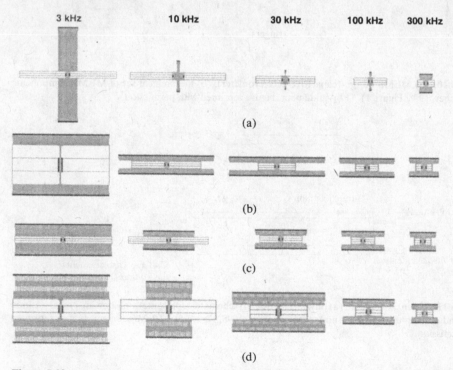

Figure 5.13. Layouts synthesized with in-plane mode separation constraints for five different frequencies and four different objective functions: (a) minimize area, (b) minimize voltage, (c) minimize area and voltage, and (d) maximize displacement at resonance (Iyer, 1998, Figure 10). © S. Iyer. Figure reprinted with permission.

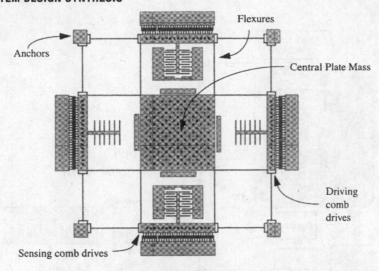

Figure 5.14. Vibratory rate gyroscope: layout of polysilicon gyroscope (Vandemeer, 1998, Figure 28b). © J. Vandemeer. Figure reprinted with permission.

of each cell (α, β and θ). [α and β are the dimensions of the rectangular hole inside each rectangular element (cell) and θ is the angular orientation of the element.] If the optimized hole dimensions α and β reach their limit values, i.e., the length and width of the cell, a hole is generated; if α and β are zero, then a solid cell is created and intermediate values give rise to porous regions. This gives the method the ability to generate *any topology, shape and size* that are optimal for given problem specifications which include applied forces, desired output displacements and the amount of material to be distributed in a prescribed design domain.

> G.K. Ananthasuresh, S. Kota, and N. Kikuchi, *Proceedings of the ASME Winter Annual Meeting* (1994)

Kris Pister and coworkers at the University of California, Los Angeles developed a parameterized synthesis procedure for electrostatic resonators (Berg et al., 1996; Lo et al., 1996).

Matt Campbell, Jon Cagan, and Ken Kotovsky at Carnegie Mellon University introduced the "A-Design" method for synthesizing novel design configurations

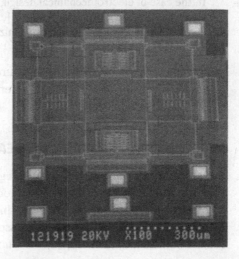

Figure 5.15. Vibratory rate gyroscope: SEM of polysilicon gyroscope (Vandemeer, 1998, Figure 28a). © J. Vandemeer. Figure reprinted with permission.

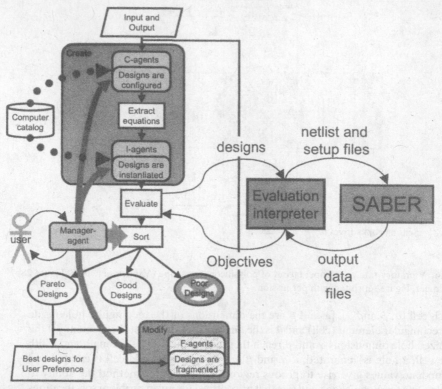

Figure 5.16. Flow chart of the A-Design method integrated with SABER dynamic simulations. (Campbell, 2000, Figure 8.8). © M. Campbell. Figure reprinted with permission.

(Campbell et al., 1999, 2000) and present an example of micromachined accelerometer design (Campbell, 2000, pp. 113–122).

> A-Design invents solutions to open-ended design problems through the interactions of a multitude of agents folded into a stochastic iterative process capable of adapting to changes in user preference (p. iv).
>
> M. Campbell (2000)

> In the case of electromechanical design, the agent [responsible for creating new configurations] searches among possible [embodiments] in the catalog to find one that leads to a new design state that maximizes the agent's evaluation function (p. 76).
>
> M. Campbell (2000)

A flowchart of the A-Design procedure for microsystems is shown in Figure 5.16. Example results of microaccelerometer layouts synthesized with the procedure are shown in Figures 5.17 and 5.18.

MICROSYSTEM SYNTHESIS BY MEANS OF STOCHASTIC EXPLORATION

An alternative approach to formalizing the design synthesis of microsystems is the use of a stochastic exploration method coupled with a computational simulation of fabrication. Initial work applying this approach has been demonstrated on multiprocess wet etching (Li and Antonsson, 1999; Lee and Antonsson; 2000; Ma and Antonsson, 2000b). This approach is shown schematically in Figure 5.19. The process

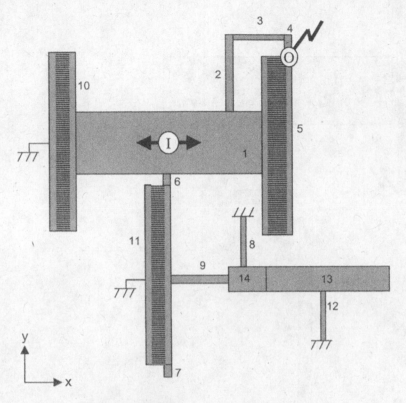

Components

1.   **MASS-25-75** h=2.5e-5 w=7.5e-5
2.   **V-BEAM-3-71** l=7.1e-5 w=3.0e-6
3.   **V-BEAM-3-15** l=1.5e-5 w=3.0e-6
4.   **H-BEAM-2-17** l=1.7e-5 w=2.0e-6
5.   **H-ES-GAP-8** w=2.0e-6 l=10.0e-6
           overlap=6.0e-6 #teeth=24
6.   **V-BEAM-2-5** l=5.0e-6 w=2.0e-6
7.   **V-BEAM-2-5** l=5.0e-6 w=2.0e-6
8.   **V-BEAM-3-71** l=7.1e-5 w=3.0e-6
9.   **H-BEAM-3-23** l=2.3e-5 w=3.0e-6
10.   **H-ES-GAP-3** w=2.0e-6 l=5.0e-6
           overlap=3.0e-6 #teeth=24
11.   **H-ES-GAP-3** w=2.0e-6 l=5.0e-6
           overlap=3.0e-6 #teeth=24
12.   **V-BEAM-2-23** l=2.3e-5 w=2.0e-6
13.   **MASS-10-50** h=1.0e-5 w=5.0e-5
14.   **MASS-10-15** h=1.0e-5 w=1.5e-5

Figure 5.17. Accelerometer created by the A-Design process with no learning (Campbell, 2000, Figure 8.9). © M. Campbell. Figure reprinted with permission.

begins with a desired three-dimensional shape. An initial population of candidate solutions (mask layouts and process parameters) is generated randomly. The features of each individual in the population are encoded into a string of characters or digits. The fabrication of each mask layout, using its associated process parameters, is then simulated to produce a three-dimensional shape. The performance of each individual in the initial population is evaluated, by comparing the three-dimensional shape it produces with the desired three-dimensional shape, and a performance

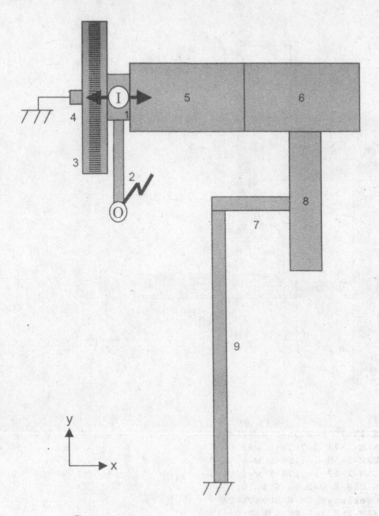

Components

```
1.    MASS-10-5 h=1.0e-5 w=5.0e-6
2.    V-BEAM-2-17 l=1.7e-5 w=2.0e-6
3.    H-ES-GAP-5 w=2.0e-6 l=10.0e-6
         overlap=4.0e-6 #teeth=12
4.    H-BEAM-3-3 l=3.0e-6 w=3.0e-6
5.    MASS-15-25 h=1.5e-5 w=2.5e-5
6.    MASS-15-25 h=1.5e-5 w=2.5e-5
7.    H-BEAM-3-17 l=1.7e-5 w=3.0e-6
8.    MASS-30-7 h=3.0e-5 w=7.0e-6
9.    V-BEAM-2-3 l=3.0e-6 w=2.0e-6
```

Figure 5.18. Accelerometer created by the A-Design process with TODO and TABOO learning (Campbell, 2000, Figure 8.10). © M. Campbell. Figure reprinted with permission.

value (called a fitness value, or FV) is calculated for each individual. Candidates that perform well (those whose corresponding shape is "close" to the desired shape) are kept in the population for the next iteration. The remaining population undergoes refinement operations according to stochastic rules, to form individuals of the next generation. With this basic stochastic exploration, the best performing candidate solutions have a higher probability of surviving into the next generation, and therefore

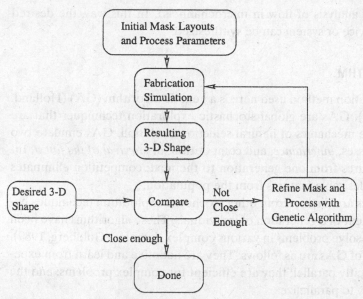

Figure 5.19. A schematic representation of a genetic algorithm MEMS synthesis technique.

are more likely to produce even better offspring for the subsequent generations. The iteration is stopped when one or more solutions sufficiently close to the desired three-dimensional shape are found.

Coupling a stochastic exploration method with a simulation of the fabrication processes can automatically synthesize a population of candidate mask layouts (and corresponding fabrication process recipes) that will produce the desired three-dimensional shape. Because for most applications the behavior or function of the microsystem is of direct interest, rather than its three-dimensional (3-D) shape, the procedure described here can be expanded (as shown in Figure 5.20) to include a simulation of the fabrication of the device, followed by a simulation of its function

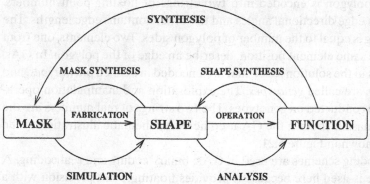

Figure 5.20. A flow chart of the microsystems design synthesis process. The central straight lines represent a prototyping process in which a candidate mask layout and process recipe are fabricated (resulting in a 3-D shape) and put in operation (resulting in a behavior or function). The curved arrows below represent engineering simulation and analysis, utilized to reduce the number of prototypes required, and to better understand the performance of the designed system. The curved arrows above represent design synthesis, where the process begins with a desired function, and from this a 3-D shape that will exhibit this function is synthesized. Starting with the desired 3-D shape, a mask layout and process recipe are synthesized that will produce the desired 3-D shape having the desired function.

(e.g., a finite-element analysis of flow in microchannels). In this way, the desired *function* of a microdevice or system can be synthesized.

THE GENETIC ALGORITHM

The stochastic exploration method used here is a genetic algorithm (GA) (Holland, 1975; Goldberg, 1989). GAs are global stochastic exploration techniques that are based on the adaptive mechanics of natural selection evolution. GAs emulate two basic evolution processes: *inheritance*, and competition, or *survival of the fittest*. Inheritance passes features from one generation to the next; competition eliminates individuals with less-desirable features from the population.

Genetic algorithms are distinct from other stochastic exploration techniques because of the involvement of a generation of individuals. These algorithms have been shown to successfully solve problems in various complex domains (Goldberg, 1989). The main advantages of GAs are as follows: They are adaptive and learn from experience; they are implicitly parallel; they are efficient for complex problems; and the computations are easy to parallelize.

In the problem addressed here, the solution space comprises the space of two-dimensional polygonal mask-layout geometries and the space of fabrication process recipes. For a multiple-process wet etching application, the process recipe space includes all the fabrication procedures available, along with parameters controlling each process and the etch time duration for each applied process.

CODING SCHEME

A coding scheme is needed to encode a two-dimensional polygon into a string that can be easily manipulated by genetic operations. Here the edge length and edge directional angle (turning angle) are chosen to describe each edge of the two-dimensional polygon (Arkin et al., 1991; Li and Antonsson, 1999). With the use of this scheme, each mask-layout polygon is encoded into two strings of floating-point numbers. One string contains edge directional angles and the other contains edge lengths. The length of each string is equal to the number of polygon sides. Two elements, one from each string with the same element position, describe an edge of the polygon. In GAs, the physical objects in the solution space that are encoded are called *phenotypes*, and the encoded strings are called *genotypes*. The exploration and manipulation operations of the GA are conducted on genotypes. The two strings of real numbers are the genotypes for the mask layouts in the GA used here. A schematic illustration of the coding scheme is shown in Figure 5.21.

Two kinds of coding schemes are used in GAs: binary coding and real coding. A real coding scheme is used here because it provides floating-point precision with a short string length.

SELECTION SCHEME

The core of a selection scheme is the method to assign a sampling rate to each individual in the generation (Grefenstette and Baker, 1989). The sampling rates determine the probability for each individual to be selected for crossover. An individual with

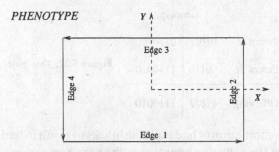

Figure 5.21. A schematic illustration of the GA coding scheme.

A Rectangle Mask Polygon With Side Length 10 & 20

GENOTYPE

	Edge 1	Edge 2	Edge 3	Edge 4
Edge Directional Angle String:	0.00	1.57	3.14	4.71
Edge Length String:	20	10	20	10

a larger sampling rate has a higher chance of surviving and passing its features on to subsequent generations. In the application addressed here, a rank-based selection algorithm is chosen to assign the sampling rates based on an individual's rank among the whole population through the following formula (Baker, 1985).

For an individual whose rank is i,

$$\text{Sampling rate} = 1.0 + \text{bias} - 2 \times \text{bias} \times i/(\text{size} - 1),$$

where

i is the individual's rank, from 0 to size -1;
size   is the population size; and
bias   is the extra fitness awarded to the top-ranked individual compared with the average-ranked individual whose fitness is 1.0.

The bias value is in general between 0.0 and 1.0 and can be used to control the selection pressure. The bias value is gradually increased as the genetic iteration proceeds to control the rate of convergence.

Crossover and Mutation. Crossover is the main operator used to produce individuals in a subsequent generation (reproduction). It combines portions of two parents to create one or more new individuals, called *offspring*, which inherit a combination of the features of the parents. There are several commonly used crossover schemes. In *one-* and *two-point crossover*, one or two gene positions are randomly selected, and crossover of the two parents is performed at those points. An example of one-point crossover is shown in Figure 5.22. The offspring is identical to the first parent up to the crossing point and identical to the second parent after the crossing point. In *uniform crossover*, each gene position is crossed with some probability, typically one-half. Crossover combines the building blocks, or *schemata*, from different solutions in various combinations. Smaller good building blocks are converted into progressively larger good building blocks over time until an entire good solution is found. Crossover is a random process, and the same process results in the

Genotype

Parent 1:	*01011*	*0110100*
Parent 2:	00110	1110010

Figure 5.22. One-point crossover of binary strings.

Offspring:	*01011*	1110010

combination of bad building blocks to result in bad offspring, but these are eliminated by the selection operator in the next generation.

Mutation is an incremental change made to each member of the population, with a small probability. Mutation enables new features to be introduced into a population so that premature convergence can be avoided.

Crossover of Polygons. The BLX-α crossover method is used here (Eshelman and Schaffer, 1991). This method produces an offspring by uniformly picking a value for each gene from the range formed by the values of two corresponding parent genes. This crossover operation is applied to the two real strings (lengths and angles) representing the mask-layout polygon. An additional step is performed to ensure that the real strings resulting after crossover are also representations of valid two-dimensional polygons (Ma and Antonsson, 2000a).

VARIABLE GENE LENGTH GA

A methodology to permit a variable number of polygon sides (and hence a variable gene length), rather than a prespecified number of sides, has been introduced to permit the GA to explore and evaluate mask-layout polygons with a range of geometric complexities (Lee and Antonsson, 2000). This approach greatly increases the likelihood that the method will synthesize a design that performs well.

A crossover operation that can accommodate parent genotypes with different gene lengths has been developed (Lee and Antonsson, 2000). An add-then-remove scheme, illustrated in Figure 5.23, is applied to extend the fixed gene length crossover to work on variable gene length cases. The crossover operation is to be applied to the two polygons shown at the top of the figure. One polygon has four sides; the other has six sides. The basic idea of the add-then-remove scheme is to add two vertices into the polygon with four sides so that this polygon can be viewed as a six-sided polygon and then crossover on these two six-sided polygons can be conducted. After crossover, some vertices of the child polygon will be removed so that the final child polygon will have a side number between the side numbers of the parent polygons.

As a way to determine the positions at which to insert additional vertices, the two parent polygons are scaled to the same perimeter length (1.0 in the example). Using a line segment from 0.0 to 1.0 to represent a polygon, each vertex of the polygon corresponds to a point on the line segment. For the square, the vertices are 0.25, 0.5, 0.75, and 1.0, and for the other polygon, the vertices are 0.2, 0.35, 0.5, 0.7, 0.85, and 1.0. The crosses in Figure 5.23 show where the new vertices are inserted.

After the insertion, the crossover operation is applied to the two polygons (with the same number of sides) to generate the initial child polygon, and then a number between the side numbers of the two parents is randomly generated as the side

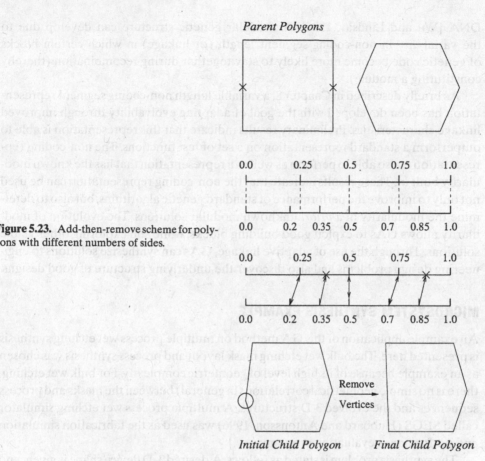

Figure 5.23. Add-then-remove scheme for polygons with different numbers of sides.

number of the final child polygon, and some vertices of the initial child polygon (the most collinear vertices) are then removed to produce the final polygon with the specified number of sides. In the example, the parents have side numbers of four and six; the initial child polygon has six sides. The randomly generated final polygon side number is five, and one vertex of the child polygon (the one in the circle) is removed to produce a final polygon with five sides.

Emergence. By this procedure, emergent properties (Stiny, 1980, 1988, 1991, 1994) are introduced into the geometric representations of the mask layouts. In this case the maximal line formed from two or more nearly collinear line segments are coalesced into a single line with the maximum length of the individual segments. The property of emergence is crucial for design systems, to avoid composition of indivisible atoms and the problems this causes for designed systems, as described in the chapters by Mitchell (Chapter 1) and Stiny (Chapter 2) in this volume.

DISCOVERY OF MODULARITY

In nature, the majority of DNA in an organism is *non-coding*, meaning that it is not transcribed into RNA for protein synthesis. It has been hypothesized that non-coding DNA prevent the destruction of good building blocks (such as genes) during recombination, since crossover is more likely to occur in a region of non-coding

DNA [Wu and Lindsay, 1996]. A modular genetic structure can develop due to the variability in non-coding segment length (or linkage) in which certain blocks of genetic code become more likely to stay together during recombination (thereby constituting a module).

As briefly described in Chapter 8, a variable length non-coding segment representation has been developed with the goal of adapting evolvability through improved linkage characteristics. Preliminary results indicate that this representation is able to outperform a standard representation on a set of test functions. The non-coding representation is also able to perform as well as a representation that has the known modularity built in. These results indicate that the non-coding representation can be used not only to improve the performance of standard genetic algorithms, but also to determine the modularity of *a priori* unknown modular solutions. The evolution of modularity allows GAs to exploit good building blocks, while still exploring other viable solutions. Through the use of adaptive linkage, GAs can synthesize solutions to engineering design problems and also discover the underlying structure of good designs.

MICROSYSTEM SYNTHESIS EXAMPLE

An example application of this GA method on multiple process wet etching synthesis is presented here. The bulk wet etching mask layout and process synthesis was chosen as an example because of its high level of geometric complexity. For bulk wet etching, there is no simple geometrical correlation (in general) between the masks and process sequences and the evolved 3-D structures. A multiple process wet etching simulator called SEGS (Hubbard and Antonsson, 1996) was used as the fabrication simulation and performance evaluation program.

The synthesis problem is stated as follows. A desired 3-D device shape is given, and three different etchants can be chosen. The fabrication procedures allowed here are two wet etching steps, which means two different etchants can be applied sequentially, each for a predetermined time duration, to generate the final shape. A multiprocess wet etching procedure can generate highly geometrically complex shapes and will require automated design tools. The goal is to find a mask layout, and the order and time duration of the two etchants, to produce a desired 3-D shape.

SHAPE COMPARISON

An approach to evaluating the closeness between the simulated evolved 3-D shapes and the desired target shape is needed. The shapes are represented here by a stack of polygonal layers. The mismatch between two 3-D shapes (contours of the target shape and a candidate evolved shape at the same vertical position) is decomposed into pure shape mismatch and size mismatch. For these two polygons, the size mismatch is calculated as the length ratio of the two polygons. The pure shape mismatch between two polygons is obtained by using the method introduced in Arkin et al. (1991) and is calculated as the *weighted* sum of the mismatch between the two polygon layers in each vertical level.

For some devices with a small z-direction dimension compared with dimensions in the x–y plane, two shapes with a similar x–y plane cross section, but different side wall slopes, may be computed to be very close by using the shape and size mismatches criteria described above, but the difference in the side wall angles of these two shapes may make one device an unacceptable replacement for the other. To overcome this difficulty, a scheme to represent side wall slope using contours of a shape is also computed.

FITNESS EVALUATION

The fitness value of each individual is evaluated in the following steps.

1. The 3-D etching simulation is run on each candidate solution. Each candidate solution consists of a mask polygon and process parameters including two etchant numbers and two etch time duration values. The simulation is run by using the mask, two etchants for those two etch time durations, and as a result an evolved 3-D shape represented by a stack of polygonal layers is produced.
2. The pure shape mismatch value, size mismatch value, and slope mismatch value between each of the evolved 3-D shapes and the specified target shape are calculated. These values are stored as the objective function values of each individual in the generation.
3. After the objective function values of all the individuals in the generation are obtained, the fitness values are calculated from the objective function values. This is a typical multiple objective optimization problem, in which an overall performance value is constructed from three different objectives. Several different ways to construct the fitness value were tried, and the noncompensating maximum-of-all scheme seems to work best:

$$\text{fitness}(i) = \max\left[\frac{\text{AShM}}{\text{ShM}(i)}, \frac{\text{ASiM}}{\text{SiM}(i)}, \frac{\text{ASlM}}{\text{SlM}(i)}\right].$$

where ShM, SiM, and SlM are shape, size, and slope mismatch values of individual i, and AShM, ASiM, and ASlM are the respective average values of the generation.

RESULT

The synthesis task was completed in 38 minutes on a Sun Ultra10 workstation with a CPU clock speed of 333 MHz. Some parameters used in the iterative synthesis are: the population size is 60; the maximum number of edges in the mask layouts is 16, and the maximum number of iterations is 60.

The results of synthesis using the approach described here are shown in Figure 5.24 and Table 5.1. In Figure 5.24, the lower-right frame shows the target shape, and the other frames show the best candidate mask layouts (the black polygons) at five different iterations during the synthesis loop, and the simulated 3-D shapes. The convergence of the evolved shapes to the target shape can be easily observed. Detailed information about these five iterations is given in Table 5.1. The shape, size, and slope mismatch values between the simulated shape and the target shape decrease in each iteration as the synthesis proceeds, and the etchant numbers converge to 2, 1, and etch times converge to 3, 3. By iteration 32, though the mask shape is nonintuitive (actually the mask shape shows corner compensation), the resulting 3-D shape

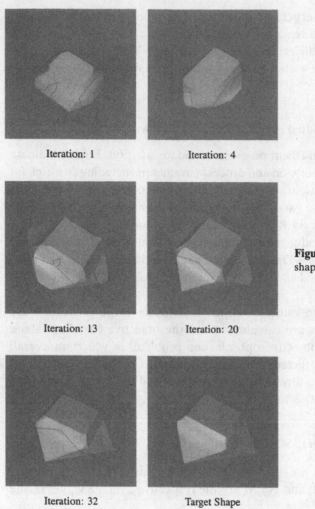

Iteration: 1   Iteration: 4

Iteration: 13   Iteration: 20

Figure 5.24. Mask layouts and evolved shapes.

Iteration: 32   Target Shape

closely approaches the target shape. During the synthesis, mask-layout polygons with different numbers of sides are explored and the final mask layout has 16 sides. Figure 5.25 shows the convergence curves of the shape, size, and slope mismatch values. Figure 5.26 shows the convergence of the side numbers of the mask layout polygons to 16. The convergence of the first and second etchant numbers to 2 and 1, and the convergence of the first etch time to 3 time units, are shown in Figures 5.27 and 5.28.

TABLE 5.1. SYNTHESIS DATA SHOWING CONVERGENCE

Iteration	Etchant Nos.	Etch Times	No. of Sides	Mismatch		
				Shape	Size	Slope
1	2, 2	4, 2	24	0.261	0.109	0.211
4	2, 1	5, 1	16	0.220	0.238	0.245
13	2, 1	4, 2	28	0.107	0.046	0.123
20	2, 1	3, 3	16	0.090	0.099	0.046
32	2, 1	3, 3	16	0.021	0.018	0.014

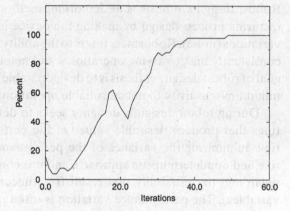

Figure 5.25. Convergence curves of shape, size, and slope mismatches.

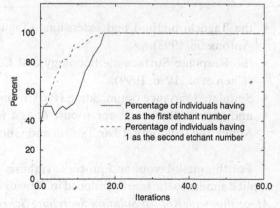

Figure 5.26. The percentage of individuals having 16 as the side number of mask-layout polygons converges to 100%.

ROBUST DESIGN SYNTHESIS

Robust design synthesis is an effort to improve the manufacturing yield. Robust process design by making the process insensitive (robust) to uncontrolled variations (noise). Robustness is measured in the ability of the synthesized process to perform consistently under operating conditions and manufacturing conditions. The primary goal of robust design is to synthesize viable products and processes that can tolerate manufacturing variations.

During robust design, performance is used to determine the design parameters that produce desirable mean performance and minimize the variance of the performance. Robust design is a multiphase and nondeterministic approach that is concerned with both the performance and its variations. The performance variation is often quantified at the cost of sacrificing the best performance, and therefore a trade-off between them is sought.

Many different robust design approaches have been developed, including Taguchi method and probabilistic approach.

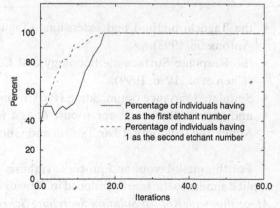

Figure 5.27. The percentage of individuals having 2 and 1 as first and second etchant numbers converges to 100%.

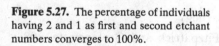

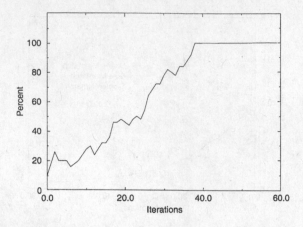

Figure 5.28. The percentage of individuals having 3 as the first etch time converges to 100%.

ROBUST DESIGN SYNTHESIS

Robust design synthesis is an important method for improving product or manufacturing process design by making the device insensitive (robust) to uncontrolled variations (noise). Robustness refers to the ability of products or processes to perform consistently under varying operational and manufacturing conditions. The primary goal of robust design synthesis is to develop stable products and processes that exhibit minimum sensitivity to uncontrollable operational and manufacturing fluctuations.

During robust design, a designer seeks to determine the control parameter settings that produce desirable values of the performance mean, while at the same time minimizing the variance of the performance. Robust design is a multiobjective and nondeterministic approach, and it is concerned with both the performance mean and the variability that result from uncertainty (represented through noise variables). The performance variation is often minimized at the cost of sacrificing the best performance, and therefore a trade-off between these two aspects cannot be avoided.

Many different robust design approaches have been developed, including:

- the Taguchi method and extensions (Taguchi, 1986; Phadke, 1989; Otto and Antonsson, 1993),
- the Response Surface methodology and Compromise Programming approach (Chen et al., 1996, 1999),
- Simulated variance optimization (Parkinson, 1998),
- and genetic algorithms for robust design (Fitzpatrick and Grefenstette, 1988; Miller and Goldberg, 1996; Tsutsui and Ghosh, 1997; Forouraghi, 2000).

For the mask-layout and process synthesis problem, a robust design scheme is applied similar to the one introduced in Tsutsui and Ghosh (1997), called the *Genetic Algorithm with Robust Solution Searching Scheme (GA/RS^3)*.

Here the mask-layout and process recipe synthesis method described above is extended to robust design. Figure 5.29 schematically shows how noise factors can be integrated into the design process by using the GA approach to synthesize a solution that is robust to uncontrolled manufacturing variations.

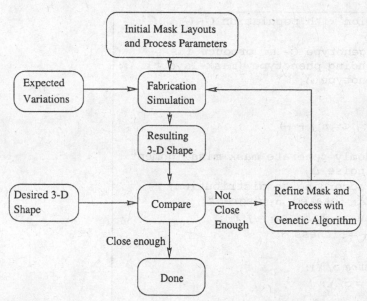

Figure 5.29. A schematic representation of a genetic algorithm MEMS synthesis technique for robust design.

ROBUST DESIGN FOR MASK MISALIGNMENT

For commercial silicon wafers, the alignment accuracy (between the flat and the crystal orientation) is usually around ±1° (Vangbo and Backlund, 1996; Madou, 1997). Additionally, the mask may not be perfectly aligned to the wafer flat. This misalignment can affect the precision of shapes fabricated. For example, the size of diaphragms formed after etching through a 500-μm thick silicon wafer can vary by 50 μm if the accuracy of the alignment is of the order of 1° (Ensell, 1996; Vangbo and Backlund, 1996; Lai, Cheng, and Huang, 1998). For this application, the optimum mask layout and process used to fabricate the target shape is synthesized by assuming perfect alignment. In actual fabrication, the misalignment of mask layout is a noise factor that affects the quality of the fabricated device, and this variation of mask-layout misalignment is taken here as the first example of an uncontrolled variation for robust design synthesis.

When robust design synthesis for mask misalignment is conducted the signal-to-noise ratio (S/N ratio) in the Taguchi method (Taguchi, 1986) is used as the performance statistic, and the sampling and exploration scheme of GA/RS^3 is utilized. For this problem, the mask-layout misalignment is considered to be uncontrolled noise with a Gaussian distribution with $\mu = 0$ and $\sigma = 1°$. For every individual in a generation, a series of randomly generated misalignments are assigned to the mask layout before the fitness values are calculated, and then the fitness values are used to construct the S/N ratio. This ratio is taken as the performance statistic to guide the search for optimum design point, and the optimum mask layout and process found will have high robustness relative to mask misalignment. The schematic model of the fitness evaluation in our robust synthesis is shown in Figure 5.30.

Given a convex protruding peg as the 3-D target shape (shown in the lower left of Figure 5.31), a robust design synthesis is conducted and a mask layout is produced

```
For each generation with population G₁, G₂, ..., Gₙ:
{
    Decode each genotype Gᵢ to produce its
        corresponding phenotype (mask-layout) Xᵢ;
    For each phenotype Xᵢ
    {
        S/N = 0;
        for (j = 0; j <= n; j++)
        {
            Randomly generate mask misalignment
                noise Δⱼ
                according to distribution;
            Yᵢ = Xᵢ rotated Δⱼ degrees;
            Evaluate fitnessᵢ = f(Yᵢ);
            S/N+ =fitnessᵢ/n;
        }
        S/N = -10log(S/N);
        fitnessᵢ = S/N;
    }
}
```

Figure 5.30. Schematic model of fitness evaluation.

that will result in a shape close to the desired 3-D shape even over a range of mask misalignment of $\pm 3°$. This result is compared with the result of synthesis without considering mask misalignment (here called *nonrobust synthesis*) in Figure 5.31. The top row shows the mask layout from nonrobust synthesis and the fabricated shapes when the mask misalignments are $0°$, $1.5°$, and $3°$. The bottom row shows the results from the robust synthesis. Both mask layouts generate shapes close to the desired 3-D target shape when the mask alignment is perfect ($0°$ misalignment). However, when the misalignment increases to $1.5°$ or $3°$, the superiority of the 3-D shape produced by the robust synthesis mask layout can be observed at the corners of the 3-D shapes. The robust mask produces square corners, even with a misalignment of $3°$. The nonrobust mask produces jagged spurs on some corners and diagonal edges on other corners when the mask is misaligned.

Figure 5.32 illustrates how the mismatch between the desired 3-D shape and the shape produced from the mask layout varies with the misalignment of the mask for two different mask layouts. The lower robust mask curve shows how little the shape mismatch grows with mask misalignment when variations in alignment are included in the mask layout synthesis procedure. The upper nonrobust mask curve shows the shape mismatch growing with misalignment, indicating that the resulting shape corresponds less well with the desired 3-D shape when variations in alignment are not incorporated in the mask-layout synthesis procedure.

SUMMARY

A mask-layout and fabrication process synthesis technique combining a genetic algorithm and forward fabrication simulation is described here, along with an application

Nonrobust Mask

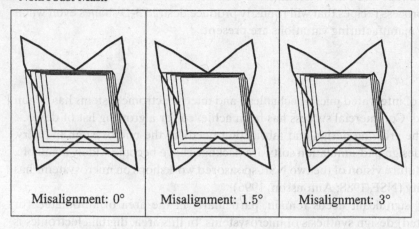

Misalignment: 0°   Misalignment: 1.5°   Misalignment: 3°

Robust Mask

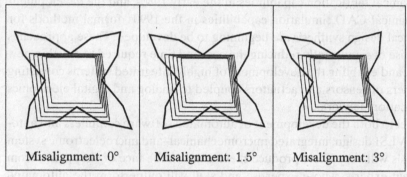

Misalignment: 0°   Misalignment: 1.5°   Misalignment: 3°

Figure 5.31. Comparison of robust and nonrobust syntheses.

of multiple process wet etching synthesis. Results demonstrate that mask layouts and fabrication process recipes can be automatically synthesized for desired 3-D shapes. An extension of this method to include expected fabrication variations, such as mask misalignment, is included in the synthesis procedure, resulting in mask layout and process sequences that are least sensitive to these variations. This robust mask layout and fabrication process synthesis procedure can readily be expanded to include

Figure 5.32. Chart showing the performance of the robust and nonrobust masks over a range of mask misalignment.

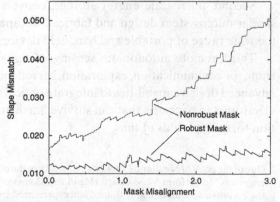

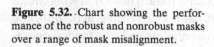

etch rate variations and other uncontrolled process variations, thus producing mask layouts and process recipes that will robustly produce desired 3-D shapes even when uncontrolled manufacturing variations are present.

OUTLOOK

The promise of integrated micromechanical–and microelectronic systems has begun to be realized. Commercial success has been achieved for a growing list of devices. To support the growing design and fabrication needs of the microsystem industry, commercial design and simulation software packages have become widely available. Much of the future vision of the two NSF-sponsored workshops on microsystems has become reality (NSF, 1988; Antonsson, 1996).

However, significant needs remain, particularly in the area of automated (or semiautomated) design synthesis of microsystems. In this area, digital electronics is far ahead of electromechanical microsystems. In a parallel to the early development of micromechanical fabrication capabilities in the early 1980s, and the development of micromechanical CAD simulation capabilities in the 1990s, formal methods for micromechanical design synthesis are beginning to be developed. These approaches hold the promise of significantly reducing the cost and time required to develop new microsystems, and enabling the development of highly integrated systems consisting of large numbers of sensors and actuators coupled to analog and digital electronics and communication networks.

In a parallel to both the development of automated software compilers and automated digital VLSI design, integrated micromechanical–and microelectronic system design synthesis will very likely produce these gains in the face of skepticism from existing domain experts, whose expertise and skill will outperform the automated design synthesis procedures in some circumstances.[4] However, expanding the range of engineers capable of performing competent microsystem design will likely significantly accelerate the pace of development in this area.

The promise of inexpensive highly integrated microsystems is significant.

First, biomedical applications, particularly for clinical diagnostics, appear to be the nearest-term development of greatest significance. These devices must integrate controlled microfluid flow with sensing and electronics. The decreases in cost and time to perform crucial tests will enhance the delivery of health care.

Second, microscale energy storage, conversion, and management will require similar microsystem design and fabrication capabilities, and will find application in the wide range of portable and handheld devices in everyday use.

Third, remote autonomous sensing, of signals across the electromagnetic spectrum, for communication, exploration, surveilance, and hazard detection will require advanced design capabilities to integrate optical systems with sensing, actuation, and electronics in packages that can survive harsh environments and unattended operation for long periods of time.

---

[4] One of the goals of Mead and Conway's early work in the digital electronics area was to permit "ordinary engineers" to perform VLSI design (Mead and Conway, 1980). Prior to their work, VLSI design was the exclusive domain of highly trained and experienced specialists.

Fourth, application of microsystems to space is in its earliest stages. These applications, because of the limited ability to test, and the severe environment in which they must operate, will place the greatest demands on microsystem design synthesis capabilities. Intercommunicating coordinated arrays of cooperating microspacecraft will be able to reconfigurably and reliably perform a host of current and future space missions.

As with earlier technical developments, the development of advanced automatic design synthesis methods will so significantly accelerate the pace of development that it is clear that they will play a role equivalent to the development of fabrication methods in the microelectromechanical systems area.

ACKNOWLEDGMENTS

The author gratefully acknowledges the contributions of Ted Hubbard, Hui (Frank) Li, Lin Ma, and Cin-Young Lee to this chapter.

This material is based upon work supported, in part, by the National Science Foundation under Grants ECS-9023646 and MIP-9529675, and the Defense Advanced Research Projects Agency (DARPA), Microsystems Technology Office (MTO), MEMS Program, under Contract N66001-97-C-8609, and the DARPA Composite CAD Program under Contract F30602-98-2-0152.

Any opinions, findings, conclusions, or recommendations expressed in this publication are those of the author and do not necessarily reflect the views of the sponsors.

APPENDIX A: NSF REPORT ON THE EMERGING FIELD OF MICRODYNAMICS

A NSF-sponsored workshop on microelectromechanical systems research held in 1987 produced a report entitled *Small Machines, Large Opportunities: A Report on the Emerging Field of Microdynamics* (NSF, 1988). A summary of the observations contained in the report relating to microsystem design is reproduced below.

DESIGN TOOLS

Microfabrication technologies, based on batch fabrication, lithography, and selective etching, impose new constraints on the design process. The conventional iterative fabrication of system components, which involves sequential refinement and modification, is widely used for testing concepts in mechanical systems design. However, the fact that a microfabricated design consists of a mask set and a process sequence makes this approach inappropriate. The entire design must be completed before fabrication is begun, and the cost of midcourse changes in process can be high because functions are interlocked so that changing one aspect of the design may affect other aspects. Making a change is a lengthy procedure; the "recipes" for process sequences of VLSI circuits can be up to 200 steps long.

Thus, simulating designs before they are fabricated, as is done in electronics and in large-scale mechanics, is highly beneficial. A set of computer-aided design tools can reduce overall cost and/or time between conception and prototype and improve designs for better performance. However, for several reasons, these simulation tools are not now readily used in microdynamical systems design.

A design, mask set, and process sequence for a microfabricated part cannot be completed until many system issues – such as system architecture and partitioning, signal processing methods – and packaging methods are worked out. Although we have good tools for mask layout, mechanical drafting, process simulation, and finite-element simulation, we do not have a merged, or compatible, set of simulation tools that permit both electrical and mechanical evaluation of device and package design, at varying levels of simulated detail.

A second, and more critical, need is to develop a database of mechanical, physical, and dynamical properties of the constituent materials. Ideally, these various properties can be tabulated and incorporated into the design frame so that a designer can anticipate the effects of process of geometry variation on the resulting structure and its performance.

Among the essential issues that microdynamical devices raise are three-dimensional modeling; large electric fields between structural components; large thermal effects; gradients in material properties on a small dimensional scale; fluid dynamics in and around small structures; and tribology (friction and lubrication). Addressing these issues may well exceed the capabilities of existing CAD tools.

APPENDIX B: STRUCTURED METHODS FOR MEMS DESIGN REPORT

A NSF-sponsored workshop on structured methods for MEMS design, held at Caltech in November, 1995 (Antonsson, 1996), produced a report that can be found on the Web.[5] The Summary of the Report is reproduced below.

Though there have been many remarkable and revolutionary advances made in microsystems design and fabrication during the past decade, the need for structured design methods remains. At present, each new MEMS development is expensive ($IM or more for each new device) and time consuming. One contributing factor is that there is not yet an equivalent to the Caltech Intermediate Form (CIF) or the other descriptive languages that are commonly used in VLSI design. MEMS fabrication processes have matured rapidly but they are still many and varied. The experience of the VLSI research community of 25 years ago in evolving design methodologies and fabrication processes should provide useful guidance.

Several elements appear to contribute to the successes in developing structured design methods for VLSI:

- a small (but growing) number of functional elements,
- a largely planar topology,
- a largely rectangular (Manhattan) geometry,
- the independence of form and function,
- conservative design rules that can eliminate complicating effects,
- a description of function exists (Boolean logic), and

[5] http://www.design.caltech.edu/NSF_MEMS_Workshop/

- "there is a clean separation between the processing done during wafer fabrication and the design effort that creates the pattern to be implemented." (Mead and Conway, 1980, p. 47).

The design of mechanical systems, more generally, appears to have none of these virtues, which largely explains the lack of structured design methods for mechanical design. However, MEMS is a hybrid of VLSI and mechanical systems, employing the materials and fabrication processes of the former, while utilizing many of the energy storage and transfer domains of the latter. This report concludes that although significant differences between digital VLSI design and MEMS design clearly exist, sufficient parallels also exist to strongly encourage research on structured design methods for MEMS.

Three areas currently in use in VLSI design have also been identified as common elements for structured MEMS design:

Languages for interchange of data among designers and between designers and fabricators,

Libraries for storing previously successful MEMS device designs for reuse, and

Simulation of desired function and of the fabrication processes.

Each of these has played a crucial role in the development of design methodologies for VLSI, and building on those prior VLSI developments will form the basis for structured design methods for MEMS. Advances in these three areas will provide the foundation for (semi-)automatic synthesis of MEMS, perhaps by compiling a schematic or language describing desired function into a set of masks and processing information to fabricate a device or system that will robustly exhibit the desired function.

Developing structured design methods for MEMS holds the promise to significantly reduce the costs and time to create new devices and systems, and to increase the complexity and robustness of devices and systems that can be designed.

REFERENCES

Abe, H., Sonobe, Y., and Enomoto, T. (1973). "Etching characteristics of silicon and its compounds by gas plasma," *Japanese Journal of Applied Physics*, 12:154–155.

Adalsteinsson, D. and Sethian, J. A. (1995a). "A unified level set approach to etching, deposition and lithography I: algorithms and two-dimensional simulations," *Journal of Computational Physics*, 120(1):128–144.

Adalsteinsson, D. and Sethian, J. A. (1995b). "A unified level set approach to etching, deposition and lithography II: three-dimensional simulations," *Journal of Computational Physics*, 122(2):348–366.

Adalsteinsson, D. and Sethian, J. A. (1997). "A level set approach to a unified model for etching, deposition, and lithography III. Re-deposition, re-emission, surface diffusion, and complex simulations," *Journal of Computational Physics*, 138(1):193–223.

Adalsteinsson, D. and Sethian, J. A. (1996). "Computational performance of level set methods for etching, deposition and photolithography development," *Proceedings of the International Conference on Simulation of Semiconductor Processes and Devices*, SISPAD96, September, 1996, pp. 79–80 IEEE, New York.

Analog Devices (2000). "ADXL 150/250 ±5 g to ±50 g, low noise, low power, single/dual axis accelerometers," Technical Datasheet, Analog Devices, One Technology Way, P.O. Box 9106, Norwood, MA. http://www. analog. com.

Ananthasuresh, G. K., Kota, S., and Gianchandani, Y. (1994a). "A methodical approach to the synthesis of compliant micromechanisms." In *Technical Digest Solid-State Sensors and Actuators Workshop*, Transducers Research Foundation, Cleveland, OH: USA, pp. 189–192.

Ananthasuresh, G. K., Kota, S., and Kikuchi, N. (1994b). "Strategies for systematic synthesis of compliant MEMS." In *Proceedings of the 1994 ASME Winter Annual Meeting*, ASME, New York, Vol. 55-2, pp. 677–686.

Angell, J. B., Terry, S. C., and Barth, P. W. (1983). "Silicon micromechanical devices," *Scientific American*, **248**:44–55.

Antonsson, E. K., Ed. (1996). *Structured Design Methods for MEMS*, Workshop Report, Caltech, Pasadena, CA; National Science Foundation, Computer and Information Science and Engineering Directorate, Microelectronics Information Processing Systems Division. http://www. design. caltech. edu/NSF_MEMS_Workshop/.

Arkin, E. M., Chew, L. P., Huttenlocher, D. P., Kedem, K., and Mitchell, J. S. B. (1991). "An efficiently computable metric for comparing polygonal shapes," *IEEE Transactions on Pattern Analysis and Machine Intelligence*, **13**:209–215.

Asaumi, K., Iriye, Y., and Sato, K. (1997). "Anisotropicetching process simulation system MICROCAD considering etch rates in total orientation of single crystal silicon." In *1997 CAD for MEMS Workshop Digest*, J.G. Korvink (ed.), Physical Electronics Laboratory, ETH Zurich, p. 16.

Athavale, M. M., Li, H. Y., and Przekwas, A. J. (1998). "Modeling of 3-D fluid flow for a MEMS laminar proportional amplifier." In *Technical Proceedings of MSM '98*, Applied Computational Research Society, Cambridge, MA, pp. 522–527.

Ayazi, F., Chen, H. H., Kocer, F., He, G., and Najafi, K. (2000). "A high aspect-ratio polysilicon vibrating ring gyroscope." In *Proceedings of the 2000 Solid-State Sensor and Actuator Workshop*, Transducers Research Foundation, Cleveland, OH, pp. 289–292.

Ayon, A. A. (1998). "Etching characteristics in a plasma etcher." In *Proceedings of the 1998 Solid-State Sensor and Actuator Workshop*, Transducers Research Foundation, Cleveland, OH.

Backus, J. W., Beeber, R. J., Best, S., Goldberg, R., Haibt, L. M., Herrick, H. L., Nelson, R. A., Sayre, D., Sheridan, P. B., Stern, H., Ziller, I., Hughes, R. A., and Nutt, R. (1967). "The FORTRAN automatic coding system." In *Programming Systems and Languages*, S. Rosen (ed.), McGraw-Hill, New York, Chap. 2A, pp. 29–47.

Baker, J. E. (1985). "Adaptive selection methods for genetic algorithms." In *Proceedings of the International Conference on Genetic Algorithms and Their Applications*, J. Grefenstette (ed.), Lawrence Erlbaum Associates, Hillsdale, NJ, pp. 101–111.

Bart, S. (1997). "CAD tools for manufacturable MEMS design." In *1997 CAD for MEMS Workshop Digest*, J.G. Korvink (ed.), Physical Electronics Laboratory, ETH Zurich, p. 17.

Bassous, E. (1978). "Fabrication of novel three dimensional microstructures by the anisotropic etching of (100) and (110) silicon," *IEEE Transactions on Electron Devices*, **ED-25**(10):1178–1185.

Bean, K. (1978). "Anisotropic etching of silicon," *IEEE Transactions on Electron Devices*, **ED-25**(10):1185–1193.

Bensøe, M. P. and Kikuchi, N. (1988). "Generating optimal topologies in structural design using a homogenization method," *Computer Methods in Applied Mechanics and Engineering*, **71**:197–224.

Berg, E. C., Lo, N. R., Simon, J. N., Lee, H.-J., and Pister, K. S. J. (1996). "Synthesis and simulation for MEMS design." In *SIGDA Physical Design Workshop*, Association for Computing Machinery ACM, New York, pp. 66–70.

Borky, J. M. and Wise, K. D. (1979). "Integrated signal conditioning for silicon pressure sensors," *IEEE Transactions on Electron Devices*, **ED-26**:1906–1910.

Bosch, D., Heimhofer, B., Muck, G., Seidel, H., Thumser, U., and Welser, W. (1993). "A silicon microvalve with combined electromagnetic/electrostatic actuation," *Sensors and Actuators A – Physical*, **37-8**:684–692.

Brooks, F. P. J. (1995). *The Mythical Man-Month*. Addison-Wesley, Reading, MA.

Bryzek, J., Petersen, K. E., and McCulley, W. (1994). "Micromachines made of silicon," *IEEE Spectrum*, **31**(5):20–31.

Buser, R. A., Crary, S. B., and Juma, O. S. (1992). "Integration of the anisotropic-silicon-etching program ASEP within the CAE-MEMS CAD/CAE framework." In *Proceedings of MEMS 92*, IEEE, New York, pp. 133–138.

Buser, R. A. and de Rooij, N. F. (1991). "ASEP: a CAD program for silicon anisotropic etching," *Sensors and Actuators A – Physical*, **28**:71–78.

Campbell, M. I. (2000). "The A-Design invention machine: a means of automating and investigating conceptual design," Ph.D. Dissertation, Carnegie Mellon University, Pittsburgh, PA.

Campbell, M., Cagan, J., and Kotovsky, K. (1999). "A-design: an agent-based approach to conceptual design in a dynamic environment," *Research in Engineering Design*, **11**(3): 172–192.

Campbell, M., Cagan, J., and Kotovsky, K. (2000). "Agent-based synthesis of electromechanical design configurations," *ASME Journal of Mechanical Design*, **122**(1):61–69.

Chakraborty, I., Tang, W., Bame, D., and Tang, T. (2000). "MEMS micro-valve for space applications," *Sensors and Actuators A – Physical*, **83**(1-3):188–193.

Chen, G., Goodson, K., Grigoropoulos, C., Hipwell, M., Liepmann, D., Majumdar, A., Maruyama, S., Thundat, T., Tien, C., and Tien, N. (1997). "Report of workshop: thermophysical phenomena in microscale sensors, devices, and structures," *Microscale Thermophysical Engineering*, **1**(4):267–274.

Chen, W., Allen, J. K., Tsui, K.-L., and Mistree, F. (1996). "A procedure for robust design: minimizing variations caused by noise factors and control factors," *ASME Journal of Mechanical Design*, **118**:478–485.

Chen, W., Wiecek, M., and Zhang, J. (1999). "Quality utility – a compromise programming approach to robust design," *ASME Journal of Mechanical Design*, **121**(2):179–187.

Chui, B., Asheghi, M., Ju, Y., Goodson, K., Kenny, T., and Mamin, H. (1999). "Intrinsic-carrier thermal runaway in silicon microcantilevers," *Microscale Thermophysical Engineering*, **3**(3):217–228.

Ciarlo, D. (1992). "A latching accelerometer fabricated by the anisotropic etching of (110) oriented silicon wafers," *Journal of Micromechanics and Microengineering*, **2**:10–13.

Crary, S. B. and Zhang, Y. (1990). "CAEMEMS: an intergrated computer-aided engineering workbench for micro-electro mechanical systems." In *Proceedings of MEMS 90*, IEEE, New York, pp. 113–115.

DeLapierre, G. (1992). "Anisotropic crystal etching: a simulation program," *Sensors and Actuators A – Physical*, **31**:267–274.

Deshmukh, A. A., Liepmann, D., and Pisano, A. P. (2000). "Continuous micromixer with pulsatile micropumps." In *Proceedings of the 2000 Solid-State Sensor and Actuator Workshop*, Transducers Research Foundation, Cleveland OH, pp. 73–76.

Dopper, J., Clemens, M., Ehrfeld, W., Jung, S., Kamper, K., and Lehr, H. (1997). "Micro gear pumps for dosing of viscous fluids," *Journal of Micromechanics and Microengineering*, **7**(3):230–232.

Elliott, J. P., Allan, G. A., and Walton, A. J. (1997). "Automatic generation from the mask layout of 3D micro-electromechanical structures that are fabricated using conformal depositions." In *1997 CAD for MEMS Workshop Digest*, J.G. Korvink (ed.), Physical Electronics Laboratory, ETH Zurich, p. 20.

Emmer, A., Jansson, M., Roeraade, J., Lindberg, U., and Hok, B. (1992). "Fabrication and characterization of a silicon microvalve," *Journal of Microcolumn Separations*, **4**(1):13–15.

Ensell, G. (1996). "Alignment of mask patterns to crystal orientation," *Sensors and Actuators A – Physical*, **53**:345–348.

Eshelman, L. J. and Schaffer, J. (1991). "Real-coded genetic algorithms and interval-schemata." In *Foundations of Genetic Algorithms I*, Morgan Kaufmann Publishers, San Mateo, pp. 187–202. Proceedings of the First Workshop on Foundations of Genetic Algorithms, International Society for Genetic Algorithms.

Fan, L., Hirano, T., Hong, J., Webb, P., Juan, W., Lee, W., Chan, S., Semba, T., Imaino, W., Pan, T., Pattanaik, S., Lee, F., McFadyen, I., Arya, S., and Wood, R. (1999). "Electrostatic microactuator and design considerations for HDD applications," *IEEE Transactions on Magnetics*, **35**(2):1000–1005.

Fan, L. S., Tai, Y.-C., and Muller, R. S. (1988). "Integrated movable micromechanical structures for sensors and actuators," *IEEE Transactons on Electron Devices*, **35**(6):724–730.

Farooqui, M. M. and Evans, A. G. R. (1991). "Polysilicon microstructures," in *MEMS '91*, IEEE, New York, pp. 187–191.

Fedder, G. K. (1999). "Structured design of integrated MEMS." In *12th Annual IEEE International Micro Electro Mechanical Systems Conference*, IEEE, New York, pp. 1–8.

Fedder, G. K., Iyer, S., and Mukherjee, T. (1997). "Automated optimal synthesis of microresonators." In *9th International Conference on Solid State Sensors and Actuators (Transducers '97)*, IEEE, New York, pp. 1109–1112.

Fedder, G. K. and Jing, Q. (1999). "A hierarchical circuit-level design methodology for microelectromechanical systems," *Transactions on Circuits and Systems II (TCAS)*, Special Issue, **46**(10):1309–1315. IEEE, New York.

Fitzpatrick, J. M. and Grefenstette, J. J. (1988). "Genetic algorithms in noisy environments," *Machine Learning*, **3**:101–120.

Flamm, D. L., Donnelly, V. M., and Ibbotson, D. E. (1990). "Basic chemistry and mechanisms of plasma etching," *J. Vac. Sci. Technol.*, **B1**(1):23–30.

Flik, M., Choi, B., and Goodson, K. (1992). "Heat-transfer regimes in microstructures" *Journal of Heat Transfer – Transactions of the ASME*, **114**(3):666–674.

Forouraghi, B. (2000). "A genetic algorithm for multiobjective robust design," *Applied Intelligence*, **12**:151–161.

Fruhauf, J., Trautman, K., Wittig, J., and Zeilke, D. (1993). "A simulation tool for orientation dependent etching," *Journal of Micromechanics and Microengineering*, **3**:113–115.

Gabbay, L., Mehner, J., and Senturia, S. (2000). "Computer-aided generation of nonlinear reduced-order dynamic macro-models – I: non-stress-stiffened case," *Journal of Microelectromechanical Systems*, **9**(2):262–269.

Gabriel, K. J. (1995). "Engineering microscopic machines," *Scientific American*, **273**(3):150–153.

Gao, R., Fang, J., Rao, B., and Warrington, R. (1997). "Miniaturized surface-driven electrostatic actuators: design and performance evaluation," *IEEE-ASME Transactions on Mechatronics*, **2**(1):1–7.

Garcia, E. and Sniegowski, J. (1995). "Surface micromachined microengine," *Sensors and Actuators A – Physical*, **48**(3):203–214.

Gilbert, J. R. (1998). "Integrating CAD tools for NEMS design," *Computer*, **31**(4):99–101.

Gogoi, B., Yeun, R., and Mastrangelo, C. H. (1994). "The automatic synthesis of planar fabrication process flows for surface micro-machined devices." In *Proceedings of the IEEE Micro Electro Mechanical Systems Workshop*, IEEE, New York, pp. 153–157.

Goldberg, D. E. (1989). *Genetic Algorithms in Search, Optimization, and Machine Learning*. Addison-Wesley, Reading, MA.

Gong, Q., Zhou, Z., Yang, Y., and Wang, X. (2000). "Design, optimization and simulation on microelectromagnetic pump," *Sensors and Actuators A – Physical*, **83**(1-3):200–207.

Goodson, K. and Asheghi, M. (1997). "Near-field optical thermometry," *Microscale Thermophysical Engineering*, **1**(3):225–235.

Goodson, K., Flik, M., Su, L., and Antoniadis, D. (1995). "Prediction and measurement of temperature-fields in silicon-on-insulator electronic-circuits," *Journal of Heat Transfer – Transactions of the ASME*, **117**(3):574–581.

Goodson, K. and Ju, Y. (1999). "Heat conduction in novel electronic films," *Annual Review of Materials Science*, **29**:261–293.

Grefenstette, J. J. and Baker, J. E. (1989). "How genetic algorithms work: a critical look at implicit parallelism." In *Proceedings of the Third International Conference on Genetic Algorithms*, Morgan Kaufmann Publishers, San Mateo, CA, pp. 20–27

Hameyer, K. and Belmans, R. (1999). "Design of very small electromagnetic and electrostatic micro motors," *IEEE Transactions on Energy Conversion*, **14**(4):1241–1246.

Hamn, G., Kahn, H., Phillips, S. M., and Heuer, A. H. (2000). "Fully microfabricated, silicon spring biased, shape memory actuated microvalve." In *Proceedings of the 2000 Solid-State Sensor and Actuator Workshop*, Transducers Research Foundation, Cleveland, OH, pp. 230–233.

Hasanuzzaman, M. and Mastrangelo, C. H. (1995). "MISTIC 1.1: a process compiler for micromachined devices." In *Transducers 95*, Vol. 1, Morgan Kaufmann Publishers, San Mateo, CA, Paper No. 38-A4.

Hasanuzzaman, M. and Mastrangelo, C. H. (1996). "Process compilation of thin film microdevices," *IEEE Transactions on Computer-Aided Design of Intergrated Circuit and Systems*, **15**:745–764.

Hirai, Y., Mori, R., Kikuta, H., Kato, N., Inoue, K., and Tanaka, Y. (1998). "Resonance characteristics of micro cantilever in liquid," *Japanese Journal of Applied Physics Part 1 – Regular Papers Short Notes and Review Papers*, **37**(12B):7064–7069.

Holland, J. H. (1975). *Adaptation in Natural and Artificial Systems*, University of Michigan Press, Ann Arbor, MI.

Horenstein, M., Bifano, T., Mali, R., and Vandelli, N. (1997). "Electrostatic effects in micromachined actuators for adaptive optics," *Journal of Electrostatics*, **42**(1-2):69–81.

Horenstein, M., Bifano, T., Pappas, S., Perreault, J., and Krishnamoorthy-Mali, R. (1999). "Real time optical correction using electrostatically actuated MEMS devices," *Journal of Electrostatics*, **46**(2–3):91–101.

Horsley, D., Wongkomet, N., Horowitz, R., and Pisano, A. (1999). "Precision positioning using a microfabricated electrostatic actuator," *IEEE Transactions on Magnetics*, **35**(2): 993–999.

Howe, R. T. and Muller, R. S. (1983). "Polycrystaline silicon micromechanical beams," *Journal of the Electrochemical Society: Solid State Science and Technology*, **130**:1420–1423.

Hubbard, T. J. and Antonsson, E. K. (1994). "Emergent faces in crystal etching," *Journal of Microelectomechanical System*, **3**(1):19–28.

Hubbard, T. J. and Antonsson, E. K. (1996). "Design of MEMS via efficient simulation of fabrication," in *Design for Manufacturing Conference*, ASME, New York.

Hubbard, T. J. and Antonsson, E. K. (1997). "Cellular automata in MEMS design," *Sensors and Materials*, **9**(7):437–448.

Huff, M. A., Mettner, M. S., Lober, T. A., and Schmidt, M. A. (1990). "A pressure-balanced electrostatically-actuated microvalve." In *Technical Digest IEEE Solid-State Sensor and Actuator Workshop*, IEEE, New York, pp. 123–127.

Hung, E. and Senturia, S. (1999). "Generating efficient dynamical models for microelectromechanical systems from a few finite-element simulation runs," *Journal of Microelectomechanical Systems*, **8**(3):280–289.

Hunt, M. and Rowson, J. A. (1996). "Blocking in a system on a chip," *IEEE Spectrum*, **33**(11):35–41.

Iyer, S. (1998). "Layout synthesis of microresonators," Master's Thesis, Carnegie Mellon University, Pittsburgh, PA.

Jaccodine, R. J. (1962). "Use of modified free energy theorems to predict equilibrium growing and etching shapes," *Journal of Applied Physics*, **33**(8):2643–2647.

Jaecklin, V., Linder, C., Derooij, N., and Moret, J. (1993). "Comb actuators for XY-microstages," *Sensors and Actuators A – Physical*, **39**(1):83–89.

Jaeger, R. C. (1988). *Modular Series on Solid State Devices: Volume V: Introduction to Microelectronic Fabrication*. Addision-Wesley, New York.

Jerman, H. (1990). "Electrically-activated, micromachined diaphragm valves." In *Technical digest IEEE Solid-State Sensor and Actuator Workshop*, IEEE, New York, pp. 65–69.

Jiang, G. Q. (2000). "Yaw sensor design optimization using finite element method." In *Proceedings of the 2000 Solid-State Sensor and Actuator Workshop*, Transducers Research Foundation, Cleveland, OH, pp. 293–295.

Ju, Y. and Goodson, K. (1997). "Thermal mapping of interconnects subjected to brief electrical stresses," *IEEE Electron Device Letters*, **18**(11):512–514.

Ju, Y. and Goodson, K. (1998). "Short-time-scale thermal mapping of microdevices using a scanning thermoreflectance technique," *Journal of Heat Transfer – Transactions of the ASME*, **120**(2):306–313.

Ju, Y. and Goodson, K. (1999). "Process-dependent thermal transport properties of silicon-dioxide films deposited using low-pressure chemical vapor deposition," *Journal of Applied Physics*, **85**(10):7130–7134.

Ju, Y., Kading, O., Leung, Y., Wong, S., and Goodson, K. (1997). "Short-timescale thermal mapping of semiconductor devices," *IEEE Electron Device Letters*, **18**(5):169–171.

Ju, Y., Kurabayashi, K., and Goodson, K. (1999). "Thermal characterization of anisotropic thin dielectric films using harmonic Joule heating," *Thin Solid Films*, **339**(1–2):160–164.

Karam, J.-M., Courtois, B., Boutamine, H., Drake, P., Poppe, A., Szekely, V., Rencz, M., Hofmann, K., and Glesner, M. (1997). "CAD and foundries for microsystems." In *Proceedings of the 34th Design Automation Conference*, ACM SIGDA, New York, Paper 42.2, pp. 674–679.

Koppleman, G. (1989). "Oyster, a simulation tool for micro electromechanical design," *Sensors and Actuators A – Physical*, **20**:179–185.

Kota, S., Ananthasuresh, G. K., Crary, S. B., and Wise, K. D. (1994). "Design and fabrication of microelectromechanical systems," *ASME Journal of Mechanical Design*, **116**(4):1081–1088.

Krishnamoorthy, S., Makhijani, V.B., Lei, M., Giridharan, M.G., and Tisone, T. (2000). "Computational studies of membrane-based test formats." In *Technical Proceedings of MSM 2000*, Applied Computational Research Society, Cambridge, MA.

Kurabayashi, K., Asheghi, M., Touzelbaev, M., and Goodson, K. (1999). "Measurement of the thermal conductivity anisotropy in polymide films," *Journal of Microelectomechanical Systems*, **8**(2):180–191.

Kurabayashi, K. and Goodson, K. (1998). "Precision measurement and mapping of die-attach thermal resistance," *IEEE Transactions on Components Packaging and Manufacturing Technology Part A*, **21**(3):506–514.

Lai, J. M., Chieng, W. H., and Huang, Y. C. (1998). "Precision alignment of mask etching with respect to crystal orientation," *Journal of Micromechanics and Microengineering*, **8**:327–329.

Lee, C.-Y. and Antonsson, E. K. (2000). "Self-adapting vertices for mask synthesis." In *MSM'2000, Modeling and Simulation of Microsystems, Semiconductors, Sensors and Actuators*, Applied Computational Research Society Cambridge, MA.

Lee, H. J., Crary, S. B., Affour, B., Bernstein, D., Gianchandani, Y. B., Woodcock, D. M., and Maher, M. (2000). "Generation of a metamodel for a micromachined accelerometer using T-Spice and the Iz-optimality option of I-OPT." In *Technical Proceedings of MSM 2000*, Applied Computational Research Society, Cambridge, MA.

Lee, J., Yoshimura, S., Yagawa, G., and Shibaike, N. (1995). "A CAE system for micromachines: its application to electrostatic micro-wobble actuator," *Sensors and Actuators A – Physical*, **50**(3):209–221.

Lewis, Jr., D. H., Janson, S. W., Cohen, R. B., and Antonsson, E. K. (2000). "Digital micropropulsion," *Sensors and Actuators A – Physical*, **80**(2):143–154.

Li, G., Hubbard, T., and Antonsson, E. K. (1998). "SEGS: on-line etch simulator." In *MSM'98, Modeling and Simulation of Microsystems, Semiconductors, Sensors, and Actuators*, IEEE, New York.

Li, H. and Antonsson, E. K. (1999). "Mask-layout synthesis through an evolutionary algorithm." In *MSM'99, Modeling and Simulation of Microsystems, Semiconductors, Sensors, and Actuators*, Applied Computational Research Society Cambridge, MA.

Li, H. Q., Roberts, D. C., Steyn, J. L., Turner, K. T., Carretero, J. A., Yagliogui, O., Su, Y.-H., Saggere, L., Hagood, N. W., Spearing, S. M., and Schmidt, M. A. (2000). "A high frequency high flow rate piezoelectrically driven MEMS micropump." In *Proceedings of the 2000 Solid-State Sensor and Actuator Workshop*, Transducers Research Foundation, Cleveland, OH, pp. 69–72.

Lien, H.-P., Funk, J., Bomholt, L., and Korvink, J. G. (1997). "SOLIDIS: an integral MEMS simulation environment." In *1997 CAD for MEMS Workshop Digest*, J. G. Korvink (ed.), Physical Electronics Laboratory, ETH Zurich, p. 18.

Lim, Y., Varadan, V., and Varadan, V. (1997). "Finite-element modeling of the transient response of MEMS sensors," *Smart Materials and Structures*, 6(1):53–61.

Lin, L. and Pisano, A. P. (1994). "Thermal bubble powered microactuators," *Microsystem Technologies Journal*, 1(1):51–58.

Ljung, P. B. and Bachtold, M. (1998). "Automatic model generation and analysis for MEMS," *Journal of Microelectro Mechanical Systems*, 66:545–551.

Ljung, P. B., Bachtold, M., and Spasojevic, M. (2000). "Analysis of realistic large MEMS devices." In *Technical Proceedings of MSM 2000*, Applied Computational Research Society, Cambridge, MA.

Lo, N. R., Berg, E. C., Quakkelaar, S. R., Simon, J. N., Tachiki, M., Lee, H.-J., and Pister, K. S. J. (1996). "Parameterized layout synthesis, extraction and SPICE simulation for MEMS." In *Proceedings of the 1996 IEEE International Symposium on Circuits and Systems*, IEEE, New York, pp. 481–484.

Ma, L. and Antonsson, E. K. (2000a). "Applying Genetic Algorithms to MEMS Synthesis." In *ASME International Mechanical Engineering Congress and Exposition*, ASME, New York.

Ma, L. and Antonsson, E. K. (2000b). "Mask-layout and process synthesis for MEMS." In *MSM2000, Modeling and Simulation of Microsystems, Semiconductors, Sensors and Actuators*, IEEE, New York.

Madou, M. (1997). *Fundamentals of Microfabrication*. CRC Press, New York.

Marcuccio, S., Genovese, A., and Andrenucci, M. (1998). "Experimental performance of field emission microthrusters," *Journal of Propulsion and Power*, 14(5):774–781.

Maseeh, F. (1997). "Intellicad MEMS computer-aided design." In *1997 CAD for MEMS Workshop Digest*, J.G. Korvink (ed.), Physical Electronics Laboratory, ETH Zurich, p. 18.

Maseeh, F., Harris, R., and Senturia, S. (1990). "A CAD architecture for MEMS." In *Transducers 90*, Institute of Electrical Engineers, New York, pp. 44–49.

May, G., Mitchell, K., Kerdoncuff, G., Parks, J., Brown, T., and Krawiecki, J. (2000). "Photolithography," http://www.ece.gatech.edu/research/labs/vc/theory/photolith.html. Virtual Cleanroom.

Mead, C. and Conway, L. (1980). *Introduction to VLSI Systems.*, Addison-Wesley, Reading, MA.

Mehner, J., Gabbay, L., and Senturia, S. (2000). "Computer-aided generation of nonlinear reduced-order dynamic macro-models – II: stress-stiffened case," *Journal of Microelectomechanical Systems*, 9(2):270–278.

Mehregany, M., Gabriel, K. J., and Trimmer, W. S. N. (1987). "Micro gears and turbines etched from silicon," *Sensors and Actuators A – Physical*, 12:341–348.

Mehregany, M., Gabriel, K. J., and Trimmer, W. S. N. (1988). "Integrated fabrication of polysilicon mechanisms," *IEEE Transactions on Electron Devices*, 35(6):719–723.

Miller, B. L. and Goldberg, D. E. (1996). "Genetic algorithms, selection scheme, and the varying effect of noise," *Evolutionary Comutation*, 4:113–131.

Mukherjee, T. and Fedder, G. K. (1997). "Structured design of microelectromechanical systems." In *Proceedings of the 34th Design Automation Conference*, ACM SIGDA, New York, Paper 42.3, pp. 680–685.

Mukherjee, T., Iyer, S., and Fedder, G. K. (1998). "Optimization-based synthesis of microresonators," *Sensors and Actuators A – Physical*, 70(1–2):118–127.

Mukherjee, T., Zhou, Y., and Fedder, G. K. (1999). "Automated optimal synthesis of microaccelerometers." In *12th Annual IEEE International Micro Electro Mechanical Systems Conference*, IEEE, New York, pp. 326–331.

Nakano, S., Maeda, R., and Yamanaka, K. (1997). "Evaluation of the elastic properties of a cantilever using resonant frequencies," *Japanese Journal of Applied Physics Part 1 – Regular Papers Short Notes and Review Papers*, **36**(5B):3265–3266.

NSF (1988). "Small machines, large opportunities: a report on the emerging field of microdynamics," Report of the NSF Workshop on Microelectromechanical Systems Research, K. Gabriel, J. Jarvis, and W. Trimmer (eds.), National Science Foundation, Salt Lake City, UT, Hyannis MA, and Princeton, NJ; research from July 27 and November 12, 1987 and January 28–29, 1988.

Ollier, E., Labeye, P., and Revol, F. (1995). "Micro-optomechanical switch integrated on silicon," *Electronics Letters*, **31**(23):2003–2005.

Ono, T., Sim, D., and Esashi, M. (2000). "Micro-discharge and electric breakdown in a microgap," *Journal of Micromechanics and Microengineering*, **10**(3):445–451.

Otto, K. N. and Antonsson, E. K. (1993). "Extensions to the Taguchi method of product design," *ASME Journal of Mechanical Design*, **115**(1):5–13.

Papavasiliou, A. P., Liepmann, D., and Pisano, A. P. (2000). "Electrolysis-bubble actuated gate valve." In *Proceedings of the 2000 Solid-State Sensor and Actuator Workshop*, Transducers Research Foundation, Cleveland, OH, pp. 48–51.

Parkinson, D. B. (1998). "Simulated variance optimization for robust design," *Quality and Reliability Engineering International*, **14**:15–21.

Petersen, K. E. (1982). "Silicon as a mechanical material," *Proceedings of the IEEE*, **70**(5):420–457.

Phadke, M. (1989). *Quality Engineering Using Robust Design*. Prentice-Hall, Englewood Cliffs, NJ.

Phillips, J. and White, J. (1997). "A precorrected-FFT method for electrostatic analysis of complicated 3-D structures," *IEEE Transactions on Computer-Aided Design of Integrated Circuits and Systems*, **16**(10):1059–1072.

Pindera, M. Z., Bayyuk, S., Upadhya, V., and Przekwas, A. J. (2000). "A computational framework for modeling one-dimensional, sub-grid components and phenomena in multi-dimensional micro-systems." In *Technical Proceedings of MSM 2000*, Applied Computational Research Society, Cambridge, MA.

Pister, K. S., Judy, M. W., Burgett, S. R., and Fearing, R. S. (1992). "Microfabricated hinges," *Sensors and Actuators A – Physical*, **33**:249–256.

Pourahmadi, F. and Twerdok, J. (1990). "Modeling micromachined sensors with finite elements," *Machine Design*, July 10: 44–60.

Przekwas, A. J., Turowski, M., and Chen, Z. (2000). "High-fidelity and reduced models of synthetic microjets." In *Technical Proceedings of MSM 2000*, Applied Computational Research Society, Cambridge, MA.

Rao, P. V. M. (1997). "Emerging CAD/CAM tools for MEMS." In *1997 CAD for MEMS Workshop Digest*, J. G. Korvink (ed.), Physical Electronics Laboratory, ETH Zurich, p. 26.

Rich, C. A. and Wise, K. D. (2000). "A thermopneumatically-actuated microvalve with improved thermal efficiency and integrated state sensing." In *Proceedings of the 2000 Solid-State Sensor and Actuator Workshop*, Transducers Research Foundation, Cleveland, pp. 234–237.

Robertson, J. K. (2000). "A vertical micromachined resistive heater for a micro gas separation column." In *Proceedings of the 2000 Solid-State Sensor and Actuator Workshop*, Transducers Research Foundation, Cleveland, OH: USA, pp. 170–174.

Robertson, J. K. and Wise, K. D. (1998). "A low pressure micromachined flow modulator," *Sensors and Actuators A – Physical*, **71**(1–2):98–106.

Rosen, S. (1967). "Programming systems and languages: a historical survey." In *Programming Systems and Languages*, S. Rosen (ed.), McGraw-Hill, New York, Chap. 1A, pp. 3–22.

Rossi, C., Esteve, D., and Mingues, C. (1999). "Pyrotechnic actuator: a new generation of Si integrated actuator," *Sensors and Actuators A – Physical*, **74**(1–3):211–215.

Rossi, C., Esteve, D., Temple-Boyer, P., and Delannoy, G. (1998a). "Realization, characterization of micro pyrotechnic actuators and FEM modelling of the combustion ignition," *Sensors and Actuators A – Physical*, **70**(1–2):141–147.

Rossi, C., Scheid, E., and Esteve, D. (1997). "Theoretical and experimental study of silicon micromachined microheater with dielectric stacked membranes," *Sensors and Actuators A – Physical*, **63**(3):183–189.

Rossi, C., Temple-Boyer, P., and Esteve, D. (1998b). "Realization and performance of thin SiO$_2$/SiNx membrane for microheater applications," *Sensors and Actuators A – Physical*, **64**(3):241–245.

Roylance, L. M. and Angell, J. B. (1979). "A batch-fabricated silicon accelerometer," *IEEE Transactions on Electron Devices*, **26**:1911–1917.

Runyan, W. R. and Bean, K. E. (1990). *Semiconductor Integrated Circuit Processing Technology*. Vol. 1, Addison-Wesley, New York.

Saif, M., Alaca, B., and Sehitoglu, H. (1999). "Analytical modeling of electrostatic membrane actuator for micro pumps," *Journal of Microelectomechanical Systems*, **8**(3):335–345.

Sakakibara, T., Izu, H., Tarui, H., and Kiyama, S. (1997). "Development of high voltage photovoltaic micro-devices for driving micro actuators," *IEICE Transactions on Electronics*, **E80C**(2):309–313.

Sammet, J. E. (1969). *Programming Languages: History and Fundamentals*," Prentice-Hall, Englewood Cliffs, NJ.

Sato, K., Shikida, M., Matsushima, Y., Yamashiro, T., Asaumi, K., Iriye, Y., and Yamamoto, M. (1997). "Orientation-dependent silicon etching database allowing process-CAD for bulk micromachining." In *1997 CAD for MEMS Workshop Digest*, J. G. Korvink (ed.), Physical Electronics Laboratory, ETH Zurich, p. 20.

Savinell, R. F. and Litt, M. H. (2000). "Proton conducting solid polymer electrolytes prepared by direct acid casting," United States Patent, 6,025,085.

Schmidt, M. A., Howe, R. T., Senturia, S. D., and Haritonidis, J. H. (1987). "A micromachined floating-element shear sensor." In *Transducers'87, The 4th International Conference on Solid-State Sensor and Actuator Workshop*, Institute of Electrical Engineers, Japan, pp. 383–386.

Schwarzenbach, H. U., Korvink, J. G., Roos, M., Sartoris, G., and Anderheggen, E. (1993). "A micro electro mechanical CAD extension for SESES," *Journal of Micromechanics and Microengineering*, **3**:118–122.

Senturia, S. (1996). "The future of microsensor and microactuator design," *Sensors and Actuators A – Physical*, **56**(1-2):125–127.

Senturia, S. (1998a). "CAD challenges for microsensors, microactuators, and microsystems," *Proceedings of the IEEE*, **86**(8):1611–1626.

Senturia, S. (1998b). "Simulation and design of microsystems: a 10-year perspective," *Sensors and Actuators A – Physical*, **67**(1–3):1–7.

Senturia, S. (2000). *Microsystem design*, Kluwer Academic Publishers, Boston/London/Dordrecht.

Senturia, S. and Schmidt, M. (1992). "Silicon micromachining," *FASEB Journal*, **6**(1): A5–A5.

Senturia, S. D., Aluru, N., and White, J. K. (1997). "Simulating the bahavior of MEMS devices: computational methods and needs," *IEEE Computational Science and Engineering*, **4**(1): 30–43.

Senturia, S. D., Harris, R. M., Johnson, B. P., Kim, S., Shulman, M. A., and White, J. K. (1992). "A computer-aided design system for microelectromechanical systems (MEMCAD)," *Journal of Microelectomechanical Systems*, **1**:3–13.

Sequin, C. H. (1992). "Computer simulation of anisotropic crystal etching," *Sensors and Actuators A – Physical*, **34**(3):225–241.

Sethian, J. A. (1996). "Fast marching level set methods for three-dimensional photolithography development." In *Proceedings, SPIE 1996 International Symposium on Microlithography*, SPIE, Bellingham, WA.

Sethian, J. A. (1999). *Level Set Methods and Fast Marching Methods*. Cambridge University Press, Cambridge, U.K.

Sethian, J. A. and Adalsteinsson, D. (1997). "An overview of level set methods for etching, deposition, and lithography," *IEEE Transactions on Semiconductor Devices*, **10**(1):167–184.

Shi, F., Ramesh, P., and Mukherjee, S. (1996). "Dynamic analysis of micro-electro-mechanical systems," *International Journal for Numerical Methods in Engineering*, 39(24):4119–4139.

Spasojevic, M., Ljung, P. B., and Bachtold, M. (2000). "Creation of 3D surface models from 2D layouts for BEM analysis." In *Technical Proceedings of MSM 2000*, Applied Computational Research Society, Cambridge, MA.

Starr, J. B. (1990). "Squeeze-film damping in solid-state accelerometers." In *Technical Digest IEEE Solid-State Sensor and Actuator Workshop*, IEEE, New York, pp. 44–47.

Steinbruchel, C., Lehmann, H. W., and Frick, K. (1985). "Mechanism of dry etching of silicon dioxide, a case of direct reactive ion etching," *Journal of the Electrochemical Society: Solid State Science and Technology*, 132 (1):180–186.

Stiny, G. (1980). "Introduction to shape and shape grammars," *Environment and Planning B: Planning and Design*, 7:343–351.

Stiny, G. (1988). "Formal devices for design." In *Design Theory '88*, S. L. Newsome, W. R. Spillers, and S. Finger (eds.), RPI, Troy, New York, pp. 173–188.

Stiny, G. (1991). "The algebras of design," *Research in Engineering Design*, 2(3):171–181.

Stiny, G. (1994). "Shape rules: closure, continuity, and emergence," *Environment and Planning B: Planning and Design*, 21.

Sun, X., Li, Z., Zheng, X., and Liu, L. (1993). "Study of fabrication process of a micro-electrostatic switch and its application to a micromechanical V-F converter," *Sensors and Actuators A – Physical*, 35(3):189–192.

Suzuki, K. and Tangigawa, H. (1991). "Single crystal silicon rotational micromotors." In *MEMS '91*, IEEE, New York, pp. 15–20.

Tabata, O. (1997). "Silicon anisotropic etching: effect of etching products and modelling." In *1997 CAD for MEMS Workshop Digest*, J. G. Korvink (ed.), Physical Electronics Laboratory, ETH Zurich, p. 210.

Taguchi, G. (1986). *Introduction to Quality Engineering*. Asian Productivity Organization, Unipub, White Plains, NY.

Tai, Y.-C. and Muller, R. S. (1989). "IC-processed electrostatic synchronous micromotors," *Sensors and Actuators A – Physical*, 20:49–50.

Tan, Z., Furmanczyk, M., Turowski, M., and Przekwas, A. J. (2000). "CFD-micromesh: a fast geometrical modeling and mesh generation tool for 3D microsystem simulations." In *Technical Proceedings of MSM 2000*, Applied Computational Research Society, Cambridge, MA.

Tang, W. C. (1997). "Overview of microelectromechanical systems and design processes." In *Proceedings of the 34th Design Automation Conference*, ACM SIGDA, New York, Paper 42.1, pp. 670–673.

Tang, W. C., Nguyen, T.-C. H., and Howe, R. T. (1989). "Laterally driven polysilicon resonant microstructures," *Sensors and Actuators A – Physical*, 20:25–32.

Tang, W. C., Nguyen, T.-C. H., Judy, M. W., and Howe, R. T. (1990). "Electrostatic-comb drive of lateral polysilicon resonators," *Sensors and Actuators A – Physical*, 21–23:328–331.

Than, O. and Buttgenbach, S. (1994). "Simulation of anisotropic chemical etching of crytalline silicon using a cellular-automata model," *Sensors and Actuators A – Physical*, 45:85–89.

Thurgate, T. (1991). "Segment based etch algorithm and modeling," *IEEE Transactions on Computer-Aided Design*, 10(9):1101–1109.

Tompsett, G., Finnerty, C., Kendall, K., and Sammes, N. (2000). "Integrated catalytic burner/micro-SOFC design and applications," *Electrochemistry*, 68(6):519–521.

Toshiyoshi, H. and Fujita, H. (1996). "Electrostatic micro torsion mirrors for an optical switch matrix," *Journal of Microelectromechanical Systems*, 5(4):231–237.

Trimmer, W. S. N., Ed. (1997). *Micromechanics and MEMS: Classic and Seminal Papers to 1990*, IEEE, New York.

Trimmer, W. S. N., and Gabriel, K. J. (1987). "Design considerations for a practical electrostatic micro-motor," *Sensors and Actuators A – Physical*, 11(2):189–206.

Tsutsui, S. and Ghosh, A. (1997). "Genetic algorithms with a robust solution searching scheme," *IEEE Transactions on Evolutionary Comutation*, 1:201–208.

Uenishi, Y., Tsugai, M., and Mehregany, M. (1995). "Micro-opto-mechanical devices fabricated by anisotropic etching of (110) silicon," *Journal of Micromechanics and Microengineering*, 5(4):305–312.

van Suchtelen, J., van Veenedaal, E., Nijdam, A. J., Elwenspoek, M., and van Enckevort, W. J. P. (1997). "Computer simulation of orientation-dependent etching of silicon." In *1997 CAD for MEMS Workshop Digest*, J. G. Korvink (ed.), Physical Electronics Laboratory, ETH Zurich, p. 26.

van Veenedaal, E., van Suchtelen, J., Nijdam, A. J., Elwenspoek, M., and van Enckevort, W. J. P. (1997). "Modeling of orientation dependent etching of silicon." In *1997 CAD for MEMS Workshop Digest*, J. G. Korvink (ed.), Physical Electronics Laboratory, ETH Zurich, p. 16.

Vandemeer, J. E. (1998). "Nodal design of actuators and sensors," Master's Thesis, Carnegie Mellon University, Pittsburgh, PA.

Vandemeer, J. E., Kranz, M. S., and Fedder, G. K. (1998). "Hierarchical representation and simulation of micromachined intertial sensors." In *Modeling and Simulation of Microsystems, Semiconductors, Sensors and Acuators (MSM'98)*, Applied Computational Research Society, MSM, Cambridge, MA.

Vangbo, M. and Backlund, Y. (1996). "Precise mask alignment to the crystallographic orientation of silicon wafers using wet anisotropic etching," *Journal of Micromechanics and Microengineering*, 6:279–284.

Verma, R., Baskett, I., and Loggins, B. (1998). "Micromachined electromechanical sensors for automotive applications," Semiconductor Application Note AN1645/D," Motorola Sensor Products Division, Phoenix, AZ. http://www.motorola.com.

Wise, K. D. (1991). "Integrated microelectromechanical systems: a perspective on MEMS in the 90s." In *MEMS '91*, IEEE, New York, pp. 33–38.

Wise, K. D. and Najafi, K. (1991). "Microfabrication techniques for integrated sensors and microsystems," *Science*, 25:1335–1342.

Wolf, S. and Tauber, R. N. (1986). *Silicon Processing*. Lattice Press, Los Angeles.

Wu, A. S. and Lindsay, R. K. (1996). "A survey of Intron research in genetics." In Ebeling, W., Rechenberg, I., Schwefel, L., and Voigt, H.-P. (eds.), *Proceedings of the 4th International Conference on Parallel Problem Solving from Nature (PPSN IV)*, Berlin, Springer-Verlag. Lecture Notes in Computer Science, Vol. 1141.

Ye, W. and Mukherjee, S. (1999). "Optimal shape design of three-dimensional MEMS with applications to electrostatic comb drives," *International Journal for Numerical Methods in Engineering*, 45(2):175–194.

Yee, Y., Bu, J., Chun, K., and Lee, J. (2000). "An integrated digital silicon micro-accelerometer with MOSFET-type sensing elements," *Journal of Micromechanics and Microengineering*, 10(3):350–358.

Zappe, S., Baltzer, M., Kraus, T., and Obermeier, E. (1997). "Electrostatically driven linear micro-actuators: FE analysis and fabrication," *Journal of Micromechanics and Microengineering*, 7(3):204–209.

Zhang, L., Bannerjee, S. S., Koo, J.-M., Laser, D. J., Asheghi, M., Goodson, K. E., Santiago, J. G., and Kenny, T. W. (2000). "A micro heat exchanger with integrated heaters and thermometers." In *Proceedings of the 2000 Solid-State Sensor and Actuator Workshop*, Transducers Research Foundation, Cleveland, pp. 275–280.

Zhang, Y., Crary, S. B., and Wise, K. D. (1990). "Pressure sensor design and simulation using the CAEMEMS-D module." In *Technical Digest of the IEEE Solid-State Sensor and Actuator Workshop*, IEEE, New York, pp. 32–35.

Zhou, Y. (1998). "Layout synthesis of accelerometers," Master's Thesis, Carnegie Mellon University, Pittsburgh, PA.

Zhu, Z. and Liu, C. (1998). "Anisotropic crystalline etching simulation using a continuous cellular automata algorithm." In *Micro-Electro-Mechanical Systems (MEMS)*, ASME, New York, pp. 577–582.

CHAPTER SIX

Function-Based Synthesis Methods in Engineering Design

State of the Art, Methods Analysis, and Visions for the Future

Kristin L. Wood and James L. Greer

OVERTURE: BACKGROUND, INTRODUCTION, AND MOTIVATION

MOTIVATION

The activity of concept generation is one of the lampposts of engineering design. It provides a forum for designers to apply creativity and contribute their personal flair. It also represents the time when technology is chosen or developed to fulfill the customer's needs.

The imaginary clay of product development is molded during concept generation. Until recently, the tools to shape this clay relied, almost entirely, on the experience and innate abilities of the designer or design team. Concept generation, fundamentally, was considered art, not science; informal, not formal.

In the past two to three decades, our tool set has changed significantly. Methods are continuously being developed, tested, implemented in industry, and taught to our engineering community (Otto and Wood, 2001). Figure 6.1 shows a simplistic view of these methods, where customer needs are first transformed to a repeatable functional representation, then to layouts and solution pieces, then to broad combinations and alternative products, and finally to an embodied realization that we can produce for the customer. With the use of these methods, our resulting abilities to develop products, and their underlying architectures, are significantly enhanced.

But is this developing tool set complete and convergent? New methods over the past decade (or less) resoundingly indicate a "no" to this question. Formal methods in engineering synthesis, although in their infancy, are emerging as new and *complimentary* possibilities for shaping the clay of product concepts. We see possibilities of formalizing that which was once thought to be informal, of systematizing that which was once thought to be purely artistic, and of understanding that which was once labeled as innate creativity.

This book seeks to convey and advance this tool set, referred to as "formal engineering design synthesis." In this chapter, an important subset of this tool set is considered: function-based synthesis methods.

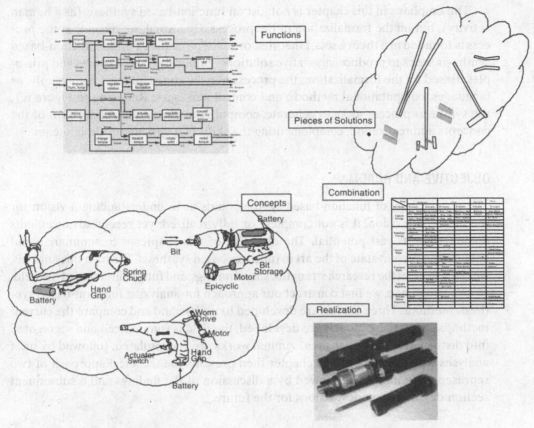

Figure 6.1. Motivation: a snapshot of functional synthesis during product development. The product example is a cordless, power screwdriver.

BACKGROUND AND ISSUES

To begin the study, let's define important terminology in function-based synthesis methods. By "function," we mean *what* a product or device must do, not *how* it will do it. The concepts of function and behavior are symbiotic; a function is the "what" for a product, and behavior results from how a function is implemented. In terms of modeling or representation, function corresponds to the *action* of a product on its inputs (materials, energies, or signals) to produce desired outputs (Stone and Wood, 1999; Otto and Wood, 2001), such as "convert torque" or "transmit electricity." Recent research in functional modeling has produced formal methods for representing product function as a vocabulary and its corresponding topology of inputs and outputs (McAdams, Stone, and Wood, 1999; Stone and Wood, 1999; Stone, Wood, and Crawford, 2000a, 2000b).

By synthesis, we mean the composition of fundamental or "atomic" elements into combinations that produce unique and desired results. The corresponding function-based synthesis process first combines functional elements, followed by structural and topological elements. The resulting combinations are realizable alternative concepts to solve a design problem.

The emphasis in this chapter is not just on function-based synthesis (as a human activity), but on the formalization of the process. By formal, we mean that the process is founded in a theory, set of theories, or set of principles. Formal function-based synthesis seeks to produce innovative solutions, guided by these theories and principles. Based on the formalization, the process may be coded, at least significantly, as languages, computational methods, and control strategies. Referring to Figure 6.1, the synthesis process seeks to generate, computationally, significant portions of the concepts and realization, complementing the skills of a designer or design team.

OBJECTIVE AND ROADMAP

The terminology of function-based synthesis aids us in understanding a vision for the field. As a vision, it is not complete or fully realized; yet recent advancements demonstrate its vast potential. The objective of this chapter is to summarize and analyze the current state of the art in function-based synthesis. Through this analysis, we hope to elicit the research strengths, shortcomings, and future direction of the field.

As a roadmap, we first construct our approach for analyzing function-based synthesis methods. Three models are developed to understand and compare the current methods. After these models are developed, the field is summarized and segregated into distinct areas. For each area, seminal works are encapsulated, followed by brief analyses with the models. The chapter then presents the technical approach of two representative methods, followed by a discussion of the findings and a subsequent section devoted to modest visions for the future.

PREAMBLE: ANALYSIS MODELS

In this section, we present the skeletal structure of our study of functional-based synthesis. Each of the contributions to this field of study could be merely summarized, reporting the basic method and the researchers' opinions regarding the inherent advantages and limitations of the method. Alternatively, we might "step back" from the methods, analyzing the basic contributions but with a common basis of comparison.

This latter approach is adopted here. Three models are developed below to assist in the analysis: a method architectural model, a design process model, and a research model. These models may be directly compared to the research reported in the literature. Through these comparisons, a number of fundamental questions may be addressed, such as the scope of a method, its coverage or niche in the design or product development process, the need that drove its creation, and its "distance" from application in industrial design practice. By answering such questions, we hope to reveal, objectively, a snapshot of the historical development of the methods and the current state of the art, as well as the avenues for future maturation of the field.

METHOD ARCHITECTURE MODEL

The area of function-based synthesis may be abstracted in terms of a general architecture representing the various generative methods, their inputs–outputs, their fundamental actions, and their layout. Figure 6.2 shows an architecture for the synthesis

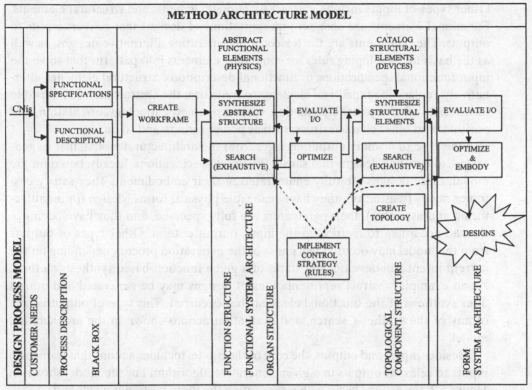

Figure 6.2. Method architecture model, in conjunction with a superimposed design process model.

methods reviewed in this chapter. This architectural model enables the analysis of the methods, including their similarities and differences. For example, considering Figure 6.1, a function-based synthesis technique might begin with a functional description of a product opportunity, as shown in the network of functions at the top of the figure. The method may then use an exhaustive search strategy of a database or repository to create a list of potential piecewise solutions to the functions. A control strategy for combining solutions may then be adopted, followed by the generation of geometry to create solutions (the alternative cordless screwdriver designs shown in Figure 6.1). These solutions may then be embodied manually or semiautomatically by using optimization or engineering modeling approaches. Together, these steps, with the requisite theory and implementation, form a function-based synthesis technique. Such a technique may be abstracted and compared with other techniques, in addition to the generic approach of Figure 6.2, to infer the comprehensiveness and depth of the technique.

Building on this brief example, the architectural model shown in Figure 6.2 includes four fundamental elements: inputs and outputs, actions or transformations performed by the method, sequential flow from one action to the next, and parallel flow representing alternative paths for a method or hybrid paths for design generation. Inputs for the method architecture begin at the fundamental level of customer needs. These needs are then expressed as either functional specifications or a functional description (functional language), initiating the generation process.

Other types of inputs include abstract functional elements and structural elements. Functional elements are abstract representations of designs that convert inputs to outputs. These elements are the lexicons for generating alternative designs, as well as the basis for developing rules for combining elements into patterns that solve the input functional specifications or functional descriptions. Structural elements, alternatively, replace the functional elements to transform the abstraction into realizable forms that include spatial information, such as geometry, topology, orientation, and position, and other characteristics, such as power ratings, and so on.

One type of primary output emerges from the architectural model, that is, generated, alternative designs that satisfy the input specifications. Ideally, based on the model, these designs are fully enumerated in their embodiment. They satisfy customer needs (feasibility), they have realizable physical forms (design for manufacturing and assembly), their parameters are fully specified, and they have been optimized according to metrics on the input–output criteria. Other types of outputs from the model may occur at any phase of the generation process, depending on the current intent and development status of a given function-based synthesis method. As an example, abstract representations of designs may be generated and output after synthesis of the functional elements has occurred. This type of output is true of any of the synthesis, search, and evaluation actions shown in the architectural model.

Besides inputs and outputs, the core of the model includes actions that transform inputs to relevant outputs in a given generative algorithm. For the model shown in Figure 6.2, the actions begin with representing the input customer needs and specifications as functional requirements or a functional language. A general workframe for generating designs is then created. This workframe is simply the media or means of representation for the designs to be synthesized, such as data structures in a computational approach or a three-dimensional reference frame for the geometric generation of kinematic mechanisms. Abstract functional elements and their associated rules are then created to form the "synthesis engine" for the method. The functional specification or functional-language representation of a design problem is fed to this synthesis engine, resulting in combinations of connected functional elements that satisfy the inputs and generation rules. A search algorithm is implemented to generate these combinations exhaustively or to some termination criterion on the complexity of the functional element chains.

The resulting designs from the synthesis process are evaluated according to their satisfaction of all input criteria (feasibility). This step is followed by a subsequent synthesis of structural elements to create physically realizable designs. Functional elements are systematically replaced with structural elements within a rule-based exhaustive search process. An important action and input into this process is the creation of a catalog of "atomic" structural elements that satisfy the input–output relationships of the functional elements. Other important actions include the optional creation of topology (position and orientation in the workframe reference axes) and control strategies to initiate, direct, and terminate the synthesis actions.

After combinations of structural elements are synthesized, the combinations are evaluated against the input–output criteria. An optimization method is then employed to refine the structural elements or to choose designs that are on the Pareto frontier (Otto and Wood, 2001) of the criteria. A subset of the generated designs emerges from this refinement process.

A given function-based synthesis method may include only a subset of the actions shown in Figure 6.2. Alternative flow paths from input to output (left to right) of the architecture model may thus exist. Alternative paths create the parallel flow structure shown in the figure, such as the creation of functional specifications *or* the creation of a functional description to begin the process. Parallel flows in this model also exist to show cyclic processing, such as the cyclic search for alternative designs in either the synthesis of functional or structural elements.

Overall, the architectural model, Figure 6.2, illustrates a metaview of the various function-based synthesis methods. We may use this model to evaluate the characteristics of the methods, such as completeness; type of synthesis; type of input information expected; and the type of strategy used in controlling the generation process, in evaluating the designs, and in optimizing the designs.

DESIGN PROCESS MODEL

The design process model presented in this chapter is a compilation based on observation from the literature (Iyengar, Lee, and Kota, 1994; Ullman, 1997; Hubka and Eder, 1998; Otto and Wood, 2001). In general, the goal of a design process is to synthesize alternative systems that perform the desired functions, meet the performance standards, and satisfy the constraints. In doing so, the design progresses through varying levels of abstraction, from the abstract concept of ascertaining what the customer wants and expects to the embodiment of the final design. At each level, the process is iterative and recursive and achieves incremental progress on a portion of the problem and its ultimate solution.

The design process considered in our analysis is superimposed on the process architecture of Figure 6.2. The superposition is intended to show how each step of the function-based-synthesis process architecture maps into the overall design process.

The overall process flow in function-based synthesis can be expressed as a hierarchical set of models, beginning with the *customer needs* (CNs), which are a representation of the customer. The CNs model the essence of the interaction between the design artifact and the customer, wherever a customer may lie along the path from manufacturing to end user.

Based on this set of CNs, the next step is to develop a *process description* for the design. Developing the process description involves an analysis of the observed or projected use patterns over the product life cycle of the design, as well as process choices used to facilitate those use patterns; for example, choosing electric power over internal combustion engine power.

The design progresses to the *black box model*, in which specific inputs and outputs to the system are determined based on the process decisions and use patterns previously stated. These inputs and outputs are the major physical flows of the system, and they are classified as energy, material, or signal (Pahl and Beitz, 1996).

The next level of the model is the *functional model*, or *function structure* (Hubka, Andreasen, and Eder, 1988; Ullman, 1993; Pahl and Beitz, 1996; Otto and Wood, 1998, 2001), which is a form-independent expression of the product design. The functional model is a domain-independent network of functions representing all the necessary working principals needed to carry out the transformation of the physical flows from input to output. These functions are selected from a set of standard functional elements known as *basis functions* (Stone and Wood, 1999). Each basis function is a primitive element that satisfies the required input–output relationship at a particular node within the network, but it is in general energy domain independent.

This network of functions spans the function space such that all input–output requirements are satisfied. The goal at this level is to refine the architecture of the network in search of a preferred functional solution for satisfaction of the CNs. This optimized (or preferred) arrangement is known as the *functional system architecture*. At this level of abstraction, the system has a specific functional architecture that is capable of carrying out the functional requirements of the design, but has no specific physical embodiment.

Further refinement of the design leads to the *organ structure* (Hubka and Eder, 1998), in which heuristics are used to identify functional elements that can be gathered together to form functional modules (Stone, Wood, and Crawford, 2000a; Otto and Wood, 2001). This organ structure is the last level where abstract functional elements are used to describe the artifact. Global product architectural decisions may be made at this point, such as modularity versus integral architectures. At this point, the design is fully described in terms of functions.

The next level of abstraction associates specific electromechanical devices with the basis functions used in the function structure. This level represents the *topological component structure* of the design. Here, structural elements, or devices, are selected to satisfy the input–output requirements at each node in the network. Because these are "real" devices, their interface specifications constrain the system configuration leading to a component topology.

The final level of the model is the *form system architecture*. This is the least abstract level in the process, where the design is now fully embodied. The goal at this level is to optimize the physical embodiment of the alternative designs that have emerged. It is at this level that the various design methods, such as function integration and design for assembly, are used to optimize the embodiment to satisfy the CNs in the most efficient and effective manner possible with the available resources.

As stated earlier, the design model presented here is a compilation of works by many authors. The goal of presenting it is to use this model as a standard against which each of the methods will be evaluated. We will look at each of the methods with regard to its overall structure and approach to design.

RESEARCH MODEL

The research model presented in this chapter is based on the observation that engineering design research is carried out in both academia and industry, but the driving force behind engineering design research must originate, or at least be targeted, in industry (Cantamessa, 2000). Industry provides the motivation for revision and extension of existing tools and methods as well as the seeds for the germination of new ideas. Development of these tools and methods has to be nurtured by both academia and industry, but validation must ultimately come from its use in a "live" product

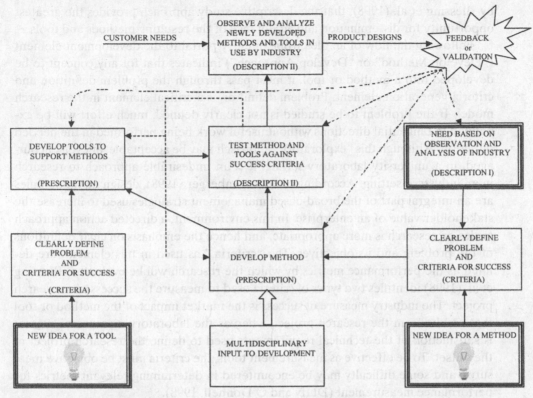

Figure 6.3. Design research model.

development environment. The mechanism for transmission of new and improved methods from research to industry has multiple paths. These paths exist both within industry alone, and between industry and academia by way of university–industry cooperation and by way of the classroom where current research results are incorporated into coursework. Each of the block elements of the proposed research model represents an area of active research in engineering design.

A schematic of the model is presented in Figure 6.3, where the parenthetical references in several of the elements, Criteria, Description I, Prescription, and Description II, are based on the work of Blessing, Chakrabarti, and Wallace (1998). These references represent a common basis for understanding design research. The elements of the model are described in more detail in the following paragraphs.

The blocks in Figure 6.3 with bold frames indicate portals of entry to the model. These entry portals represent opportunities for research contributions to the field of function-based design research. Clearly, there are two classes of opportunities, those originating from observation and analysis of industrial practices and processes, and those originating from perceived or expected needs based on original thought (the light bulb). The former are based on descriptive studies of the way product development is done in industry, and the latter are based on intuition, experience, and logical reasoning. In this context, the phrase *descriptive study* is meant as a study that is in the form of or based upon a coherent theory, not the more common meaning of being vivid or graphic in nature. Both approaches have merits, but it is proposed by Blessing et al. (1998), that the descriptive study approach provides the greatest opportunity for dissemination and acceptance of the resulting methods and tools.

Following the flow of an idea from an entry portal to the development element ("Develop Method" or "Develop Approach") indicates that for any concept to be developed into a method or tool, it must pass through the problem definition and criteria generation element. Problem definition is a critical element in the research model. If the problem to be studied is not clearly defined, much effort will be expended in tangential directions without useful work being performed in the needed direction. Although this "exploratory" approach may be acceptable, or even encouraged, in a university laboratory setting, it is an undesirable approach to research in an industrial setting. According to Schregenberger (1998), design methodologies are an integral part of the broad-based management strategies used to increase the stakeholder value of an enterprise. In this environment, a directed action approach to design research is more appropriate, and hence the emphasis on clear definitions of the problem and its objectives. The "criteria," as used in this element, are defined as the performance metrics by which the research will be evaluated. Blessing et al. (1998) identifies two types of criteria used to measure the success of a research project. The industry measure of success is the market impact of the method or tool that results from the research project, whereas the laboratory measure of success is satisfaction of the technical requirements used to define the research project at the outset. To be effective as measurement tools, the criteria must be objective measures, and some difficulty may be encountered in determining relevant metrics for performance measurement (Duffy and O'Donnell, 1998).

The *method-development* element ("Develop Method") of the model is one of the two primary areas of research in function-based synthesis (tool development is the other). Here we begin to discuss the prescriptive aspects of the engineering design research model. This element is prescriptive in nature, that is, prescribed by the innovation and thought processes of the researcher, because the outcome is in the form of instructions, heuristics, guidelines, theory, or advice for practice and application of a method. The essence of this element is the synthesis of useful electromechanical design methods based on the results of descriptive studies. Note the inclusion of the multidisciplinary element, which feeds directly into development of the method. This aspect of method development has been highlighted to bring attention to the need for involvement of disciplines other than mechanical engineering in the process of developing methods. The reasons for this comment are based, first of all, on the fact that method development is, in and of itself, a design effort, where the artifact being designed is related to the creative process of synthesis, and must involve the sciences that are best suited for understanding such efforts. A second reason for involving experts from other disciplines is the fact that the resulting method must be integrated into a much larger enterprise of product development in the business domain. Pahl (1998) highlights a chronology of collaborative efforts in engineering design research. Schregenberger (1998) pleads the case for involving these other disciplines because they provide expertise in areas well outside the domain of engineering, areas such as business and industrial management, cognitive psychology, organizational psychology, and sociology to name a few. Antonsson (1987) highlighted the need for a more scientific approach in engineering design research.

Several disciplines have expertise in clinical experiments using the scientific method, which is invaluable in researching the behavior and internal processes of designers who are engaged in synthesis. In order to make the developed methods more usable in engineering practice, there must be an interface between the method and the engineer; that interface is provided by the tools that are developed to support the methods.

Tool development is a prescriptive effort that leads to instructions, directions, worksheets, or procedures relating a design method to its practical application. Full-scale development of tools is typically a commercial venture carried out by software firms and consulting groups, but the seed that leads to commercial exploitation invariably comes from research. There are two possible paths leading to the tool development research element; the first is as an extension of method development just discussed. The second path is independent of any current method-development research. As an extension of *method development, tool development* is seen as a continuation of the process in which applying the method to engineering practice is the primary consideration. The independent path approach is the result of perceived or expected needs and is based on original thought (the light bulb). This independent path approach is typically based on the recognition or supposition of the need for a new or improved tool that is associated with an existing design method. Once a method or tool has been developed, its suitability for use must be evaluated. As such, this path implicitly includes an element of peril; that is, if the perceived need does not exist, the resulting method will likely not be accepted or used in actual product development. This inherent danger calls for more industry data to validate research efforts, as shown in the model; however, the independent path approach is still needed to cause bifurcations and leaps in the practice of engineering design.

The methods and tools that result from design research must be tested and validated before they are promoted to industry for implementation. The bottom line is that as researchers, we must ensure that the criteria developed based on descriptive analysis of industry data are used to guide the research and thus lead to useful results. The next two elements of the model are associated with the Description II nomenclature proposed by Blessing et al. (1998), and they represent the testing and validation process for a method or tool. As the *description* moniker suggests, this is a descriptive study element. Blessing et al. highlight two principal difficulties with the validation of design methods and tools: "(i) to identify whether the method or tool has the expected effect on the influencing factors that are addressed directly; and (ii) to identify whether this indeed contributes to success."

The element of *test method and tools against success criteria* is designed to address the first of the issues highlighted by Blessing et al. In this element, the method or tool is tested and evaluated against the success criteria determined at the outset of the research project. A model or tool that successfully satisfies these criteria will proceed to the next level of validation, whereas an unsuccessful candidate will be evaluated to determine what further research and development is needed to satisfy the success criteria (iteration–feedback).

Ultimately, the goal of engineering design research is to develop methods and tools that contribute to the success of designers in industry. In order to understand the appropriateness and usefulness of the research, the methods or tools that result must be observed and evaluated in an industrial setting (Blessing et al., 1998). The *observe and analyze newly developed methods and tools in use by industry* element is designed to address this issue by validating the results of design research in an industrial setting, with the preferred setting being a "live" design project. The designs that result from this element are evaluated to determine the efficacy of the method or tool used to execute it.

A second level of feedback is initiated at this level. The resulting designs are analyzed, and the results are fed back to several levels of the design research model. This feedback allows researchers to fine tune their efforts toward more effective methods and tools. In addition, a new round of descriptive studies is begun, thus allowing the studies to evolve to higher levels of refinement. The motivation for this entire effort comes from the fact that the customer of any engineering design research effort is the designer who uses the results, and our goal as researchers, and designers of design methods and tools, is to delight the customer (Otto and Wood, 2001). These final two elements are critical to the ultimate success of integrating engineering design research into industry, as the research must satisfy some need and it must be usable in real-world design situations faced by design engineers.

The purpose of presenting an overall research model is to provide a frame of reference for the various works discussed later in the chapter. We identify the entry portal, the location of the research work within the model, and the level of progress toward validating each of the works presented.

TENOR: STATE OF THE ART, BODIES OF RESEARCH

OVERVIEW OF THE RESEARCH FIELD

With the method architecture, design process, and research models defined, this section seeks to analyze the field of function-based synthesis research. To begin this analysis, let's consider a generic overview. As shown in Figure 6.4, the field may be segregated into a number of focal areas: Synthesis of Dynamic Systems (Bond Graph Chunks, Function-Based Bond Graph Chunks, Impedance Methods); Agent-Based Methods; and Catalog Design Methods (Component Composition, Set-Based, Grammars). All of these areas share the common goal of generating mechanical or electromechanical designs based on an input functional model or functional specification. They also share the common goal of formal synthesis, that is, developing, ultimately, a mathematical, grammar based, or lexicon language for transforming functional representations to physical designs or products. In the following sections, we study selected research work from each of the areas shown in Figure 6.4, applying the models developed above.

SYNTHESIS OF DYNAMIC SYSTEMS

The bulk of the research work carried out in synthesizing dynamic systems focuses on the use of bond graph methods (Paynter, 1961; Karnopp, Margolis, and Rosenberg, 1990), or equivalent, as the formal schematic description of a system being designed. Concentrating on the schematic description without regard to a physical description

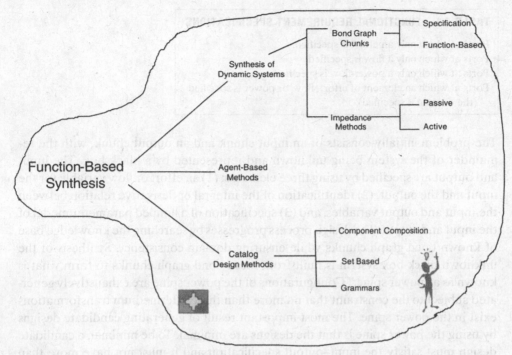

Figure 6.4. Overview of function-based synthesis research.

forces the designer to abstract the functional behavior before worrying about instantiation. Ulrich and Seering (1989) define a schematic description as a "graph of functional elements," and schematic synthesis is defined as "generating a schematic description in response to a specification of desired device behavior." The use of schematic descriptions in function-based synthesis of dynamic systems effectively separates functional issues from structural issues. Hence, synthesis of dynamic systems has come to mean the generation of bond graphs (or equivalent) that represent the functional relationships within a dynamic system.

Bond graphs are nondomain specific representations of the exchange of energy in systems composed of lumped parameter elements. Systems are represented as interconnected components with power flows across their interfaces (ports). The ports are specified in terms of effort and flow variables in various domains. The effort variables are force, torque, voltage, and pressure, for example, whereas the flow variables are velocities, current, magnetic flux, and volume flow. A distinct advantage of using bond graphs as the schematic description is that the equations of motion are generated as a byproduct of their use. The effort and flow variables can be represented as generalized variables that are nondomain specific, thus allowing the use of this schematic graph to describe multiple energy domains. The use of bond graphs in function-based synthesis has evolved into two distinct fields.

Bond Graph Chunks Methods

The first field is closely related to the work of Ulrich (Ulrich, 1988; Ulrich and Seering, 1989) in which "chunks" (or modules) of bond graphs are used to specify function.

TABLE 6.1. FUNCTIONAL REQUIREMENT SPECIFICATIONS

Ports at which only an effort is specified.
Ports at which only a flow is specified.
Ports at which only a power-flow is specified.
Ports at which an element of effort, flow, or power is specified
   (the domain is specified).

The problem initially consists of an input chunk and an output chunk, with the remainder of the system being unknown and represented by a black box. The input and output are specified by using three elements: (1) an effort or flow variable for the input and the output, (2) identification of the integral or derivative relation between the input and output variables, and (3) specification of a lumped parameter model of the input and output. The design process progresses by searching the knowledge base of known bond graph chunks while ensuring domain consistency. Synthesis of the unknown black box system is built from these bond graph chunks to form what is known as a "power spine." Configurations of the power spine are exhaustively generated subject to the constraint that no more than three intermedium transformations exist in the power spine. The most important result of generating candidate designs by using the power spine is that the designs are minimal. To be minimal, a candidate design must satisfy the input–output specification and it must not have more than three intermedium transformations. The approach of Ulrich and Seering has been characterized as a design and debug strategy, and one limitation of the approach is that it only considers SISO (single input, single output) systems.

Another work that utilizes the representational power of bond graphs is that of Prabhu and Taylor (1989). Prabhu and Taylor treat the process of designing systems, given functional requirements, as a process of mapping the requirements onto physical artifacts. Functional requirements, shown in Table 6.1, consist of a set of power, effort, and flow variables that exist in the requirements space. A library of predefined components is used to represent the physical artifacts that exist in the design space. The authors define a vector representation of the functional requirements as a combination of the elements from Table 6.1, consisting of the following variables:

> i is the port index,
> e is the effort variable value,
> f is the flow variable value,
> $\delta \equiv 1$ if power flow is out of the system and
> $ \equiv -1$ if power flow is into the system, and
> D is the domain (translational, rotational, etc.)

The function-based synthesis process begins with this vector specification of the input and output functions attached to a black box representing the unknown system. The next step in the process is to reduce the vector problem to a scalar problem and solve it by selecting a spanning set of functional primitives that satisfies the input–output specifications without regard to power flow direction (Prabhu and Taylor, 1988). The result of the scalar problem solution is a design topology representation of the black box that is bond graph based. This scalar topological solution will in general not satisfy the vector requirements; thus a "vector tuning" operation is implemented to satisfy the magnitude, orientation, and position requirements of the input–output requirement vector.

Analysis: Architecture of "Bond Graph Chunks" Methods. The methods by Ulrich and Seering and Prabhu and Taylor are quite innovative. They use the physics underlying power-flow systems to design a correct abstract schematic, followed by a physical instantiation.

Considering Figure 6.2, we see that these methods contributed significantly to many of the early activities needed for a fully developed function-based approach. However, as with the methods reviewed for grammar-based conceptual design (Cagan, Chapter 3), bond graph chunks methods are short of completeness. Customer needs are again implicitly treated as an input leading to the development of a black box model. Input to the method consists of the type of junction, one or zero; the relationship between the input and output, derivative, integral, proportional, and so on; and the bond graph chunks that represent the input and output functions. The method synthesizes a set of candidate schematic descriptions from which all possible solutions can be generated, and it allows multiple solutions to be evaluated. The number of solutions is controlled by the power spine concept and the limitation of three energy transformations. Beyond this control, the remainder of the control strategy is based in the user of the method. The designer must exercise considerable judgment in selecting which designs will be pursued. There is no optimization scheme per se; once again this activity falls to the user.

Analysis: Design Model. The bond graph chunks methods use a portion of the fundamental design method that is proposed in this chapter (Figure 6.2). The method begins at the black box level and stops just after the synthesis of abstract functional elements. The workframe of the method consists of bond graph schematic synthesis. The method has great potential to be expanded and advanced, especially with respect to coverage of the design process and domains of application.

Analysis: Research Model. The bond graph chunks methods enter the model at the New Idea for a Method portal (Figure 6.3). There is not a documented need for the method, and its applicability to industry is not expressed. The works reviewed in this section are preliminary and have spawned other research efforts. In this context, the work can be considered fundamental in nature; yet, there is a tremendous need to obtain industry data and develop the methods toward specific industrial sectors.

Function-Based Bond Graph Chunks Methods

The furthest evolution of the bond graph chunks approach to function-based synthesis is represented by the work of Malmqvist (1994). The foundation of this work is the technical systems theory of Hubka (Hubka, 1982; Hubka and Eder, 1988), and bond graph theory as previously discussed. Hubka's technical systems theory is summarized as a general characterization of the purpose of machines. A machine can be represented as four separate models: the process structure, the function structure, the organ structure, and the component structure. In addition, an extension of the function vocabulary developed by Krumhauer (1974), Table 6.2, is used to develop a library of elementary functions. These two fundamental modeling approaches along

TABLE 6.2. FUNCTION VOCABULARY

Elementary Functions	Functions with Effects	Refinement of Functions	
Change	Change physical domain	Rotation → translation Rotation → electricity Rotation → hydraulics ⋮ ⋮ Translation → electricity Translation → hydraulics ⋮ ⋮	
	Change causality	Flow → effort Effort → flow	
Vary magnitude	Noncontrolled	Steady-state response	Proportional Integrate Derivate Complex dependency Frequency Relationships
	Controlled by signal	Dynamic response	
Connect	Join	Join effort Join flow	
	Separate	Separate effort Separate flow	
Channel	Change place	Change position Change orientation	
	Stop	Translation support Rotation support Prevent leakage Isolate	
Store	Store kinetic energy	Effort input Flow input	
	Store potential energy	Effort input Flow input	

Source: Malmqvist, 1994.

with the function vocabulary are melded together with the design methodology proposed by Pahl and Beitz (1996) to produce the dynamic systems synthesis method proposed by Malmqvist.

The input to the synthesis procedure is a black box model of the artifact to be designed. Output from the method consists of a set of alternative function structures that correspond to the black box model, and a set of matching organ structures for each alternative function structure. The design problem is specified by the flows at the input and output ports with the unknown system represented by a black box model. The flows are represented as bond graph chunks that are specified by effort and flow variables and are positioned and oriented in space. The specifications are given as port type (input or output), physical domain, power value, effort or flow, time operator (integral, derivative, constant), position, and orientation.

Once specification of the model is complete, the method synthesizes alternative function structures by first constructing a "minimal" function structure by applying decomposition rules. This function structure is then varied systematically by applying variation rules. In the context of this work, "minimal" has essentially the same meaning as it does for Ulrich and Seering (bond graph chunks method, above). A minimal function structure consists of a set of elementary functions, any of which if removed would cause the system to no longer perform the required global function. The minimal function structure is generated by searching the library of elementary functions for a minimal set that satisfies the input and output specifications of the black box. The method is equipped to deal with SISO systems with a constant time operator, SISO dynamic systems, and MIMO (multiple input, multiple output) systems.

Synthesis of constant power SISO systems is a straightforward search for elementary functions that satisfy the difference between input and output specifications. For dynamic functions, the straight mapping to a single elementary function does not hold. This difficulty arises because the "requirements on causality and time operators can not be decomposed separately, since physical realizations of dynamic functions always involve two physical effects." (Malmqvist, 1994). The first effect is realized through the use of capacitive or inertial elements, but the side effect of adding an inertial or capacitive element is that the causality of the system is changed. This result can have a positive or negative side effect, depending on whether the desired change involves the time operator and the causality, or just the time operator. In either case, techniques are available to exploit or counter the effect. The method treats MIMO systems by either joining or separating functions. The system strategy is to first identify *common* and *individual* functions, and then the *joining* and *separating* functions are inserted between the common and individual functions, respectively. Once a minimal function structure is established, function structure variants are synthesized by transforming the minimal function structure in such a way that the overall specification is preserved. To achieve these various configurations, variation rules are applied.

Organ structures are now created by using the results of the function structure variation. To create an organ structure, organs that match the functions within the function structure are selected from the library by using a strategy of "one function at a time." The organ structure is constructed such that the overall function of the organ structure matches that of the function structure. Alternative organ structures are generated when more than one organ matches the function in question.

The method is applied to the synthesis of a design for an accelerometer. The input and output functions used to specify the problem are twofold. The input flow will be translation, and the output flow will be rotation of a dial; the time operator of the input will be derivative, and the output time operator will be integral (refer to Figure 6.5). The results are encouraging, as the method produces a solution consisting of a rack and pinion, two grounded torsional springs, two rotational dashpots, and a dial to indicate acceleration. The equations of motion of the synthesized system are then exercised by using a smoothly varying function that reaches a steady-state value.

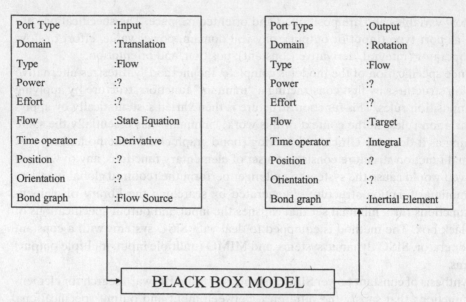

Port Type	:Input		Port Type	:Output
Domain	:Translation		Domain	: Rotation
Type	:Flow		Type	:Flow
Power	:?		Power	:?
Effort	:?		Effort	:?
Flow	:State Equation		Flow	:Target
Time operator	:Derivative		Time operator	:Integral
Position	:?		Position	:?
Orientation	:?		Orientation	:?
Bond graph	:Flow Source		Bond graph	:Inertial Element

BLACK BOX MODEL

Figure 6.5. Example input–output model for Malmqvist (1994).

The result is a set of state equations that produce the desired result–proportionality between an input acceleration and an output displacement.

Analysis: Architecture of Function-Based Dynamic Systems Methods. Overall, the function based bond graph method of Malmqvist (1994) is the most complete approach of all those reviewed in the bond graph chunks class of methods. Input to the model is given as shown in Figure 6.5. The method begins with a black box model and addresses all aspects of the proposed process architecture up through synthesis of structural elements with topology. The final result is a set of alternative designs containing components with functional and spatial characteristics that satisfy the overall design goal set forth in the initial specification. Control of the generation process is accomplished by constraining the synthesized function structure to be minimal. The generation process is guided by the decomposition and variation rules as well as the criteria for the organ structure search.

Analysis: Design Model. The design model for this method is based on similar fundamentals as the design model proposed for this chapter. The primary difference lies in the absence of the customer voice. Final output is at the level of a topological component structure with embodiment of the design represented in a limited way. No optimization of the embodiment is done; hence no form system architecture is achieved. Benefits of this method include the generation of multiple concept variants with global geometric and functional constraints satisfied.

Analysis: Research Model. Malmqvist's hypothetical basis for the research is "that computational conceptual design systems may be improved by matching the function vocabulary used to the modeling concepts of a physical systems modeling language" (Malmqvist, 1994). Entry to the research model must then be through the New Idea for a Method portal. Method development and tool development are achieved (in part), as is testing against success criteria through an example, where an accelerometer with a known output response is designed. Success is claimed based on the example, with the caveat that considerable work remains before "real" problems can be treated. Regardless of the shortcomings identified, considerable resonance exists between this method and the models proposed in this chapter. Industrial sectors should be sought, with "real" problems identified.

Impedance Methods

The second field of research that has evolved out of bond graph analysis is the use of impedance techniques for the synthesis of dynamic systems. Impedance techniques use many of the same concepts and functional models as the bond graph chunks approach. The system is still specified as a set of input and output functional criteria, with the unknown system lying between the two. The fundamental difference between the methods is the approach to synthesizing the unknown system.

The basis for the impedance method harkens back to the network synthesis work of Foster (1924), who developed the technique, and Cauer (1958), who extended the work to apply to RL and RC circuits. The method employs partial fraction expansion to separate the impedance relation until the fractional components are recognizable as representing physical artifacts. According to Redfield and Krishnan (1993), network synthesis is the generation of electrical circuits and networks based on filter specifications in the frequency domain. Early work formulated the impedance synthesis problem in which an impedance relationship between a voltage and current at a single power port is represented as a rational function. Mathematically, impedances are either positive real, corresponding to passive systems, or negative real, corresponding to active systems. This distinction results in two distinct areas of impedance methods research.

IMPEDANCE METHODS I: PASSIVE SYSTEMS

Synthesis of passive systems by using impedance bond graph techniques is based on the work of Redfield (1993) and Redfield and Krishnan (1991, 1993). Passive systems are systems that require no active control or external source of power. The network synthesis of *passive* functions typically separates an impedance specification into smaller pieces in a process known as *reticulation*. Each of the smaller parts is in turn reticulated until the smaller elements are recognizable as representations of physical circuit elements.

The work of Redfield and Krishnan (1993) adopts the network synthesis problem to the bond graph domain and extends it to allow the generation of multiple configurations for a given impedance. In this context, impedance is defined as the ratio of effort and flow at a single port. In addition, impedance is the performance specification for the system. The goal of the work is to synthesize bond graphs and subbond graphs from impedances, and then associate the bond graphs with physical artifacts. At the fundamental level, this method uses a mapping of reticulated impedances to simple bond graph elements. As an example, Table 6.3 contains a basic set of bond graph elements and the associated impedances, but it is possible to generate much more sophisticated bond-graph-to-impedance mappings. At a more sophisticated level, this technique can be automated to include libraries of mappings to be accessed during an automated design search process (Redfield and Krishnan, 1991).

TABLE 6.3. BOND-GRAPH-TO-IMPEDANCE MAPPINGS

$\xrightarrow[f]{e}$ R	$\dfrac{f(s)}{e(s)} = \dfrac{1}{R}$	$\xrightarrow[f]{e}$ C	$\dfrac{e(s)}{f(s)} = C$	$\xrightarrow[f]{e}$ I	$\dfrac{f(s)}{e(s)} = \dfrac{1}{Is}$
$\xrightarrow[f]{e}$ R	$\dfrac{e(s)}{f(s)} = R$	$\xrightarrow[f]{e}$ C	$\dfrac{f(s)}{e(s)} = \dfrac{1}{Cs}$	$\xrightarrow[f]{e}$ I	$\dfrac{e(s)}{f(s)} = Is$

Note: These mappings are basic one-port impedences.

Synthesis of dynamic systems, using passive elements, is demonstrated in Redfield (1993). In this work, the design of a velocity sensor is synthesized. The functional specification is given as an initial frequency specification for the system. This frequency specification is given as a ratio of the input velocity to some measurable output such as displacement or relative velocity. The schematic description of the system is as shown in Figure 6.6. In this figure, D represents the unknown system in both the bond graph and the schematic.

The impedance is given by the ratio of the effort on bond 1 to the flow on bond 1. This impedance can be represented by the transfer function:

$$\frac{F_s(s)}{V(s)} = \frac{e_1(s)}{f_1(s)} = \frac{Is}{CIs^2 + D(s)Is + 1} = A,\tag{6.1}$$

where A represents a performance specification based on the assumption that all frequencies are important; therefore a constant relationship is required for all frequencies. Here s represents the Laplacian operator (or more generally, an operator that converts a differential equation to an algebraic equation), and $F_s(s)$, the spring force, is the desired output. This ratio can be solved for $D(s)$ and reticulated to produce the following relationship:

$$D(s) = \frac{Is - CIAs^2 - A}{AIs} = \frac{1}{A} - Cs - \frac{1}{Is}.\tag{6.2}$$

Using these simple ratios and Table 6.3, one can construct a bond graph by using the synthesized representation for $D(s)$. This bond graph is the schematic description of the design for a velocity indicator. Clearly, this example illustrates a simplified overview of the process, and much more complex frequency-response specifications can be imposed on the system. For further review of this technique, consider the work of Redfield and Krishnan (1993), and Redfield (1993). For a more advanced application of this method, consider further work by Redfield (1999), in which the

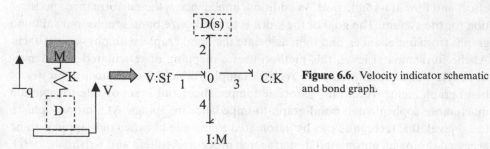

Figure 6.6. Velocity indicator schematic and bond graph.

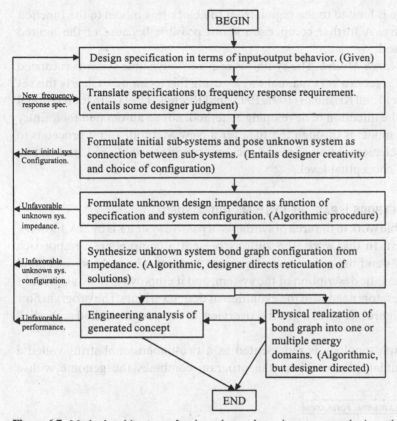

Figure 6.7. Method architecture of an impedance, dynamic systems synthesis methods (Redfield, 1993).

method has been applied to the synthesis of novel design concepts for a constantly variable transmission (CVT).

Analysis: Architecture of Impedance Methods I. Redfield (1993) provides a clear picture of the architecture of this method. Figure 6.7 illustrates a reproduction of the method as the author envisioned it.

Clearly, customer needs are assumed to exist, but they aren't treated explicitly; rather, the input to the method is a design specification. Design specifications are in the form of a desired system response and an unknown impedance representing the system to be designed. Figure 6.7 also highlights the flow and feedback mechanism of the method. Overall, the architecture is similar to that proposed in this chapter, starting with the black box model and proceeding to search and synthesis of abstract structures. The control mechanisms of this method intimately involve the designer who directs the process from concept generation to optimization.

Although the method has these features, more automation of the technique is required. The method is algorithmic and procedural in nature, but it has yet to be effectively coded as a testbed tool. In turn, industrial applications, such as with the CVT, are currently being pursued; yet, descriptive studies are needed to provide clear conduits to design practice.

Analysis: Design Model. The correlation between the design approach of this method and the one proposed is good for the region of the model covered by the method. Coverage is limited to the region from the black box model to the function system architecture. A further comparison is not possible because of the limited scope of the method.

Analysis: Research Model. The synthesis of passive dynamic systems is carried out algorithmically, yet in a very manual way by using this technique. Clearly this set of works by Redfield and Krishnan (1991, 1993) and Redfield (1993) are foundational in this field, with the intention of developing a method, not an automated tool. Entry into the research model is through the Idea for a New Method, and it proceeds to the Test Method element. Limited use of success criteria is established, and testing is done at a more conceptual level.

IMPEDANCE METHODS I: PASSIVE SYSTEMS (CONTINUED)

Another notable work in the area of synthesis of passive systems is by Tay, Flowers, and Barrus (1998). In this work, the authors present a comprehensive approach to design synthesis and the generation of concept variants. The method uses bond graphs as the schematic description of the system, and it employs genetic algorithms to provide the operators needed in the evolution of design variants. The program flow chart, shown in Figure 6.8, gives a general overview of the design method embodied by the program.

The bond graph topology is represented as a two-dimensional string called a *genome*. To formulate the problem, the program combines the genome with a

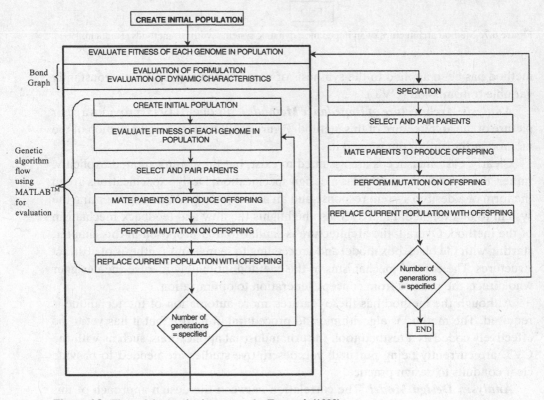

Figure 6.8. Flow of the synthesis program by Tay et al. (1998).

TABLE 6.4. INPUT QUANTITIES FOR TAY ET AL.		
Model Parameters	First-Stage Genetic Algorithm Control	Second-Stage Inputs
1. Max. no. of storage elements	No. of generations	No. of generations
2. Max. no. of dissipative elements	Size of population	Size of population
3. Range of parameter values	Probability of crossover	Probability of crossover
4. Nominal parameter values	Probability of mutation	Probability of mutation
5. Input quantity	Proportion for replacement	
6. Output quantity	Sharing function used	

Source: Tay et al., 1998.

binary-to-decimal string representing the parameter values to form an incidence matrix. Once the problem is formulated, the mutation, selection, and combination operators of the genetic algorithm transform the population of randomly generated two-dimensional binary strings to a valid bond graph with the desired dynamic characteristics. A valid bond graph satisfies the conditions of Kirchhoff's current and voltage laws, geometric compatibility, Newton's laws, and conservation of energy. In all, three programs are presented: *Topology Generator, Performance Generator, and Variant*. For a more detailed account of this method, refer to the referenced article by Tay et al.

Analysis: Architecture of Tay's Passive Method. Figure 6.8 is a good representation of the process model used in this method. Input to the model is a black box model, with a genetic algorithm used to direct and control the concept generation process. A sample input vector is as shown in Table 6.4.

Model parameter inputs are the most interesting elements of the table for this discussion. A note of absence in the input specifications is geometric content. Energy is the only global type parameter being specified, thus allowing the method to generate concepts free of any geometric constraints but forcing the designer to invest significant effort embodying the conceptual results of the method.

Analysis: Design Model. A method for concept generation is developed, and as a method, the work attempts to be comprehensive in nature. There are still omissions to be noted, first of which is the voice of the customer. Additionally, the method stops short of actual embodiment of a design by terminating the process at the functional model variant level. Initial specification of the artifact is again at the black box level, with the generation of function-structure concept variants being carried out by the application of a genetic algorithm.

Analysis: Research Model. Tay et al. describe the research as being "situated comfortably in the computational-conceptual design area." They note its contribution in two areas: (1) it uses operators rather than rules allowing the designer to explore areas with which he or she is not familiar, and (2) it contributes to the application of genetic algorithms to dynamic systems. The synthesis of dynamic systems using bond graph chunks is extended considerably by the inclusion of commercial software in the analysis (MATLAB) and the inclusion of a genetic algorithmic approach. Although the authors are affiliated with industry, the influence of industry as a prime mover is still missing. Several example designs are generated and then compared to known good solutions to test the effectiveness of the model, resulting in novel concepts for a pumping machine that uses components from multiple energy domains.

IMPEDANCE METHODS II: ACTIVE SYSTEMS

The limitation of dynamic system synthesis using passive elements is that it cannot generate designs that include active elements, thus restricting the generality of the method. The second classification of dynamic systems synthesis, using impedance methods and bond graphs, expands this approach to include active elements in the schematic description of the design.

The final method to be treated in this section is the synthesis of dynamic systems with active elements. Active elements are elements that require a power source that is external to the system in which they operate. The work of Connolly and Longoria (2000) is representative of work in this area. The method proposed by Connolly and Longoria uses negative impedances, generally a nonrealizable physical phenomenon, as a mathematical modeling tool to represent the behavior of active elements. This approach is structurally the same as that presented by Redfield and Krishnan (1993), but the inclusion of negative impedances opens the door to designs with actively controlled elements. At the point in the process where the system is fully reticulated and the resulting impedance relationships are mapped to bond graph form, the system is fully designed from a functional perspective (at least in terms of dynamic behavior functions). At this point, there is still no physical or topological form to the design. Connolly and Longoria propose a four-step process to realize, physically, the active elements.

The first step in physically realizing the active elements is to separate the active from the passive elements in the bond graph structure. In general, all active and passive elements are not separable. The next step is to decide whether the separable elements should be realized individually or if they should be incorporated into the design of the active elements (function sharing). Next, a physical realization of the negative bond graph elements is created. Finally, a control strategy for the active element must be designed. This control strategy is unique to the active system synthesis method. At this point, a tuning phase is entered as the model of the system is tested against the system specifications.

Connolly and Longoria test their synthesis method in an example, where an active vehicle suspension is redesigned. The vehicle suspension model is a quarter vehicle model that includes a tire, an unsprung mass, and a sprung mass. Between the sprung and unsprung masses exists a spring and an unknown suspension system specified to have some impedance relationship $Z(s)$. The input specification to this method is in the form of a transfer function of the desired response, in the form of either experimental data, or it can be prescriptive in nature. With the use of a bond graph, which represents the system model with the unknown impedance of the suspension system as $Z(s)$, a transfer function for the system is generated based on this bond graph. The specified and generated transfer functions are set equal, and the resulting equality is solved for the unknown impedance. The resulting rational function is then reticulated to find the recognizable impedances that represent physical elements. Embodiment of the design follows this step, with the replacement of these elements with actual components. This last stage is not part of the synthesis procedure–architecture.

Experimental or prescribed data are also used as a basis for evaluation of the performance of the dynamic system synthesized by using the method. The authors claim good correspondence between the experimental data and the behavior of the synthesized active suspension.

Analysis: Architecture of Active Methods and Design Model. In general, the approach to the design taken in this method is identical to that of the bond graph chunks method developed by Redfield and Krishnan (1993). With regard to completeness, this method stops short of emulating the entire method architecture proposed in this chapter. Once again, the method begins at the black box level and carries forward to the synthesis of abstract structures. Some effort is put forth to develop embodiments of the design, but this is done through the experience of the designer and is very much dependent on bond graph analysis experience of the designer. In addition, the designer, with no automated synthesis, carries out control of the process. To the authors' credit, this is a very new approach to dynamic systems synthesis that has not shown up in the literature prior to this work.

With regard to the design model (Figure 6.2), the middle portion of the model is covered from black box model to the organ structure level (in part). Likewise, entry into the research model (Figure 6.3) is through the New Idea for Method path, with an analysis corresponding to that of Redfield and Krishnan.

AGENT-BASED METHODS

Another focal area of function-based synthesis is agent-based methods. Agent-based systems implement computational methods founded in the analogy of basic artificial life elements that are coded with certain reflexive operations. When combined through a control strategy, these elements, or agents, can create what appear to be quite complicated behaviors.

Of particular note in this area is the work by Campbell, Cagan, and Kotovsky (1999, 2000). In this work, a four-prong strategy is adopted for generating concepts to particular design domains. This strategy is embodied, as a computational design method, through what is termed *basic subsystems* of an A-Design theory.

The subsystems of this theory are (1) an agent architecture for creating and improving design alternatives, (2) a functional representation of a design problem that is synergistic with the agents, (3) a multiobjective decision-making approach for searching the design space and for retaining or discarding potential solutions, and (4) an algorithmic approach, based on iteration, that seeks to improve generated solutions (Campbell et al., 1999). Table 6.5 illustrates a generic functional specification, known as functional parameters (FPs), for electromechanical design. This specification language represents the required input–output states of the machine being designed. Table 6.6 shows the input and output FPs for a human weighing machine application.

Based on the functional specification, concepts and architectures are generated by using catalog electromechanical elements, such as generic springs, gears, analog dials, and the like. Agents in the system are coded with operations to combine the physical elements according to the demands stated in the input–output FPs. These

TABLE 6.5.   FP (SPECIFICATION) STRUCTURE AND CONTENTS
FOR INPUT–OUTPUT REPRESENTATION

Function Variables (Flows/Constraints)	Possible States/Values
Through	{any real number, bounded, unbounded}
Across–Integral	{any real number, bounded, unbounded}
Across–None	{any real number, bounded, unbounded}
Across–Differential	{any real number, bounded, unbounded}
Class	{power, signal, material}
Domain	{trans, rotate, electric, hydraulic}
Position	{x, y, z}
Orientation	{θ, ϕ, ψ}
Interface	{standard size, e.g., 9/16 in. (1.42 cm) bolt}
Direction	{nil, source, sink}

Source: Campbell et al., 2000.

agents use composition rules, in concert with the physics underlying the physical elements and design domain. Other agents modify concepts as they are created, through fragmentation and iteration. In turn, still other agents manage the search process through Pareto optimization strategies to retain or discard concept paths in the MIMO design space.

Figure 6.9 illustrates a design (one of many) generated as a concept for the human weighing machine. Notice that the weighing machine includes the expected functional elements, at least in "stick figure" form. It also calculates (estimates) the results of design objectives, such as cost, mass, and accuracy, which is the basis for its search strategy.

In addition to this domain, Campbell et al.'s work has been applied to MEMS design. Intriguing results have been created through this application, even with a very small catalog of physical elements to initiate the design search. This use in another domain illustrates the potential efficacy of the method.

TABLE 6.6.   INPUT–OUTPUT FPS FOR A HUMAN WEIGHT
MACHINE APPLICATION

Input		Output	
Function Variables	States/Values	Function Variables	States/Values
Through	{[0300]}	Through	nil
Across–Integral	{goal 0}	Across–Integral	{goal [0 5]}
Across–None	{goal bounded}	Across–None	{goal bounded}
Across–Differential	nil	Across–Differential	nil
Class	power	Class	signal
Domain	translation	Domain	rotation
Position		Position	
Orientation		Orientation	
Interface	feet	Interface	dial
Direction	source	Direction	

Source: Campbell et al., 2000.

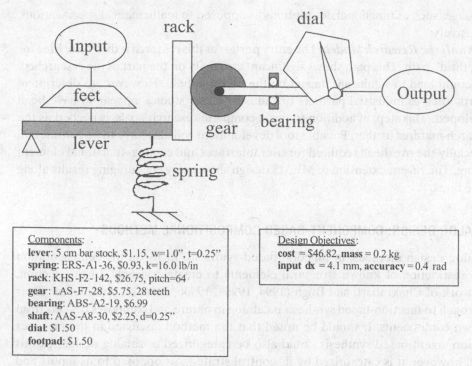

Figure 6.9. Example weighing machine concept (Campbell et al., 2000).

Analysis: Architecture of Campbell et al.'s Agent-Based Synthesis Method. With considerations of Figure 6.2, the agent-based functional synthesis method of Campbell et al. shows significant coverage of the general architectural model. Their method includes a subset of customer needs, translated into FPs and objective functions. Inputs include functional specifications (again the FPs), functional elements (generic springs, etc.), and catalog structural elements (coded part numbers with associated data). The workframe for the method is the set of data structures for the methods FPs and components, given in Figure 6.9, where the geometry is not fully created. This agent-based method also includes synthesis and search strategies, as well as control, evaluation, and optimization.

Overall, the method shows a remarkable level of generative capability. It "learns" and formalizes strategies for fragmentation and component combination that are known (but only at an informal and ad hoc level) design principles and rules of thumb. However, most of the effort relates to the stick-figure concept generation. Significant functions of the designs are not represented, nor is the geometric layout fully captured. These basic capabilities will need to be explored further to satisfy the research goals. In general, the strategy seems to be inherently flexible for such extensions.

Analysis: Design Model. Again, Campbell et al.'s method shows a good coverage of the design space. Physical architecture (organ structures), layout, geometry, and embodiment structures are necessary future extensions to create more competitive designs. In addition, the functional representation could be advanced. The use of the functional parameters is innovative and very powerful. However, adding product (organ) architectural and geometric issues may necessitate a more rich representation language, such as functional descriptions as opposed to mathematical specifications, exclusively.

Analysis: Research Model. The entry portal for this research is the "New Idea for a Method" path. This path shows significant creativity on the part of the researchers and could lead to significant impact on the research field. However, no descriptive efforts, such as industrial partners or industry sector studies, appear to have been developed. This step, in addition to more complete research goals, is needed as the research matures further. Possible tool development will also have to be considered, especially the overhead required for user interfaces and catalog–functional element coding. The recent extension to MEMS design shows very encouraging results along these lines.

CATALOG DESIGN: COMPONENT-BASED COMPOSITIONAL METHODS

Catalog design methods, in function-based synthesis, focus on the representation and abstraction of known structural elements to create concepts. In this section, the work of Chakrabarti and Bligh (1994, 1996a, 1996b, 1996c) is considered. Their approach to function-based synthesis is catalog in nature and composes designs from known components. It should be noted that the method discussed in the previous section, agent-based synthesis, could also be categorized as catalog design. In this case, however, it is categorized by its control strategy, as opposed to its inputs and representational strategy.

The component-based compositional method of Chakrabarti and Bligh represents design problems according to an input–output structure, comparatively to Campbell et al. This representation is in the form of a functional specification (mathematical) as opposed to a more general functional description. It also focuses entirely on the mechanical domain, such as kinematics and mechanical power flow.

To compose a design based on the input–output specification for a machine, Chakrabarti and Bligh abstract mechanical components as input–output elements with motion transmission and motion constraints on each end. Example elements include abstracted versions of generic shafts, tie rods, cams, levers, screws, and so on. Their library of elements may be extended by using their general representational approach.

The method then includes a control strategy for synthesizing the range of mechanical devices from SISO to MIMO. This strategy, referred to as *kind synthesis*, exhaustively searches for chains of elements that satisfy the input–output functional specification. This computational search ends with predetermined bounds on the number of elements or complexity of the chains. The designer is then left to interpret the chains as design concepts, embodying the concepts with geometry and interfaces to realize a physical structure. The chain concepts do, however, provide layout information in terms of orientations and positions in three-dimensional space.

Analysis: Architectural, Design, and Research Models. The analysis of the component-based composition method of Chakrabarti and Bligh is very similar to the agent-based approach of Campbell et al., Chakrabarti and Bligh have completed some basic descriptive studies with designers, but industry data are needed, as with the agent-based method. Similar comparisons exist for the activities in the architectural model, the steps in the design process, and the elements of the research model. The main difference occurs, however, in the architectural model. Component-based composition does not include as sophisticated a control strategy as the agent-based method. Likewise, very little optimization is considered, whereas Pareto frontiers are considered in Campbell et al.'s method. On the flip side, however, component-based composition includes more complete spatial information, with the generation of some very unique, as well as historical significant, concepts.

SET-BASED CATALOG DESIGN METHODS: MECHANICAL COMPILERS

Beyond component composition and agent-based synthesis, other catalog design strategies exist. The most significant is the strategy implemented in set-based methods. These methods use mathematical representations of the design problem and consider the search for solutions as an identification of the feasible subset of designs that satisfy a functional specification.

Most notable of the set-based methods is the work of Ward and Seering (1988, 1989a, 1989b). This method takes a schematic input of a mechanical system. Besides this primary input, other inputs include mathematic representations of the functional specification for the system, as well as a catalog of elements that may be "synthesized" into the schematic to solve the specifications. The outputs from the method are sets of components that combine to satisfy the design problem.

Key features of the method are a high-level set based language for representing a systems design problem and an interval calculus for transforming the specification into lower-level structural elements. These features provide the machinery and formalism necessary to "compile" design solutions (thus the term *mechanical compilers*). They also provide a control strategy for the search of design solutions, without the need for exhaustive search or unneeded enumeration of alternatives that are infeasible.

This method has been applied to relatively small-scale mechanical and sensing design problems, such as an ice cream machine, referred to as Toscanini's Problem. The set-based method by Ward and Seering has also been extended significantly to include more powerful set-based operations, such as vector-space mathematics. It has also been extended, at least in principle, by other researchers, such as Bradley and Agogino (1994), to include uncertainty information during the catalog design synthesis.

Analysis: Architecture of the Set-Based Method of Ward and Seering. With a review of Figure 6.2, Ward and Seering's set-based synthesis method begins with a mathematical functional specification (derived, at least in spirit but not explicitly, from CNs), in addition to a workframe composed of a system schematic (layout). Interval calculus and constraint propagation techniques are used to control the search for structural elements that satisfy the functional specification. No activities are devoted to abstract functional elements (besides the mathematical representation), creation of topology, or optimization and embodiment. In fact, the initial schematic plays a large role in setting the physical architecture (general topology) of the system. Alternative layouts (product architectures–organ structures) are not considered.

Analysis: Design Model. Coverage of the design process (Figure 6.2) by the method is focused. Because of the initial schematic input, the method concentrates on a controlled search for solutions as prescribed by the schematic. The focus is thus concept generation for a given organ structure, with little follow-up in terms of detailed topological or form (embodiment, geometry, etc.) structures.

Analysis: Research Model. The entry portal for Ward's and Seering's set-based method is the New Idea for a Method path. Clearly, the researchers seek an innovative approach to synthesis through compilation, a noteworthy and seminal work. However, the criteria for success are not completely defined. The method generates designs as combinations of catalog part numbers, where feasibility is the primary criterion to note success. This criterion is met very well; yet no criteria for realizing the method in realistic scale problems are provided. Thus, industry data and studies are needed (descriptive approaches) to mature the method, realize actual designs, and identify the critical research issues.

CATALOG DESIGN METHODS USING GRAMMARS

A method that is related to that of Campbell et al. (1999, 2000) is one that uses grammars to transform abstract representations of function to less abstract representations of form. This approach can be associated with the techniques developed and presented by Pahl and Beitz (1996), whereby design artifacts are represented using energy, material, and signal flows connected by means of a hierarchical network spanning input to output. As the design process progresses, this representation is refined from a simple black box model to a fully developed function structure (functional model) that represents all primary and ancillary functions of the artifact. A more detailed description of this representation is described in the design model section of this chapter.

In this context, grammars represent a vocabulary of valid symbols and rules used to describe the function of an artifact, using the rules to guide combination of the symbols to form meaningful expressions. The essential symbolic content of a function grammar is the representation of energy conversion relationships and the rules for combining elements that perform the conversions. The energy conversion relationships of the grammar may consist of input–output relationships such as conversion of flows: rotation → translation, human → mechanical, pressure → force, and so on. The rules of the grammar will define the conditions under which these conversions may be combined to satisfy the overall functional requirement of the artifact being designed. Function grammars are currently found in association with optimally directed approaches to design synthesis (Schmidt and Cagan, 1996). These approaches use high-level optimization techniques, such as simulated annealing, within the design space search algorithm. Simulated annealing directs the search through the design space to end on the state that represents a design that is optimal, or near optimal, in terms of quantifiable design and performance attributes.

To date, design-method work, using function grammars, has focused on machine design, the transformation of a description of the function to be performed by a machine to a description of the machine that will perform the function (Finger and Rinderle, 1989a; 1989b; Hoover and Rinderle, 1989; Schimdt and Cagan, 1995, 1997,

1998). Machine design is clearly a function to form application, one that seems to fit the goals of function grammar research. The following subsections highlight a representative set of works presented thus far, where each is evaluated against the architecture, design, and research models developed above.

Abstraction Grammar Methods

The work of Schmidt and Cagan (1995) highlights the abstraction grammar, the first function grammar to be discussed. This work is a predecessor of the A-Design (agent-based) synthesis method discussed above. In Schmidt and Cagan's approach, abstraction grammars are developed as a basis for the larger framework of the function to form recursive annealing design algorithm (FFREADA), which is an optimally directed search approach to design synthesis. Within this technique, abstraction grammars are defined as "a production system for the representation and generation of function and form layouts." The abstraction grammar representation system uses a library of function and form entities from which selections are made to satisfy input–output specifications at the current level of abstraction. Several suggestions are made for the system of representation, such as bond graphs (Paynter, 1961) or the energy, material, signal approach of Pahl and Beitz (1996), but none are prescribed, the authors preferring to allow the user to select a representation system appropriate to the domain of the problem. The representation system will manifest itself in the library of function, form, and rule entities developed for the application at hand.

In general, the function and form of the artifact are initially represented at a very high level of abstraction; the authors use the energy–material–signal representation of Pahl and Beitz (1996) to represent a black box model in their examples. Initial specifications are in the form of target values for the output function. In this example, system cost is to be minimized, and the output torque is to be at least 1.36 N m. The functional specifications are given, so the left side of the arrow is the input, and the right side is the output. The drill power train example is thus specified as shown in Table 6.7. This initial specification is built upon and refined as the synthesis process progresses.

This specification indicates energy input is zero, and the output is *Rotation* that is *Continuous*. Zero material (0) is input, and a drill bit (1) is output. A trigger signal is input as indicated by the 1, and no signal is output.

The rules of the grammar are applied to ensure that compatibility between form and function entities is maintained. Abstraction grammars are string grammars; relationships contained within the grammar are sequential in nature. The rules then must

TABLE 6.7. INITIAL SPECIFICATION FOR FFREADA

Constraints	Energy	Material	Signal
Torque ≥ 1.36 N m	$0 \rightarrow R$	$0 \rightarrow 1$	$1 \rightarrow 0$
Minimize cost	$0 \rightarrow C$		

Source: Schmidt and Cagan, 1995.

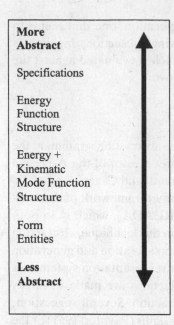

More
Abstract

Specifications

Energy
Function
Structure

Energy +
Kinematic
Mode Function
Structure

Form
Entities

Less
Abstract

Figure 6.10. FFREADA design model (Schmidt and Cagan, 1995).

ensure that each element in a string is compatible with the previous and subsequent elements so that generated strings represent valid machine elements or assemblies. Having satisfied the rules at one level of abstraction, the abstraction grammar approach proceeds to the next lower level of abstraction, considering Figure 6.10. The essence of moving to a lower level of abstraction is that the form and function entities contain more detailed information. Designs at the next lower level of abstraction are instantiations of the previous level, with several designs being possible valid instantiations for the higher level. This process is repeated through *m* levels of abstraction until level 0 is reached. Level 0 represents a component topology and is the final design or set of designs generated by FFREADA.

The authors highlight several observations regarding the approach. The sequential nature of string grammars limits the types of machines that can be represented by using this approach. Sequential machines are still an acceptable way of representing an artifact such as a transmission, but there are clearly limitations. The rule set results in designs biased toward minimizing the number of components in a machine. Finally, the designs are functionally segmented; that is, each function is discretely instantiated by a single machine component. This observation leads to two further consequences.

1. Synthesis of explicit function sharing is not possible.
2. Mapping of performance metrics of specific components up through the function structure, to the functions they satisfy, is possible.

Analysis: Architecture of Abstraction Grammar Methods. With consideration of Figure 6.2, this method provides an interesting synthesis strategy to abstract elements. However, in many respects, it is incomplete. An overview of the architecture model of Figure 6.2 reveals that customer needs are not addressed explicitly by the method, but they are implied in the functional specification. The method promotes the use of functional specifications as the initial description of the artifact.

Abstract functional elements are represented by the high-level function and form elements, and an optimally directed search is performed on the library of primitives in order to synthesize the abstract structure. At each level of abstraction, the grammar rules are used to ensure the input–output relationships are satisfied. The method represents structural elements at abstraction level 0 and uses them in an optimally directed search of the design space to synthesize variants on the component topology. At each level of abstraction, application of the grammar rules ensures the input–output relationships are satisfied. In addition, the embodiment is optimized by using the rules of the grammar. These rules do not address optimization of the topology, but they do address minimization of the number of elements used to satisfy the specifications.

Analysis: Design Model. A black box description is used as the starting point of the design process in this method. In this method, the function structure is limited to series configurations based on the nature of the string grammar used to describe function and form. As a result, the functional system architecture is SISO. The algorithm used does not have sufficient detail to develop an organ structure; hence issues such as function sharing are beyond the capabilities of FFREADA. Function sharing is one of the identified shortcomings of the approach. A topological component structure is treated in a rudimentary way as the result of abstraction level 0, component level design. There are no specific geometric details that result from this abstraction level, only hierarchical relationships between the components. Finally, the form system architecture is not developed, as the authors have chosen to limit the scope of the project to abstract and structural elements. Overall, the FFREADA method represents a high-level conceptual design approach that produces multiple concepts without significant detail about instantiation.

Analysis: Research Model. This method enters the model through the New Idea for a Method model portal. The method is "a computational model for optimally directed conceptual design of machines in which the transformation of function to form occurs iteratively along an abstraction continuum on a number of defined levels of abstraction" (Schmidt and Cagan, 1995). The development approach taken by the authors is limited primarily to the mechanical engineering and mathematical or statistical domains. Tangential reference to the cognitive processes of designers during the act of synthesis is made, but no references are made to research in that area. Development of the method is based on an assumed need for computationally based conceptual design in industry, but it does not elucidate the foundation for this assumption. There are no references to descriptive observations or analysis of industry practices and needs. An academic tool is developed, FREADA, to support the method, and this tool is tested against a limited set of success criteria through an example design of a power supply. In addition, the search and simulated annealing aspects of the tool are exercised to determine the benefit of these techniques over a purely random search of the design space. The results indicate that the optimally directed nature of the tool allows it to produce optimum solutions in fewer iterations than is required to achieve the same results by using a purely random exploration of the design space. Because the method does not take the design process to embodiment, no attempt is made to observe and analyze the results in an industrial setting; thus no feedback mechanism is generated for a descriptive analysis of the method.

Abstraction Grammar Methods II: Extension of FFREADA

Further evolution of this method is carried out in Schmidt and Cagan (1998). The unique contribution of this work is the application of the technique to a design problem using drill power trains as an example. The drill power train is a sequential machine type of design problem, one for which the FFREADA method is well suited. The results are encouraging. The method produces results that are consistent with commercially available products from Black & Decker and DeWalt. In addition, a specific catalog of machine components is specified and searched during the design process. The catalog is limited to energy, material, and signal entities, and the costs associated with the components that realize each of the functions. The method results in designs that satisfy the functional specifications, but it doesn't go any further than the original work toward defining a specific topology in the final design. This work is a move toward realizing a descriptive test of the method against industry-based success criteria, with the logical next step being the use of the method in a live product design effort.

In summary, the work done in developing FFREADA leads to some significant results. The first result is that an algorithmic method can generate feasible serial designs given a library of components that represent the domain of the desired design, drill motors and their drivetrains in this case. These designs are optimal or near optimal in configuration as compared with that of products available on the market today.

Graph Grammars

In response to the limitations imposed by the use of string grammars in the two previous works, Schmidt and Cagan (1997) developed a method for treating a basic level of function sharing in the design artifacts that result from application of the method. In this method, the string grammar is replaced by a graph grammar.

The method is called the graph grammar recursive annealing design algorithm. According to the authors, "GGREADA uses a graph-based abstraction grammar to create designs and a recursive simulated annealing process to select a near-optimum design from those created" (Schmidt and Cagan, 1997). A graph grammar is a mathematical formalism for representing graph vertices and edges. The graph grammar representation allows nonserial component arrangements and functional relationships, thus including the possibility of function sharing in the resulting designs. The overall approach to design is similar between the string grammar and graph grammar implementations, with the grammar consisting of both form and function entities and a set of rules. The design evolves from an abstract black box model to a component level topology through various levels of abstraction, with each level being an instantiation of the previous level (refer to Figure 6.10 of the previous method). The rules of the grammar ensure that the final design satisfies the initial functional specifications.

TABLE 6.8. FUNCTION SPECIFICATIONS FOR GGREADA, MECCANO ERECTOR SET CARTS

	Function Specifications	Function Applicability
1	Create rolling	N/A
2	Support load	N/A
3	Mount 2 parallel wheels	1
4	Mount 1 wheel	1
5	Provide surface area	2

At this point in its development, GGREADA's rules do not include any type of geometric reasoning. The implication of this omission is that the resulting design satisfies the functional requirements, but still requires interpretation from the designer to ensure the design physically provides the desired behavior. Some uncertainty about the behavior is reduced by the use of a library of components. This library includes geometric content, resulting in a limited number of possible component combinations. Additional reduction of uncertainty can be achieved through the implementation of more sophisticated rules for geometric reasoning, but designer interpretation will still be required.

As an example of the power of GGREADA, the method is applied to the design of Meccano Erector Set Carts. A sample of the required input specification is shown in Table 6.8. Functional applicability refers to the primary functions supported by the subfunctions in rows 3, 4, and 5. With the use of a limited set of components (12), an example is undertaken to design a cart given the functional requirement to "create rolling." Given this simple functional requirement, the combinatorial nature of the catalog design approach becomes quite obvious, as over 130,000 designs are possible with the design space growing as more components are included in the catalog. This issue highlights the need for a technique to reduce the size of the design space that is searched. Simulated annealing is the technique chosen for GGREADA. Simulated annealing directs the search through the design space, leading to an end state representing optimal or near-optimal results based on quantifiable performance attributes.

Analysis: Architecture of Graph Grammar Methods. Again, customer needs are not explicitly addressed in this work; they are implicitly represented in the functional specifications used as the initial description of the artifact for the method. Abstract functional elements are represented by the high-level function and form elements, and an optimally directed search is performed on the library of primitives in order to synthesize the abstract structure. At each level of abstraction, the grammar rules are used to ensure the input–output relationships are satisfied. As with FFREADA, GGREADA represents structural elements at abstraction level 0 and uses them in an optimally directed search of the design space to synthesize variants on the component topology. The embodiment is optimized by using the rules of the grammar, but these rules do not address optimization of the topology; in fact, a multitude of topological configurations are generated and left to the designer to sort out.

Analysis: Design Model. The process description is not explicitly addressed in this method; rather it is assumed to have been established prior to initiation of the design effort at the conceptual level. Based on that assumption, the black box description is used as the starting point of the design process in this method. Functional models (function structures) within GGREADA are multipath in nature, allowing function sharing and MIMO systems to be synthesized. A topological component structure is generated as a result of the geometric data contained in the component library and is the output at abstraction level 0, component level design. A basic level of form system architecture results from the geometric data. Designing with Meccano Erector Set Carts in the example made the resultant architecture especially apparent.

Analysis: Research Model. Similarly to FFREADA, this method enters the model through the New Idea for a Method portal. The method is "a computational model for optimally directed conceptual design of machines in which the transformation of function to form occurs iteratively along an abstraction continuum on a number of defined levels of abstraction." The development approach taken by the authors is limited primarily to the mechanical engineering and mathematical or statistical domains. The method is developed based on the limited success of FFREADA, and it seeks to extend the work toward a tool that can be validated in industry. There are currently no references to descriptive observations or analysis of industry practices, but the work using the Meccano Erector Set Carts is a step toward validation and the establishment of a feedback mechanism for descriptive analysis of the method.

In summary, GGREADA overcomes, at least in part, the inability of a string grammar to deal with function sharing. A comparison of the two methods is difficult because the examples presented are very different in function and form. An interesting comparison might be to use the GGREADA method to design a serial device such as a drill motor, and compare the results. The method and example clearly demonstrate the feasibility of the graph grammar design process and the efficacy of the simulated annealing process, albeit on a limited scope.

SUPPLEMENTARY FUNCTION-BASED SYNTHESIS METHODS

The analysis of function-based synthesis methods, in this chapter, focuses on the areas of dynamic systems synthesis, agent-based, and catalog design. Specific methods and research are considered in each of these areas. These particular methods are chosen based on their seminal (original) nature or their unique contribution to one or more of the activities in the method architecture (Figure 6.2).

Other methods and contributions exist in this field, besides those analyzed above. Such methods and contributions include the works of Carlson-Skalak, White, and Teng (1998), Finger and Rinderle (1989), Finger and Dixon, 1989); Frecker et al. (1997), Hoover and Rinderle (1989), Kannapan and Marshek (1987, 1991), Kota and Chiou (1992), Navinchandra, Sycara, and Narasimhan (1991), Welch and Dixon (1992, 1994), and Schmidt, Shetty, and Chase (1998). These works fall within the areas of function-based synthesis, as defined in Figure 6.4. They also follow, generally, the model analysis results of the particular methods reviewed above.

BASS: TECHNICAL DESCRIPTIONS OF REPRESENTATIVE FUNCTION-BASED METHODS

The previous section develops "scaffolding" for understanding the body of research in function-based synthesis. Significant data exist in this scaffolding; however, before inferences are made from these data, the technical depth of the field should be investigated further. Representative methods from the field (Figure 6.4) will aid us in this investigation.

Two methods are presented below. The choice of these methods is based on a number of factors: (1) the most recent methods typically build upon principles learned in the original research, so they are the most sophisticated; (2) while presenting depth, the breadth of the field (Figure 6.4) must not be sacrificed; and (3) it is important to understand the range of activities in the architectural model (Figure 6.2). The impedance synthesis method by Connolly and Longoria (2000) and the A-Design by Campbell et al. (1999, 2000) address these factors.

DYNAMIC SYSTEMS SYNTHESIS: THE CONNOLLY AND LONGORIA IMPEDANCE SYNTHESIS METHOD

The impedance synthesis method of Connolly and Longoria seeks to advance classical synthesis techniques for electrical network theory (Foster, 1924; Brune, 1931; Baher, 1984), focusing on systems with active elements. Active elements, in contrast to passive elements, provide power flow into a system. Such elements are typically realized by actuators, in conjunction with appropriate physical control of the power source. The synthesis of such elements represents a significant advancement of dynamic systems synthesis methods. The inclusion of active elements can potentially simplify the complexity of devices, reduce weight and cost, and increase the level of automation (active control).

The driving impetus of the method is to synthesize a collection of resistive, inductive, capacitive, transformer, gyrator, and power source functional elements to satisfy a desired dynamic response of a system. This synthesis must follow physical power laws and is applicable to multiple energy domains. After synthesizing the functional elements, the method then seeks a physical realization of these elements. The goal is to create alternative networks of functional elements to satisfy the desired response. Physical simulation of the system drives the creation and selection of these networks.

For context, let's consider the method of Connolly and Longoria with respect to the architectural model in Figure 6.2, building on the summary and analysis in the previous section. Their method focuses on a subset of the activities in the architecture: functional specification, creation of a workframe, synthesis of functional element structures, evaluation and search, and control of the synthesis. Functional specifications are in the form of a desired frequency response, stated as an impedance function. The workframe is an abstract functional space, that is, a power flow network in terms of bond graph elements. The initial workframe commences with an initial bond graph of the partial passive system. Synthesis of these elements occurs through mathematical manipulation of the input functional specification to create an equivalent bond graph system (functional element structure). Alternatives are generated through directed replacement of the functional elements with combinations of passive, active, and control elements. These alternatives are created, selected, and controlled based on simulation of the bond graph representations and physical rules for replacement.

After physically realizable functional designs are synthesized, a "catalog" approach is taken to replace the functional elements with structural elements. Physical components replace the functional elements in the bond graph. This process is directed and controlled by the designer, where no formal representation of the structural elements is provided. The result of this process is a "skeletal" schematic of the structural system. No embodiment is carried out as part of the method.

Procedure – Algorithm

Based on this architectural summary, Figure 6.11 presents the organization of the Connolly and Longoria synthesis method. The following sections detail the steps shown in Figure 6.11. A motivating design example, that is, an active vehicle suspension, is used to illustrate the application and results generated for each step (Connolly and Longoria, 2000).

STEP 1: CREATE INPUT FUNCTIONAL SPECIFICATION

The first step involves developing a functional specification for the design. In this case, the functional specification should be in the form of a frequency-response function: magnitude (decibels) and phase (degrees) versus frequency (Figure 6.11). This response function may be prescribed as an ideal relationship from the customer needs, or, alternatively, it may be constructed empirically through experiments and curve fitting.

Application. To demonstrate the method, the system under study is a quarter vehicle configuration that consists of a tire, an unsprung mass, and a sprung mass, between which there is an original electromechanical suspension system. This suspension system consists of an air spring and an electric motor, coupled with a rack and pinion gearing system controlling the position of the sprung mass. Figure 6.12 shows a double-exposure photograph of an experimental setup for this system.

The parameters of the system are as follows: sprung mass, $m_{sm} = 613$ kg; unsprung mass, $m_{um} = 81$ kg; suspension spring stiffness, $k_p = 35,550$ N/m; effective tire stiffness, $k_t = 250,000$ N/m; and effective tire damping coefficient, $b_t = 4000$ Ns/m. Experiments are run with this system to plot a frequency response (not shown here). It is desired to replace the existing electromechanical actuator with an actuator that involves another combination of energy domains, such as hydraulic-mechanical. Thus, the active element synthesis method of Connolly and Longoria is applied, in this case, to an adaptive redesign problem. Figure 6.13 illustrates a schematic of the synthesis problem with an unknown suspension system.

STEP 2: TRANSFORM FUNCTIONAL SPECIFICATION TO AN UNKNOWN IMPEDANCE REPRESENTATION

The next step of the method involves casting the desired frequency response into an unknown configuration with an impedance given by $Z(s)$. This process

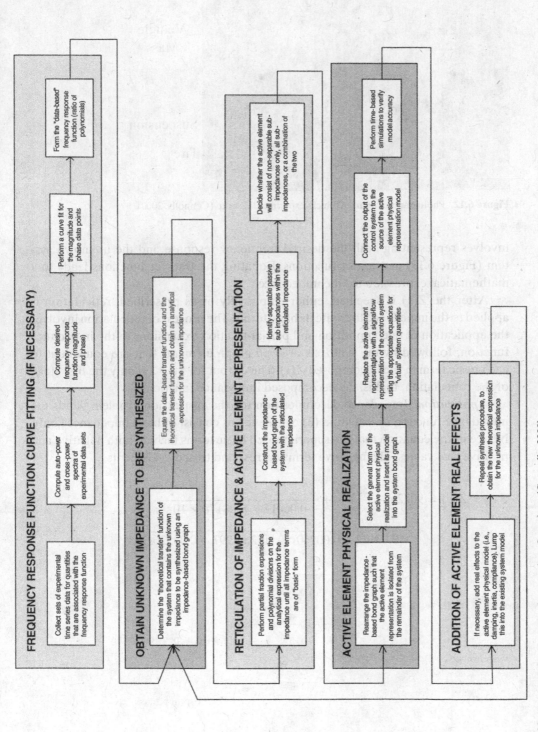

Figure 6.11. Active element synthesis procedure flow chart (Connolly, 2000).

Figure 6.12. Photograph of quarter vehicle experimental setup (Connolly, 2000).

involves representing both the desired frequency response and the unknown system (Figure 6.13) as transfer functions. Equating the transfer functions leads to a mathematical expression for the impedance.

After the $Z(s)$ is specified, either empirically or as prescribed, reticulation is applied to the impedance to create its basic terms. The process of reticulation involves the application of decomposition and partial-fraction expansions to the impedance function, followed by the synthesis of bond graph elements that are equivalent to each basic term in the decomposed $Z(s)$. The output for Step 2 is only the first stage of the reticulation, that is, the decomposed impedance function.

Application. Impedance decomposition for the suspension problem yields the impedance function in Equation (3), whose terms consist of basic impedances only. Notice in the equation that the superimposed basic terms include both negative and positive real values. The negative terms correspond to active elements.

$$Z(s) = 9072.4 - \frac{35550}{s} + \frac{1}{-0.00011 - 6.08178 \times 10^{-7}s}$$
$$+ \frac{1}{-0.03122 - 0.00960s + [1/2.64997 - 0.93526s]} \tag{6.3}$$

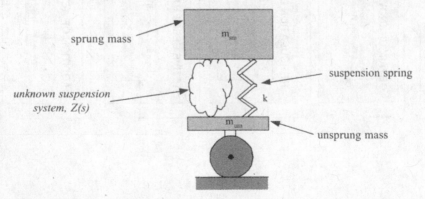

Figure 6.13. Actuator replaced by an unknown suspension system (Connolly, 2000).

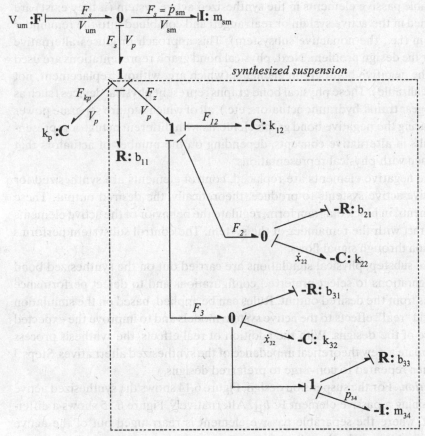

Figure 6.14. Bond graph representation of synthesized suspension system (Connolly, 2000).

STEP 3: SYNTHESIZE ACTIVE FUNCTIONAL ELEMENT REPRESENTATION

The second stage of reticulation is applied to the decomposed impedance function to synthesize a functional element representation of the desired dynamic system. The elements in the representation include both passive and active elements, depending on the basic terms in $Z(s)$. This synthesis procedure involves the one-to-one mapping of the basic terms to energy-domain-independent resistors, capacitors, and the like. These elements are connected with power bonds following the procedures of bond graph modeling.

Application. Figure 6.14 shows the bond graph structure that represents the synthesis of the suspension system, equivalent to Equation (3). Active elements in this representation are illustrated as bond graph elements with prefaced negative signs.

STEP 4: SYNTHESIZE ACTIVE FUNCTIONAL ELEMENT PHYSICAL REALIZATIONS (INCLUDING CONTROL)

The second stage of synthesis is now executed, that is, the composition of functional elements that are physically realizable. This step involves a number of substeps.

First, separable passive elements in the synthesized active system (if they exist) are either retained in the active system or rearranged and combined into the remainder of the system (i.e., the nonactive subsystem). This approach produces alternative concepts for the design problem. Next, physical bond graph representations are used to replace the negative bond graph elements (which are, without replacement, not physically realizable). These physical bond graphs represent physical devices (such as motors and gear trains, hydraulic actuators, etc.), all of which require separate power input. Replacing the negative bond graph elements with different actuator representations results in alternative concepts, depending on the number of actuators that are catalogued with physical representations.

After the negative elements are replaced, control elements are synthesized for the alternative active systems to produce, theoretically, the desired output. These control elements, in block diagram form, regulate the behavior of the active elements as they interact with the remainder of the system. The control subsystem performs this regulation through signal flows.

As a final substep, physical simulations are carried out on the synthesized bond graph configurations to select preferred configurations and to detect performance that deviates from the desired output. Rules can be applied, based on the simulation results, to add "real" effects to the active system models and to improve the expected performance of the designs. With the addition of real effects, the synthesis process iterates to create a new theoretical impedance of the synthesized alternatives. Steps 3 and 4 are then repeated to converge to preferred designs.

Application. For the suspension design, Figure 6.14 shows the synthesized active system, retaining a passive element R: b_{11}. Alternatively, Figure 6.15 shows a different concept, where the separable passive element is rearranged out of the active subsystem, creating a hybrid configuration.

We may simulate these alternative configurations, using parameters for the elements, as derived from the impedance function. The resulting force-velocity profiles that are due to random excitation for the fully active and hybrid cases are shown in Figure 6.16. The profile for the hybrid case is notably flattened, indicating that the power requirements for the actuator in this case are comparatively lower. Alternatively, the range of the actuator force for the fully active case is much wider than that of the hybrid case. In many cases, an actuator requiring lower power is more desirable. However it is important to investigate both cases, because there may be a situation in which a design constraint, such as available space, may rule out a hybrid system.

We now synthesize the physical realizations of the negative bond graph elements. This step entails replacing the negative elements with physical actuator models. We also generate the control elements for the actuator. Considering the fully active configuration, for example, and choosing a hydraulic actuator as the active element, we show the synthesized bond graph of the system, along with signal flow for the controlled effort source of the actuator, in Figure 6.17.

For the concept shown in Figure 6.17, real effects may now be added. Figure 6.18 illustrates the addition of real effects associated with the actuator piston, such as piston mass and viscous damping, to the bond graph. A piston mass, $m_p = 1$ kg, and a piston viscous damping coefficient, $b_p = 100$ N s/m, are selected.

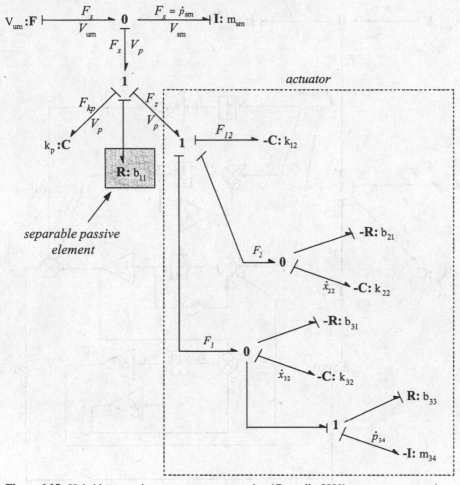

Figure 6.15. Hybrid suspension actuator representation (Connolly, 2000).

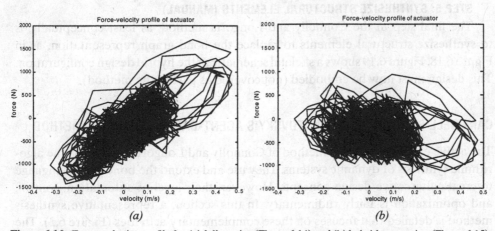

Figure 6.16. Force-velocity profile for (a) fully active (Figure 6.14) and (b) hybrid suspension (Figure 6.15) actuators (Connolly, 2000).

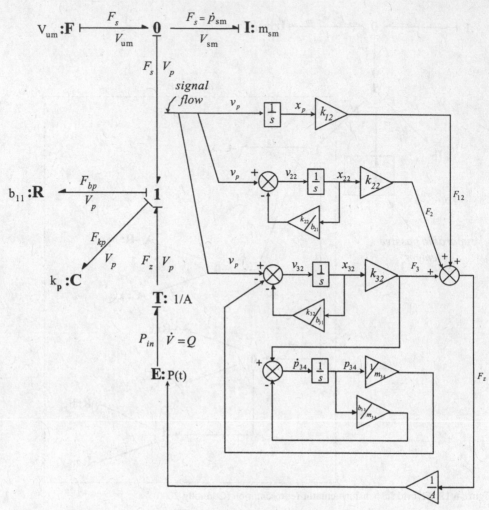

Figure 6.17. System bond graph with signal flows and controlled effort source (Connolly, 2000).

STEP 5: SYNTHESIZE STRUCTURAL ELEMENTS (MANUAL)

The final step in the Connolly and Longoria method, as least conceptually, is to synthesize structural elements to replace the bond graph representation, as in Figure 6.18. Figure 6.19 shows a skeletal schematic of the hybrid design configuration. This design must now be embodied (not covered as part of the method).

CAMPBELL'S, CAGAN'S, AND KOTOVSKY'S AGENT-BASED SYNTHESIS METHOD

The active element synthesis method of Connolly and Longoria focuses on the algorithmic synthesis of dynamic systems. They use and extend the bond graph language to create alternative design concepts. However, their level of automation, control, and optimization is fairly rudimentary. In this section, a representative synthesis method is detailed that focuses on these complementary activities (Figure 6.2). The chosen method for this study is the agent-based method (A-Design) of Campbell et al. (1999, 2000).

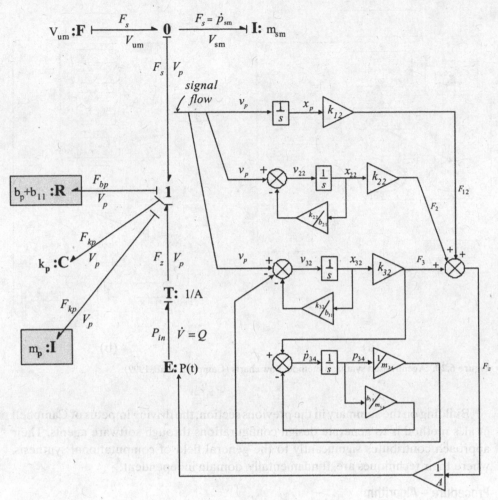

Figure 6.18. Bond graph illustrating real actuator effects (Connolly, 2000).

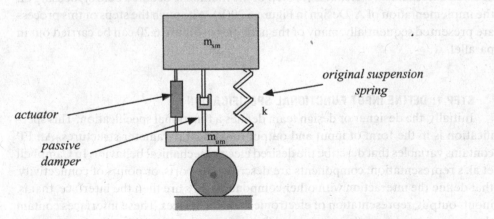

Figure 6.19. Hybrid configuration for a synthesized suspension system (Connolly, 2000).

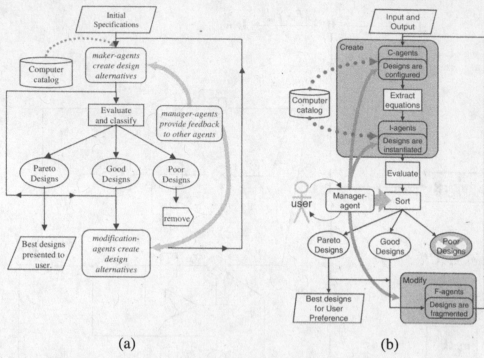

Figure 6.20. Agent-based synthesis method: flow charts (Campbell et al., 1999).

Building on the summary in the previous section, the driving impetus of Campbell et al.'s method is to generate design configurations through software agents. Their approach contributes significantly to the general field of computational synthesis, where their techniques are, fundamentally, domain independent.

Procedure – Algorithm

Figure 6.20(a) illustrates the general approach of Campbell et al.'s agent-based method. Figure 6.20(b), in turn, refines this approach for the application domain of electromechanical systems. Each step of the method is detailed below, focusing on the implementation of A-Design in Figure 6.20(b). Although the steps of this process are presented sequentially, many of the activities in Figure 6.20 can be carried out in parallel.

STEP 1: DEFINE INPUT FUNCTIONAL SPECIFICATION

Initially, the designer or design team defines a functional specification. This specification is in the form of input and output functional parameter structures. An FP contains variables that describe the desired electromechanical behavior. In Campbell et al.'s representation, components are described by ports, or points of connectivity, that define the interaction with other components. FPs are then the interface, that is, input–output, representation of electromechanical devices. These interfaces contain the constraints, energy flow, and signal flow information describing the components. Table 6.5 presents the contents of a generic FP structure. Table 6.9, in turn, lists the

TABLE 6.9. DEFINITIONS OF FP THROUGH AND ACROSS VARIABLES FOR DIFFERENT ENERGY DOMAINS

Variable	Translational	Electrical	Rotational	Hydraulic
Through	Force f (Newtons)	Current I (amps)	Torque T (N m)	Flow Rate m (kg/s)
Across	Velocity v (m/s)	Voltage v (volts)	Angular speed Ω (rad/s)	Pressure P (Pa)
Through $\propto$ Across	Damper friction	Resistor	Damper friction	Valve viscous drag
Through $\propto d$(Across)/dt	Mass	Capacitor	Rotational inertia	Tank
Through $\propto \int$ (Across) dt	Spring	Inductor coil	Rotational spring	Long piping

Source: Campbell et al., 2000.

specific relationships for the Through and Across variables of an FP for different energy domains.

Application. To illustrate the agent-based synthesis method, the design of a machine to weigh humans is considered. Figure 6.21 and Table 6.6 define the input specification, according to the FP formal representation. The input is a downward force, supplied by a specimen's weight, and the desired output is an angular displacement of a dial indicator.

STEP 2: CONFIGURE (SYNTHESIZE) FUNCTIONAL DESIGNS WITH C-AGENTS

After defining the functional specification, the configuration agents, shown in Figure 6.20(b), synthesize design configurations by adding one component (represented with FPs) at a time to a given configuration until the desired input–output specification is functionally satisfied. Configuration, or C, agents compose the design configurations through a reasoning scheme, where they add a component embodiment, either in series or parallel, to existing components at a given state of the configuration development. The goal is to add components that fully satisfy the input–output specification. If this is not possible, components are added to input–output ports that further the design toward its requirements. A combination of decomposition (problem simplification), goal-oriented, and random-search strategies are adopted by the agents to synthesize the design configurations.

An important input into this synthesis process is a catalog of components (referred to as embodiments–EBs–by Campbell et al.), from which the C agents create

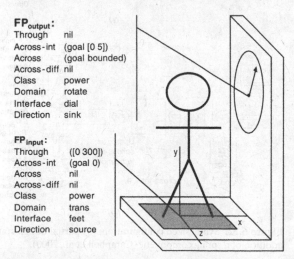

Figure 6.21. Weighing machine design problem: functional specification (Campbell et al., 2000).

FP_{output}:
Through   nil
Across-int   (goal [0 5])
Across   (goal bounded)
Across-diff   nil
Class   power
Domain   rotate
Interface   dial
Direction   sink

FP_{input}:
Through   ([0 300])
Across-int   (goal 0)
Across   nil
Across-diff   nil
Class   power
Domain   trans
Interface   feet
Direction   source

TABLE 6.10. CURRENT COMPONENTS REPRESENTED IN THE A-DESIGN TESTBED FOR ELECTROMECHANICAL SYSTEMS

Battery	cable	capacitor	electrical valve
Spur gear	inductor coil	lever (class 1)	lever (class 2)
Lever (class 3)	motor	pipe	piston
Potentiometer	pulley	rack	relay
Resistor	rotational bearing	rotational damper	rotational valve
Shaft	solenoid	spring	sprocket
Stopper	switch	tank	torsional spring
Transistor	translational bearing	translational damper	worm gear

Source: Campbell et al., 1999, 2000.

configurations. This catalog contains functional representations (FPs) of physical components, such as gears, springs, levers, and dials. A given catalog depends on the domain of application, but covers, in general, multiple energy domains and atomic (one-function) behavior. Table 6.10 lists the current components (embodiments) implemented in the A-Design testbed.

Application. Figure 6.22 illustrates the synthesis of a weighing machine configuration using C agents. A spring component is added to the configuration to make it complete, satisfying the input–output specification (Figure 6.21).

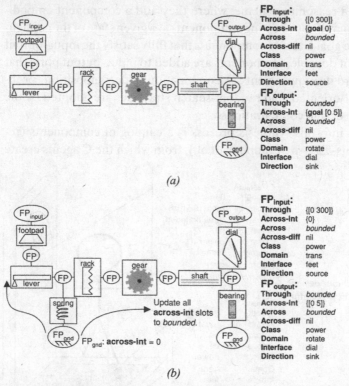

Figure 6.22. Synthesized configuration: (a) partial design state with FPs; (b) completed design with the addition of a spring component (Campbell et al., 2000).

STEP 3: EXTRACT BEHAVIORAL EQUATIONS

Once a design configuration is completed, behavioral equations are extracted, in symbolic form, for the goal states of the system. Equation extraction is completed by recursively computing a system state by starting at a goal state and working backward to given data at ground FPs (Campbell et al., 1998). The extracted equations are used to evaluate how well a configuration satisfies the design specification. The agent-based method uses these equations, and their evaluated values, to synthesize, iteratively, updated and improved configurations.

Application. The extracted equations for a weighing machine configuration are given by the goal dial displacement and goal input displacement, as shown in Equations (4) and (5).

$$\theta_{\text{dial}} = \frac{1}{r_{\text{gear}}k_{\text{spring}}} \frac{d_1}{(d_1 + d_2)} F_{\text{weight}}, \tag{6.4}$$

$$x_{\text{input}} = \frac{1}{k_{\text{spring}}} \frac{d_1^2}{(d_1 + d_2)^2} F_{\text{weight}}, \tag{6.5}$$

where $\theta_{\text{dial}} = [0\ 5]$, $x_{\text{input}} = 0$, and $F_{\text{weight}} = [0\ 300]$.

STEP 4: SYNTHESIZE ACTUAL COMPONENTS USING I AGENTS

For each completed alternative configuration, instantiation agents, or I agents, replace the functional components with actual catalog components that may be purchased or fabricated. The I agents perform the replacement operations through the use of the extracted system equations. I agents choose preferred actual components by using different preference strategies, as a function of the calculated states from the extracted equations.

Application. Figure 6.9 shows the instantiation of an example configuration for the weighing machine problem.

STEP 5: EVALUATE, SORT, AND MODIFY DESIGNS

Instantiated configurations are next evaluated and sorted by using a strategy based on Pareto optimization frontiers. The evaluation and sorting operations lead to the categorization of designs into Pareto optimal, good, and bad populations. These categories are determined based on the input specifications. Poor designs are eliminated from the iterative synthesis process, and Pareto optimal and good designs are carried forward for refinement and the synthesis of further configurations.

The remaining configurations enter the modification stage of the process. Fragmentation agents (F agents) are implemented to improve the states of the configurations. These agents selectively remove and add components to configurations based on a number of preference strategies. Components that don't affect the system states (or negatively affect the states) are early candidates for removal. Function sharing is currently not part of this modification approach, but it is viewed as an important research extension.

STEP 6: REPEAT PROCESS TO TERMINATION WITH M-AGENTS

Manager agents (M agents) monitor the emerging configurations and control the overall process. These agents control other agents and the design configurations through the Pareto optimality strategy. These agents also group component chunks (subsystems of modules) that are successful at meeting the input specification. These chunks are used in constructing new alternatives as the process continues. The process terminates when no improvement is seen along the Pareto frontier of the design space.

Application. Figure 6.9 shows a Pareto design that emerges from the A-Design process. The design objective values, calculated for the configuration, are shown in the figure, in addition to the component descriptions.

DISCUSSION OF THE STATE OF THE ART: GOLDEN NUGGETS

Considering the body of function-based synthesis research, the methods, their analysis, and their applications demonstrate compelling results. Concepts may be generated from formal methods, beginning with a functional description or specification. The generated concepts are intriguing. They follow fundamental theories of physics and design. They are also innovative and broad in number and in scope.

On the flip side, function-based synthesis methods have only considered relatively small-scale applications and only a fraction of the possible set of customer needs. They are currently not used widely in industrial practice; they do not show great coverage of our current function vocabulary or design process, and they are far from the generation of realizable geometries and structural topologies.

Even with these limitations, however, great strides have been taken in the research field. We can learn much from this state of the art. Let's consider a brief summary of the important features of the current research. This summary is based on observations from the research review and technical descriptions, where no particular order or priority is implied.

Functional description or functional specification languages drive function-based synthesis methods. Further research is needed to determine preferred types of representations, especially as more functions of electromechanical designs are used in the synthesis process. Hybrid representations, integrating both descriptions (function-basis language) and specifications have yet to be investigated.

Search techniques include exhaustive, agent based, set based, and optimization (or a hybrid).

Embodiment techniques should be advanced as part of the synthesis process. Few methods consider the generation of detailed geometry and topology. Perhaps a hybrid of shape grammars and function-based synthesis methods will help to achieve this capability.

The methods studied in this chapter are just beginning to be tested with industrial-type problems, such as active dynamic systems (vehicle design), and MEMS components. This component of the research model (Figure 6.3) must be explored further. Methods must be tested in industry, and, in parallel, descriptive studies in industry must be carried out to enter the research activities through the Need Based on Observation and Analysis of Industry portal. Currently, all of the synthesis methods (in this chapter) began as "New Idea . . ." work, focusing on basic research issues.

Agent-based, grammars, and other strategies exist for the synthesis and control activities of the method architecture (Figure 6.2). Comparisons of these techniques are needed to determine their scope for formal representations.

Synthesis methods are now exploring the ability for component integration and function sharing as part of the synthesis process. This area of research must continue to aid the efficacy of the methods for industrial practice.

Synthesis methods are useful even if actual geometry is not generated as an output. Most of the Dynamic Systems methods create abstractions as alternative designs. It is left to the designer to interpret the results and embody the concepts. This process is nonlinear and not unique (Otto and Wood, 2001), but it does greatly assist designers in creating a broad range of creative solutions.

User interface issues are not treated in any of the work reviewed thus far. As the research transforms or develops from methods to tools, this issue will become paramount. It may even "make or break" a given method's use in practice.

Algorithmic approaches to design, as reviewed in this chapter, are generally at the concept design level.

Issues of ergonomics are not treated by the "automated" design methods found in this review. This result is similar to the mechanical design executed at the beginning of the industrial revolution, where machine function was the dominant factor and human-machine-interface functions were not given much if any consideration. A review of the dominant functions, with respect to customer needs, shows that ergonomic-type functions, such as "import/export hand," are the most critical to achieve customer satisfaction (Little and Wood, 1997; Stone and Wood, 1999).

Building on this thought, an important fraction of typical customer needs (or classes of customer needs) are covered by current function-based synthesis methods. This fraction focuses on the physics and functions related to power flow, in addition to limited kinematics. Although this fraction is important, many other types of customer needs are apparent in all devices and products (Urban and Hauser, 1993; Otto and Wood, 1998; McAdams et al., 1999). The synthesis research field will need to grow to subsume these addition types of customer needs and the functions they generate in a design.

Many technologies will converge to standardized organs (architectures), such as motors, fasteners, and so on. Catalog design methods should be able to exploit this principle of electromechanical design in future research.

Common functional languages (descriptive or specifications) require further research. Through common or formal languages, repositories of functional and structural elements may be created, filtered, and shared between the competing synthesis methods.

Scale is a critical and unresolved issue of the research field. All of the methods have only considered relatively small-scale design problems. What will happen as the scale increases, for example, from household consumer products to medium-scale power equipment (lawnmowers, etc.) or large-scale products such as automobiles and aircraft?

The current function-based synthesis methods focus on component combinations with a serial type of architecture. What about other types of architecture, such as the various modular types, integral, mass customization, tunable, and so on (Otto and Wood, 2001; Stone et al., 2000a, 2000b)? The research in product architecture has advanced significantly over the past decade. Synthesis methods should integrate these research advancements and formalisms.

Formal engineering design synthesis has been compared to the formalisms and level of automation in VLSI design (Calvez, Heller, and Bakowski, 1993; Sander and Jantsch, 1999). In particular, a debate exists regarding the analogous possibilities of compilation in mechanical design with that of VLSI design (Whitney, 1996; Antonsson, 1997). No attempt is made in this chapter to answer, explicitly, the fundamental issues in this debate. It is clear, however, that new and significant data exist from recent synthesis methods (in the mechanical and electromechanical domains). These data, although far from complete, should be strongly considered in reviewing the opinions of Antonsson and Whitney.

Design principles and theories underlie the synthesis and control strategies of the methods reviewed here. Research into principle development (for specific domains) will greatly assist the future development of the synthesis methods. Better formalisms and representations will create better methods.

ANALYSIS FINALE: AN EXAMPLE VISION FOR THE FUTURE

The field of function-based synthesis has advanced significantly over the past decade, based on the contributions of a number of talented researchers around the world. The analysis and discussion in the previous sections provide many points or opinions regarding the future of function-based synthesis. In particular, the most important point concerns the research model (Figure 6.3). Descriptive industrial studies (to create new research needs) and industrial tests are needed to advance the field to the next stage. In the subsections below, an example approach to this point is described. This approach might be used as an analogy for similar efforts.

WALK BEFORE WE RUN

In order for function-based synthesis to gain industry acceptance, successes must be demonstrated in a real engineering design setting. One method of achieving this goal is to develop a full-scale working tool that embodies one of the methods reviewed in this chapter. Unfortunately, none of the present set of methods is ready for full embodiment and application in industry. The research results just do not appear to be mature enough to move forward. A more appropriate approach may be to start with a much more focused method and embody that method for use by industry. The benefits of this approach are multifaceted. The first benefit is that valuable descriptive information can be gathered on the effectiveness of the method in an industrial setting, thus leading to refinement of the method and establishment of the feedback mechanism discussed in the research model (Figure 6.3). A second and equally important benefit is the establishment of credibility for this type of design tool with industry. These methods must be shown to be either advantageous over or synergistic with traditional approaches to design; otherwise, they will remain academic exercises without industrial relevance.

An example of a generative method that takes this focused approach is *effort flow analysis* (EFA), presented by Jensen et al. (2000). Effort flow analysis is the embodiment of a method that is based on a descriptive analysis of an industrial need. The method evolved out of industrial design problems in which the goal was to redesign automotive components to reduce complexity. The EFA method has been applied to the redesign of an auxiliary sun visor, for example, for a major automotive manufacturer with dramatic bottom line results. An overview of the technique is presented in the following subsection.

EFFORT FLOW ANALYSIS

Effort-flow analysis, as presented here, is a redesign method. As a redesign method, EFA starts with an existing product with an established set of customer needs, a known functional model (functional structure), and an existing topology for the artifact. The goal of the method is to reduce product complexity, thus leading to a reduction in production costs. In this case, product complexity is reduced through the synthesis of function sharing within the system. EFA highlights opportunities for function sharing by developing a diagram depicting the flow of effort as it is transmitted through the components that embody the known functional model. This diagram is appropriatley called an *effort flow diagram*, and it represents the workframe of the method. Effort flow diagrams represent components using nodes (circles or squares), and the interfaces between components are drawn as links (directed arrows) connecting the components. The links are the critical element, as it is through the interface where the effort flow takes place. Each of the links is then characterized based on the presence of relative motion (R-Link) at the interface represented by the link. The relative motion in the link has been classified at two levels:

- Class I: A link used to represent effort flow between components having relatively small ratios of displacement to characteristic dimension. A characteristic dimension, as noted here, is a distinguishing spatial property necessary to achieve the R-link function between components. A small ratio of displacements to characteristic dimension implies that the resulting strains and associated stresses are small enough to keep the members from experiencing a mechanical failure mode over the design life of the device.
- Class II: A link used to represent effort flow between components that have relatively large ratios of displacement to characteristic dimension or that must perform multiple functions.

The most obvious opportunities for component of combination are made apparent by looking for links where there is no relative motion, but a more sophisticated approach involves the analysis of the links that do have relative motion. These links present the opportunity for novel designs based on the generation of new hybrid components that satisfy the original functional and geometric requirements of the device. Effort flow diagrams also seek to represent a general topological arrangement of the components so there is a geometric components to the diagram as well as a functional component. The existing geometry of the artifact must be considered as function sharing is synthesized. The goal is to generate combined components that interface with the existing geometry, thus reducing the impact on the overall system

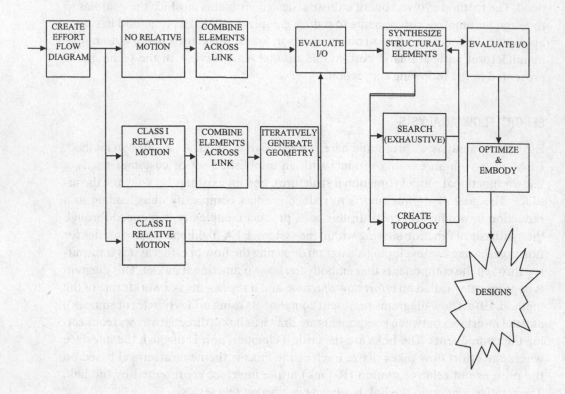

Figure 6.23. Method architecture for EFA synthesis.

design. An extension to this method, which will lend itself to computational application, is the automated generation of geometric forms that satisfy the geometric constraints while combining components through a link with relative motion. The method architecture for this approach is shown in Figure 23. Finite elements analysis is a likely candidate tool for this application. Examples of generating geometry can be found in the compliant mechanism literature (Frecker et al., 1997; Nishiwaki et al., 1998).

CLOSURE

Concept generation represents the time when product function and architecture are transformed to actual geometry. This stage in product development is exciting and

Figure 6.24. Consumer product examples (Otto and Wood, 2001).

Figure 6.25. Mechanical breadboard prototype (left) and a production version (right) of a printer product (Otto and Wood, 2001).

challenging. It is the time when creativity and design principles are used to create innovative solutions. It is also the time when the first glimpses of a realized product appear from the design teams' inner thoughts and dreams.

Formal function-based synthesis methods are emerging as a potential tool set for concept generation. These methods currently focus on the areas of dynamic systems, agent-based, and catalog design synthesis (Figure 6.4). Within these areas, significant advancements have been made, demonstrating basic research results as well as early applications.

With consideration of the household consumer products in Figure 6.24, the current methods are not capable of generating the full set of functions or embodiments shown in these products. Skeletal structural elements or abstract functional representations of such products are the state of the art.

This limitation should not deter us, however. For illustration purposes, Figure 6.25 shows two versions of a concept for a printer product. The left-hand figure is a mechanical breadboard prototype of the product. The right-hand figure shows the production version of the product. Based on the analysis of this chapter, the prototype product, shown in Figure 6.25, could very likely be generated with the current methods. Although many functions and embodiment issues must be solved to arrive at the production version of the product, the abilities to generate a design like the prototype is significant and exciting. What was a mere dream 10 years ago is realizable today!

ACKNOWLEDGMENTS

The authors offer a special thank you to Jon Cagan, Matthew Campbell, Erik Antonsson, Tom Connolly, and Raul Longoria for their advice, contributions of knowledge, and investment of time to discuss their research. The work reported in this document was made possible, in part, by various grants from the National Science Foundation. The authors also acknowledge the support of the United States Air Force Academy, Fluor-Daniel, Ford Motor Company, Texas Instruments, Desktop Manufacturing Corporation, and the UT June and Gene Gills Endowed Faculty Fellow. Any opinions, findings, or recommendations are those of authors and do not necessarily reflect the views of the sponsors.

REFERENCES

Antonsson, E. K. (1987). "Development and testing of hypotheses in engineering design research," ASME *Journal of Mechanisms, Transmissions, and Automation in Design* (currently the *Journal of Mechanical Design*), **109**:153–154.

Antonsson, E. K. (1997). "The potential for mechanical design compilation," *Research in Engineering Design*, **9**:191–194.

Baher, H. (1984). *Synthesis of Electrical Networks*. Wiley, New York.

Bradley, S. R. and Agogino, A. M. (1994). "An intelligent real time design methodology for component selection: an approach to managing uncertainty," *Journal of Mechanical Design*, **116**:980–988.

Blessing, L. T. M, Chakrabarti, A., and Wallace, K. M. (1998). "An overview of descriptive studies in relation to a general design research methodology." In *Designers: The Key to Successful Product Development*, Springer-Verlag, New York.

Brune, O. (1931). "Synthesis of a finite two-terminal network whose driving-point impedance is a prescribed function of frequency," *MIT Journal*.

Carlson-Skalak, S., White, M. D., and Teng, Y. (1998). "Using an evolutionary algorithm for catalog design," *Journal of Research in Engineering Design*, **10**:63–83.

Calvez, J. P., Heller, D., and Bakowski, O. (1993). "Functional-level synthesis with VHDL," Paper No. 0-8186-4350-1/93, *IEEE*, New York, pp. 554–559.

Campbell, M. I., Cagan, J., and Kotovsky, K. (1998). "A-Design: theory and implementation of an adaptive, agent-based method of conceptual design." In *Artificial Intelligence in Design '98*, J. Gero and F. Sudweeks (eds.), Kluwer Academic Publishers, Dordrecht, Lisbon, pp. 579–598.

Campbell, M. I., Cagan, J., and Kotovsky, K. (1999). "An agent-based approach to conceptual design in a dynamic environment," *Journal of Research in Engineering Design*, **11**:172–192.

Campbell, M. I., Cagan, J., and Kotovsky, K. (2000). "Agent-based synthesis of electromechanical design configurations," *ASME Journal of Mechanical Design*, **122**:1–9.

Cantamessa, M. (2000). "Design methods and tools with respect to industrial practice." Presented at the WDK Design Society Workshop, Rigi, Switzerland, March 23. (Contact: Marco Cantamessa, Politecnico di Torino, Italy.)

Cauer, W. (1958). *Synthesis of Linear Communication Networks*. Translated from German by G. E. Kanausenberger and J. N. Warfield (eds.), McGraw-Hill, New York.

Chakrabarti, A. and Bligh, T. P. (1994). "An approach to functional synthesis of solutions in mechanical conceptual design. Part I: Introduction and knowledge representation," *Journal of Research in Engineering Design*, **6**:127–141.

Chakrabarti, A. and Bligh, T. P. (1996a). "An approach to functional synthesis of solutions in mechanical conceptual design. Part II: Kind synthesis," *Journal of Research in Engineering Design*, **8**:52–62.

Chakrabarti, A. and Bligh, T. P. (1996b). "An approach to functional synthesis of solutions in mechanical conceptual design. Part III: Kind synthesis," *Journal of Research in Engineering Design*, **8**:116–124.

Chakrabarti, A. and Bligh, T. P. (1996c). "An approach to functional synthesis of mechanical design concepts: theory, applications, and emerging research issues," *Artificial Intelligence for Engineering Design, Analysis, and Manufacturing*, **10**:313–331.

Connolly, T. J. and Longoria, R. G. (2000). "A simulation-based synthesis method for engineering systems that contain active elements." In *Proceedings of the 2000 ASME International Mechanical Engineering Congress and Expo*, ASME, New York.

Connolly, T. J. (2000). "Synthesis of multiple-energy active elements for mechanical systems," Ph.D. Dissertation, The University of Texas at Austin.

Duffy, A. H. B. and O'Donnell, F. J. (1998). "A model of product development performance," in *Designers: The Key to Successful Product Development*, Springer-Verlag, New York.

Finger, S. and Dixon J. R. (1989). "A Review of Research in Mechanical Engineering Design, Part II: Representation Analysis and Design for the life Cycle," *Research in Engineering Design*, **1**(2):121–137.

Finger, S. and Rinderle, J. R. (1989). "A transformational approach for mechanical design using a bond graph grammar." In *Proceedings of the ASME Design Theory and Methodology Conference*, ASME, New York, pp. 107–116.

Foster, R. (1924a). "Theorems on the driving point impedance of two-mesh circuits," *Bell Systems Technical Journal*, **2**:651.

Foster, R. (1924b). "A Reactance Theorem," *Bell System Technical Journal*, **3**:259–267.

Frecker, M. I., Ananthasuresh, F. K., Nishiwaki, S., Kikuchi, N., Kota, S. (1997). "Topological synthesis of compliant mechanisms using multi-criteria optimization," *ASME Journal of Mechanical Design*, **119**(2):238–245.

Hoover, S. P. and Rinderle, J. R. (1989). "A synthesis strategy for mechanical devices," *Research in Engineering Design*. **1**:87–103.

Hubka, V. (1982). *Principles of Engineering Design*. Butterworth, London.

Hubka, V. and Eder, W. E. (1988). *Theory of Technical Systems*. Springer-Verlag, Berlin.

Hubka, V., Andreasen, M. M., and Eder, W. E. (1988). *Practical Studies in Systematic Design*. Butterworth, London.

Hubka, V. Eder, W. E. (1998). "Theoretical approach in design methodology." In *Designers: The Key to Successful Product Development*, Springer-Verlag, New York.

Iyengar, G., Lee, C., and Kota, S. (1994). "Towards an objective evaluation of alternate designs," *ASME Journal of Mechanical Design*, **116**(2):487–492.

Jensen, D. L., Greer, J. L., Wood, K. L., and Nowack, M. L. (2000), "Force flow analysis: opportunities for creative component combination." In *Proceedings of the 2000 ASME International Mechanical Engineering Congress and Expo*, ASME, New York.

Kannapan, S. and Marshek, K. M. (1987). "Design synthetic reasoning: a program for research," Technical Report 202, Department of Mechanical Engineering, The University of Texas at Austin.

Kannapan, S. M. and Marshek, K. M. (1991). "Design synthetic reasoning: a methodology for mechanical design," *Research in Engineering Design*, **2**:221–238.

Karnopp, D. C., Margolis, D. L., and Rosenberg, R. C. (1990). *Systems Dynamics: A Unified Approach*. 2nd ed., Wiley Interscience, New York.

Kota, S. and Chiou, S. J. (1992). "Conceptual design of mechanisms based on computational synthesis and simulation, *Research in Engineering Design*, **4**:75–87.

Krumhauer, P. (1974). "Rechnerunterstützung für die Konzeptphase der Konstruktion," Ph.D. Dissertation, TU Berlin, Germany.

Little, A. D. and Wood, K. L. (1997). "Functional analysis: a fundamental empirical study for reverse engineering, benchmarking, and redesign." In *ASME Design Theory and Methodology Conference*, ASME, New York, DETC97/DTM-3879.

Malmqvist, J. (1994). "Computational synthesis and simulation of dynamic systems," *ASME Proceedings of the Design Theory and Methodology Conference*, ASME, New York, pp. 221–230.

McAdams, D. A., Stone, R. B., and Wood, K. L. (1999). "Functional interdependence and product similarity based on customer needs," *Journal of Research in Engineering Design*, **11**(1):1–19.

Navinchandra, D., Sycara, K. and Narasimhan, S. (1991). "A transformational approach to case-based synthesis," *Artificial Intelligence in Engineering Design, Analysis, and Manufacturing*, **5**:31–45.

Nishiwaki, S., Frecker, M. I., Min, S., and Kikuchi, N. (1998). "Topology optimization of compliant mechanisms using the homogenization method," *Int. J. Numer. Meth. Eng.*, **42**:535–559.

Otto, K. N. and Wood, K. L. (1998). "Product evolution: a reverse engineering and redesign methodology," *Journal of Research in Engineering Design*, **10**(4):226–243.

Otto, K. N. and Wood, K. L. (2001). *Product Design: Techniques in Reverse Engineering and New Product Development*. Prentice-Hall, Englewood Cliffs, NJ.

Pahl, G. and Beitz, W. (1996). *Engineering Design: A Systematic Approach*. 2nd ed., Springer-Verlag, London.

Pahl, G. (1998). "Keynote." In *Designers: The Key to Successful Product Development*, Springer-Verlag, New York.

Paynter, H. M. (1961). *Analysis and Design of Engineering Systems*, MIT Press, Cambridge, MA.

Prabhu, D. R. and Taylor, D. L. (1988). "Some issues in the generation of the topology of systems with constant power-flow input-output requirements." In *Advances in Design Automation – 1988, Proceedings of the 1988 ASME Design Automation Conference*, DE, ASME, New York, DE 14, pp. 41–48.

Prabhu, D. R. and Taylor, D. L. (1989). "Synthesis of systems from specifications containing orientations and positions associated with flow variables." In *Advances in Design Automation – 1989, Proceedings of the 1989 ASME Design Automation Conference*, ASME, New York, DE 19-1, pp. 273–280.

Redfield, R. C. and Krishnan, S. (1991). 'Automated conceptual design of physical dynamic systems using bond graphs," in *Proceedings of 1991 International Conference on Engineering Design (ICED '91)*, ASME, New York, Zurich, Switzerland.

Redfield, R. C. and Krishnan, S. (1993). "Dynamic system synthesis with a bond graph approach: part I – synthesis of one-port impedances," *ASME Journal of Dynamic Systems, Measurement, and Control*, **115**(3):357–363.

Redfield, R. C. (1993). "Dynamic system synthesis with a bond graph approach: part II – conceptual design of an inertial velocity indicator," *ASME Journal of Dynamic Systems, Measurement, and Control*, **115**(3):364–369.

Redfield, R. C. (1999). "Bond graphs in dynamic systems design: concepts for a continuously variable transmission," *International Conference on Bond Graph Modeling and Simulation*, **31**(1):225–230.

Sander, I. and Jantsch, A. (1999). "Formal system design based on synchrony hypothesis, functional modls, skeletons." In *Proceedings of the 12th International Conference on VLSI Design, IEEE*; New York, Paper No. 0-7695-0013-7/99, pp. 318–323.

Schmidt, L. and Cagan, J. (1995). "Recursive annealing: a computational model for machine design," *Research in Engineering Design*, **7**(1):102–125.

Schmidt, L. and Cagan, J. (1996). "Grammars for machine design." In *Artificial Intelligence in Design*, J. Gero and F. Sudweeks (eds.), Kluwer Academic, The Netherlands, pp. 325–344.

Schmidt, L. and Cagan, J. (1997). "GGREADA: a graph grammar-based machine design algorithm," *Research in Engineering Design*, **9**(2):195–213.

Schmidt, L. and Cagan, J. (1998). "Optimal configuration design: an integrated approach using grammars," *ASME Journal of Mechanical Design*, **120**(1):2–9.

Schmidt, L. Shetty, H., and Chase, S. (1998). "A graph grammar approach for structure synthesis of mechanisms." In *Proceedings of the 1998 ASME Design Engineering Technical Conferences*, ASME, New York, DETC98/DTM-5668.

Schregenberger, J. W. (1998). "The further development of design methodologies." In *Designers: The Key to Successful Product Development*, Springer-Verlag, New York.

Stone, R. B. and Wood, K. L. (1999). "Development of a functional basis for design." In *Proceedings of DETC99, 1999 ASME Design Engineering Technical Conferences*, ASME, New York, DETC99/DTM-8765. (Also to appear in *ASME Journal of Mechanical Design*.)

Stone, R. B., Wood, K. L., and Crawford, R. H. (2000a). "A heuristic method to identify modules from a functional description of a product," *Design Studies*, **21**(1):5–31, January.

Stone, R. B., Wood, K. L., and Crawford, R. H. (2000b). "Using quantitative functional models to develop product architectures," *Design Studies*, V, **21**(3):239–260.

Tay, H. E., Flowers, W., Barrus, J. (1998). "Automated generation and analysis of dynamic system designs," *Research in Engineering Design*, **10**(1):15–29.

Ullman, D. G. (1993). "A new view of function modeling." In *International Conference on Engineering Design, ICED '93*, ASME, New York.

Ullman, D. G. (1997). *The Mechanical Design Process*. McGraw-Hill, New York.

Ulrich K. T. (1988). "Computation and pre-parametric design," MIT Artificial Intelligence Lab Technical Report 1043, MIT, Cambridge, MA.

Ulrich K. T. and Seering W. P. (1989). "Synthesis of schematic descriptions in mechanical design," *Research in Engineering Design*, **1**(1):3–18.

Urban, G. and Hauser, J. (1993). *Design and Marketing of New Products*. 2nd ed., Prentice-Hall, Englewood Cliffs, NY.

Ward, A. C. (1988). "A theory of quantitative inference applied to a mechanical design compiler," Ph.D. Dissertation, MIT, Cambridge, MA.

Ward, A. C. and Seering, W. (1989a). "The Performance of a mechanical design compiler," AI Memo No. 1084, MIT, Cambridge, MA.

Ward, A. C. and Seering, W. (1989b), "Quantitative inference in a mechanical design compiler," AI Memo No. 1062, MIT Cambridge, MA.

Welch, R. V. and Dixon, J. R. (1992), "Representing behavior and structure during conceptual design," *Proceedings of the ASME Design Theory and Methodology Conference*, ASME, New York, pp. 11–18.

Welch, R. V. and Dixon, J. R. (1994). "Guiding conceptual design through behavioral reasoning," *Research in Engineering Design*, **6**:169–188.

Whitney, D. (1996). "Why mechanical design cannot be like VLSI design," *Research in Engineering Design*, **8**:125–138.

CHAPTER SEVEN

Artificial Intelligence for Design

Thomas F. Stahovich

INTRODUCTION

Artificial Intelligence (AI) is the study of knowledge representations and inference mechanisms necessary for reasoning and problem solving.[1] AI encompasses a wide variety of topics such as logic, planning, machine vision, and natural language processing. This chapter focuses on those topics that are the most useful for design synthesis: search, knowledge-based systems, machine learning, and qualitative physical reasoning.

This chapter is divided into four main sections, one for each of these topics. Each section discusses the theory behind the techniques, provides examples of their application to synthesis, and summarizes the current understanding about their usefulness. To the extent possible, the sections are independent of one another, and thus can be read in any order without loss of continuity. This chapter draws examples primarily from mechanical engineering; however, the techniques are suitable for many kinds of design problems. The earlier sections describe more mature technologies, whereas the later sections describe technologies with much of the potential left to be explored. The chapter concludes with thoughts about future directions.

SEARCH

Search is one of the oldest AI techniques and was the basis of much of the early work in AI. For example, the Logic Theorist of Newell and Simon (Newell, Simon, and Shaw, 1963) which was one of the first implemented AI systems, used search to prove theorems in propositional calculus (logic). Search has continued to be an important AI tool and has contributed to a number of significant accomplishments. For example, Deep Blue, a chess-playing computer program based on search, recently defeated a grand master (Hamilton and Garber, 1997).

In some applications, search is the primary problem-solving tool. However, even when it is not, search often still has an important role to play. For example, in

---

[1] Philosophers and scientists have long struggled to define what constitutes both (human) intelligence and artificial intelligence. The definition used here takes the pragmatic view of AI as a set of problem-solving tools. Russell and Norvig (1994) lists several alternative definitions of AI.

rule-based systems (second section), search can be used to implement rule chaining; in machine learning (third section), it is used to find the best hypothesis to explain a set of observations; and in qualitative physical reasoning (fourth section), it is used to find a consistent set of values for the state variables.

This section provides an overview of common search techniques and describes their application to design synthesis. It also describes the key issues in search-based design, including the problem of exponential explosion and the benefits of abstraction. Note that this section covers only a representative sampling of search techniques to illustrate the key issues involved. For a more detailed discussion of individual algorithms, refer to Korf (1988) or any introductory AI text such as that by Winston (1992) or by Russell and Norvig (1994).

SEARCH TECHNIQUES

Path-Finding Problems. A search problem is characterized by a "search space" or "problem space" consisting of states and operators. States are possible solutions or possible partial solutions to the problem. Operators map from one state to another. A particular instance of a search problem is characterized by the initial and goal states, and the search task is to identify a sequence of operators that map from one to the other.

For some problems the goal state is known, and the task is to find a path (sequence of operators) from the initial state to the known goal state. An example is finding an interference-free path by which tubing can be routed from one hydraulic component to another. In this case the location of one of the hydraulic components is the initial search state and the location of the other is the goal state. For many search problems, however, the goal state is not known explicitly, but rather is described implicitly by a test. An example is searching for a layout for a set of objects that will allow them to fit inside a small container (the packing problem). In this case the final state is defined implicitly by a test that determines if the objects fit without interference.

A search space is typically represented as a tree. The nodes in the tree are the search states and the arcs connecting them are the operators. The root of a search tree is the initial state of the search problem. The next level deeper in the tree consists of states that can be reached by applying a single legal operator to the root. Similarly, the nodes n levels deep in the tree are the states that can be reached by applying legal operators to the states at level $n-1$. Figure 7.1 shows a generic example of a search tree.

The size of a search space is characterized by the branching factor (b) and the depth of the tree (d). The branching factor is the average number of child nodes that can be reached by applying the legal operators to a node. The depth is the distance – number of nodes – from the root to a solution. A good estimate of the total size of the search space is b^d, and thus the time required to exhaustively search the tree is exponential in the tree depth.

Search techniques differ in terms of the order in which they visit the nodes in the search tree. The remainder of this section considers three classes of techniques: brute-force, heuristically informed, and stochastic.

Brute-force techniques search the tree systematically, without using any knowledge of the search space. The two classic techniques are breadth-first and depth-first

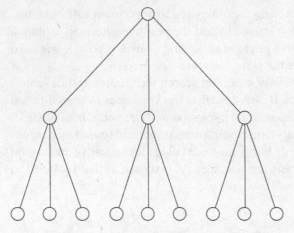

Figure 7.1. A search tree with the branching factor and depth both equal to three.

search. Breadth-first search visits all of nodes at one level of the tree before proceeding to the next level. This requires exponential storage and exponential time; both are proportional to b^d. Depth-first search attempts to extend a single line of reasoning from the root all the way to a solution. Whereas breadth-first search examines the tree row by row, depth-first search examines it branch by branch. Depth-first search selects one child of the root, then one child of that child, and so on, until a solution is found or a leaf of the tree is reached. In the latter case, the search backtracks to the last decision point and selects a different child node. Depth-first search still requires exponential time because on average it visits the same nodes as breadth-first search, but it simply does so in a different order. The benefit of depth-first search is that the storage requirements are only linear in the depth of the tree rather than exponential as in the case of breadth-first search.

If the search tree is infinitely deep, depth-first search may not terminate. For example, if the first branch does not contain the solution and the tree is infinite, the approach will continue along that branch forever. Breadth-first search avoids this problem, but at the expense of exponential storage. Depth-first iteratively deepening (DFID) is a brute-force search technique that avoids both of these problems (Korf, 1985). In DFID, the tree is searched as in depth-first search, except that each branch is terminated when a preselected cutoff depth is reached. Initially, a small cutoff depth is selected. If no solution is found, the cutoff depth is increased and the search is repeated. This process continues in this fashion until a solution is found. Although DFID ends up visiting some nodes over and over again, it can be shown that this does not have a significant adverse affect on performance.

Because search is exponentially expensive, there will always be problems that are too large to solve with brute-force methods, even when the fastest computers are used. Imagine, for example, that the task is to synthesize a device by brute-force combination of components selected from a library. If there were only 10 different kinds of components in the library and we consider devices composed of only 20 components, the size of the search space would be 10^{20}. It is clearly not feasible to search a space this large by brute force. The remedy is to use problem-specific knowledge to guide the search so as to avoid unnecessarily searching large portions of the space. Search methods that do this are called *heuristically informed*. They use a heuristic measure of the remaining distance to the goal, or some other measure of quality, to decided which nodes in the search tree to visit first. For example, hill climbing expands the best child of the root node, the best child of that node, and so on until a solution is reached. In this sense, hill climbing is analogous to gradient descent in numerical optimization.

Heuristics are often quite effective at reducing the amount of search required to find a solution. Additionally, if the heuristics are known to be an underestimate of the remaining distance to the goal, they can provide a guarantee that the solution is optimal.[2] For example in A^* search, each unexpanded node is assigned a cost equal to the sum of the cost of the solution so far plus an underestimate of the remaining cost of reaching the goal (Hart, Nillson, and Raphael, 1968). A^* expands all nodes whose cost is less than the known cost of the current best solution. When all of the nodes yet to be expanded have a cost greater than the current best solution, that solution is guaranteed to be optimal.

Another way to avoid the exponential cost of search is through abstraction: Instead of directly solving a complex problem, solve a simpler abstraction of it and then fill in the missing details with a second problem-solving effort. For example, in mechanical design, it is often easier to first solve a problem at the functional level and then map each of the functions to an embodiment. (Several examples of this are discussed following, under Search Applications.) Because there are often multiple ways to implement a given function, the search space for functional design is much smaller than for embodiment design. Although abstraction reduces the cost of finding a solution, it may sacrifice optimality: even if each step is solved optimally, there is no guarantee that the final solution is globally optimal.

If heuristics and abstractions are unable to reduce the search space to a tractable size, stochastic search methods are useful. Rather than searching systematically, these methods stochastically sample a large number of points distributed throughout the search space. By pursuing many diverse solutions, these approaches avoid being trapped at local maxima (or minima). Common stochastic methods include simulated annealing and genetic algorithms (Chapter 8).

Constraint Satisfaction. The discussion thus far has focused on search problems that can be characterized as path-finding problems because the task is to find a path (sequence of operators) from the initial search state to the goal state. Another major class of search problems is constraint-satisfaction problems (CSPs). These problems are characterized in terms of a set of variables, a set of possible values for each variable, and a set of constraints on the variables. The task is to select a value for each variable such that the constraints are satisfied. A common example is map coloring in which the goal is to assign a color – selected from a small set of colors – to each country on the map such that no adjacent countries have the same color.

Constraint satisfaction problems can be solved with the brute-force approaches used for path-finding problems, but there are better techniques. The simplest of these is backtracking. This approach begins by assigning an order to the variables. Then, similar to depth-first search, values are assigned to the variables, in order, one at a

---

[2] There are other optimal search techniques such as exhaustive search and branch-and-bound search that do not rely on heuristics, but these approaches are less efficient.

time. After each assignment, the constraints are evaluated, and if any are violated, the approach backtracks to the first variable for which there are other choices. Another choice is selected and the method continues on in this fashion until either a consistent solution is found or all combinations of assignments have been explored.

Backtracking attempts to resolve constraint violations by returning to the most recent decision point. Often, however, the violation is a result of a much earlier decision and a substantial amount of backtracking is required to resolve the conflict. Dependency-directed backtracking (Stallman and Sussman, 1976) attempts to identify which variable is actually responsible for the conflict so that the solution process can directly backtrack to that variable. Dependency-directed backtracking is thus typically far more efficient than plain backtracking, although in the worst case, both are exponentially expensive.

An even more efficient way to solve constraint satisfaction problems is to pre-process the search space and eliminate any local inconsistencies before searching for a globally consistent solution. The techniques for doing this are called *arc consistency* (Waltz, 1975; Mackworth, 1977). Consider a pair of variables x and y that are related by a constraint. All those values of x that do not have a corresponding legal value in y can be pruned, and vice versa. For example, if x and y are constrained to be equal, then the possible values for x and y can be reduced to the intersection of the initial sets of possible values. To solve a problem, arc consistency is repeatedly applied to each of the constraints until no more variable values can be eliminated. Then, either a backtracking or brute-force approach is used to find a globally consistent solution from the choices remaining. Often arc consistency results in an enormous reduction in the search space so that comparatively little search has to be performed.

SEARCH APPLICATIONS

This section describes synthesis systems that rely on search as the primary problem-solving method.[3] Search has been used for three main types of synthesis problems: (1) synthesis as the combination of standard components, (2) synthesis as repair, that is, synthesis as the application of modification operators to transform an initial design into a working design, and (3) synthesis as the selection of parameter values for a parametric design. This section provides examples of each of these three types of problems.

Synthesis as Combination. Ulrich (1988) uses search combined with the bond graph representation to synthesize single-input, single-output devices. Bond graphs are used to generate schematic designs that are then mapped to implementations. Bond graphs provide a form of abstraction allowing problems to be solved at the functional level before considering embodiments. Bond graphs are a modeling formalism for describing devices composed of networks of lumped-parameter

---

[3] Many of the systems described later in this chapter also employ search in one form or another. For example, the CADET system described at the beginning of the third section uses search to synthesize a design by selecting a sequence of cases from a case base. Similarly, the LearnIT II system described later in that section uses a form of hill climbing to construct decision trees.

elements including generalized capacitors, resistors, and inductors (Karnopp and Rosenberg, 1975). These elements can be mechanical, electrical, and fluidic. Bond graphs can also model elements such as transformers and gyrators that convert power in one medium to power in another.

The synthesis problem is specified in terms of two bond graph chunks, one describing the input to the device, the other describing the output. Each chunk is associated with a variable. The designer specifies the design requirements in terms of a desired relationship between the input and output variables. Schematic designs are generated by using search: Bond graph elements are chained together until the input chunk is connected to the output chunk. To limit the search, restrictions are placed on the number of bond graph elements a solution can contain. The initial schematic solutions are then evaluated to determine if they provide the desired relationship between the input and output variables. If not, debugging rules are used to modify the bond graph.

Each of the successfully debugged bond graphs is mapped to an implementation by using a library of embodiments for the different types of bond graph elements. The resulting designs are inefficient because each element is mapped to its own embodiment. To produce more efficient designs, a function-sharing procedure is used. This procedure eliminates a component from the embodiment and then uses a set of rules to identify other components that could be modified to provide the missing functionality.

Figure 7.2 shows an example concerning the design of a device for measuring the rate of change of pressure. The input is pressure and the output is a linear displacement that is required to be proportional to the time rate of change of the input. The bottom of the figure shows the final schematic design and an initial implementation. When the function-sharing procedure is applied, the fluid resistance in the bottom branch of the circuit is eliminated and replaced by using an undersized piston that allows leakage.

Williams' IBIS system (Williams, 1990) synthesizes devices by searching through a network of qualitative interactions. The desired behavior is specified as a desired relationship between two or more quantities. For example, the sign of the derivative of one quantity may be specified to be equal to the sign of the difference between two other quantities (a fluid level regulator). IBIS searches through its network of interactions to identify a set of interactions that could connect the specified quantities and achieve the desired relationships between them. Once it has found such a set, IBIS uses a library to map each interaction to a structure that implements it.

Subramanian and Wang (1993, 1995) use search to synthesize mechanisms that transform specified input motions, such as continuous rotation, into specified output motions, such as translational oscillation. They first find a sequence of primitive mechanisms that can achieve the desired transformation. They then produce a detailed design by selecting implementations for the primitive mechanisms from a library as shown in Figure 7.3. Their search algorithm is recursive and works backward from the specified output motion toward the specified input motion. The algorithm begins by identifying all primitive mechanisms that could produce the desired output and selects one at random. If the input of this primitive mechanism is compatible with the specified device input motion, the process terminates. If not, the input of

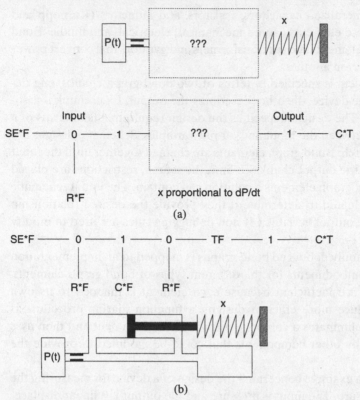

Figure 7.2. A schematic synthesis problem from Ulrich (1988): (a) specification and (b) solution.

this primitive mechanism is treated as if it were the specified device output and the algorithm recurses. In this fashion, the approach is able to chain together components to construct single-input, single-output (SISO) mechanisms.

The approach can also be used to design single-input, multiple-output (SIMO) devices. SIMO devices are initially designed as multiple SISO devices operating in parallel. Then, if any of the SISO solutions have primitive mechanisms in common, those solutions are merged so that the common mechanisms are shared.

Synthesis as Repair. Joskowicz and Addanki (1988) use search to design kinematic pairs (Figure 7.4). The desired behavior of a kinematic pair is described as a desired configuration space (C space). The C space represents the configurations in which the pair of parts interpenetrates (blocked space), the configurations in which they do not touch (free space), and the configurations in which they just touch (boundaries between free and blocked space). Only the latter two types of configurations are legal kinematic states.

Each of the two interacting parts is described as a two-dimensional contour composed of line and arc segments. They begin with an initial contour for each part. They compute the corresponding C space and compare it to the desired C space. If the two do not match, the part contours must be modified. If the actual C space contains a boundary not found in the desired C space, one or more of the contour segments responsible for that boundary must be removed. If, in contrast, the actual C space is lacking boundaries contained in the desired C space, then one or more segments

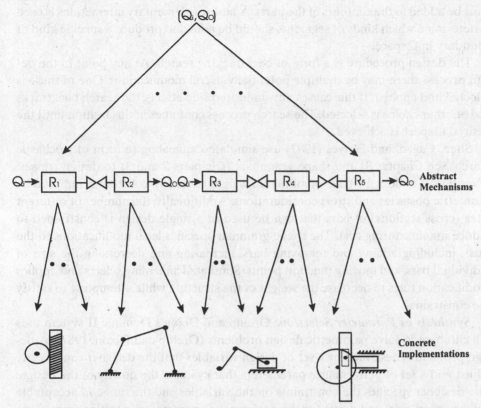

Figure 7.3. Mechanism synthesis using search from Subramanian and Wang (1993). Here Q_i and Q_O are the specified input and output motions; R_j are primitive mechanisms each of which has multiple possible embodiments. The last row shows examples of embodiments.

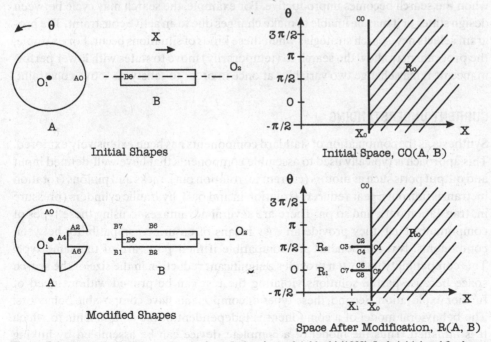

Figure 7.4. The design of kinematic pairs from Joskowicz and Addanki (1988). Left: initial and final part contours; right: initial and final C spaces.

must be added to the contours of the parts. A table of elementary interactions is used to determine which kinds of segments should be added to produce a specific kind of boundary in C space.

The design procedure is a form of backtracking search. At any point in the design process there may be multiple potentially useful modifications. One of these is selected and applied. If this causes any undesired side effects, the search backtracks and another choice is selected. The search process continues in this fashion until the desired C space is achieved.

Shea, Cagan, and Fenves (1997) use simulated annealing (a form of stochastic search; see Chapter 8) and shape grammars (Chapters 2 and 3) to design trusses. Their task is to design a truss of minimum weight subject to constraints imposed by geometric obstacles and stress considerations. Additionally, the number of different sizes (cross sections) of bars that can be used in a single design is constrained to reduce manufacturing cost. The shape grammar specifies legal modifications to the truss, including adding and removing bars, increasing and decreasing the size of individual bars, and moving junction points. Simulated annealing selects and applies modification rules to decrease the weight of the structure while attempting to satisfy the constraints.

Synthesis as Parameter Selection. Orelup and Dixon's Dominic II system uses hill climbing to solve parametric design problems (Orelup and Dixon, 1987). A design problem is described by a set of design variables that the designer can directly adjust and a set of performance parameters that evaluate the quality of the design. The designer specifies the constraints on the variables and the range of acceptable values (good, fair, and poor) for the performance parameters. Starting from some initial design state, Dominic II uses hill climbing to adjust the variables and improve the performance parameters. The program monitors its own performance and detects when the search becomes unproductive. For example, the search may cycle between design states or it may be unable to make changes due to an active constraint. The program selects new search strategies when these kinds of situations occur. For example, the program may allow the search to (temporarily) move to states with lower performance or it may change two variables at once in order to satisfy an active constraint.

CURRENT UNDERSTANDING

Synthesis as the combination of standard components has been extensively explored. This approach is typically used to assemble components that have well-defined input and output ports such as motors (current in, rotation out), racks and pinions (rotation in, translation out), gear reducers (rotation in and out), hydraulic cylinders (pressure in, translation out), and so on. There are several advantages to using these types of components. First, they provide an easy means of ensuring compatibility between components: two components are compatible if their ports are of the same type. This compatibility test often provides a significant reduction in the size of the search space because partial solutions violating the test can be pruned without need of further exploration. Second, these types of components have composable behaviors: The behavioral model of a component is independent of the components to which it is attached. Thus the model of a complete device can be assembled by linking together predefined component models. Third, for these types of devices, the desired function can be conveniently described as a desired qualitative relationship between scalar parameters. For example, the desired function of a pressure gauge is for a displacement to be proportional to a pressure.

The main limitation of synthesis as combination, as the name suggests, is combinatorial explosion: the size of the search space is n^m where n is the number of available components and m is the maximum number of components allowed in the design. The exponential problem size often prohibits brute-force search. Abstraction is commonly used to help manage problem size. For example, the systems described above use bond graphs, abstract mechanisms, and qualitative interactions to synthesize an abstract functional design. This functional solution is then used as the starting point for embodiment design. This two-step process replaces one large exponential with the sum of two much smaller exponentials. Although abstraction is often quite effective at reducing problem size, it is often still necessary to use heuristics to guide the search process.

Synthesis by repair is another common application of search to synthesis. This approach also suffers from combinatorial explosion, however, heuristics are often available in the form of explicit debugging knowledge. For example, Joskowicz and Addanki (1988) repair shape by using explicit knowledge of which modifications are likely to produce particular boundaries in C space.

Heuristically informed search techniques such as hill climbing have been used for selecting parameter values in parametric design. However, numerical optimization techniques often perform better for this application. One of the deficiencies of hill climbing is that the solution can get trapped at a local maxima before reaching the solution. Stochastic optimization techniques (Chapter 8) are particularly good at avoiding local maxima.

KNOWLEDGE-BASED SYSTEMS

Knowledge-based systems (KBSs) have been widely used in design. These types of computer systems are often called *expert systems* because they solve problems by using knowledge obtained from experts and because they can often achieve expert-level performance. A knowledge-based system consists of a knowledge base and a compatible inference engine. There are a variety of different knowledge representations for constructing knowledge bases, and thus a variety of different inference engines. The representations differ in the types of inferences they support and how they describe facts about the world. Davis, Shrobe, and Szolovits (1993) provide a comprehensive analysis of existing knowledge representation technologies. In this section, we discuss the technologies most commonly used in design: rules and frames.

Rule-based systems describe knowledge in the form of production rules (Davis, Buchanan, and Shortliffe, 1977; Davis and Lenat, 1982; Hayes-Roth, Waterman, and Lenat, 1983; Buchanan and Shortliffe, 1984; Dym and Levitt, 1991). A rule is composed of an "if" part, called an antecedent, and a "then" part, called a consequent. The antecedent is a set of patterns or clauses that indicate when the rule is applicable. The consequent describes the deductions to be made or the actions to be taken when the rule is executed. Figure 7.5 shows an example of a rule from R1,

IF:

- The most current active context is putting unibus modules in the backplanes in some box

- It has been determined which module to try to put in a backplane

- That module is a multiplexer terminal interface

- It has not been associated with panel space

- The type and number of backplane slots it requires are known

- There are least that many slots available in a backplane of the appropriate type

- The current unibus load on that backplane is known

- The position of the backplane in the box is known

THEN:

- Enter the context of verifying panel space for a multiplexer

Figure 7.5. A rule for configuring computer systems from R1 (McDermott, 1981).

a rule-based system that translates a customer's requirements for a computer system into a detailed configuration of components.

Individual rules are small chunks of knowledge. Solving a complete problem typically requires chaining together multiple rules with a rule-chaining engine. In forward-chaining systems, the rule antecedents are matched against the known facts to determine which rules are applicable. If more than one rule applies, a conflict resolution strategy is used to determine which should be executed first. Common strategies include picking the most specific rule (the one with the most clauses in the antecedent) or picking the rule whose antecedent is satisfied by the most recently deduced facts. The rule chainer continues to execute all applicable rules until there are none remaining. Note that in contrast to traditional procedural programs, rule-based systems do not provide a means for explicitly controlling the order of the program's execution. The rule chainer, rather than the system designer, determines the order in which the rules are used.

Frame-based systems describe knowledge in terms of taxonomic hierarchies (Bobrow and Winograd, 1977; Brachman and Schmoze, 1985).[4] A frame can be a "class frame" describing an entire class of objects or an "instance frame" describing a particular instance of a class. For example, one frame could represent the entire class of trucks, whereas another could represent a particular red truck (i.e., an instance). Frames contain slots for describing the attributes of a class or instance. For example, a frame describing a truck may have a slot for the cargo capacity. Class frames are

---

[4] Frames grew out of work in semantic networks. See Brachman (1979).

organized in a superclass–subclass hierarchy in which subclasses inherit attributes from their superclasses. Inferences are made about objects (instance frames) by knowing the classes to which they belong.

Rules and frames are often combined to form a hybrid representation (Stefik et al., 1983; Kehler and Clemenson, 1984; Fikes and Kehler, 1985). This is accomplished by associating sets of rules with individual frames. The frame taxonomies serve to partition the rules and define their scopes of application. This helps the system designer control when and for what purposes different rules are used. Frames also provide a language for describing the objects referred to in the rules. Additionally, frames provide a means for making certain inferences about objects, based on class membership, without need for explicit rules. This hybrid representation is perhaps the most common knowledge representation for design tools.

KNOWLEDGE-BASED SYSTEMS APPLICATIONS

The origins of knowledge-based systems are generally traced to the DENDRAL system (Feigenbaum, Buchanan, and Lederberg, 1971), which used heuristic knowledge to interpret mass-spectroscopy data and infer the structure of an unknown compound. The MYCIN system (Shortliffe, 1974; Davis et al., 1977), which used production rules to select antibiotic therapies for bacteremia, was the first rule-based system with a separable knowledge base and inference engine. MYCIN's inference engine, called EMYCIN (van Melle, Shortliffe, and Buchanan, 1984), was used to implement a variety of other rule-based systems including SACON, a system that assisted an analyst in the use of a complicated finite-element analysis tool (Bennett and Englemore, 1984).

Knowledge-based systems have been used widely in design. For example, they have been used to design computer systems (McDermott, 1981), V-belts (Dixon, Simmons, and Cohen, 1984), VLSI devices (Subrahmanyam, 1986), pneumatic cylinders (Brown and Chandrasekaran, 1986), paper transport systems (Mittal and Dym, 1986), dwell mechanisms (Kota et al., 1987; Rosen, Riley, and Erdman, 1991), and electrical transformers (Garrett and Jain, 1988). Here we review two of these systems to illustrate the issues involved in building knowledge-based systems for design.

R1. The R1 systems (also called XCON) is one of the best known rule-based systems (McDermott, 1981; Bachant and McDermott, 1984; van de Brug, Brachant, and McDermott, 1986; Barker and O'Connor, 1989). It was developed by Digital Equipment Corporation to configure built-to-order computer systems. R1 took as input a list of computer components that a customer had ordered and produced as output a set of diagrams showing how those components should be assembled. The program also determined what other components were needed to complete the order and produce a functional computer.

R1 decomposed the configuration task into a set of loosely coupled, temporally ordered subtasks. The program imposed a partial order on the set of components to be configured, such that if the components were configured in that order they could be configured without need of backtracking. The program determined the partial ordering dynamically, based on the characteristics of the problem at hand.

R1 had a knowledge-base of approximately 10,000 rules and a database of approximately 30,000 computer components. Figure 7.5 shows a typical rule. The rules were placed into groups on the basis of the subtasks to which they were relevant. This allowed each rule to presuppose certain things about the current state of the configuration without need of explicit clauses in the antecedents. In performing a subtask, the program applied all applicable rules relevant to the subtask. When multiple rules were applicable, the program used generic conflict resolution strategies to determine which to apply first. When all of the applicable rules had been used, the subtask was completed and no additional rules were needed to verify success.

In contrast to traditional procedural software, the R1 system required a substantial amount of ongoing maintenance. As a result of new product releases and new configuration techniques, 40% of R1's rules changed each year. Because of the large number of rules in R1's rule base, adding new rules presented a significant technical challenge. The main difficultly was in bounding the potential relevance of a piece of knowledge and controlling which piece to apply when more than one was relevant. To overcome these difficulties, R1's developers created the Rime problem-solving approach (van de Brug et al., 1986). Rime defines six different roles that a rule could serve. These included proposing which configuration operator to apply, rejecting clearly inferior operators, selecting the best of the remaining operators, applying a selected operator, recognizing success, and recognizing failure. By providing the programmer with a precise way to specify the role of any new piece of knowledge, Rime greatly simplified the task of maintaining R1's large scale, evolving rule base.

The R1 system was used at Digital Equipment on a daily basis for over a decade. The program was used to configure hundreds of thousands of computer systems and was estimated to have saved the company $40 million annually (Barker and O'Connor, 1989).

PRIDE. PRIDE is a knowledge-based system for designing paper transport systems in copy machines (Mittal and Dym, 1986). A design problem is described to the system in terms of the required locations of the paper entrance and exit, the properties of the paper to be used, the timing requirements, the desired entrance and exit speeds, and so on. The program specifies a design solution in terms of the number and locations of the pinch rolls, the materials for the rolls, the values of various geometric parameters, and the like.

In PRIDE, the plan for designing a transport system is expressed as a hierarchy of design goals that decompose the design process into simpler steps. To begin a new design, the top-level goal of designing a paper transport is instantiated. This, in turn, results in the instantiation of subgoals representing subproblems, such as deciding how many roll stations are needed and where they should be located. Figure 7.6 shows a snapshot of the goals that exist when a new design has begun.

Each goal is an autonomous specialist responsible for designing some subset of the design parameters. A goal contains all of the alternative ways or "methods" for making a decision about the values of the goal's design parameters. There are a variety of different kinds of methods: A method can be a "design generator" that explicitly chooses parameter values; a method can be a set of production rules whose consequents are themselves methods; and a method can be a new set of goals.

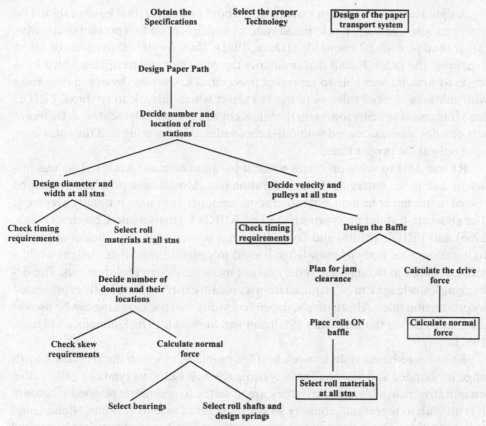

Figure 7.6. A snapshot of the goals that exist when PRIDE (Mittal and Dym, 1986) begins a new paper transport design problem.

The latter type of method is the mechanism by which the program traverses the hierarchical design plan.

Each goal contains a list of constraints that verify that the design is satisfactory. If a constraint is violated, PRIDE uses a form of dependency-directed backtracking to try to resolve the conflict. The violated constraint itself can direct the backtracking by sending advice to the solver. This advice is provided to the system by an expert and is explicitly included in the representation of the constraints.

The PRIDE system is used daily by a copy machine manufacturer for performing feasibility studies of new copier designs. The program performs competently and can do in 30 minutes what previously took weeks (Mittal and Dym, 1986; Dym, 1994).

CURRENT UNDERSTANDING

As knowledge-based systems have been extensively used, a number of specialized development techniques have emerged. These have been documented in a variety of sources, including Davis and Lenat (1982), Hayes-Roth et al. (1983), and Buchanan and Shortliffe (1984). These techniques include specialized software engineering techniques and methods for obtaining knowledge from experts (knowledge engineering).

A significant part of design knowledge is about procedure, that is, what should be done and when. However, rule-based systems were intended to represent declarative rather than procedural knowledge (Davis, 1982). Thus, special efforts must be taken to express the latter. Recall that ordinarily the rule chainer determines which rules are used first. R1 was able to represent procedural knowledge by associating rules with subtasks. Special rules were used to select which subtask to try next. PRIDE used frames to explicitly represent the hierarchy of subgoals to be achieved. Different sets of rules were associated with different frames, thus ensuring that the rules were used only at the proper time.

R1 was able to solve problems without iteration or backtracking. This was primarily due to the nature of the configuration task. Most design problems cannot be solved in this linear fashion, but rather some amount of iteration is usually necessary. The generate-test-debug approach used by AIRCYL (Brown and Chandrasekaran, 1986) and PRIDE (Mittal and Dym, 1986) has proven to be an efficient approach to iteration. One body of knowledge is used to generate candidate designs while a second is used to debug them if they do not meet the design requirements. The debugging knowledge can be acquired from a domain expert and explicitly represented by production rules. Alternatively, dependency-directed backtracking can be used to attempt to debug the deficiency (Stallman and Sussman, 1976; Simmons and Davis, 1987).

Knowledge-based systems work best for problems in which the relevant knowledge is bounded and known. These systems are well suited to symbolic rather than quantitative reasoning. Similarly, they are ill suited to geometric reasoning because it is difficult to represent geometry with a small set of axioms (Forbus, Nielsen, and Faltings, 1991a). Given these considerations, knowledge-based systems are best suited to routine design problems such as selecting components from a library or selecting parameter values in a parametric design problem. As R1 and PRIDE demonstrated, knowledge-based systems can achieve expert-level performance. Furthermore, knowledge-based system techniques can be used to construct robust design tools suitable for production use in real-world industrial applications.

MACHINE LEARNING

Machine Learning (ML) is the subdiscipline of AI concerned with collecting knowledge computationally. A program is said to have learned if its performance at a task improves as a result of previous experiences (Mitchell, 1997). There are several different classes of machine-learning methods. Our discussion is limited to the three most common: inductive methods, instance-based methods, and analytical methods.

Inductive methods draw generalizations from a set of examples by identifying regularities. Most inductive methods operate by computing an approximation to an unknown target function. Common inductive methods include decision-tree learning (Quinlann, 1986; Quinlann, 1993), version-space learning (Mitchell, 1977; Mitchell, 1982), and neural networks (Parker, 1985; Rumelhart and McClelland, 1986). Instance-based methods do not compute an explicit generalization and instead directly match new problems to the most similar training examples. Common instance-based approaches include the k-Nearest-Neighbor algorithm (Cover and Hart, 1967) and case-based reasoning (Kolodner, 1993).

Similar to inductive methods, analytical methods explicitly generalize from training examples (Winston et al., 1983; Mitchell, Keller, and Kedar-Cabelli, 1986; Kedar-Cabelli and McCarty, 1987). However, unlike inductive methods that identify empirical (statistical) regularities, analytical methods use a domain model to construct the generalizations. The domain model is used to determine which properties of the example are significant, and hence which should be the basis of the generalization. Analytical methods are also referred to as "explanation-based learning" because they generalize by constructing an explanation (proof) for the example.

Learning is often used to approximate design spaces in order to accelerate the search for a satisfactory or optimal design. For example, Ivezic and Garrett (1998) have used neural networks to predict the performance of parametric designs. Their program is trained on a small number of sample designs and can then predict the performance of new designs without incurring the cost of expensive simulations and analyses. Similarly, Jamalabad and Langrana (1998) use an instance-based approach to accelerate numerical optimization by learning sensitivity (derivative) information. Finally, Schwabacher, Ellman, and Hirsh (1998) use decision-tree learning algorithms to learn how to select starting prototypes for numerical optimization problems.

As these examples illustrate, machine learning has a number of uses in design. In the sections that follow, we focus on the two applications that have the most impact on automated synthesis: case-based reasoning and learning and reusing design strategies.

CASE-BASED REASONING

When solving new problems, designers frequently rely on previous experience. Case-based reasoning (CBR) is a learning technique intended to assist in the reuse of previous problem-solving experience. There are two main tasks a case-based system must perform: (1) identifying relevant previous design cases and (2) adapting them as necessary to satisfy the new design requirements. Some case-based systems focus on just the first of these tasks and rely on the designer to perform case adaptation. Other systems perform both tasks.

An important consideration in case-based design is how the cases should be represented. The requirements for a suitable representation depend on the nature of design domain and whether or not the system performs case adaptation in addition to retrieval. For retrieval-only systems, it is often adequate to characterize a case in terms of a fixed set of attributes. This allows for efficient indexing and retrieval of cases. (The cases can be organized in a taxonomic hierarchy to further accelerate the retrieval process.) Other types of data, such as text and CAD models, must be included for the designer to be able to adapt the case to new problems; however, this additional information need not be described in a machine-understandable form. For systems that automate case adaptation, the representation must support a broader range of inferences. Adapting a case typically involves reasoning about structure, behavior, function, and the like, and thus the representation must facilitate this reasoning.

Case-based reasoning has been used in a variety of design domains, including architectural design (Domeshek and Kolodner, 1992), structural design (Maher,

Balachandran, and Zhang, 1995), and mechanical design (Goel, Bhatta, and Stroulia, 1997; Chandra et. al., 1992). Here, we focus on just three systems that illustrate the basic issues in case-based reasoning for design.

CASECAD is a case-based system for the conceptual design of structural systems for buildings (Domeshek and Kolodner, 1992). CASECAD uses a multimedia representation for cases. Each case contains references to CAD drawings, images, and text that are used by the designer to understand the case information. Each case also contains a list of attribute-value pairs that are used for indexing and retrieval. The attributes describe function, such as the desired dead load and wind load capacities; behavior, such as the displacement and the cost; and structure, such as the maximum span and material types. CASECAD has a browsing mode that allows the designer to interactively navigate through the case base. It also has a retrieval mode that allows the designer to retrieve designs by specifying desired attribute values.

Kritik2 is a case-based system for the conceptual design of physical devices (Goel et al., 1997). Conceptual design can be viewed as the task of mapping function to structure. Kritik2 performs this task by using the structure-to-function maps of previous designs to adapt them to new functional specifications. A case in Kritik2 is represented with a structure-behavior-function (SBF) model that explains how the structure of the device accomplishes its functions. The structure of a device in the SBF language is expressed in terms of its constituent components and substances, and the interactions between them. Components include things like batteries and light bulbs, whereas substances include things like electricity and light. The behavior of a device is described as a sequence of causal transitions between states. A state is defined in terms of the existence of a substance, such as the existence of electricity at a light bulb. The function of a device is characterized by the input and output states. Figure 7.7 shows the SBF model describing a red light bulb circuit.

To begin a new design problem with Kritik2, the designer specifies the desired function with an SBF model. The program uses this as a probe into the case memory and identifies all cases that at least partially match the desired function. These are then heuristically ordered by their ease of adaptation. Once the program has selected a design case, it checks if the new design requirements are satisfied, and repairs the design if necessary. The program is able to repair failed designs by modifying components and by changing substances. For example, when adapting the design in Figure 7.7 to a problem with a larger required light output, the program identifies that the voltage of the battery is responsible for the failure of the retrieved design. The program then replaces the battery with a higher voltage part.

Whereas Kritik2 adapts a single case, CADET (Chandra et al., 1992) constructs conceptual designs by combining snippets accessed from multiple previous design cases. Each of CADET's cases is a complete design of some physically realizable device. The cases are described in terms of function, behavior, and structure. The behavior, which is perhaps the most important part of the representation, is defined by a set of qualitative differential relations ("influences") relating the input and output variables. For example, Figure 7.8 shows a case in which the flow (Q) out of a water tap monotonically increases with the displacement (X) of the gate valve.

The desired artifact is described to CADET as a set of input and output variables related by qualitative influences. CADET's task is to synthesize a structure that can implement the desired influences. The program first searches the case base

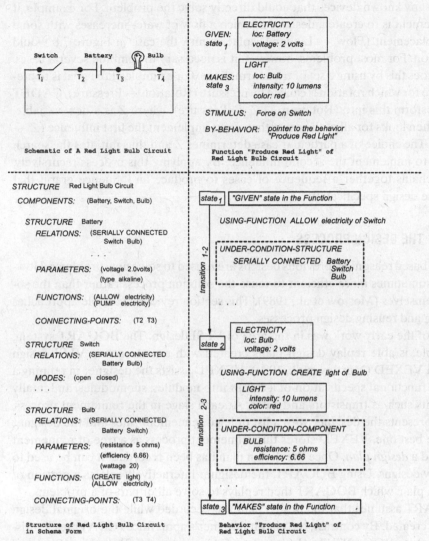

Schematic of A Red Light Bulb Circuit

```
GIVEN:      ELECTRICITY
state 1     loc: Battery
            voltage: 2 volts

MAKES:      LIGHT
state 3     loc: Bulb
            intensity: 10 lumens
            color: red

STIMULUS:   Force on Switch

BY-BEHAVIOR:  pointer to the behavior
              "Produce Red Light"
```

Function "Produce Red Light" of
Red Light Bulb Circuit

```
STRUCTURE    Red Light Bulb Circuit
COMPONENTS:  (Battery, Switch, Bulb)

STRUCTURE    Battery
  RELATIONS: (SERIALLY CONNECTED
              Switch Bulb)
  ...
  PARAMETERS: (voltage 2.0volts)
              (type alkaline)

  FUNCTIONS:  (ALLOW electricity)
              (PUMP electricity)

  CONNECTING-POINTS:  (T2 T3)

STRUCTURE    Switch
  RELATIONS: (SERIALLY CONNECTED
              Battery Bulb)
  MODES:     (open closed)
  ...

STRUCTURE    Bulb
  RELATIONS: (SERIALLY CONNECTED
              Battery Switch)
  PARAMETERS: (resistance 5 ohms)
              (efficiency 6.66)
              (wattage 20)
  FUNCTIONS:  (CREATE light)
              (ALLOW electricity)

  CONNECTING-POINTS:  (T3 T4)
```

Structure of Red Light Bulb Circuit
in Schema Form

```
state 1 ──  "GIVEN" state in the Function

        USING-FUNCTION ALLOW electricity of Switch

        ┌─────────────────────────────────┐
        │ UNDER-CONDITION-STRUCTURE        │
        │   SERIALLY CONNECTED   Battery   │
        │                        Switch    │
        │                        Bulb      │
        └─────────────────────────────────┘

state 2 ──  ELECTRICITY
            loc: Bulb
            voltage: 2 volts

        USING-FUNCTION CREATE light of Bulb

            LIGHT
            intensity: 10 lumens
            color: red

        ┌─────────────────────────────────┐
        │ UNDER-CONDITION-COMPONENT        │
        │   BULB                           │
        │     resistance: 5 ohms           │
        │     efficiency: 6.66             │
        └─────────────────────────────────┘
        ...

state 3 ──  "MAKES" state in the Function
```

(transition 1-2)

(transition 2-3)

Behavior "Produce Red Light" of
Red Light Bulb Circuit

Figure 7.7. The SBF model of a red light bulb circuit from Kritik2 (Goel et al., 1997).

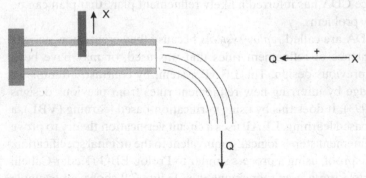

Figure 7.8. The water tap case from CADET (Chandra et al., 1992). Left: physical structure; right: influence graph.

to identify any known devices that could directly solve the problem. For example, if the requirement is to create a design for which a flow of water increases with some linear displacement (Flow ← Linear Displacement), the case in Figure 7.8 would be a solution. For most problems, however, it is necessary to combine several cases. CADET does this by using a set of transformations. For example, if the goal is to create a device for which rotation is caused by pressure (Rotation ← Pressure), CADET might transform this into (Rotation ← Z ← Pressure), where Z is a new variable. CADET then looks for all of the cases that could implement the first influence (Z ← Pressure). The choice of a particular case determines Z and thus initiates the search for a case to implement the second influence. By applying this process recursively, CADET chains together a sequence of cases to produce an influence graph that satisfies the design specification.

LEARNING THE DESIGN PROCESS

With case-based reasoning, previous designs are reused to solve new problems. However, it is sometimes more efficient to reuse the solution process rather than the solutions themselves (Mostow et al., 1989). This section reviews available approaches to learning and reusing design processes.

Much of the early work was in the area of VLSI design. The BOGART system, for example, is able replay design plans created with an interactive VLSI design tool called VEXED (Mostow et al., 1989). VEXED assists the designer in refining a high-level functional specification of a circuit into modules, submodules, and finally components such as transistors and gates. At each stage in the refinement process, VEXED presents the designer with a list of legal refinement rules and the designer selects the best one. VEXED stores this refinement process as a tree of refinement rules called a *design plan*. Once the design plan has been recorded, it can be used to create new designs. Using BOGART, the designer interactively selects a portion of the design plan, which BOGART then replays to solve all or part of a problem.

BOGART assumes that the design plan was recorded while the original design was being created. By contrast, the circuit designer's apprentice (CDA) uses heuristics to generate BOGART-like design plans from existing VLSI designs (Britt and Glagowski, 1996). CDA's approach is called *reconstructive derivational analogy*. CDA uses heuristics to determine which refinement rules might have been used to construct the circuit. Once CDA has inferred a likely refinement plan, that plan can be replayed to solve new problems.

BOGART and CDA are called *replay systems* because they create new designs by replaying the sequence of refinement rules that was used, or may have been used, to construct a previous design. The LEAP system, by contrast, attempts to reuse design knowledge by inferring new refinement rules from previous designs (Mahadevan et al., 1993). It does this by using verification-based learning (VBL), a form of explanation-based learning. LEAP uses a circuit verification theory to prove that a given circuit refinement step is logically equivalent to the original specification. It then generalizes the proof, using a process similar to Prolog-EBG (Kedar-Cabelli and McCarty, 1987), to form a new refinement rule. Figure 7.9 shows an example that concerns the design of a module whose output is specified to be the conjunction

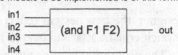

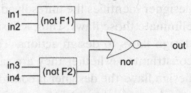

Figure 7.9. Top: specification for a module in LEAP (Mahadevan et al., 1993); middle: the result of manual refinement of the module; bottom: the new refinement rule LEAP learned from this example.

If the module to be implemented is of this form:

Then refine it into the following circuit:

(AND) of two disjunctions (ORs). The designer chose to implement this module by using three NOR (Negated OR) gates as shown in the figure. LEAP first proves that the refined circuit is valid. It then generalizes this proof into the refinement rule shown at the bottom of the figure. This rule states that the conjunction of any two binary Boolean functions (not just ORs) can be implemented by negating each Boolean function and combining their outputs with a NOR gate.

The IDeAL system uses a model-based method for learning generic teleological mechanisms (GTM's) such as cascading, feedback, and feedforward (Bhatta and Goeal, 1997). IDeAL is built on the KRITIK system, an earlier version of the Kritik2 system described above. To learn a GTM, IDeAL requires SBF models of two similar devices, one embodying the GTM, the other lacking it. IDeAL is able to infer the GTM by comparing the two SBF models. For example, when comparing a light bulb circuit with two batteries in series, to one with a single battery, IDeAL learns the notion of cascading: using multiple components in series to achieve a greater output. IDeAL is able to abstract this notion into a domain-independent principle. For example, it can use cascading to increase the temperature drop in a heat exchanger.

LearnIT is a computer program that learns a designer's design strategy by observing how he or she solves a design problem (Stahovich, 2000). The program's domain is iterative parametric design: design problems that are solved by iteratively adjusting a set of parameters until the design requirements, expressed as algebraic constraints, are satisfied. LearnIT is a transparent software layer that sits on top of the usual modeling and analysis software (Figure 7.10). It unobtrusively observes the sequence of design iterations and from this generates a set of design rules describing the designer's strategy. These rules can then be used to automatically create new designs satisfying new design requirements. Because the rules are learned from the

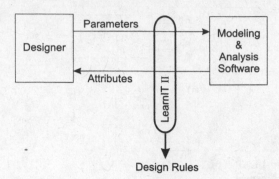

Figure 7.10. LearnIT (Stahovich, 2000) un-obtrusively observes the interaction between the designer and the CAD software and learns the design strategy employed.

designer, the new designs reflect the designer's engineering judgment, knowledge of implicit constraints, and overall familiarity with the problem.

The program's learning technique is based on two insights. The first is that iterative solutions to design problems are typically a form of debugging. At each iteration the designer identifies the unresolved flaws in the design and chooses a design action to eliminate those flaws. Thus, a design strategy can be thought of as a mapping from design flaws to design actions. The second insight is that the states of the design constraints – whether they are satisfied or not – are often a good indicator of the design flaws the designer is considering at any given time.

These insights lead to a specialized instance-based learning technique. Because the technique is instance based, the learning consists simply of recording observed design iterations. These are recorded in the form of design rules. A rule describes a particular state of the design and the corresponding action to take. The design state is defined by the states of the constraints and the action is a modification (increase or decrease) of a specific parameter.

Figure 7.11 shows a sample rule from the LearnIT system. This rule comes from the design of a circuit breaker. The original design task was to find parameter values to make the device trip after 5 seconds of a 15-amp overload. The sample rule indicates that when the device does not trip at the specified current, the preferred action is to decrease the thickness of each layer of the bimetallic strip. This rule embodies, but does not explicitly represent, several of the designer's insights into this particular design problem. For example, it reflects the fact that reducing the thickness of the bimetallic layers is the most efficient way to increase the hook deflection and thus make the device trip. (The fact that this rule is the most preferred way to repair this flaw is represented by the rule's low rule number.) This modification is particularly effective because it both increases the electrical resistance so that the hook heats faster and decreases the bending resistance. Similarly, the rule's limit embodies the insight that making the hook too thin will make the hook fragile and increase the risk of accidental tripping. Designs created with this rule automatically reflect these implicitly captured insights.

Because LearnIT's approach is instance based, the bulk of the work occurs when a new situation must be matched to a previous rule. LearnIT's rule matching procedure considers the states of the design constraints, the designer's preferences for particular rules, the rule limits, and the likely outcomes of the rules. To create new

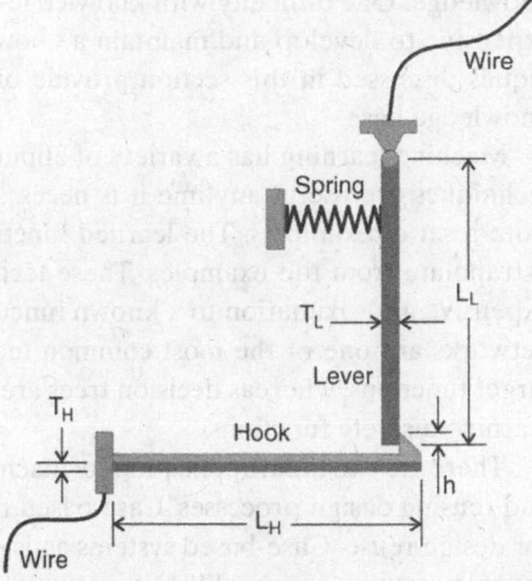

Figure 7.11. Top: a parameterized circuit breaker model; bottom: a design rule from LearnIT (Stahovich, 2000). Each C_i is a design constraint. δ is the final hook deflection; ΔT is the trip time; σ is the hook stress.

Rule 1:
If:
    C_1: δ too small
    C_2: ΔT too large
    C_3: σ less then yield (SAT)
Then:
    decrease T_H with a limit of 0.1 mm
Expected outcome:
    $\delta - h$ will decrease
    ΔT will not change
    σ will increase

designs, LearnIT applies its rule base in an iterative fashion. It evaluates the state of the design, identifies the best rule, and changes the corresponding parameter. This process repeats until all of the constraints are satisfied or until there are no remaining rules. The new designs that are created in this fashion are similar to those the designer would have created because the rules are learned from the designer.

The LearnIT II system performs the same task as the LearnIT system, but it uses an inductive learning technique rather than an instance-based one (Stahovich and Bal, 2001). LearnIT assumes that the designer's actions are determined primarily by the states of the design constraints. LearnIT II, in contrast, uses decision-tree learning to explicitly determine which properties of the design, or of the design history, best indicate which design actions the designer will take. LearnIT II is able to learn a much broader range of design strategies than LearnIT could. For example, it is able to learn strategies that depend strongly on the design history as well as those that depend strongly on the state of the design itself.

CURRENT UNDERSTANDING

The previous section on knowledge-based systems demonstrated the high level of performance that can be achieved by programs using expert problem-solving knowledge. One difficulty with knowledge-based systems, however, is that it can be expensive to develop and maintain a knowledge base. The machine-learning techniques discussed in this section provide one means of automatically generating a knowledge base.

Machine learning has a variety of applications in design and engineering. These techniques are useful anytime it is necessary to infer an unknown target function from a set of examples. The learned function can be used to both interpolate and extrapolate from the examples. These techniques can also be used to learn an inexpensive approximation to a known function that is expensive to evaluate. Neural networks are one of the most common techniques for learning continuous-valued target functions, whereas decision trees are one of the most common techniques for learning discrete functions.

There are two main applications of machine learning for synthesis: reusing designs and reusing design processes. Case-based reasoning is the most common technique for design reuse. Case-based systems assist in identifying previous design cases that are relevant to a new problem. Some systems can automatically adapt the previous cases to the new problem. Doing this requires a model of behavior and function. Most current systems rely on hand-generated models. The qualitative physical reasoning techniques described in the next section may provide a means of automatically generating these models, thus greatly extending the range of capabilities of case-based design systems.

The learning and reuse of design processes is an emerging application of machine learning. There is some evidence that suggests it is often more efficient to reuse the solution process rather than the solutions themselves. For this application, some success has been achieved by using learning techniques that identify empirical regularities in the training data. However, the next advances will likely come from explanation-based learning techniques. Here again, the qualitative physical reasoning techniques described in the next section may provide the necessary tools for using explanation-based learning for this application.

As more of design is performed electronically, there are more opportunities to apply machine learning. Much useful information generated during the design process is lost. Machine learning has the potential to capture and preserve this information for future use.

QUALITATIVE PHYSICS

Qualitative physical reasoning (QPR) techniques allow a computer program to perform commonsense reasoning about the physical world. This set of techniques is intended to provide computers with some of the same kinds of physical reasoning abilities that human designer's use in problem solving.

As the name suggests, qualitative reasoning techniques work from qualitative rather than quantitative problem representations. Qualitative representations capture the significant characteristics of a problem and abstract away the rest. For example, for a problem involving fluid flow into and out of a tank, a qualitative representation might describe whether the inflow was less than, equal to, or greater than the outflow. Even this abstract representation would allow useful predictions to be made. For example, this information is adequate for determining if the tank will eventually empty, although it would not be adequate for predicting how long that would take. The benefit of working from a qualitative representation is that predictions of behavior can be made before the quantitative details have been determined. For example, the IBIS and CADET systems described earlier (in the first and third sections) use networks of qualitative influences to predict the behavior of a candidate design prior to selecting actual physical components.

This section first reviews the two main classes of qualitative reasoning techniques: those suitable for lumped parameter systems and those that consider geometry. Next, two important applications are considered: causal explanation and design generalization. Causal explanations are descriptions of how a device achieves its behavior. These explanations are useful for adapting a design to new applications. Design generalization techniques take a single design, construct an explanation of how it works, and then generate new alternatives that work the same way.

LUMPED PARAMETER SYSTEMS

Much of the early work in qualitative physics focused on devices that could be described with lumped parameter models, such as electric circuits, hydraulic systems, and thermodynamic systems. In a quantitative world these devices are described with ordinary differential equations and algebraic constraint equations. In a qualitative world they are described with qualitative versions of these equations. Typically, these are in the form of qualitative constraints. Behavior is predicted by propagating qualitative values through these constraints.

Among the earliest work was de Kleer's QUAL program, which could produce causal explanations of the small signal behavior of electric circuits (de Kleer, 1979). Small signal behavior is the behavior that occurs within a single operating mode of a device. Each component is modeled with a qualitative constraint equation that relates the signs of the derivatives of the inputs and outputs. The program uses a constraint propagation technique to perform "incremental qualitative" analysis and determine how changes in the circuit's inputs propagate through the circuit.

Williams (1984) created a program that could reason about both the small signal and the large signal behavior of an electric circuit, greatly extending the ability of a program to reason about changes in operating mode. The program computed the small signal model that applied in a particular operating mode, then predicted which parameters changed, possibly causing a transition to another operating mode. The program then used constraint analysis to determine which of the possible transitions would actually happen first.

The ENVISION program, which built upon QUAL, is another system that could reason about changes in operating mode (de Kleer and Brown, 1984). With ENVISION, the behavior of each component is characterized by a set of qualitative states or operating modes. The behavior in each state is characterized by a set of "confluences," describing how changes in variables propagate to other variables. Confluences are concerned with the directions of change rather than the magnitudes. For example, flow through a valve (or other orifice) could be modeled with this confluence: $\partial P - \partial Q = 0$. This represents the fact that an increase in pressure results in an increase in flow, but it says nothing about the magnitudes of the changes.

ENVISION models the passage of time as a sequence of episodes. Within an episode each component remains in the same state. The behavior within an episode is determined by identifying a consistent direction of change ($+$, 0, or $-$) for each variable. The task is a constraint satisfaction problem (see the first section) because the confluences form constraints on the directions of change. ENVISION identifies all possible device states by identifying the consistent variable assignments for each combination of component states. It then reasons about the directions of change to identify legal transitions between device states. Suppose, for example, that device state **A** is valid only when some variable X is less than C, and state **B** is valid when $X \geq C$. If X is increasing when the device is in state **A**, then a transition from **A** to **B** is possible.

Kuipers later formalized the qualitative mathematics for predicting behavior from qualitative constraint equations, resulting in QSIM (Kuipers, 1986). With QSIM, a physical system is described by a set of symbols representing the physical parameters. The value of a parameter is specified in terms of its relationship with a set of ordered landmark values. A set of constraint equations describe how the parameters are related to each other. Some of the constraints are qualitative analogs of common mathematical relationships such as DERIV (velocity, acceleration), MULT (mass, acceleration, force), ADD (net flow, outflow, inflow), and MINUS (forward, reverse). Others specify that one parameter is a monotonically increasing or decreasing function of another: M^+(size, weight) and M^-(resistance, current).

With QSIM, the initial state of a system is defined by a set of active constraints and a set of qualitative values for the parameters. The simulation proceeds by first enumerating all possible qualitative values each parameter can have next, ignoring the constraints. For example, if a parameter is at a landmark and steady, in the next step it may be beyond the landmark and increasing, it may be below the landmark and decreasing, or it may still be at the landmark and steady. The constraints are then applied to identify the legal combinations of next parameter values. It is common for there to be multiple legal combinations and thus the simulation often branches.

Figure 7.12 shows an example of a QSIM simulation of the flow of water between two tanks. Figure 7.12(b) shows the constraints relating the parameters. Initially the levels in the two tanks are the same; then additional water is added to tank **A**. Figure 7.12(c) shows the simulation results. As the water is added to **A**, the pressure at the bottom of the tank increases, causing water to flow into **B**. As the water flows from **A** to **B**, the pressure in **A** decreases while that in **B** increases. As the pressure difference decreases, so does the flow between the tanks. Eventually, the pressure difference and flow become zero. Notice that QSIM is able to predict this behavior without knowing anything quantitative about the system: the specific amounts of water, the density of water, the size of the tanks, and so on are unspecified.

Much of the work in QPR has been device centric, meaning that device models are constructed by assembling models of the individual components. This approach is well suited to systems such as electrical and hydraulic circuits, but it cannot easily handle phenomena such as boiling, which do not involve a fixed collection of "stuff." To handle those kinds of problems, Forbus developed qualitative process theory that takes a process-centric perspective (Forbus, 1984). The representation of a process

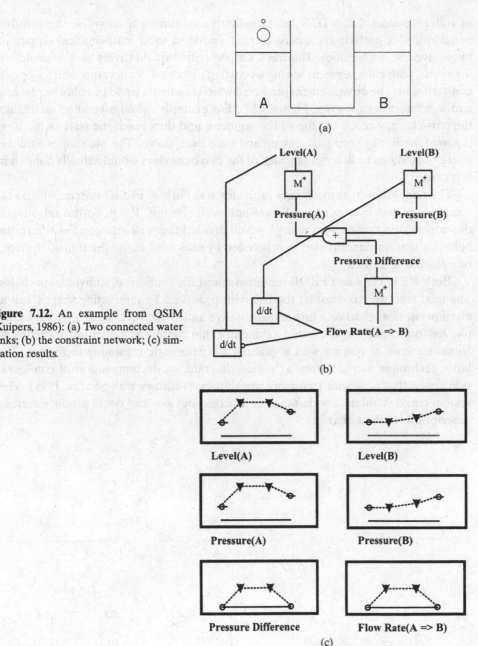

Figure 7.12. An example from QSIM (Kuipers, 1986): (a) Two connected water tanks; (b) the constraint network; (c) simulation results.

includes the set of objects that must exist and the preconditions that must be satisfied for the process to occur. The representation also includes a set of "influences" (similar to confluences) describing how the parameters change when the process is active.

QUALITATIVE PHYSICS WITH GEOMETRY

The first attempt at qualitative reasoning about force and geometry was de Kleer's NEWTON program (de Kleer, 1977), whose domain was the one-dimensional world of roller coasters. NEWTON used qualitative reasoning to envision the possible behaviors of a particle on a curved track and then used mathematical equations to resolve the ambiguities. The track was described qualitatively as a sequence of segments, with each segment having a constant direction of curvature and a slope of constant sign. The envisionment predicted which segments could possibly be reached and in what order (see e.g., Figure 7.13). For example, when ascending an incline, the particle may reach the top of the segment, and thus reach the start of the next segment, or it may stop part way up and slide back down. The program would use energy equations to.determine which of the two behaviors would actually happen in this case.

The next attempt at mechanics problems was Forbus' FROB system, which reasoned about particles in a two-dimensional well (Forbus, 1980). Forbus introduced the notion of a "place vocabulary," which divided the well into regions where the behavior was similar. Simulation proceeded by reasoning about the transitions from one place to the next.

Both NEWTON and FROB reasoned about the motions of individual particles. The next step was to consider the behavior produced by interacting shapes. Initial attempts at this relied on a mix of quantitative and qualitative reasoning. For example, Joskowicz and Sacks created a simulator that combined a qualitative version of Newton's laws of motion with a quantitative geometric reasoning technique. The latter technique worked from a "region diagram," a decomposition of configuration space that is similar to a place vocabulary (Joskowicz and Sacks, 1991). This system could simulate a wide range of practical devices and could produce concise descriptions of the behavior.

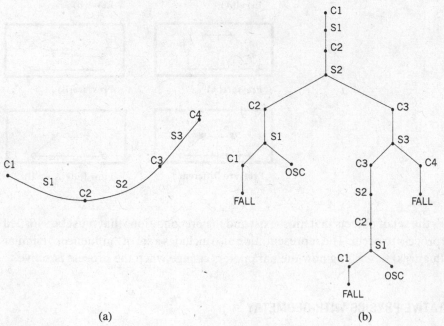

(a)   (b)

Figure 7.13. An example from NEWTON (de Kleer, 1977): (a) one-dimensional roller coaster with labeled segments; (b) envisionment (tree) indicating possible sequences of motions.

Faltings (1990) developed a system that used place vocabularies to produce qualitative kinematic simulations. The place vocabulary is a qualitative geometric representation that is constructed using quantitative analysis. Thus, this approach, too, is partially quantitative. This work was later combined with qualitative techniques for reasoning about forces and mechanical constraints, producing a system capable of qualitatively predicting rigid-body dynamics (Forbus et al., 1991). This system works as a total envisioner: it computes all possible states of the device and all legal transitions between them. It then computes which of these states are actually visited for a specific set of initial conditions and external inputs.

Stahovich, Davis, and Shrobe (1997, 2000) developed a qualitative rigid-body dynamic simulator that requires no quantitative information. This simulator works directly from a qualitative geometric representation called qualitative configuration space (Qc space). In Qc space a mechanical interaction between a pair of part faces is represented as a qualitative configuration space curve (Qcs curve) describing the configurations of the device for which the pair of faces touch each other (Figure 7.14). Each Qcs curve is a family of monotonic curves all having the same qualitative slope. The end points of the Qcs curves are marked with landmarks. The landmarks are ordered relative to one another, but there are no quantitative distances in Qc space. A special form of Qcs curve, called a *boundary*, is used to represent the neutral positions of springs and the motion limits of actuators.

To compute a step of simulation, the simulator first computes the net force on each body. Summing qualitative quantities often leads to ambiguity. To avoid this, a special qualitative force representation was used. This representation considers the projections of the forces onto the degrees of freedom of the device and describes each force by its magnitude, direction, and the type of constraint it imposes. Once the net forces are computed, quasi-static assumptions are used to determine the direction of motion of each part.

A simulation step ends when an event changes the nature of the forces. Events include two bodies colliding, two objects separating, a spring passing through its neutral position, and a motion source turning on or off. The simulator performs the geometry-intensive task of identifying the next event by working directly from the Qc-space representation. As the parts of the device move, the configuration traces out a qualitative trajectory through Qc space. An event is detected by geometrically determining when the trajectory leaves or reaches a Qcs curve or boundary in Qc space.

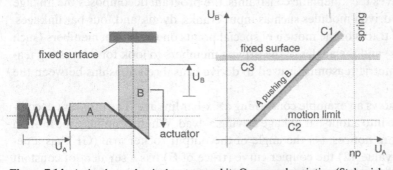

Figure 7.14. A simple mechanical system and its Qc space description (Stahovich et al., 2000).

Sometimes there is more than one possible next event. When this occurs, the simulator branches to consider all of the possibilities. However, unlike an envisioner, which must first compute all possible states of the device, this simulator directly computes just those states of the device that could be visited for the given initial conditions and external inputs.

CAUSAL EXPLANATIONS

This section describes a set of qualitative reasoning techniques that allow a program to produce causal explanations of a device's behavior. These explanations are useful for a variety of tasks including diagnosis and design reuse. Currently, the explanations these techniques generate are intended for human use. However, it may be possible to extend these techniques to create tools for automated design reuse. For example, case-based reasoning systems can adapt previous design solutions to solve new problems, but doing this requires models of behavior and function. Most current systems rely on manually constructed models such as the one in Figure 7.7. Causal reasoning techniques may provide a means of automatically generating behavioral and functional models.

There are a variety of causal reasoning techniques suitable for devices that can be described by ordinary differential equations and algebraic constraints (i.e., lumped parameter models). Many of the qualitative simulation techniques described in the fourth section naturally produce causal explanations. For example both de Kleer's QUAL system (de Kleer and Brown, 1984) and Williams' qualitative circuit simulator (1984) produce causal explanations of behavior. These techniques identify causality by examining the order in which quantities propagate through the qualitative equations. Another technique, called *causal ordering*, produces causal explanations by examining the order in which the equations must be solved; that is, which equations can be solved first, which can be solved once those are solved, and so on (Iwasaki and Simon, 1986; Gautier and Gruber, 1993; Gruber and Gautier, 1993).

Although there are a number of causal reasoning techniques for lumped parameter systems, there are relatively few techniques for reasoning about geometric interactions. One such system is Shrobe's linkage understanding program, which examines kinematic simulations of linkages in order to construct explanations for the purpose of the parts (Shrobe, 1993). The program first numerically simulates the behavior of the linkage by using a kinematic simulator. By examining the order in which the simulator solves the kinematic constraints, the program decomposes the linkage into driving and driven modules such as input cranks, dyads, and four-bar linkages. It then examines traces of the motion of special points on the driven members (such as coupler points) and the angles of the driving members to look for interesting features. Next, geometric reasoning is used to derive causal relationships between the features.

Figure 7.15 shows an example concerning a six-bar linkage. The program decomposes this linkage into a four-bar linkage driving a dyad. When examining the motion traces the program notices: (1) the angle of the output rocker arm (GF) has a period of constant value, (2) the coupler curve (trace of E) has a segment of constant curvature, and (3) the radius of curvature of this segment is nearly equal to the length

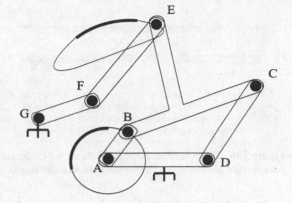

Figure 7.15. Shrobe's linkage understanding systems determines that the purpose of this device is to cause dwell (Shrobe, 1993).

of the driven arm of the dyad (EF). From this and other similar facts, the program hypothesizes that the purpose of the coupler moving in a circular arc is to cause the output to dwell.

Stahovich and Raghavan's ExplainIT program is another system that can produce causal explanations of behavior for devices that depend on geometry. ExplainIT is a computer program that computes the purposes of the geometric features on the parts of a device (Stahovich and Raghavan, 2000). ExplainIT identifies purpose by simulating how removing a geometric feature alters the behavior of a device. Ordinarily simulations describe what happens but not why. Thus, a simulation does not directly indicate which of a device's many behaviors are caused by a given feature on a given part. To identify those behaviors, ExplainIT compares a simulation of the nominal device to a simulation with the feature removed. The differences between them are indicative of the behaviors the feature ultimately causes.

The primary challenge in implementing this "remove and simulate" technique is accurately identifying the differences between the nominal and modified simulations. Direct numerical comparison of the state variables is not useful because there are likely to be differences in force magnitudes, velocities, accelerations, and so on, at every instant of time. Many of these differences are insignificant, such as those resulting from the small change in mass that occurs when the feature is removed. To avoid this problem, ExplainIT abstracts the simulation results into a qualitative form that reveals the essential details. The program then identifies the first point at which the two simulations begin to qualitatively differ. This is the point when the feature must perform its intended purpose to make the device end up in the correct final state. The program uses the laws of mechanics to construct a causal explanation for how the feature causes the behavior observed at this point in the simulation. It then translates this causal explanation into a human-readable description of the feature's purpose. Figure 7.16 shows an example of the kind of explanation the program can provide.

Stahovich and Kara's ExplainIT II system uses a different version of the remove and simulate technique to compute purpose (Stahovich and Kara, 2001). The program starts with two simulations of a device, one with the feature and one without. It then rerepresents the two simulations as a set of processes with associated causes, that is "causal processes." To identify the purpose of the removed feature, the program identifies all causal processes unique to one or the other of the two simulations.

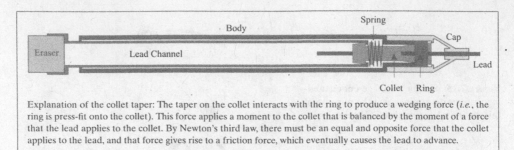

Figure 7.16. Top: A mechanical pencil. Bottom: An explanation for the purpose of the taper on the collet paraphrased from ExplainIT's output (Stahovich and Raghavan, 2000).

ExplainIT II's causal-process representation allows the program to reliably determine when a piece of behavior from one simulation is the same as a piece from the other. By a process of elimination, this allows the program to accurately determine when a piece of behavior is unique to one or the other of the simulations. It is common for a particular behavior to repeat multiple times during a simulation, especially if the device operates cyclically. By explicitly considering the causes of behavior, ExplainIT II can determine which instance of behavior from one simulation is the same as that from the other: two similar behaviors are the same if they have the same cause.

DESIGN GENERALIZATION

One of the challenges in automating design synthesis is specifying what is desired. It is often difficult to provide a completely abstract description of the design requirements. The problem is not simply the lack of a suitable specification language. The more challenging problem is thinking about an as yet unrealized design in abstract terms. This section describes an alternative means of describing requirements that avoids this problem. Rather than working from an abstract description, the programs described here work from a concrete example. The designer provides a specific example of the kind of device that is desired, and the programs generalize it to provide alternative designs. These programs could be described as performing "design by example" rather than the traditional "design by specification."

The first example of this approach is 1stPRINCE, which uses a methodology called dimensional variable expansion (DVE) to generalize the design of a structure in the context of numerical optimization (Cagan and Agogino, 1991a, 1991b).[5] The program generalizes a design by expanding the variables representing the critical dimensions (i.e., the critical "dimensional variables") into multiple regions, each of which can have its own set of material properties. This provides additional degrees of freedom to the optimization process and allows the program to innovate new designs that are substantially better with respect to the objective function than the

---

[5] DVE is technically not a qualitative physics technique in that it reasons primarily about the form of the governing equations. DVE is included in this section because of the kind of task it performs rather than the approach it uses.

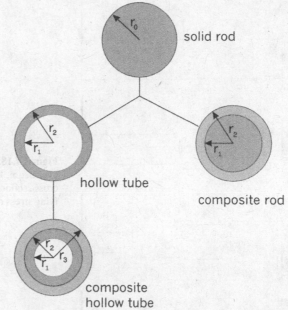

Figure 7.17. Applying DVE to a solid shaft under torsion, 1ˢᵗPRINCE generates a hollow tube, a composite rod, and a composite hollow tube (Cagan and Agogino, 1991a).

original. Figure 7.17 shows an example in which the objective is to design a torsion shaft of minimum weight. DVE generalizes the initial solid rod into a hollow tube, a composite rod, and a composite hollow tube.

The critical dimensional variables are identified by examining the nature of the design constraints. The solution to a constrained optimization problem is typically determined by a set of "active" constraints. In satisfying the active constraints, the solution naturally satisfies the other inactive constraints, which are still far from their limits. For example, in structural optimization, a constraint imposed by the ultimate tensile strength would be naturally satisfied if a constraint imposed by the yield strength were already satisfied. 1ˢᵗPRINCE uses monotonicity analysis (Papalambros and Wilde, 1988) to identify the possible sets of constraints that may be active in the optimal solution. Each such set is called a *prototype*. The program applies DVE to the dimensional variables associated with the active constraints in each prototype.

After 1ˢᵗPRINCE applies DVE and splits a critical dimensional variable into two regions, it repeats the monotonicity analysis. If each of the new regions is limited by the same set of active constraints as the original prototype, the program applies DVE to the two new regions, resulting in a total of four regions. If DVE can be applied for three consecutive iterations, 1ˢᵗPRINCE uses induction to generalize to an infinite number of infinitesimal regions, resulting in a continuous function. For example, Figure 7.18 shows an example in which the objective is to minimize the weight of a cantilever beam. The initial design has a uniform circular cross section. In the final design, the cross section is a continuous function of the position along the beam.

The second example of "design by example" is Stahovich et al.'s SKETCHIT system, which can transform a stylized sketch of a mechanical device into multiple

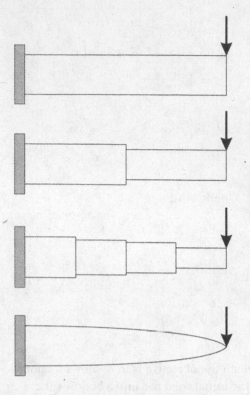

Figure 7.18. Applying DVE to the uniform cantilever beam, 1st PRINCE induces a beam with variable cross section as a means to reduce weight while satisfying stress requirements (1991b).

families of new designs (Stahovich et al., 1996, 1998). SKETCHIT uses a paradigm of abstraction and resynthesis as shown in Figure 7.19. During the abstraction process, the program reverse engineers and generalizes the original design by using the qualitative configuration space (Qc space) representation described above. Qc space enables the program to identify the behavior of the individual parts of the design while abstracting away the particular geometry used to depict those behaviors.

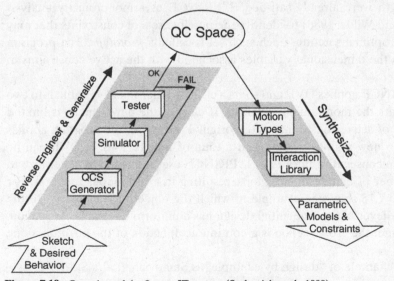

Figure 7.19. Overview of the SKETCHIT system (Stahovich et al., 1998).

To interpret a sketch, the program begins by directly translating it into an initial Qc space. The program then uses Stahovich's qualitative rigid-body dynamic simulator described above (in a previous subsection) to simulate the behavior of the Qc space and determine if the sketch as drawn is capable of producing the desired behavior. The desired behavior is a specific sequence of kinematic states that this particular geometry should achieve (i.e., the description is concrete rather than abstract). The sequence is described with a state transition diagram.

If the initial Qc space works correctly, the program is done with the abstraction process. If not, the program modifies the Qc space and repeats the process. Once it has found a working Qc space, the program uses it as a specification from which the program synthesizes new implementations. SKETCHIT uses a library of geometric interactions to map the individual parts of the Qc space (Qcs curves) back to geometry. Each library entry is a chunk of parametric geometry with constraints that ensure it implements a particular kind of behavior. By assembling these chunks, the program produces a behavior-ensuring parametric model (BEP model) for the device. A BEP model is a parametric model augmented with constraints that ensure the device produces the desired behavior. Each BEP model is a family of solutions that are all guaranteed to work correctly. Different members of the family are obtained by selecting different parameter values that satisfy the constraints of the BEP model.

SKETCHIT's library contains multiple implementations for each kind of behavior (Qcs curve), and thus the program is able to generate multiple families of implementations (BEP models). Additionally, the Qc space representation allows SKETCHIT to identify when it is possible to replace rotating parts with translating ones and vice versa. This ability provides SKETCHIT with another means of generating design alternatives.

Figures 7.20 and 7.21 show the kind of stylized sketch and state transition diagram that SKETCHIT takes as input. This particular example concerns the design of a circuit

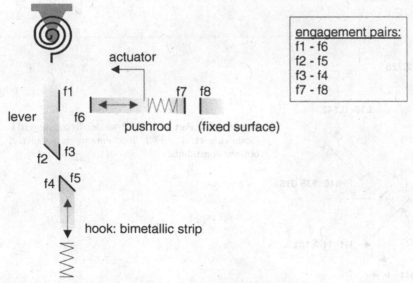

Figure 7.20. A stylized sketch of a circuit breaker given to SKETCHIT as input (Stahovich et al., 1998). Engagement faces are bold lines. The sketch is created in a mouse-driven sketching environment.

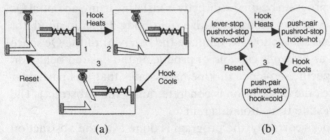

Figure 7.21. The desired behavior of a circuit breaker (Stahovich et al., 1998): (a) physical interpretation; (b) state transition diagram.

breaker similar to those found in a residential electrical system. From this input, the program generates several families of designs. Figure 7.22 shows a portion of the BEP model for one of these families (the one that is most similar to the original sketch). Figure 7.23 shows a different design the program created by selecting new geometry for the interacting faces, and Figure 7.24 shows another design created by replacing a rotating part with a translating one.

QUALITATIVE PHYSICS: CURRENT UNDERSTANDING

Qualitative physical reasoning (QPR) techniques have been used extensively for performing qualitative simulation. They are also frequently used for generating explanations of behavior and for performing diagnosis. The most common application to design is the use of qualitative simulation to reason about the behavior of a device before concrete implementations have been selected for the parts. The IBIS, Kritik2, and CADET systems described in earlier sections all use QPR for this purpose.

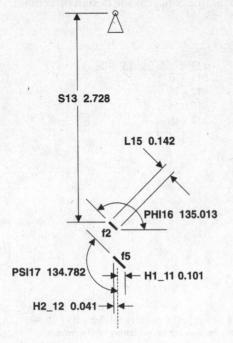

Figure 7.22. Part of a BEP model from SKETCHIT (Stahovich et al., 1998). Top: Parametric geometry; bottom: constraints.

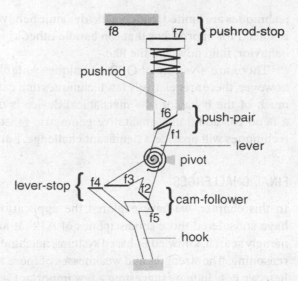

Figure 7.23. A design variant SKETCHIT obtained by using different implementations for the engagement faces (Stahovich et al., 1998).

Current applications do not tap the full potential of QPR as a synthesis tool. There is clearly much work left to be done. Interestingly, some of the areas that are ripe for progress may be the areas where QPR intersects the other techniques described in this chapter. For example, the first section described search-based "design by repair" techniques. These techniques synthesize by searching for a sequence of modification operators to repair an initial candidate design. Qualitative physical reasoning may provide tools for understanding why a design fails so that suitable modification operators can be identified with a minimum of search.

There are a number of ways that QPR can be used in conjunction with machine learning. For example, the third section described systems that learn and reuse design strategies. Qualitative physical reasoning might be able to extend explanation-based learning to this problem area. Recall that explanation-based approaches often perform better than approaches that rely solely on empirical regularities. As mentioned earlier, QPR might also provide a means of extending the capabilities of case-based reasoning systems. The techniques described above in the previous subsection might provide the necessary tools for creating the behavioral and functional models needed for case adaptation.

Another emerging application of QPR is design generalization. A common difficulty with automated synthesis techniques is specifying what is desired. It is often easier to generate an example of what is required than to provide an abstract description of it. Design generalization techniques enable a program to automatically derive the abstract specification from a specific example. The current generalization

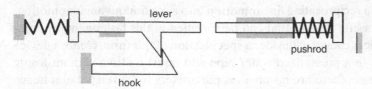

Figure 7.24. A design variant SKETCHIT obtained by replacing a rotating lever with a translating part (Stahovich et al., 1998).

techniques are limited to kinematic/dynamic behavior and structural behavior. There is a need for approaches that can handle other kinds of behavior, including thermal behavior, fluid flow, and the like.

There are a variety of QPR techniques suitable for lumped parameter models; however, there are relatively few techniques that can reason about geometry. Because much of the behavior of a mechanical device is determined by geometry, there is a need for additional qualitative geometric reasoning techniques. Creating these techniques will provide a significant challenge, but the potential benefits are vast.

FINAL CHALLENGES

In this chapter, we have explored the application of AI to design synthesis. We have considered those subdisciplines of AI that are the most relevant to synthesis, namely search, knowledge-based systems, machine learning, and qualitative physical reasoning. The strengths and weaknesses of these techniques were discussed earlier; here we conclude by suggesting a few important areas for future work.

It has been common in AI research to consider devices that are composed of linear sequences of idealized components connected at well-defined ports. Examples include chains of power transmission components and chains of simple mechanisms. These types of devices provide a number of computational advantages. The linear topology helps to restrict the size of the search space, and the discrete nature of the components and interconnections facilitates the construction of behavioral models.

Although these types of devices are used in practice, there are many common, real-world devices that cannot be described as a linear sequence of discrete parts. Many real-world devices have components that are highly interconnected, and the connections between the components often change as the device operates. Furthermore, a given function may be split across multiple components, or a single component may have multiple functions. Consider, for example, a car door, which must provide a means of entering, exiting, and securing the vehicle. The door must also provide structural integrity, aesthetic shape, controls for power mirrors and windows, ducts for defrosters, channels for wires, and so on. In this case, there are no obvious linear sequences of functions or components. In fact, much of the functionality is distributed in a few large pieces of sheet metal. Scaling existing synthesis techniques to handle these kinds of complexities will provide a significant challenge. Part of the solution may be to choose a new ontology, viewing devices as collections of interactions (Williams, 1990; Stahovich et al., 1998) rather than as collections of components.

Many of the automated synthesis techniques described in this chapter are restricted to functions that can be expressed as a desired relationship between two scalar parameters. Examples include transforming a voltage into an angular deflection or transforming a reciprocating linear motion into a continuous angular motion. Although there are useful devices that can be specified in this fashion, many common, real-world devices cannot. Consider a specification of this form: create a device that can bore a hole in a pressurized water pipe and insert a fitting without losing any water.[6] In this case, there are no obvious parameters to be related and hence traditional specification languages would be unsuitable.

---

[6] Such devices are used to tap into pressurized water mains.

Future work on specification languages should also address the fact that devices typically have multiple states or operating modes, and hence have different functions at different times. For example, a camera has modes for loading film, winding film, cocking, shooting, rewinding, and so on. Work in this area should also be driven by the design of devices with highly interconnected parts and integrated functionality (as described above).

The traditional approach to specification is to describe a desired function in terms of a simple, identifiable piece of behavior, such as a quantity remaining constant or one quantity being proportional to another. It may be possible to extend this approach to more complicated kinds of functions by developing better techniques for characterizing behavior. Alternatively, the "design by example" approach described above may provide a means of specifying a broader range of functions. Rather than attempting to categorize different types of standard behaviors, the"design by example" approach relies on methods for identifying when two behaviors are the same. Identifying similar behaviors may prove more general than creating definitions of behaviors.

Perhaps the most important area for future work is geometric reasoning. Most of the automated synthesis techniques described in this chapter have only limited geometric reasoning abilities. However, for mechanical design, geometry is of primary importance. Much of the behavior of a mechanical device is determined by the geometry of its parts. The geometric reasoning techniques of Joskowicz and Addanki (1988) and Stahovich et al. (1998) are a starting point for this work, but there is much left to be done. Geometric reasoning will likely pose the biggest challenges in applying AI to synthesis. However, success in this area will lead to substantially more powerful and more general synthesis tools.

REFERENCES

Bachant, J. and McDermott, J. (1984). "R1 revisited: four years in the trenches," *The Artificial Intelligence Magazine*, **5**(3):21–32.

Barker, V. E. and O'Connor, D. E. (1989). "Expert systems for configuration at digital: XCON and beyond," *Communications of the ACM*, **32**(3):298–318.

Bennett, J. S. and Englemore, R. S. (1984). "SACON: a knowledge-based consultant for structural analysis." In *AAAI-84*, AAAI Press, Menlo Park, CA.

Bhatta, S. R. and Goel, A. (1997). "Learning generic mechanisms for innovative strategies in adaptive design." *Journal of the Learning Sciences*, **6**(4):367–396.

Bobrow, D. G. and Winograd, T. (1977). "An overview of KRL, a knowledge representation language," *Cognative Science*, **1**(1):3–46.

Brachman, R. J. (1979). "On the epistemological status of sematic networks." In *Associative Networks: Representation and Use of Knowledge by Computers*, N. V. Finder (ed.), Academic Press, New York.

Brachman, R. J. and Schmoze, J. G. (1985). "An overview of the KL-ONE knowledge representation system," *Cogntive Science*, **9**(2):171–216.

Britt, D. B. and Glagowski, T. (1996). "Reconstructive derivational analogy: a machine learning approach to automated redesign," *Artificial Intelligence for Engineering Design*, **10**:115–126.

Brown, D. C. and Chandrasekaran, B. (1986). "Knowledge and control for a mechanical design expert system," *IEEE Computers*, **19**(7):92–100.

Buchanan, B. G. and Shortliffe, E. H., Eds. (1984). *Rule-Based Expert Systems. The MYCIN Experiments of the Stanford Heuristic Programming Project*, Addison-Wesley, Reading, MA.

Cagan, J. and Agogino, A. M. (1991a). "Dimensional variable expansion – a formal approach to innovative design," *Research in Engineering Design*, **3**:75–85.

Cagan, J. and Agogino, A. M. (1991b). "Inducing constraint activity in innovative design," *Artificial Intelligence in Engineering Design, Analysis, and Manufacturing*, **5**(1):47–61.

Chandra, K. S. D. N., Guttal, R., Koning, J., and Narasimhan, S. (1992). "CADET: a case-based synthesis tool for engineering design," *International Journal of Expert Systems*, **4**(2): 157–188.

Cover, T. and Hart, P. (1967). " Nearest neighbor pattern classification," *IEEE Transactions on Information Theory*, **13**:21–27.

Davis, R. (1982). "Expert systems: Where are we? and Where do we go from here?" *Artificial Intelligence Magazine*, **3**(2):3–22.

Davis, R., Buchanan, B., and Shortliffe, E. (1977). "Production rules as a representation for a knowledge-based consultation program," *Artificial Intelligence*, **8**:15–45.

Davis, R. and Lenat, D. B. (1982). *Knowledge-Based Systems in Artificial Intelligence*, McGraw-Hill, New York.

Davis, R., Shrobe, H., and Szolovits, P. (1993). "What is a knowledge representation?" *Artificial Intelligence Magazine*, **14**(1):17–33.

de Kleer, J. (1977). "Multiple representations of knowledge in a mechanics problem solver." In *Proceedings IJCAI-77*, Morgan Kaufmann, San Francisco, CA, pp. 290–304.

de Kleer, J. (1979). "Causal and teleological reasoning in circuit recognition," Ph.D. Dissertation, Massachusetts Institute of Technology, Cambridge.

de Kleer, J. and Brown, J. S. (1984). "A qualitative physics based on confluences," *Artificial Intelligence*, **24**:7–83.

Dixon, J. R., Simmons, M. K., and Cohen, P. R. (1984). "An architecture for the application of artificial intelligence to design." In *21st Design Automation Conference*, IEEE Computer Society Press, pp. 634–640.

Domeshek, E. A. and Kolodner, J. L. (1992). "A case-based design aid for architectural design." In *Artificial Intelligence in Design '92*, J. S. Gero (ed.); Kluwer Academic, Dordrecht, The Netherlands, pp. 497–516.

Dym, C. L. (1994). *Engineering Design: A Synthesis of Views*. Cambridge University Press, Cambridge, MA.

Dym, C. L. and Levitt, R. E. (1991). *Knowledge-Based Systems in Engineering*. McGraw-Hill, New York.

Faltings, B. (1990). "Qualitative kinematics in mechanisms," *Artificial Intelligence*, **44**: 89–119.

Feigenbaum, E. A., Buchanan, B. G., and Lederberg, J. (1971). "On generality and problem solving: a case study using the DENDRAL program." In Meltzer, B. and Michie, D., editors, *Machine Intelligence 6*, B. Meltzer and D. Michie (eds.); Edinburgh University Press, pp. 165–189.

Fikes, R. and Kehler, T. (1985). "The role of frame-based representation in reasoning," *Communications of the ACM*, **28**:904–920.

Forbus, K. D. (1980). "Spatial and qualitative aspects of reasoning about motion." In *Proceedings AAAI-80*, AAAI Press, Menlo Park, CA.

Forbus, K. D. (1984). "Qualitative process theory," *Artificial Intelligence*, **24**:85–168.

Forbus, K. D., Nielsen, P., and Faltings, B. (1991). "Qualitative spatial reasoning: the clock project," *Artificial Intelligence*, **51**(3):417–471.

Garrett, Jr., J. H. and Jain, A. (1988). "ENCORE: an object-oriented knowledge-based system for transformer design," *Artificial Intelligence in Engineering Design, Analysis and Manufacturing*, **2**(2):123–134.

Gautier, P. O. and Gruber, T. R. (1993). "Generating explanations of device behavior using compositional modeling and causal ordering." In *Eleventh National Conference on Artificial Intelligence*, AAAI Press, Menlo Park, CA.

Goel, A., Bhatta, S., and Stroulia, E. (1997). "Kritik: an early case-based design system." In *Issues and Applications of Case-Based Reasoning in Design*, M. L. Maher and P. Pu (eds.), Lawrence Erlbaum, Mahwah, NJ, pp. 87–132.

Gruber, T. R. and Gautier, P. O. (1993). "Machine-generated explanations of engineering models: A compositional modeling approach." In *1993 International Joint Conference on Artificial Intelligence*, Morgan Kaufmann, San Francisco.

Hamilton, S. and Garber, L. (1997). "Deep Blue's hardware-software synergy," *Computer*, **30**(10):29–35.

Hart, T. P., Nillson, N. J., and Raphael, B. (1968). "A formal basis for the heuristic determination of minimum cost paths," *IEEE Transactions on Systems Science and Cybernetics*, **SSC-4**(2):293–326.

Hayes-Roth, F., Waterman, D. A., and Lenat, D. B. (1983). *Building Expert Systems*. Addison-Wesley, Reading, MA.

Ivezic, N. and Garrett, J. H. Jr. (1998). "Machine learning for simulation-based support of early collaborative design," *Artificial Intelligence for Engineering Design, Analysis, and Manufacturing*, **12**(2):123–140.

Iwasaki, Y. and Simon, H. A. (1986). "Causality in device behavior," *Artificial Intelligence*, **29**:3–32.

Jamalabad, V. R. and Langrana, N. A. (1998). "A learning shell for iterative design (L'SID): concepts and applications," *Journal of Mechanical Design*, **120**(2):203–209.

Joskowicz, L. and Addanki, S. (1988). "From kinematics to shape: an approach to innovative design." In *Proceedings AAAI-88*, AAAI Press, Menlo Park, CA, pp. 347–352.

Joskowicz, L. and Sacks, E. (1991). "Computational kinematics," *Artificial Intelligence*, **51**: 381–416.

Karnopp, D. and Rosenberg, R. (1975). *System Dynamics: A Unified Approach*. Wiley, New York.

Kedar-Cabelli, S. and McCarty, T. (1987). "Explanation-based generalization as resolution theorem proving." In *Proceedings of the Fourth International Workshop on Machine Learning*, Morgan Kaufmann, San Francisco, CA, pp. 383–389.

Kehler, T. P. and Clemenson, G. D. (1984). "An application development systems for expert systems," *System Software*, **3**(1):212–224.

Kolodner, J. L., Ed. (1993). *Case-Based Reasoning*, Morgan Kaufmann, San Francisco.

Korf, R. E. (1985). "Depth-first iterative deepening: an optimal admissible tree search," *Artificial Intelligence*, **27**(1):97–109.

Korf, R. E. (1988). "Search: a survey of recent results." In *Exploring AI*, H.E. Shrobe (ed.), Morgan Kaufmann, San Francisco, pp. 197–237.

Kota, S., Erdman, A. G., Riley, D. R., Esterline, A., and Slagle, J. (1987). "An expert system for initial selection of dwell linkages." In *ASME Design Automatiion Conference*, ASME, New York.

Kuipers, B. (1986). "Qualitative simulation," *Artificial Intelligence*, **29**:289–388.

Mackworth, A. K. (1977). "Consistency in networks of relations," *Artificial Intelligence*, **8**(1):99–118.

Mahadevan, S., Mitchell, T. M. Mostow, J., Steinberg, L., and Tadepalli, P. V. (1993). "An apprentice-based approach to knowledge acquisition," *Artificial Intelligence*, **64**: 1–52.

Maher, M. L., Balachandran, B., and Zhang, D. M. (1995). *Case-Based Reasoning in Design*, Lawrence Erlbaum, Mahwah, NJ.

McDermott, J. (1981). "R1, the formative years," *Artificial Intelligence Magazine*, **2**(2): 21–29.

Mitchell, T. M. (1977). "Version spaces: a candidate elimination approach to rule learning." In *Fifth International Joint Conference on AI*, Morgan Kaufmann, San Francisco, CA, pp. 305–310.

Mitchell, T. M. (1982). "Generalization as search," *Artificial Intelligence*, **18**(2):203–226.

Mitchell, T. M. (1997). *Machine Learning*. McGraw-Hill, New York.

Mitchell, T. M., Keller, R., and Kedar-Cabelli, S. (1986). "Explanation-based generalization: a unifying view," *Machine Learning*, **1**(1):47–80.

Mittal, S. and Dym, C. L. (1986). "PRIDE: an expert system for the design of paper handling systems," *IEEE Computers*, **19**(7):102–114.

Mostow, J., Barley, M., and Weinrich, T. (1989). "Automated reuse of design plans," *Aritificial Intelligence in Engineering*, **4**(4):181–196.

Newell, A., Simon, H. A., and Shaw, J. C. (1963). "Empirical explorations with the logic theory machine: a case study in heuristics." In *Computers and Thought*, E. Feigenbaum, and J. Feldman (eds.), McGraw-Hill, New York.

Orelup, M. F. and Dixon, J. R. (1987), "Dominic II: More progress towards domain independent design by iterative redesign." In *Proceedings Intelligent and Integrated Manufacturing Analysis and Synthesis*, Winter Annual Meeting of the ASME, 67–80.

Papalambros, P. and Wilde, D. J. (1988). *Principles of Optimal Design*. Cambridge University Press, Cambridge.

Parker, D. (1985). "Learning logic," Technical Report TR-47, MIT, Cambridge; MA.

Quinlann, J. R. (1986). "Induction of decision trees," *Machine Learning*, **1**(1):81–106.

Quinlann, J. R. (1993). *C4.5: Programs for Machine Learning*, Morgan Kaufmann, San Mateo, CA.

Rosen, D., Riley, D., and Erdman, A. (1991). "A knowledge based dwell mechanism assistant designer," *Journal of Mechanical Design*, **113**:205–212.

Rumelhart, D. E. and McClelland, J. L. (1986). *Parallel Distributed Processing: Exploration in the Microstructure of Cogntion*. Vols. 1 and 2, MIT Press, Cambridge, MA.

Russell, S. and Norvig, P. (1994). *Artificial Intelligence: A Modern Approach*. Prentice-Hall, Englewood Cliffs; NJ.

Schwabacher, M., Ellman, T., and Hirsh, H. (1998). "Learning to set up numerical optimizations of engineering designs," *Artificial Intelligence for Engineering Design, Analysis, and Manufacturing*, **12**(2):173–192.

Shea, K., Cagan, J., and Fenves, S. J. (1997) "A shape annealing approach to optimal truss design with dynamic grouping of members," *ASME Journal of Mechanical Design*, **119**(3): 388–394.

Shortliffe, E. H. (1974). "MYCIN: a rule-based computer program for advising physicians regarding antimicrobial therapy selection," Ph.D. Dissertation, Stanford, University, Stanford, CA.

Shrobe, H. (1993). "Understanding linkages." In *Proceedings AAAI-93*, AAAI Press, Menlo Park, CA, pp. 620–625.

Simmons, R. and Davis, R. (1987). "Generate, test, and debug: combining associatational rules and causal models." In *International Joint Conferences on Artificial Intelligence*, Morgan Kaufmann, San Francisco, pp. 1071–1078.

Stahovich, T. F. (2000). "LearnIT: an instance-based approach to learning and reusing design strategies," *Journal of Mechanical Design*, **122**(3):249–256.

Stahovich, T. F. and Bal, H. (2001). "LearnIT II: an inductive approach to learning and reusing design strategies," *Research in Engineering Design*, submitted.

Stahovich, T. F., Davis, R., and Shrobe, H. (1996). "Generating multiple new designs from a sketch." In *Proceedings of the Thirteenth National Conference on Artificial Intelligence*, AAAI Press, Menlo Park CA, pp. 1022–1029.

Stahovich, T. F., Davis, R., and Shrobe, H. (1997). "Qualitative rigid body mechanics." In *Proceedings of the Fourteenth National Conference on Artificial Intelligence*, AAAI Press, Menlo Park, CA.

Stahovich, T. F., Davis, R., and Shrobe, H. (1998). "Generating multiple new designs from a sketch," *Artificial Intelligence*, **104**:211–264.

Stahovich, T. F., Davis, R., and Shrobe, H. (2000). "Qualitative rigid-body dynamics," *Artificial Intelligence*, **119**:19–60.

Stahovich, T. F. and Kara, L. B. (2001). "A representation for comparing simulations and computing the purpose of geometric features," *Artificial Intelligence in Engineering Design, Analysis and Manufacturing*, **15**(2):189–201.

Stahovich, T. F. and Raghavan, A. (2000). "Computing design rationales by interpreting simulations," *Journal of Mechanical Design*, **122**(1):77–82.

Stallman, R. M. and Sussman, G. J. (1976). "Forward reasoning and dependency-directed backtracking in a system for computer-aided circuit analysis," Technical Report Memo 380, AI Lab, MIT, Cambridge, MA.

Stefik, M., Bobrow, D. G., Mittal, S., and Conway, L. (1983). "Knowledge programming in LOOPS: report on an experimental course," *Artificial Intelligence*, **4**(3):3–14.

Subrahmanyam, P. A. (1986). "Synapse: An expert system for VLSI design," *IEEE Computers*, **19**(7):78–89.

Subramanian, D. and Wang, C.-S. E. (1993). "Kinematic synthesis with configuration spaces." In *The 7th International Workshop on Qualitative Reasoning about Physical Systems*, pp. 228–239.

Subramanian, D. and Wang, C.-S. E. (1995). "Kinematic synthesis with configuration spaces," *Research in Engineering Design*, **7**:193–213.

Ulrich, K. T. (1988). "Computation and pre-parametric design," Technical Report 1043, AI Lab, MIT, Cambridge, MA.

van de Brug, A., Brachant, J., and McDermott, J. (1986). "The taming of R1," *IEEE Expert*, **1**(3):33–39.

van Melle, W., Shortliffe, E. H., and Buchanan, B. G. (1984). "Emycin: a knowledge engineer's tool for constructing rulebased expert systems." In *Rule-Based Expert Systems: The MYCIN Experiments of the Stanford Heuristic Programming Project*. E. H. Shortliffe and B. G. Buchanan (eds.), Addison-Wesley, Reading, MA.

Waltz, D. (1975). "Understanding line drawings of scenes with shadows." In *Psychology of Computer Vision*, P. H. Winston (ed.), MIT Press, Cambridge, MA.

Williams, B. C. (1984). "Qualitative analysis of MOS circuits," *Artificial Intelligence*, **24**(1–3):281–346.

Williams, B. C. (1990). "Interaction-based invention: designing novel devices from first principles." In *AAAI-90*, AAAI Press, Menlo Park, CA.

Winston, P., Binford, T., Katz, B., and Lawry, M. (1983). "Learning physical descriptions from functional definitions, examples, and precedents." In *Proceedings of the National Conference on Artificial Intelligence*, AAAI Press, Menlo Park, CA, pp. 433–439.

Winston, P. H. (1992). *Artificial Intelligence*. 3rd ed., Addision-Wesley, Reading, MA.

CHAPTER EIGHT

Evolutionary and Adaptive Synthesis Methods

Cin-Young Lee, Lin Ma, and Erik K. Antonsson

The simplest scheme of evolution is one that depends on two processes; a generator and a test. The task of the generator is to produce variety, new forms that have not existed previously, whereas the task of the test is to cull out the newly generated forms so that only those that are well fitted to the environment will survive.

— Herbert A. Simon, *The Sciences of the Artificial*, 1969, MIT Press

INTRODUCTION

The synthesis of novel engineering designs can take on the character of an exploration, especially when the exploration has the capability to experiment, either in a random or directed manner, not simply with the values of parameters that describe a single design configuration, but with a wide range of different configurations.

Since the 1960s, there has been an increasing interest in simulating the natural evolutionary process to solve optimization problems, leading to the development of adaptive and stochastic exploration techniques (Fogel, 1962, 1994, 1998; Holland, 1962; Rechenberg, 1965; Schwefel, 1965; Goldberg, 1989; Hammel, Bäck, and Schwefel, 1997) that can often outperform conventional optimization methods when applied to challenging real-world problems. Evolutionary and adaptive search methods, such as genetic algorithms and simulated annealing, stochastically refine or alter individual candidate solutions in a population, evaluate the fitness or performance of these new candidates, and keep only those with good fitness values for the next iteration. These methods have been demonstrated to successfully synthesize novel design configurations of the following:

- VLSI layouts (Wong, Leong, and Liu, 1989; Chatterje and Hartley, 1990; Gen, Ida, and Cheng, 1995; Naito et al., 1996; Salami, Murakawa, and Higuchi, 1996; Lienig, 1997a, 1997b; Mihaila, 1997; Drechsler, 1998),
- structures (Goldberg, 1987; Bensøe and Kikuchi, 1988; Chapman, Saitou, and Jakiela, 1994; Reddy and Cagan, 1995a, 1995b; Chapman and Jakiela, 1996; Duda and Jakiela, 1997; Shea, Cagan, and Fenves, 1997; Shea and Cagan, 1997; Campbell, 2000; Jakiela et al., 2000),
- mechanical systems (Gupta and Jakiela, 1994; Schmidt and Cagan, 1995; Saitou and Jakiela, 1995; Husbands et al., 1996),
- electromechanical systems (Iffenecker and Ferber, 1992),

- artificial lifeforms (Lipson and Pollack, 2000),
- optical coatings (Wiesmann, Hammel, and Back, 1998),
- three-dimensional component configurations Szykman and Cagan, 1995, 1997; Cagan et al., 1996; Campbell, Amon, and Cagan, 1997; Cagan, Degentcsh, and Yin, 1998; Szykman, Cagan, and Weisser, 1998; Yin and Cagan, 2000), and
- MEMS mask-layouts (Li and Antonsson, 1999; Lee and Antonsson, 2000b; Ma and Antonsson, 2000a, 2000b; Campbell, 2000).

These applications have demonstrated the practicability of the approach, and they have shown that novel engineering designs and design configurations can be automatically synthesized.

Although the definition of optimality varies from case to case, for any feasible design synthesis, a mapping that allows the comparison of different design candidates must exist. This mapping, be it cardinal, ordinal, quantitative, or otherwise, creates a "performance landscape" of the design space that can be explored to find the best design. In many instances, the performance landscape has one or more of the following characteristics: (i) high nonlinearity, (ii) multiple performance optima, and (iii) nonanalyticity (i.e., cannot be written explicitly in a functional form). These characteristics preclude the use of gradient-like search techniques. Perhaps more troubling is that typical design spaces are so large that traditional search techniques, such as branch and bound, linear programming, and the like; are ineffective. As a result, a variety of *adaptive* search methods have been developed to synthesize good solutions in a short amount of time.

To describe adaptive search methods, it is useful to first review the basic framework of iterative searches. A search starts with one or more at least marginally feasible candidate solutions from which new trial solutions are created by means of search operators. These search operators are chosen, or guided, by a selection mechanism that attempts to ensure that large performance gains are achieved with the desired result being convergence to an optimal solution. The search continues by repeatedly finding new solutions until termination criteria are met. For example, in discretized gradient descent, the search operator samples points within a given radius or step size of the current solution. The selection mechanism chooses the sampled point with the best performance (i.e., the one in the approximate gradient direction) to replace the current solution; and, finally, the search terminates when the improvement is less than some specified value. Returning to the framework here, an *adaptive* search method adapts its selection mechanism, search operators, or both in response to the structure of the performance landscape.

Four adaptive search methods have received considerable attention recently. These are evolutionary computation, simulated annealing, Tabu search, and multiagent systems. The first two are stochastic iterative methods; the third is also iterative but can be deterministic or stochastic. Multiagent systems are not in general iterative (they often operate asynchronously), nor stochastic; although they can exhibit random behavior akin to chaotic systems. Furthermore, multiagent systems, in contrast to the other methods, have more diverse implementations and do not have search as their primary application.

Of the four methods, evolutionary computation is the oldest, as it was introduced independently in Fogel (1962), Holland (1962, 1975), Rechenberg (1965, 1973), Schwefel (1965), and Goldberg (1989). A collection of the seminal papers in the area can be found in Fogel (1998). Multiagent systems have their origins in Hewitt (1977), whereas Tabu search was introduced in Glover (1986). Simulated annealing is one of the younger methods having been developed in 1983 by Kirkpatrick, Gellatt, and Vechi (Kirkpatrick et al. 1983), although traces of simulated annealing can be found in Metropolis et al. (1953).

The focus of this chapter is on design synthesis using evolutionary computation, though for completeness, the three other methods are briefly reviewed and discussed. Following these reviews, an example synthesis of a cantilever beam is presented, followed by synthesis of a fuzzy system to approximate unknown functions. The chapter closes with some thoughts on future directions for the field.

EVOLUTIONARY COMPUTATION

Evolutionary computation (EC) encompasses a broad range of search techniques, with the three most prevalent being *genetic algorithms*, *evolutionary programming*, and *evolution strategies*. The common thread that ties these disparate techniques to EC is the use of simulated evolution as the search procedure.

Simulated evolution is the attempt to mimic evolutionary processes in a virtual setting. In regards to search algorithms, simulated evolution is a rough imitation of biological evolution in which the search operators take their inspiration from genetic adaptive operators (e.g., mutation and recombination), and the selection mechanisms are derived from Darwinian selection, or "survival of the fittest". EC further distinguishes itself from other search methods in employing sets of solutions, called *populations*, rather than single solutions.

Another characteristic typical of EC implementations is the representation of feasible solutions as vectors of objects. Any particular solution can be written as a string of objects called a *chromosome*, with each object referred to as a *gene*. A *genome* is the set of all possible chromosomes, or equivalently, the search space. The most commonly used search operators in EC are mutation and crossover – recombination. Mutation creates new candidate solutions by perturbing a gene(s) according to a random distribution. The effect of mutation is similar to a discretized gradient descent search operator, useful for hill climbing when the step size is small or for escaping local minima when the step size is large. Conversely, crossover obtains new solutions by swapping genes between two or more "parent" chromosomes. The premise of crossover is to preserve genes common to both parents that are presumably good "building blocks." These search operators are guided implicitly through selection of offspring that exhibit better performance as compared with other offspring.

EC implementations differ primarily in the solution representation, search operators, and selection mechanisms. For example, a subset of genetic algorithms called genetic programming (GP) evolves computer programs rather than one-dimensional strings of objects. In most GP applications, solutions are represented as two-dimensional LISP-like trees.

The general format of EC methods is to initialize a population of trial candidates, then to iterate by creating offspring by using mutation and crossover and selecting the offspring that perform the best for survival to the next iteration. Iteration continues until termination criteria are met. This format changes slightly depending on the selection mechanism.

SIMULATION OF NATURAL EVOLUTION

There are currently three main avenues of research in *evolutionary algorithms* (Fogel, 1994; Bäck et. al., 1997): *genetic algorithms* (GAs; Holland, 1975; Goldberg, 1989), *evolutionary programming* (EP; Fogel, 1962, 1964), and *evolution strategies* (ESs; Schwefel, 1965; Rechenberg, 1973). Among them, GAs are perhaps the most widely known type of evolutionary algorithm today and have been successfully applied to many science and engineering problems in various domains (Goldberg, 1989).

Genetic Algorithms. Genetic algorithms are global stochastic optimization techniques that are based on the adaptive mechanics of natural selection evolution. They were introduced in Holland (1975), and they were subsequently made widely popular by Goldberg (1989). These algorithms use two basic processes from evolution: *inheritance*, or the passing of features from one generation to the next, and competition, or *survival of the fittest*, which results in the weeding out of individuals with bad features from the population. In general, GAs are nonproblem specific techniques that can be applied to virtually any optimization problem if its fitness function is available.

Genetic algorithms are particularly good at finding maxima where the fitness function is nonlinear, multimodal, and dependent on several parameters simultaneously. Although in many cases a GA is unlikely to produce a globally optimal solution to a problem, it can produce a solution that is within a few percent of the optimum in a time that is orders of magnitude less than a full solution takes by using other algorithms such as exhaustive search.

General Structure. A genetic algorithm maintains a population of solution candidates and works as an iteration loop. First, an initial population is generated randomly. Each individual in the population is an encoded form of a solution to the problem under consideration, called a *chromosome*, which is usually a string of characters or symbols, e.g., a string of 0s and 1s (a binary string). The chromosomes evolve through successive iterations, called *generations*. During each generation, the chromosomes are evaluated by a fitness evaluation function and selected according to the fitness values by using a selection mechanism, e.g., fitness proportionate selection, so that fitter chromosomes have higher probabilities of being selected. New chromosomes, called *offspring*, are formed by either merging two selected chromosomes from the current generation by using a crossover operator, or modifying a chromosome by using a mutation operator. Crossover results in the exchange of genetic material between relatively fit members of the population, potentially leading to a better pool of solutions. Mutation randomly introduces new features into the population to ensure a more thorough exploration of the search space. A new generation is created by selecting chromosomes from the parents and the offspring. The population's average fitness will improve as this procedure continues, and the algorithm will converge to

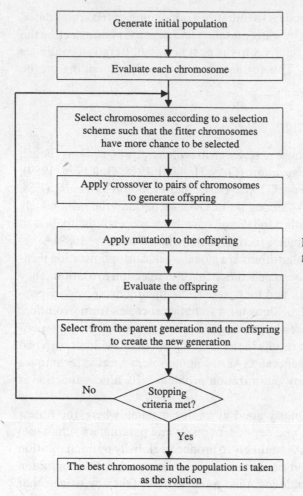

Figure 8.1. The general structure of genetic algorithms.

a best chromosome approaching the optimal or near-optimal solution. The general structure of genetic algorithms is shown in Figure 8.1 in flow-chart form.

For GAs to be used, each of the following must be developed.

1. Encoding scheme: In GAs, a population of candidate solutions is maintained and manipulated by genetic operators. The solutions are encoded as chromosomes (usually strings of characters or symbols, e.g., binary strings, real-number strings, or symbol strings) to which genetic operators can be applied. An encoding scheme is needed to map candidate solutions into coded strings.

2. Initialization of population: The initialization is usually done randomly to sample the search space uniformly without bias. A well-initialized population can improve the algorithm's robustness and effectiveness in finding an optimal solution, whereas a poorly initialized population may trap the algorithm in local optima and make it impossible to reach the global optimum.

3. Evaluation function: During the operation of GAs, all chromosomes are evaluated to see how fit they are as solutions to the problem. An evaluation function is required to assign a fitness value to each chromosome.

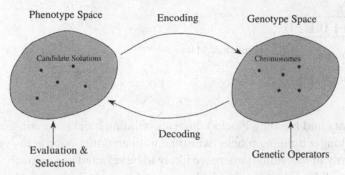

Figure 8.2. The relation of phenotype space and genotype space.

4. Genetic operators: In general the well-studied typical canonical genetic operators cannot be directly applied, because they only work with particular encoding
schemes. Additionally, problem-specific information should be considered when
efficient, meaningful operators are designed. Genetic operators suitable to the
problem and the encoding scheme must be developed.

The terminology used in genetic algorithms is described in Appendix A.

Encoding. Before a GA can be applied to a problem, a suitable *encoding* (or
representation) for the problem must be developed. The encoding scheme is the
linkage between the solutions in the phenotype space and the chromosomes in the
genotype space, as illustrated in Figure 8.2, and selecting an appropriate encoding
scheme is a key issue for GAs.

The two most common encoding schemes are binary and real (Holland, 1975;
Goldberg, 1989; Davis, 1991; Michalewicz, 1994). In binary coding, each chromosome
is a string of binary bits that, with a suitable scaling, describe candidate solutions. A
real-number coding uses a vector of k real numbers to encode the solution vector
(Goldberg, 1990; Janilow and Michalewicz, 1991; Wright, 1991). One advantage of
real-number coding over binary coding is improved precision. Another advantage is
that meaningful problem-specific genetic operators can be more easily defined.

Selection. The key principle of Darwinian natural evolution theory is that fitter
individuals have a greater chance to reproduce offspring, and it is by this principle of
"survival of the fittest" that species evolve into better forms. In genetic algorithms,
the bias toward fitter individuals is achieved through selection. The objective of any
selection scheme is to statistically guarantee that fitter individuals have a higher
probability of selection for reproduction.

In a GA, selection is carried out in two different stages: parent selection and
generational selection. Parent selection is the step in which individuals from the
parent generation are selected as parents to create offspring. Generational selection
is carried out after a specified number of offspring are generated. In general, the new
generation is created by selecting individuals from both the parent generation and
the offspring generation. Details of selection schemes in common use are described
in Appendix C.

Crossover. Once two chromosomes are selected, the crossover operator is applied to generate two offspring. Crossover combines the schemata or building blocks
from two different solutions in various combinations (see Appendix B for more

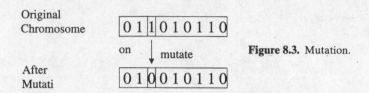

Original
Chromosome

on   mutate

Figure 8.3. Mutation.

After
Mutati

information about schemata and building blocks). Shorter building blocks are converted into progressively longer building blocks over time until an entire good solution is found. Because highly fit individuals are more likely to be selected as parents, the GA examines more candidate solutions in good regions of the search space and fewer candidate solutions in other regions. Crossover is the most important genetic operator for a GA, and it is the driving force for exploration of the search space. The performance of the GA depends to a great extent on the performance of the crossover operator used (Holland, 1975). Appendix D describes crossover operators in more detail.

Mutation. After new individuals are generated through crossover, mutation is applied with a low probability to introduce random changes into the population. In a binary-coded GA, mutation may be done by flipping a bit of a binary string, as shown in Figure 8.3, whereas in a nonbinary-coded GA, mutation involves randomly generating a new value in a specified position in the chromosome. In GAs, mutation serves the crucial roles of replacing gene values lost from the population during the selection process so that they can be tried in a new context, and of providing gene values that were not present in the initial population. By the introduction of random changes into the population, more regions of the search space can be evaluated, and premature convergence can be avoided.

Although the crossover operator is the most efficient search mechanism and the most important genetic operator, by itself, it does not guarantee the reachability of the entire search space with a finite population size. It is mutation that makes the entire search space reachable. Mutation is described in detail in Appendix E.

LINKAGE

In nature, the majority of DNA in an organism is *non-coding*, meaning that it is not transcribed into RNA for protein synthesis. The reason for non-coding DNA is not well known (Wu et al., 1996). It has been hypothesized that non-coding DNA prevent the destruction of good building blocks (such as genes) during recombination, since crossover is more likely to occur in a region of non-coding DNA (Wu and Lindsay, 1996). A modular genetic structure can develop due to the variability in non-coding segment length (or linkage) in which certain blocks of genetic code become more likely to stay together during recombination (thereby constituting a module). This premise of modularity, a hallmark of increased evolvability, has led to several studies on the utility of non-coding segments in genetic algorithms.

Examples of adaptive linkage are presented in (Goldberg et al., 1989, Levenick, 1995, Lobo et al., 1998, Rosenberg, 1967). An overview of this topic is presented in (Harik, 1997).

The concept of *linkage* is an integral part of GAs and is also pertinent in the discussion of GAs applied to design problems. Linkage is defined as the probability that two genes will be separated after recombination. In a canonical simple GA, adjacent genes have tighter linkages than non-adjacent genes, *i.e.*, there are more crossover points between non-adjacent genes such that there is a greater probability of disruption of non-adjacent genes than adjacent genes during crossover and recombination.

If linkages are allowed to adapt in a GA with non-coding segments, some sets of genes will become tightly linked as a result of the survival of only the fittest solutions. These tightly linked genes can be considered to be a 'module' since they are highly likely to remain together after recombination. The implication of adaptive linkages is that any inherent modularity of a design can be found by the GA.

Adaptation of the lengths of the non-coding segments can be accomplished in the following manner. Prior to recombining two chromosome, the non-coding genes of the first parent are randomly perturbed (lengthened or shortened). As the genetic information of individuals recombines in subsequent generations, the success of valuable attributes or features will increase the likelihood that certain coding genes will stay together. This allows the GA to exploit good building blocks or modules, while still exploring other regions with loose linkage.

A variable length non-coding segment representation has been developed with the goal of adapting evolvability through improved linkage characteristics. Preliminary results indicate that this representation is able to outperform a standard representation on a set of test functions. The non-coding representation is also able to perform as well as a representation that has the known modularity built in. These results indicate that the non-coding representation can be used not only to improve the performance of standard genetic algorithms, but also to determine the modularity of *a priori* unknown modular solutions. The evolution of modularity allows GAs to exploit good building blocks, while still exploring other viable solutions. Through the use of adaptive linkage, GAs can synthesize solutions to engineering design problems and also discover the underlying structure of good designs.

EVOLUTIONARY PROGRAMMING AND EVOLUTION STRATEGIES

In addition to genetic algorithms, two other evolutionary algorithms are commonly used: evolutionary programming and evolution strategies. GAs, EP, and ESs, all simulate natural biological evolution to solve optimization problems, and the principles and implementations of these three methods are similar. Each method maintains a population of contending candidate solutions. New solutions are created by stochastically altering the existing solutions. An objective measure of performance is used to assess the "fitness" of each trial solution, and a selection mechanism determines which solutions to maintain as "parents" for the subsequent generation. In spite of the many similarities, there are distinct differences among GAs, EP and ESs.

SYNTHESIS EXAMPLES USING EVOLUTIONARY COMPUTATION

Design synthesis has received considerable attention in the EC literature. The majority of design synthesis problems can be roughly divided into combinatorial problems, which are typically *NP* complete, and noncombinatorial problems.

The set of combinatorial design synthesis problems consists of routing, networking, placement or layout, packing, timetabling, jobshop scheduling, catalog design, compression code books, and so on. A large body of work exists for these applications. An even larger body of work exists for the quadratic assignment problem (QAP) and traveling salesman problem (TSP), which are *NP* complete and isomorphic. Given that nearly all combinatorial design synthesis problems are variants of QAP or TSP, this literature is relevant here.

A variety of hybrid methods have been developed to address design synthesis problems. A hybrid search – evolutionary programming approach to placement design is presented in Burke, DeCausmaecker, and VandenBerghe (1998). Genetic algorithms and local search have been implemented together for jobshop scheduling in Cai, Wu, and Yong (2000). Hybrids of simulated annealing and genetic algorithms can be found in Wong and Wong (1994).

Many of the design synthesis problems to which EC have been successfully applied are listed below.

Timetabling, Jobshop Scheduling. Timetabling is the problem of scheduling a set of events to specific time slots such that no person or resource is expected to be in more than one location at the same time and that there is enough space available in each location for the number of people expected to be there. These two main fundamental constraints, and many others, combine to make timetabling a classically hard problem to solve.

For jobshop scheduling, the task is to efficiently allocate a set of resources (machines, people, rooms, facilities) to carry out a set of tasks, such as the manufacture of a number of batches of machine components. There are obvious constraints: for example, the same machine cannot be used for doing two different things at the same time. The optimum allocation has the earliest overall completion time, or the minimum amount of "idle time" for each resource.

Research applying EC to timetabling and jobshop scheduling problems has been carried out in Corne, Ross, and Fang (1994), Cartwright and Tuson (1994), Paechter (1994), Ross and Corne (1995), Fang, Corne, and Ross (1996), Corne and Ross (1997), Zhang, Zhang, and Lee (1997), Tezuka et al. (2000), and Urquhart, Chisholm, and Paechter. Other examples include the scheduling of the maintenance of a power system (Langdon, 1995); task scheduling for multiprocessor computers (Tsujimura and Gen, 1996; Yue and Lilja, 1997), and manufacturing procedure optimization (Dimopoulos and Zalzala, 2000).

Layout, Placement, and Networks. In this category, the problems are to optimize layouts (or placement of components) of different systems, for example, the VLSI electronic chip layout design problem (Wong et al., 1989; Chatterje and Hartley, 1990; Gen, Ida, and Cheng, 1995; Drechsler, 1998) and the automated synthesis of topology and sizing of analog electrical circuits (Lienig, 1997a, 1997b; Mazumder and Rudnick, 1999). Field programmable gate arrays (FPGA), or reconfigurable hardware, have been the subject of much research with Mihaila (1997), Naito et al. (1996), and Salami et al. (1996), being a small sample of such work. Another example is the pipe network layout design of branched hydraulic networks (Savic and Walters, 1995). EC is also applied to solve constrained placement problems that require solutions to problems involving arrangements of discrete objects (Schnecke and Vorngerger, 1997).

Bin packing, the task of determining how to fit a number of objects into a limited space, has many applications in industry, and it has been widely studied (Dighe and Jakiela, 1995). Actually, the layout problem of VLSI-integrated circuits can be viewed as a particular example of the bin-packing problem. Another example of the constrained placement problem is the spatial layout planning problem in civil engineering (Lunn and Johnson, 1996, Gero, Kazakov, and Schnier, 1997). Bin packing has also been addressed by other methods, including simulated annealing, and will be discussed in the next section of this chapter.

Neural network topologies have been synthesized by numerous researchers (Fogel, Fogel, and Porto, 1990; Zhao, 1996; FigueiraPujol and Poli, 1998; Hancock, 1990; Kaise and Fujimoto, 1998; Skourikhine, 1998; Wicker, Rizki, and Tamburino, 1998). Similarly, many have synthesized fuzzy systems (Murta et al., 1998, Garcia, Gonzalez, and Sanchez, 1999). Fuzzy systems are a type of rule-based system, which have also received much attention in evolutionary computation. Most evolutionary approaches to rule-based systems attempt to determine the optimal rule set and knowledge base. Some recent work in this field can be found in Brotherton and Chadderdon (1998) and in Fogarty (1994).Control systems also have not been neglected (Tan, Lee, and Khor, 2000).

Structure Design. In this category, EC is used to synthesize structure shape and topology. EC has been demonstrated to efficiently solve structure design problems such as truss structure optimization (Goldberg, 1987, Bensøe and Kikuchi, 1998), structural configuration and topology design (Chapman et al., 1994; Chapman and Jakiela, 1996; Duda and Jakiela, 1997; Jakiela et al., 2000), and mechanical component design (Gupta and Jakiela, 1994; Husbands et al., 1996; Obayashi et al., 2000). Automated design of mechanical conformational switches is reported in Saitou and Jakiela (1995), and multilayer optical coating in Wiesmann et al. (1998).

Routing, Sequencing, Synthesizing. Many engineering problems belong to this category, for example, the mobile robot path planning problem (Xiao et al., 1997). This problem is typically formulated as follows: given a robot and a description of an environment, plan a path between two specified locations that is collision free and satisfies certain optimization criteria. EC has been utilized to solve this problem.

An example of data routing can be found in Zhang and Leung (1999). Design of optimal codebooks for data compression, vector quantization, and clustering, which are all variants of the QAP, have received much attention (Fogel and Simpson, 1993; Kim and Ahn, 1999; Lee and Antonsson, 2000a; Meng, Wu and Yong, 2000).

Another application is the synthetic route design problem in chemistry, that is, the development of efficient methods to synthesize chemicals (Cartwright and Hopkins, 1996). Most computer-driven synthesis programs incorporate a database of chemical reactions and expert system rules. These are used to propose a reaction sequence that transforms readily available starting materials into the desired product. Many possible routes for such transformations may exist, depending on the number of intermediates and the complexity of the product, and this presents synthesis design programs with various difficulties. EC is well suited to handle this type problem.

The design of sequences for stacking composite laminates is another application in which EC has been successfully used (Riche and Haftka, 1997; Liu et al., 2000). Laminated composite structures are stacked from layers made up of fibers embedded in a matrix of isotropic material. Fibrous composites are usually manufactured in the form of layers of fixed thickness. Designing composite structures involves finding the number of layers and the fiber orientations of each layer inside each laminate that maximizes the performance of the structure under constraints such as failure, geometry, cost, and so on.

Software. Genetic programming (a variant of GAs) is itself the automatic design of computer programs for a specified function (Koza, Andre, and Keane, 1999). One example of software design is the evolution of C programs (O'Neill and Ryan, 1999). A large portion of genetic programming has been geared toward analog circuit design such as bandpass filters. Reviews of the research done with genetic programming for analog circuit design can be found in Banzhaf et al. (1998) and in Koza et al. (1999).

Robot Evolution. One of the most recent developments in the field of evolutionary design (and artificial life) was reported in Lipson and Pollack (2000). This project, called the GOLEM (Genetically Organized Lifelike Electro Mechanical) project, has been described in review articles in national newspapers and magazines. In the project, Lipson and Pollack were able to evolve robots with rudimentary locomotive abilities. One significant aspect of this achievement is that the design syntheses of the control system and the mechanical structure of the robot are performed simultaneously. This is analogous to evolving a lifeform's neurological and physiological structures simultaneously. Lipson and Pollack accomplished this feat by means of a specialized genetic algorithm whose design space consisted of mechanical linkages, actuators, and artificial neurons.

SIMULATED ANNEALING

Annealing is the process of heating a material to a high temperature and then allowing it to cool slowly. The purpose of annealing is to let a material reach its lowest energy, or equilibrium state. High temperatures impart energy to the material's atoms, such that the material is able to jump out of metastable states. The larger the energy barrier, the more difficult the jump. Thus, as the temperature cools, the material, with less energy, will be trapped into the deepest energy "well," which is the equilibrium state. Of course, cooling cannot occur too rapidly or the material will freeze in a metastable state. The annealing process is effectively a physical search process that finds a material's equilibrium state. Simulated annealing (SA) was inspired by this observation.

The standard implementation of simulated annealing is composed of four parts: an energy function, a perturbation mechanism, an acceptance function, and a cooling schedule. The energy function maps a solution to a real value, which is to be minimized by simulated annealing. The perturbation mechanism is the search operator and simply moves the solution in a random direction. The acceptance function determines whether or not the new solution replaces the old solution. If the new solution has a lower energy, it is always accepted. Otherwise, the solution is accepted with finite probability according to a decreasing function of the variable T that is analogous to temperature in real annealing. In most cases the acceptance function is $\sim e^{-\Delta E/T}$, where ΔE denotes the change in energy. The cooling schedule is the method by which T is decreased from iteration to iteration. The standard schedule is logarithmic and has been shown to guarantee convergence to global optima because any point can be sampled infinitely often in asymptotic annealing time (Geman and Geman, 1984; Szu and Hartley, 1987). However, "these results say that SA . . . retains enough traces of exhaustive search to guarantee asymptotic convergence, but if exhaustive search were a realistic option, we would not be using SA anyway" (Neal, 1993).

Standard SA is often quite slow, as useful moves are only taken in the final temperature steps. A variety of methods have been implemented to speed up standard SA, with most simply increasing the cooling rate, typically without claims of ergodicity (convergence to global optima). One of the more prominent of these new techniques is Ingber's adaptive simulated annealing, in which the cooling schedule depends on the status of the current search (Ingber, 1993). An extended pattern search algorithm has shown 10–100 times improvement over standard SA for layout synthesis (Cagan et al., 1998; Yin and Cagan, 2000).

SYNTHESIS EXAMPLES USING SIMULATED ANNEALING

Because simulated annealing is a general search technique, it is not surprising to find that many of the design synthesis problems tackled by using evolutionary computation are topics of interest for simulated annealing application as well. A sampling of design synthesis using simulated annealing follows.

Trusses and mechanical structures have been designed by using simulated annealing in Reddy and Cagan (1995a, 1995b), Shea et al. (1997), Shea and Cagan (1997), and Campbell (2000). Communication network topologies have been designed in DeSanctis, Benjamin, and Taylor (1998) and in Mukherjee (1996). Routing and scheduling have been addressed in Purohit, Clark, and Richards (1995). Many VLSI layout and chip partitioning programs utilize simulated annealing for design (Brady, 1994; Wong et al., 1989; Chatterje and Hartley, 1990; Wu and Sloane, 1992). Work has also been done in developing hardware annealing for in situ modification of analog neurocomputing VLSI (Lee and Sheu, 1991). Design of lookup tables and vector quantization (typically used for data compression) can be found in Lech and Hua (1991; 1992). Filters for image reconstruction have been designed successfully in Bilbro (1991) and in Raittinen and Kaski (1990). Artificial neural network design can be found in Ingber (1992), Niittylahti (1992) and Peterson and Soderberg (1992). Although perhaps it is not an engineering design application, the design of hybrid financial securities has been attempted with some success in Marshall and Bansal (1992).

Layout problems for three-dimensional (3-D) mechanical and electro mechanical products have been solved by using extensions of simulated annealing methods originally developed for 2-D VLSI layout (Szykman and Cagan, 1995, 1996, 1997). This approach utilizes blocks and cylinders with rotations constrained to multiples of 90°. A perturbation-based approach was used in which infeasible states with component overlap and constraint violations are allowed and penalized. The move set includes translation, rotation, and swap moves. An adaptive annealing schedule (Huang, Romeo, and Sangiovanni-Vincentelli, 1986) is used to control the temperature, and a probabilistic move selection strategy (Hustin and Sangiovanni-Vincentelli, 1987) is used to choose moves based on their prior performance. The

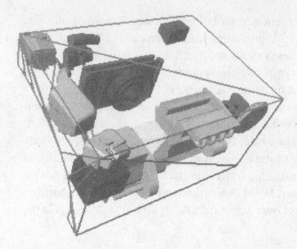

Figure 8.4. Constrained layout of portion of an automobile engine compartment with SA (Cagan et al., 1998).

objective function consists of a weighted sum of multiple performance measures, including packing density and tube layout and routing (Szykman and Cagan, 1996). Several types of spatial constraints that are characteristic of layout problems are defined, including the global and relative constraints on component locations and orientations.

The work of Szykman and Cagan was extended in Kolli, Cagan, and Rutenbar (1996) and in Cagan et al. (1998) by relaxing the restrictions on component geometry and rotations. The result is an algorithm able to layout reasonably large numbers of components of arbitrary geometries in any orientation. Applications include the packaging of several components from a passenger automobile (shown in Figure 8.4), and the component layout and tube routing of a heat pump layout problem (Cagan et al., 1998).

Although effective, the major problem with SA remains speed. Simulated annealing wastes a significant amount of time during the more random, earlier portions of the algorithm (Cagan et al., 1998; Yin, 2000).

An extended pattern search algorithm is introduced in Cagan et al. (1998) that balances focused search with stochastic jumps out of local optima. In the extended pattern search algorithm, components are moved through translations and rotations in predefined patterns at a given, shrinking step size, and only better objective states are accepted. The algorithm is extended to a stochastic technique by randomly selecting the order of components, allowing for jumps in step size and strategically swapping components when search gets stuck in local optima. Yin and Cagan have applied the method to the layout of heat pumps where finite-element method (FEM) simulators are called to analyze the natural frequency of the tubes (Cagan et al., 1998), automatic transmission layout where the clutches are resized based on their position in the layout (shown in Figure 8.5), truck chassis configuration where certain components are selected during execution, and automobile trunk packing where luggage pieces are selected to maximize cargo space packing (Yin et al., 1999; Yin, 2000, Yin and Cagan, 2000). In all applications the extended pattern search algorithm was shown to be 1–2 orders of magnitude faster than the robust SA algorithm highlighted above, for the same quality solutions. The heat pump example presented in Cagan

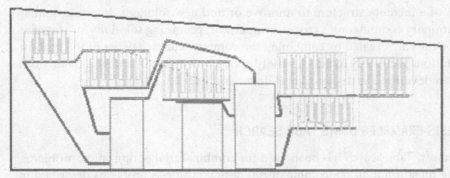

Figure 8.5. Axially symmetric cross section of automatic transmission configured by an extended pattern search (Yin et al., 1999).

et al. (1998), without FEM analysis, took only 1.2 seconds (versus 2 minutes with SA), whereas the heat pump layout with FEM analysis of six tubes took 5 minutes. The automobile engine compartment layout shown in Figure 8.4 took only 5 seconds with an extended pattern search (versus 1.25 minutes with SA).

TABU SEARCH

Tabu search, as with evolutionary computation and simulated annealing, is based on a natural phenomenon, that of memory. Hence, Tabu search methods are often grouped under the heading of adaptive memory programming. Glover, the originator of Tabu search, believed that human beings are capable of intelligent problem solving because of their ability to incorporate new knowledge with previous experience, or memory (Glover and Laguna, 1997). In particular, people can avoid past mistakes and/or imitate past successes.

The original Tabu search, regarded now as a naïve approach, only accounted for the avoidance of past mistakes. A Tabu search is conducted by moving in the direction of steepest ascent or mildest descent (for a maximization problem) while being restricted from moving to previously visited solutions. These restricted solutions are kept in a "Tabu" list whose length can change depending on the desired intensification and diversification (similar to exploitation and exploration in evolutionary computation). For example, if the list is long, meaning that the most recently visited points cannot be revisited, the search is forced to diversify by moving to a new neighborhood. The usefulness of such a memory approach is that, at subsequent search iterations, the search neighborhood is reduced significantly, often speeding the search up orders of magnitude in comparison to blind search. The primary difficulties of implementing Tabu search lie in maintaining efficient Tabu lists and choosing a good list length. Much work has been done in optimizing Tabu lists by using hash tables (Glover and Laguna, 1997).

Tabu list management (i.e., the size and number of lists) is the the most important issue and is still open to research. Most recently, reactive Tabu search (RTS) has been introduced that uses feedback mechanisms to determine the length of the Tabu list (Battiti and Tecchiolli, 1994). A more recent development in Tabu search is the addition of a memory structure to improve or find new solutions. In a vein similar to evolutionary computation's crossover operator, promising solutions for stagnant Tabu searches are created by combining the common characteristics of the best, or elite, solutions previously found. A variety of other memory structures and functions have been developed as discussed in (Glover and Laguna, 1997).

SYNTHESIS EXAMPLES USING TABU SEARCH

Traditionally, Tabu search has been used for combinatorial optimization problems, including most if not all the combinatorial design synthesis problems described in the evolutionary computation section. This is primarily a side effect of the Tabu list. Because the Tabu list keeps track of previously visited solutions, it is easier to prohibit moves if the solutions are discrete, as in combinatorial optimization problems. However, this does not mean continuous optimizations cannot be solved by using Tabu search. As with canonical genetic algorithms, which use binary strings, Tabu search can handle continuous optimizations by treating continuous values as binary strings, although the effectiveness of Tabu search under these conditions has yet to be thoroughly investigated. An illustrative list of design synthesis applications using Tabu search follows.

A large amount of Tabu search work has been devoted to various jobshop scheduling and planning problems. Resource allocation is addressed in Mazzola and Schantz (1995), and scheduling in Nowicki and Smutnicki (1995), Skorin-Kapov and Vakharia (1993), and Vaessens, Aarts, and Lenstra (1995), and process planning for machining centers in Veeramani and Stinnes (1996). Also along these lines was work done for task scheduling in parallel computing (Chakrpani and Skorin-Kapov, 1995; Porto and Ribeiro, 1995a, 1995b). Partitioning for VLSI circuits is accomplished in Andreatta and Ribeiro (1994). Optimal routing and network design are discussed in Chiang and Kouvelis (1994), Domschke, Forst, and Voss (1992), Gendreau, Soriano, and Salvail (1995), Rochat and Semet (1994), Rochat and Taillard (1995), and Skorin-Kapov and Labourdette (1994, 1995). Furthermore, Tabu search has been used for structure optimization and design. For example, in Fanni, Giacinto, and Marchesi (1996), electromagnetic structures are optimized; in Kincaid and Berger (1993), damper placement for space structures are determined; and in Kincaid, Martin, and Hinkley (1995), straight polymers are discovered. Neural network design is another recent application, with Battiti and Tecchiolli (1994, 1995) as examples. Many of these papers are comparative in nature, showing the results for Tabu search and other techniques (simulated annealing, evolutionary computation, etc.).

MULTIAGENT SYSTEMS

Although a simplification, the premise of multiagent systems (MAS) is to effect a behavior through the interaction of distributed autonomous "agents." As an example of real-world agents, consider the design process of an automobile. Because no single person is an expert in all aspects of automobile design, a distributed approach is taken in which experts in certain aspects collaborate with one another to accomplish the design task. Such a distributed approach provides "modularity (which reduces complexity), speed (due to parallelism), reliability (due to redundancy) and flexibility (i.e., new tasks are composed more easily from the more modular organization)" (Nwana and Ndumu, 1997).

The definition of agents varies greatly, but they often have the following characteristics: (i) an autonomous behavior (i.e., agents do not need constant instruction) (ii) the ability to interact with other agents, (iii) the ability to act and react to an environment such as stored data, and (iv) the capability to learn new behaviors. Thus, the design of a MAS essentially is to specify agent behavior such that the desired "global" behavior emerges as a result of agent interaction in the environment. Obviously, this is a difficult task and is suitable for global search techniques. One example of a global search technique for MAS design (using evolutionary computation in this instance) is presented in Haynes, Wainwright, and Sen (1995).

SYNTHESIS EXAMPLES USING MULTIAGENT SYSTEMS

An introduction to MASs can be found in Ferber (1999). A tutorial on such systems based on social insects is presented in Bonabeau, Dorigo, and Theraulaz (1999).

Design synthesis problems requiring combinatorial optimization have been attempted by using MASs, including the QAP of assembling useful electromechanical devices (Iffenecker and Ferber, 1992; Campbell, Cagan, and Kotovsky, 1999). Multiagent systems, as a result of their interactive abilities, are often endowed with conflict resolution mechanisms allowing them to accommodate constrained combinatorial problems. Some examples are flow shop control and scheduling (Morley, 1993; Balasubramanian and Norrie, 1995; Daouas, Ghedira, and Muler, 1995; Fischer, Mueller, and Pischel, 1995). Expert systems have been designed in Itoh, Watanabe, and Yamaguchi (1995); electromechanical systems have been synthesized in Campbell et al. (1999; 2000), and general combinatorial problems have been addressed in de Souza and Talukdar (1995).

A great deal of the design work using MASs is based on social insects, in particular ants, that are often referred to as *ant systems* (ASs). These systems have been used predominantly for network design and routing, as in Caro and Dorigo (1998) and Schoonderwoerd et al. (1996). These problems are similar to the TSP (i.e., finding shortest paths, least congestion) that can be transformed in polynomial time to QAPs such as VLSI layout and jobshop scheduling. This transformation approach has been used with ACSs to address jobshop scheduling in Colorni et al. (1994).

TRUSS DESIGN USING GENETIC ALGORITHMS

As mentioned in the introduction to this chapter, one example of engineering design synthesis using evolutionary computation that has received considerable attention is the automated design of mechanical structures (Goldberg, 1987; Bensøe and Kikuchi, 1988; Reddy and Cagan, 1995a, 1995b; Chapman and Jakiela, 1996; Dadu and Jakiela, 1997; Shea et al., 1997; Shea and Cagan, 1997; Jakiela et al., 2000). The approach to this problem presented in Chapman et al. (1994) is summarized here.

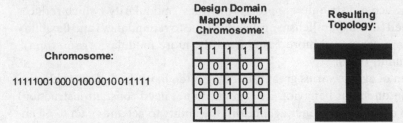

Figure 8.6. Mapping chromosome into design domain (Chapman et al., 1994, Figure 3). © 1994 ASME. Figure reprinted with permission.

The problem is to design the 2-D cantilevered plate with the highest ratio of stiffness to weight. The design space (usually a rectangular area, as illustrated in Figure 8.8) is discretized into binary square elements where the value of each bit determines whether or not its corresponding element is empty or full of material. This representation lends itself to a traditional genetic algorithm approach where solutions are encoded as binary strings. Every chromosome is a binary string with as many bits as the number of discretized elements in the design domain. The chromosome is mapped to the design topology by simply filling in the discretized domain from left to right and top to bottom with each successive bit in the chromosome. Figure 8.6 depicts one such mapping.

Although a chromosome can always be mapped to a topology, it is possible, and quite likely, that a randomly created chromosome will result in an unstable topology. For example, some element may be unconnected to any support, or "seed," elements except at the corners. Because these corner-attached elements cannot support torques, certain loads cannot be supported. To remedy this problem, all disconnected elements, which have no edge neighbors, are changed to empty elements. Note that this change is only to the design domain element and not to the chromosome, which remains unchanged. Figure 8.7 shows examples of connected and disconnected elements.

The fitness function is the stiffness to weight ratio of the 2-D cantilevered beam. The stiffness is taken as the inverse of beam deflection that is due to an applied load (applied at the same point for all topologies). Beam deflection is calculated by transforming each design domain element into four triangular finite elements and performing a finite-element analysis on the resulting structure.

The remainder of the GA implementation utilizes a random initialization of the population, single-point crossover, bit-flip mutations, and an "elitist" stochastic universal sampling selection strategy. Crossover probability is chosen as 0.95, mutation probability as 0.01, and population size as 30.

Figure 8.7. Elements that are (a) connected and (b) disconnected (Chapman et al., 1994, Figure 4). © 1994 ASME. Figure reprinted with permission.

(a)   (b)

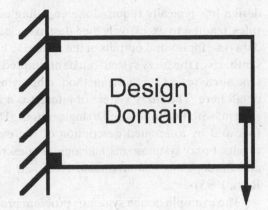

Figure 8.8. Design domain (Chapman et al., 1994, Figure 5). © 1994 ASME. Figure reprinted with permission.

The specific problem addressed is the design of a cantilevered plate subject to a vertical load as shown in Figure 8.8. As reported in Chapman et al. (1994), results of the GA for domain discretizations of 10×16, 15×24, and 20×32 grids are shown in Figure 8.9. The 10×16 design ran 225 generations, whereas the other two designs required 600 generations. Thus, the 20×32 discretization searched roughly 18,000 different solutions (600 generations $\times$ 30 topologies per generation), which is considerably smaller than the $2^{20 \times 32}$ total number of topologies required for exhaustive search.

A DETAILED DESIGN SYNTHESIS EXAMPLE

Recently, fuzzy systems have been used in a variety of applications, most notably in robust nonlinear control. Although fuzzy systems have had much success, their

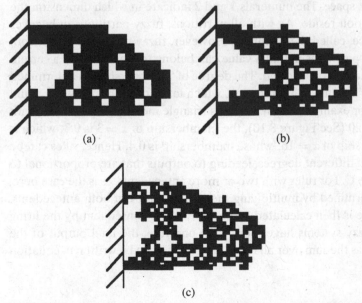

Figure 8.9. Results of (a) 10×16, (b) 15×24, and (c) 20×32 discretizations (Chapman et al., 1994, Figure 6). © 1994 ASME. Figure reprinted with permission.

design has typically required the encoding of expert knowledge as a set of if/then rules, leading to excessively long design cycles. However, if a set of suitable training data (i.e., inputs and outputs of the system to be controlled) is available, then a direct synthesis of the fuzzy system can be attempted by using an adaptive search technique. One such design synthesis method, employing evolution strategies, is presented in detail here. The fuzzy system of interest is a zero-order Takagi–Sugeno (TS) fuzzy system, with two inputs and a single output. TS fuzzy systems are briefly introduced, followed by a detailed description of the evolution strategy implementation and results. Fuzzy systems, sets, and logic are described in more detail elsewhere (Dubois and Prade, 1980; Zimmermann, 1985; Klir and Folger, 1998; Klir and Yuan, 1995; Ross, 1995).

The example design synthesis problem presented here is to synthesize a TS fuzzy system to approximate an unknown function based on a small number of data points. Basically the problem is to create (or synthesize) a set of rules to best approximate the unknown function. Neither the number of data points nor the number of rules are known in advance. This example is analogous to an engineering configuration synthesis, in which establishing an appropriate number of rules is analogous to creating a topology or configuration of subsystems, and setting the parameter values for each rule is analogous to parameter design for an engineering configuration.

TAKAGI–SUGENO FUZZY SYSTEMS

The zero order, two input, single output, TS fuzzy system to be synthesized has rules, R_i, of this form: *if x_1 is $A_{1,i}$ and x_2 is $A_{2,i}$, then $y_i = C_i$.* The inputs, x, are real numbers, as is the output, y, which is equal to a constant, C, if the rule antecedent is satisfied. Here i denotes the rule number and the As are so-called fuzzy partitions of the input space. The numerals 1 and 2 indicate in which dimension the input and fuzzy partition reside. As with all partitions, fuzzy partitions only cover part of the input space, called their *support*. However, fuzzy partitions differ from conventional, "crisp" partitions in that a value can belong to a partition to a certain degree, rather than the binary *yes* or *no*. The degree of "membership" is determined by the fuzzy partition's membership function, which takes on values between 0 and 1 over the support. For example, for an isosceles triangle membership function with a support of $0 \leq x \leq 20$ (See Figure 8.10), the membership of $x = 3$ is 0.3, which is less than the membership of $x = 16$, whose membership is 0.4. Hence, rules can be applied (or fired) with different degrees, leading to outputs that are proportional to C rather than equal to C. For rules with two or more antecedents, as is the case here, firing strength is determined by multiplying the memberships of both antecedents. The output of the rule is then calculated by multiplying its consequent by the firing strength. Because fuzzy systems have more than one rule, the final output of the fuzzy system is taken as the sum over all rule outputs. This can be written in equation form as

$$y = \sum_{i=1}^{N} \mu_{1,i} \mu_{2,i} C_i,$$

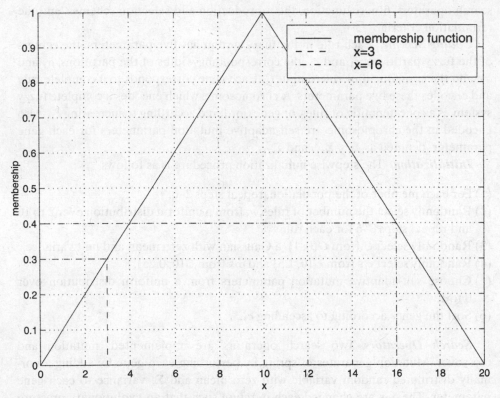

Figure 8.10. Example fuzzy partition and membership function.

where i denotes the rule number, N is the number of rules, $\mu_{j,i}$ is the membership, or firing strength, of input's jth dimension to the ith rule's jth dimension partition, and C_i is the ith rule's consequent.

EVOLUTION STRATEGY IMPLEMENTATION

Here, the fuzzy synthesis problem is stated as finding the optimal number of rules and the corresponding antecedents and consequents such that each rule has its own unique set of antecedents, or partitions, and consequents. The fuzzy system abides by the descriptions given above, with multiplicative rule firing and summation of all rule outputs. However, Gaussian rather than triangular membership functions are used. In this case, the firing strengths, or membership, of an input to the ith rule is written as

$$\mu_i = \mu_{1,i}\mu_{2,i} = \text{strength} = e^{\frac{-(x_1 - c_{1,i})^2}{\sigma_{1,i}^2}} e^{\frac{-(x_2 - c_{2,i})^2}{\sigma_{2,i}^2}},$$

where $c_{j,i}$ is the center of the Gaussian and $\sigma_{j,i}$ is its variance. It should be noted that this type of fuzzy system is equivalent to a radial basis function network (Jang and Sun, 1993).

The evolution strategy implementation for synthesizing such a fuzzy system is detailed in the following by discussing the coding scheme, initialization procedure, search operators, fitness function, the reproduction and selection schemes, and the termination criteria.

Coding Scheme. Each rule in the fuzzy system has five parameters: the centers of the fuzzy partitions, c_1 and c_2, the corresponding widths of the partitions, σ_1 and σ_2, and the consequent or rule output C. Each gene corresponds to a complete rule and encodes these five parameters. A chromosome, which encodes a complete fuzzy system, is a concatenation of rules, or genes, ordered according to increasing c_1. Also encoded in the chromosome are self-adaptive mutation parameters for each gene parameter, denoted as Σ_{c_j}, Σ_{σ_j}, and Σ_C.

Initialization. The stepwise initialization procedure is as follows.

(1) For each member of the population, repeat steps 2–6.
(2) Randomly select the number of rules, n, from a uniform distribution over 2 to 16 and repeat steps 3–6 for each rule.
(3) Randomly select C from $G(0, 1)$, a Gaussian with zero mean and unit variance.
(4) Randomly select c_is from $G(0, 1.5)$ and σ_is from $G(0, 0.25)$.
(5) Choose self-adaptive mutation parameters from a uniform distribution over 0 to 1.
(6) Sort the genes according to ascending c_1.

Search Operators. Two search operators are implemented: mutation and crossover. Mutation generates offspring by perturbing each gene by adding a normally distributed random variable with zero mean and Σ_i variance to each gene parameter. The Σ_is are changed each iteration such that an evolutionary pressure acts to determine good Σ_is. The mutation operator can be thought of as a discretized gradient-descent operator with self-adaptive step sizes.

Crossover produces offspring by swapping genes between two different parent chromosomes. The swap range is randomly selected as a range of c_1 values, and any genes lying within this range are swapped between parents. This type of crossover allows the search to move to different numbers of rules, because the length, or number of rules, can change after crossover. For example, if the first parent has no genes in the given range and the second parent has one, then after crossover, the first parent has increased by one gene and the second parent has decreased by one.

Fitness Function. The fitness function is the standard mean-squared error used in most function approximation methods. Moreover, no penalization term is implemented to prevent overfitting. Written explicitly, the fitness is

$$\text{fitness} = \frac{1}{N}\sum_{i=1}^{N}(y_i - Y_i)^2$$

where N is the number of training data, y_i is the fuzzy system output for the ith input data, and Y_i is the actual output for the ith input data.

Reproduction, Selection, and Termination. Six offspring are generated per parent chromosome, with two from mutation and the other four from crossover. A $(\mu + \lambda)$ truncation selection strategy is chosen, where μ is the number of parents and λ is the number of offspring. The plus sign indicates that the best μ individuals from the combined pool of parents and offspring are selected for survival to the next generation. Because the parents are included in this competition, population fitness never decreases, a so-called elitist approach. In this implementation, μ is 25 and λ is $6\mu = 150$. Termination occurs when the number of generations exceeds 250, which was chosen empirically.

RESULTS

Results are presented below for two test cases. For both cases, 150 training data are randomly generated from the functions (to be described later) over the intervals $-2.5 \leq x_1, x_2 \leq 2.5$. The first test case is a proof-of-concept test case that determines whether or not the synthesis algorithm can be used to synthesize a two-rule fuzzy system. The two rules are that $c = (-1, -1)$, $\sigma = (0.5, 1)$, $C = 1$ and that $c = (1, 1)$, $\sigma = (1.5, 1)$, $C = -2$. The second test case is the nonlinear function $y = \sin(3x_1) + \cos(2x_2)$.

Two-Rule Fuzzy System. The fuzzy partitions and resultant surface of the chosen two-rule fuzzy system are shown in Figures 8.11 and 8.12. The synthesis algorithm was quite effective in finding the two-rule fuzzy system. For a typical run of the evolution strategy, the fitness convergence is shown in Figure 8.13, the evolved fuzzy partitions in Figure 8.14 (in fact these are the same as in Figure 8.11), the evolved surface in Figure 8.15, and the error surface in Figure 8.16. The final evolved fuzzy system is

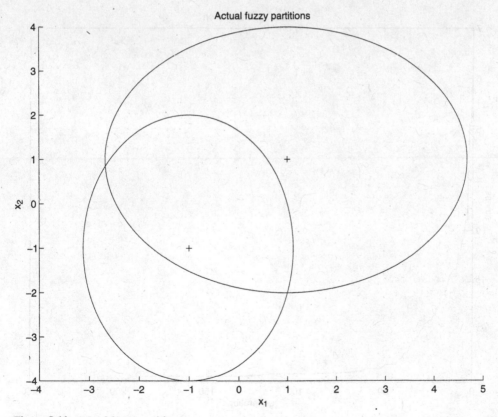

Figure 8.11. Actual fuzzy partitions.

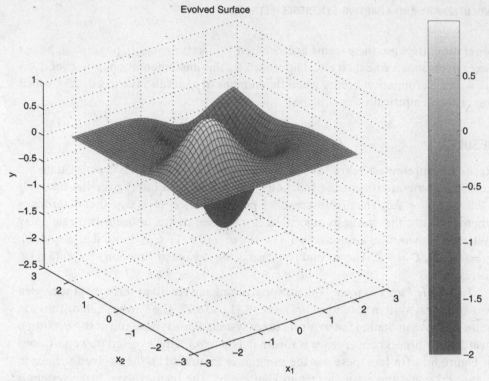

Figure 8.12. Actual surface.

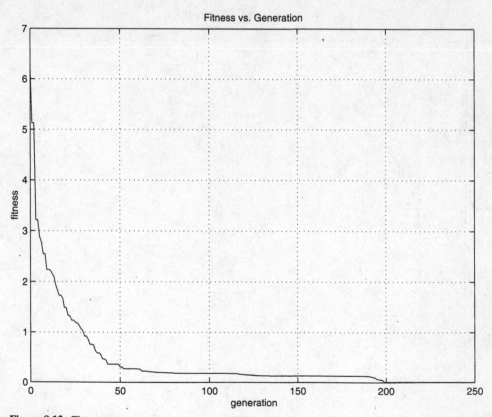

Figure 8.13. Fitness convergence.

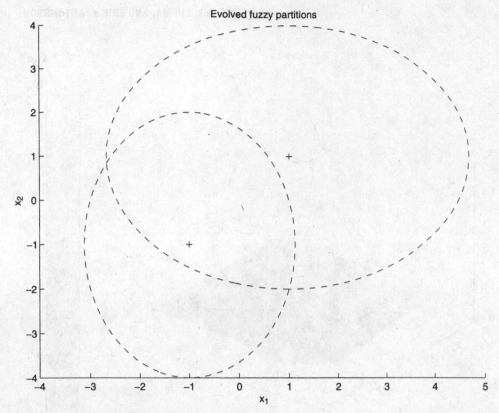

Figure 8.14. Evolved fuzzy partitions.

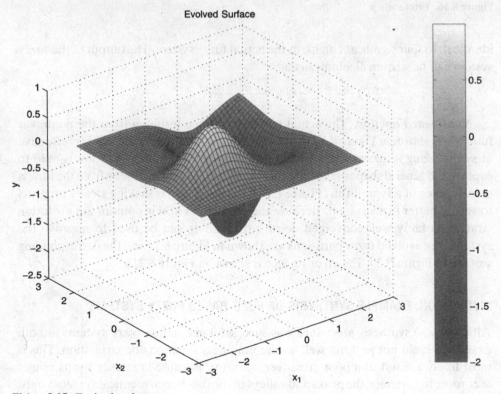

Figure 8.15. Evolved surface.

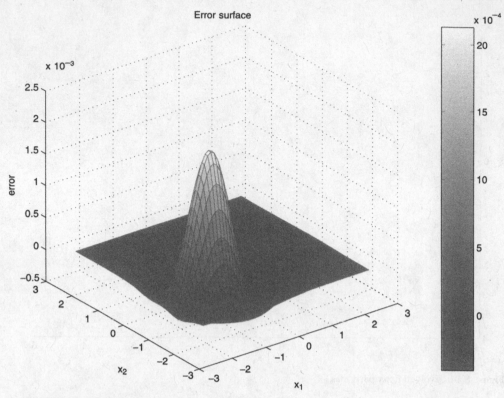

Figure 8.16. Error surface.

identical, to four significant digits, to the actual fuzzy system. The output of the fuzzy system can be written algebraically as

$$y = 1 \times e^{-\frac{(x_1+1)^2}{0.5}} e^{-\frac{(x_2+1)^2}{1}} - 2 \times e^{-\frac{(x_1-1)^2}{1.5}} e^{-\frac{(x_2-1)^2}{1}}$$

Nonlinear Function. The actual response surface resulting from the nonlinear function is shown in Figure 8.17. Although the synthesis algorithm was less effective at synthesizing a fuzzy system with a matching response, the algorithm is able to capture the general shape of the function. Furthermore, as can be seen from the fitness convergence of a typical run (Figure 8.18), the algorithm is making steady progress toward a better fuzzy system. Because the population at this point in the evolution strategy is fairly well converged, local hill-climbing can be used to improve the system. The evolved fuzzy partitions are shown in Figure 8.19 and the corresponding surface in Figure 8.20. The error surface is shown in Figure 8.21.

DISCUSSION: EXAMPLE SYNTHESIS OF RULE-BASED FUZZY SYSTEMS

Although the synthesis procedure was successful in creating fuzzy systems in both examples, it did not perform well on the nonlinear function approximation. This is most likely a result of a poor crossover operator. Because crossover swaps ranges according to x_1 ranges, the peaks and valleys of the function at identical x_1 values must

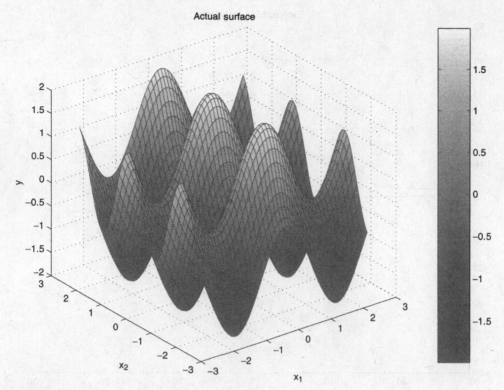

Figure 8.17. Actual response surface.

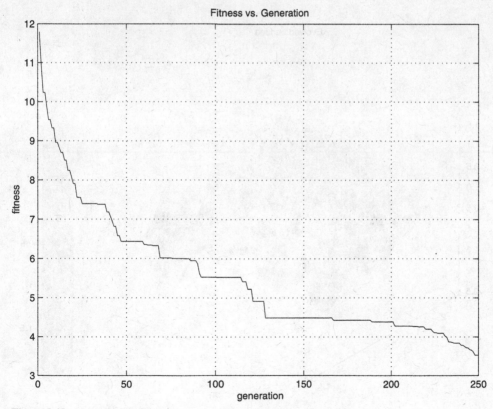

Figure 8.18. Fitness convergence.

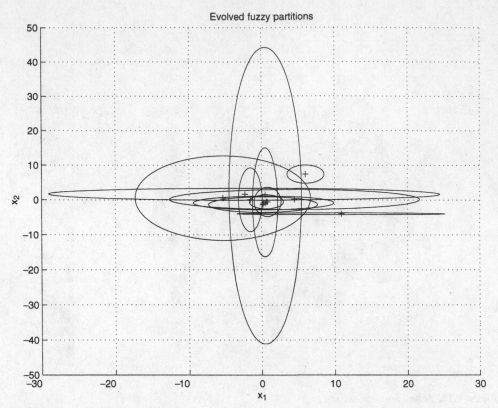

Figure 8.19. Evolved fuzzy partitions.

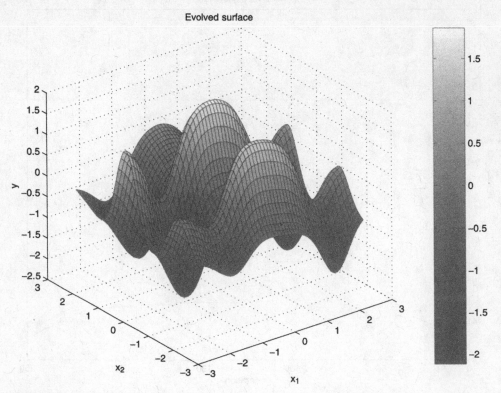

Figure 8.20. Evolved surface.

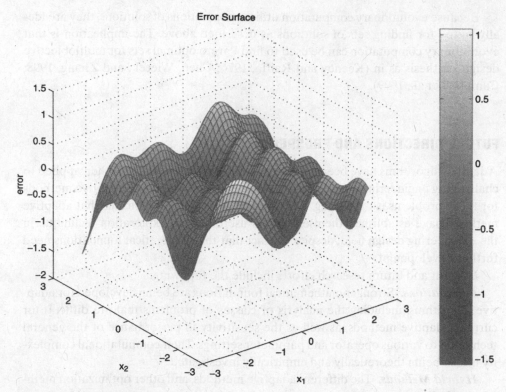

Figure 8.21. Error surface.

evolve simultaneously (as they will almost always be swapped together). Of course, this is highly unlikely to occur, so the results are actually better than expected. A more suitable crossover operator would be to find similar fuzzy rules and to combine them in an averaging manner while keeping other fuzzy rules with some probability, allowing each fuzzy rule to evolve independently, if necessary. This problem is a hallmark of adaptive methods in that search operators have to be tailored to each problem for rapid convergence to the global optimum.

However, the nonlinear example shows that variable dimensionality or complexity searches are amenable to adaptive search. In this case, variable complexity is manifested by the variable number of rules. Traditional search techniques, such as gradient descent, cannot address variable complexity search spaces. Given the fact that a large variety of topological design problems are of variable complexity (i.e., truss design requires determination of the number of truss elements), the utility of adaptive search becomes more apparent.

Furthermore, determining the optimal complexity in a variable complexity exploration is a crucial question. For example, consider a fuzzy system that has as many fuzzy rules as data inputs. Then, if each fuzzy rule has a point support corresponding to one of the data inputs and a consequent equal to the respective output, the approximation error is identically zero. This is obviously an overly complex model that has no real use as it is unable to generalize to unknown data inputs. A simple way to avoid this problem is to find the best performing fuzzy systems for a variety of different complexities or, for engineering examples in which complexity imposes costs, to penalize complex solutions in proportion to their cost.

Because evolutionary computation utilizes populations of solutions, they are ideally suited for finding sets of solutions as described above. The implication is that evolutionary computation can be used to find Pareto optimal sets for multiobjective design synthesis as in (Keeney and Raiffa, 1976; Chen, Wiecek, and Zhang, 1998; Campbell et al., 1999).

FUTURE DIRECTIONS AND PROSPECTS

Adaptive algorithms have been demonstrated to be of practical use when applied to challenging engineering design synthesis problems, and the use of adaptive methods for these problems is expanding. The amount of work itself indicates that adaptive methods have established themselves as a useful design exploration technique in the engineering design field, despite the fact that their theoretical foundations need further development.

Current and future research efforts include the following.

Foundations. Stronger mathematical foundations are being developed for adaptive algorithms, including the analysis of classes of problems that are difficult for current adaptive methods, as well as the sensitivity to performance of the general technique to various operator and parameter settings. Their computational complexity is also being theoretically and empirically investigated.

Hybrid Methods. The different adaptive methods, and other optimization methods are being compared, and the manner in which they can be enhanced by incorporating additional procedures is being examined. Combining adaptive algorithms with more traditional search techniques, or combining multiple adaptive algorithms, has been demonstrated to be able to achieve improved performance. For example, the robustness of EC can be greatly enhanced when it is hybridized with other optimization methods such as local search techniques, simulated annealing, Tabu search, neural networks, or fuzzy systems. The number of papers introducing hybrid systems is growing, indicating that there is a trend toward this direction.

Design Representation. How are solutions in the design space represented such that the different design synthesis methods can efficiently find good design solutions? This in particular has been a significant hurdle for evolutionary approaches that implement crossover. The challenge of developing representations for design problems is particularly difficult, because each design synthesis problem is qualitatively different.

Design Space Size. Engineering design spaces are typically large, with many tens to hundreds of dimensions. Unfortunately, it is well known that as the number of parameters in a design space increases, the effectiveness of adaptive operators decreases. The problem of large design spaces can be addressed by decomposing the exploration into different subspaces (similar to branch and bound or divide and conquer methods) that can be explored in parallel. These distributed searches are ideal for multiagent systems, which, in turn, may consist of various adaptive exploration methods.

Concurrent Computation. EC is being developed for implementation on parallel processing machines. The greatest potential for the application of EAs to real-world problems is likely to occur as a result of their implementation on multi-processor parallel computers.

Applications. Adaptive algorithms are being adapted to a wide range of fields, including engineering problems, computer programming, machine learning, rule-based classifier systems, neural network design, and so on.

Mass Customization. Tailored designs will become more and more a part of everyday life, and they are already appearing in limited forms in some consumer products in a trend called *mass customization*. Adaptive methods can be envisioned that would synthesize multiple design alternatives in response to user input. The user would then interactively choose the more preferable designs to further guide the automated engineering design synthesis.

CONCLUSIONS

Engineering design synthesis, by way of adaptive exploration, has been demonstrated to create a wide range of novel configurations. These methods simulate the natural evolutionary process to explore large spaces of potential design solutions. The character of these spaces can be quite complex, including variable topologies and variable complexities, as well as wide ranges of parameter values. The computational cost of adaptive exploration can be significant; however, the demonstrated successes of adaptive exploration methods in many design synthesis application areas confirm the practicability of this approach and suggest that further research in this area will be fruitful.

APPENDIX A: GENETIC ALGORITHM TERMINOLOGY

In this section, the terminology used in genetic algorithms is described. Because genetic algorithms simulate natural evolution, some terminology used in GA literature is borrowed from natural genetics.

Genetic algorithms work on a *population*, or a collection of candidate solutions to the given problem. Solutions are encoded as *chromosomes*, which are usually strings of characters and symbols, and the individual characters or symbols in the strings are referred to as *genes*. Candidate solutions constitute the *phenotype space*, and chromosomes constitute the *genotype space*. An *encoding* scheme is the mapping function between the phenotype and genotype space. Each chromosome is evaluated by an *evaluation function* (or *fitness function*) to determine the *fitness* value. The evaluation function is usually user defined and problem – specific.

Individuals are selected from the population for reproduction, with the *selection* biased toward fitter individuals. Different *selection schemes* have been developed to ensure that survival of the fittest occurs. The selected individuals form pairs, called *parents*, and a genetic operator called *crossover* is applied to two parents to create two new individuals, called *offspring*. Crossover combines portions of two parents so that the offspring inherit a combination of the features of the parents. For each pair of parents, crossover is usually performed with a high probability P_C, which is called the *crossover probability*. With probability $1 - P_C$, crossover is not performed, and the offspring pair is the same as the parent pair. *Mutation* is another genetic operator applied to parents. Mutation randomly changes the chromosome in some way to introduce new features into a population and make the whole search space reachable. Mutation is usually applied with a low probability, P_M, called *mutation probability*.

A new *generation* is created by selecting individuals from the parents and offspring. The *generation gap* is the fraction of individuals in the population that are replaced from one generation to the next. The generation gap is equal to 1 for *simple GAs* (also referred to as the *total replacement algorithm*) in which all individuals in the parent generation are replaced by the new offspring. In *steady-state GAs* (Syswerda, 1989), at each iteration only one pair of parents is selected to create offspring and then the two offspring replace the worst individual in the current population. Obviously in steady-state GAs, the generation gap is minimal, as only two offspring are produced in each generation.

Figure 8.2 illustrates the terminology. It should be noted that genetic operators (crossover and mutation) work on *genotype* space, whereas evaluation and selection work on *phenotype* space.

Selection pressure is the term used to describe the bias of a selection scheme toward the reproduction of fitter individuals. This pressure is the driving force that determines the rate of convergence of a GA to an optimal solution. Low selection pressure allows the algorithm to explore more freely, whereas high selection pressure focuses the algorithm in the neighborhood of the highly fit individuals leading to exploitation of the most promising search region. Hence, it is imperative to choose a selection scheme with adequate selection pressure such that premature or slow convergence is avoided.

APPENDIX B: SCHEMA

A schema (plural *schemata*) is a fixed template describing a subset of strings with similarities at certain defined positions. Thus, strings that contain the same schema contain, to some degree, similar information. Here only binary alphabets will be considered, allowing templates to be represented by the ternary alphabet {0, 1, #}. The metasymbol # is called the "don't care" element, and within any string the presence of the metasymbol # at a position implies that either a 0 or a 1 could be present at that position. A particular chromosome is said to be an instance of a particular schema, or contain a particular schema if it matches that schema, with the # symbol matching anything. So for example, 101010 and 010010 are both instances of the schema 1##010. Conversely, two examples of schemata that are contained within 101010 are 1010#0 and 1#10##. A schema is a partial solution and represents a set of possible fully specified solutions. A schema with m specified elements and $(n - m)$ #s can be considered to be an $(n - m)$ dimensional hyperplane in the solution space. All points on that hyperplane are instances of the schema.

In a typical binary-coded GA, where the chromosomes are binary strings, each string in the population is an instance of 2^L schemata (or say each string contains 2^L schemata), where L is the length of each individual string. Therefore a population of N chromosomes could contain between 2^L and $N2^L$ possible schemata, because there may be duplications. The total number of different schemata contained in all possible strings is 3^L, because each gene in a schema may be 0, 1, or #. In general, for an alphabet of cardinality (or distinct characters) k, there are $(k+1)^L$ schemata. For a population of N chromosomes, there could be between k^L and Nk^L possible schemata. Thus, although there are only N chromosomes in the population, a much greater number of schemata are processed in parallel, and this property is called *implicit parallelism* (Holland, 1975).

A schema represents a region of the search space and the area of the search space represented by a schema and the location of this area depend on the number and location of the metasymbols within the schema. The regions of the search space represented by schemata such as 1#### are much larger than schemata such as 1110#. Schemata are typically classified by their *defining length* and their *order*. The order o of a schema S is the number of positions within the schema that are not defined by a metasymbol, that is,

$$o(S) = L - m,$$

where m is the number of metasymbols and L is the schema length. In other words, the order is the number of fixed positions within the schema:

$$S = \#1\#0\#, \qquad o(S) = 2.$$

The defining length d specifies the distance between the first and last nonmetasymbol characters within the schema:

$$S = \#1\#0\#, \qquad d(S) = 4 - 2 = 2.$$

In general, low-order schemata cover large regions of space and high-order schemata cover much smaller regions, which is illustrated in Figure B.1. In Figure B.1, binary strings with four bits are used to represent integers in the range [0, 15]. Low-order schema 1### covers half of the space, [8, 15], and high-order schema 01#0 covers a much smaller region (it actually represents only integers 4 and 6).

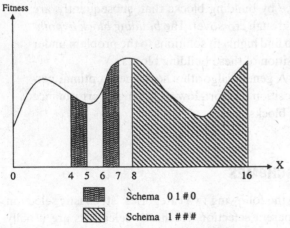

Schema 0 1 # 0

Schema 1 # # #

Figure B.1. Visualization of regions of schemata.

SCHEMA PROCESSING

When genetic operators such as selection, crossover, and mutation are applied to the population, the chromosomes are modified and the distribution of the schemata is changed accordingly. For any particular chromosome (string) within a GA, it can get fragmented by crossover, attacked by mutation, or simply thrown away by the selection operator. Because the selection mechanism favors chromosomes with high fitness, then fitter parents, which are expected to contain some good schemata, will produce more offspring. Therefore, the number of instances of good schemata tends to increase, and the number of instances of bad schemata tends to decrease. The change of the number of instances of a particular schema during a GA run can be roughly estimated and this estimation throws light on how GAs work.

The combined effect of selection, crossover, and mutation gives the so-called *reproductive schema growth equation*.

For a particular schema S, if $\Phi(S, g) > 0$ is the number of instances of S within the population at generation g,

$$\Phi(S, g+1) \geq \frac{f(S, g)}{\bar{f}(g)} \Phi(S, g) \left[1 - P_c \frac{d(S)}{L-1} - o(S) P_m \right],$$

where $f(S, g)$ is the average fitness of all instances of S at generation g, and $\bar{f}(g)$ is the average fitness of generation g. Here P_c and P_m are the crossover probability and mutation probability, and $d(S)$ and o(S) are the defining length and order of S. The equation above indicates the expected number of strings matching a schema S in the next generation as a function of the actual number of strings matching the schema in the current generation, the relative fitness of the schema, and its defining length and order.

The final result of the growth equation can be stated as follows.

> *Schema Theorem*: Short (defining length), low-order, above-average schemata receive exponentially increasing trials in subsequent generations of a genetic algorithm.

The short, low-order, above-average schemata are termed *building blocks* by Goldberg (1989). An immediate result of the schema theorem is that GAs explore the search space by building blocks that, subsequently, are combined into larger blocks through crossover. The *building block hypothesis* states that GAs attempt to find highly fit solutions to the problem under consideration by the juxtaposition of these building blocks:

> *Building Block Hypothesis*: A genetic algorithm seeks near-optimal performance through the juxtaposition of short, low-order, high-performance schemata, called the building blocks.

APPENDIX C: SELECTION SCHEMES

Most selection schemes belong to the following two categories: stochastic selection and deterministic selection. For parent selection, stochastic selections are usually applied, and for generational selection, deterministic selections are usually used.

Fitness proportionate selection and tournament selection are two of the most popular stochastic selection algorithms.

FITNESS PROPORTIONATE SELECTION

Fitness proportionate selection is typically implemented as a probabilistic operator for parent selection, using the relative fitness to determine the selection probability of an individual. In fitness proportionate selection, the selection probability for each chromosome is proportional to its fitness value. For chromosome k with fitness value F_k, its selection probability p_k is calculated as follows:

$$p_k = F_k \bigg/ \sum_{j=1}^{\text{pop\_size}} F_j.$$

For example, roulette wheel selection is a proportionate selection scheme in which the slots of a roulette wheel are sized according to the fitness of each individual in the population. An individual is selected by spinning the roulette wheel and noting the position of the marker. The probability of selecting an individual is therefore proportional to its fitness. Another well-known proportionate selection scheme is stochastic universal selection (Baker, 1985). Stochastic universal selection is a less noisy version of roulette wheel selection in which N equidistant markers are placed around the roulette wheel, where N is the number of individuals in the population. Here N individuals are selected in a single spin of the roulette wheel, and the number of copies of each individual selected is equal to the number of markers inside the corresponding slot.

Proportionate selection methods assign selection probability to an individual according to its fitness, and this can be problematic. Proportionate selection depends upon positive values, and, simply adding a large constant value to the objective function can eliminate any selection pressure, with the algorithm then proceeding as a purely random search. There are several heuristics that have been devised to compensate for these issues. With the use of *fitness scaling*, the fitness of all parents can be scaled relative to some reference value, and proportionate selection then assigns selection probability according to the scaled fitness values. *Rank-based selection* methods utilize the indices of individuals when ordered according to fitness to calculate the corresponding selection probabilities, rather than using absolute fitness values (Baker, 1987). Rank-based selection also eliminates problems with functions that have large offsets. Linear as well as nonlinear fitness mappings have been proposed for fitness scaling and rank-based selection.

Fitness Scaling

If raw fitness values are used for proportionate selection, without any scaling or normalization, then one of two things can happen. On one hand, if the fitness range is too large, then only a few good individuals will be selected. This will tend to fill the entire population with similar chromosomes and will limit the ability of the GA to explore the search space. On the other hand, if the fitness values are too close to each other, then the GA will tend to select one copy of each individual, with only random variations in selection. Consequently, it will not be guided by small fitness variations and will be reduced to random search. In early generations, there is a tendency for a few superchromosomes to dominate the selection process, and in later generations, when the population is largely converged, competition among chromosomes is less strong and a random search behavior will emerge. Fitness scaling is used to scale the raw fitness values so that the GA sees a reasonable amount of difference in the scaled fitness values of the best versus the worst individuals (Mazumder and Rudnick, 1999). Thus, fitness scaling controls the selection pressure or discriminating power of the GA.

Beginning with De Jong (1975), scaling of objective function values has become a widely accepted practice and several scaling mechanisms have been proposed. In general, the scaled fitness F'_k derived from the raw fitness F_k for chromosome k can be expressed as follows:

$$F'_k = G(F_k),$$

where the mapping function $G(\cdot)$ transforms the raw fitness into scaled fitness. The function $G(\cdot)$ may take different forms to yield different scaling methods, such as linear scaling, sigma truncation, power law scaling, and so on. Some of them are illustrated briefly below. For detailed description, see Gen and Cheng (1996).

Linear Scaling. When the mapping function takes the form of a linear transformation, the following linear scaling method applies:

$$F'_k = a \times F_k + b,$$

where parameters a and b are normally selected such that an average chromosome receives one offspring copy on average, and the best receives the specified number of copies (usually two). Linear scaling adjusts the fitness values of all chromosomes in such a way that the best chromosome gets a fixed number of expected offspring and thus prevents it from reproducing too many offspring. This method may give negative fitness values that are usually taken as zero.

Sigma Truncation. For sigma truncation (Goldberg, 1989),

$$F'_k = F_k - (\bar{F} - c \times \sigma),$$

where c is a small user-defined integer called the sigma scaling factor, σ is the standard deviation of the fitness of the population, and $\bar{F}$ is the average raw fitness value. Negative scaled fitnesses F'_k are set to zero.

This scales the fitness such that, if the raw fitness is $\pm k$ standard deviations from the population average, the fitness is

$$F'_k = (c \pm k)\sigma.$$

This means that any individual worse than c standard deviations from the population mean ($k = c$) is not selected at all. The usual value of c reported in the literature is between 1 and 5.

Boltzmann Selection. Boltzmann selection (Michalewicz, 1994) is a nonlinear scaling method for proportionate selection, using the following scaling function:

$$F'_k = e^{(F_k/T)},$$

where T is a user-defined control parameter. The selection pressure can be adjusted by assigning T high or low.

Rank-Based Selection

An alternative way to control the selection process is to use ranking. Individuals are sorted according to their fitness values, and the selection probability of each chromosome is assigned according to its rank instead of its raw fitness. After individuals are sorted in order of raw fitness, the reproductive fitness values are assigned according to rank. Two methods for mapping rank into reproductive fitness values are in common use: linear ranking and exponential ranking.

Let F_k' be the reproductive fitness value for the kth chromosome in the ranking of population; the linear ranking takes the following form:

$$F_k' = q - (k-1) \times \frac{q - q_0}{\text{pop\_size} - 1},$$

where parameter q is the reproductive fitness for the best chromosome, and q_0 is the reproductive fitness for the worst chromosome. Fitness values of intermediate chromosomes are decreased from q to q_0, proportional to their rank. When q_0 is set to 0, it provides the maximum selective pressure.

Michalewicz proposed the following *exponential ranking* method (Michalewicz, 1994):

$$F_k' = q(1-q)^{k-1}.$$

A larger value of q implies stronger selective pressure. Hancock proposed the following exponential ranking method:

$$F_k' = q^{k-1},$$

where q is typically ~ 0.99. The best chromosome has a fitness of 1, and the last one receives $q^{\text{pop\_size}-1}$.

Controlling the variance of the fitness values is one of the frequent problems of GAs. Ranking ensures that the variance is constant throughout the optimization process. Mapping ranks to reproductive fitness for proportionate selection gives a similar result to fitness scaling, in that the ratio of the maximum to average fitness is normalized to a particular value. However, it also ensures that the remapped fitnesses of intermediate individuals are regularly spread out. Because of this, the effect of one or two extreme individuals will be negligible, irrespective of how much greater or less their fitnesses are than the rest of the population.

TOURNAMENT SELECTION

In binary tournament selection (Goldberg and Deb, 1991), two individuals are taken at random, and the better individual is selected from the two. If binary tournament selection is being done without replacement, then the two individuals are set aside for the next selection operation, and they are not replaced into the population. Because two individuals are removed from the population for every individual selected, and the population size remains constant from one generation to the next, the original population is restored after the new population is half-filled. Therefore, the best individual will be selected twice, and the worst individual will not be selected at all. The number of copies selected of any other individual cannot be predicted except that it is either zero, one or two. In binary tournament selection with replacement, the two individuals are immediately replaced into the population for the next selection operation.

Binary tournament selection was generalized to *Tournament selection*, which works by taking a random uniform sample of a certain size $q > 1$ from the population, selecting the best of these q individuals to survive for the next generation. This method has gained increasing popularity because it is easy to implement, computationally efficient, and allows for fine-tuning of selection pressure by increasing or decreasing the tournament size q.

DETERMINISTIC SELECTION

Deterministic selection schemes select individuals according to their fitness values in a deterministic way.

Deterministic selection schemes are usually used in generational selection to select individuals from both the parent generation and offspring generation to create the next generation. An example of deterministic selection is the *generational replacement*, which replaces the entire parent generation by their offspring (i.e., the offspring generation is taken as the new generation, and the parent generation is discarded after the offspring generation is created).

Elitist selection (Eshelman, 1991) is another popular deterministic selection scheme that ensures that the best chromosome is passed onto the new generation if it is not selected through another process of selection. Fitness proportionate selection does not guarantee the selection of any particular individual, including the fittest. Thus with fitness proportionate selection the best solution to the problem discovered so far can be regularly thrown away. Sometimes this is counterproductive. For many applications the search speed can be greatly improved by not losing the best, or *elite*, member between generations. Ensuring the propagation of the elite member is termed *elitism*, and by retaining the best chromosome in the population, elitist selection guarantees asymptotic convergence.

The (μ, λ) evolution strategy (Bäck, 1994) uses a deterministic selection scheme that has been introduced to genetic algorithms. The notation (μ, λ) indicates that μ parents create $\lambda > \mu$ offspring by means of recombination and mutation, and the best μ offspring individuals are deterministically selected to replace the parents. Notice that this mechanism allows that the best member of the population at generation $t + 1$ might perform *worse* than the best individual at generation t (i.e., the method is not *elitist*), thus allowing the strategy to accept temporary deteriorations that might help to leave the region of attraction of a local optimum and reach a better optimum. In contrast, the $(\mu + \lambda)$ evolution strategy (Bäck and Hoffmeister, 1991) selects the μ survivors from the union of parents and offspring, such that a monotonic course of evolution is guaranteed.

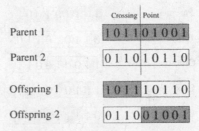

Figure D.1. One-point binary crossover.

APPENDIX D: CROSSOVER

The implementation of crossover depends on the encoding scheme used. Most canonical GAs use binary coding, and therefore the crossover schemes for binary codings (called *binary crossover* here) have been studied most thoroughly. Several real crossover schemes also have been proposed and applied to optimization problems utilizing real-number encoding. These crossover operators are discussed below.

BINARY CROSSOVER

The "traditional" GA uses one-point crossover, in which the two mating chromosomes (binary strings) are each cut once at corresponding points, and the sections after the cuts exchanged. The two offspring each inherit some genes from each parent. See Figure D.1 for details.

Many different binary crossover schemes have been devised, often involving more than one cut point. In two-point crossover (and multiple-point crossover in general), chromosomes are regarded as loops formed by joining the ends together. To exchange a segment from one loop with that from another loop requires the selection of two cut points, as shown in Figure D.2. Obviously, one-point crossover can be viewed as two-point crossover with one of the cut points fixed at the start of the string, and it's fair to say that two-point crossover is more general than one-point crossover, although they both perform the same task (exchanging single segment of two parents). Researchers now agree that two-point crossover is generally bette than one-point crossover. De Jong (1975) investigated the effectiveness of multiple-point

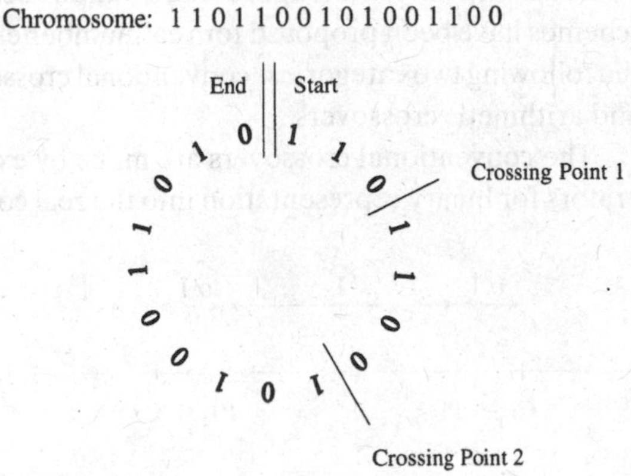

Chromosome: 1 1 0 1 1 0 0 1 0 1 0 0 1 1 0 0

Figure D.2. Two-point binary crossover.

Parent 1   1 0 1 1 0 1 0 0 1

Parent 2   0 1 1 0 1 0 1 1 0

Crossover Mask   1 1 0 0 1 0 0 1 1   **Figure D.3.** Uniform binary crossover.

Offspring 1   1 0 1 0 0 0 1 0 1

Offspring 2   0 1 1 1 1 1 0 1 0

crossover and concluded that two-point crossover gives the best performance, but that adding more crossover points reduces the performance of the GA. The problem with adding additional crossover points is that building blocks are more likely to be disrupted. However, an advantage of having more crossover points is that the solution space may be searched more thoroughly.

Uniform crossover is another binary crossover scheme that is radically different from one-point crossover. In uniform crossover, after a pair of parents are selected, a crossover mask, which is also a binary string with the same length as the chromosomes, is randomly generated. An offspring is created by assigning each gene in the offspring by copying the corresponding gene from one or the other parent, chosen according to the crossover mask. Where there is a 1 in the crossover mask, the gene is copied from the first parent, and where there is a 0 in the mask, the gene is copied from the second parent, as shown in Figure D.3. The process is repeated to produce the second offspring. The offspring therefore contain a mixture of genes from each parent, and the number of effective crossing points is not fixed, but will average $L/2$ where L is the chromosome length.

A comparison of different binary crossover operators, including one-point, two-point, multipoint, and uniform crossover, was undertaken in Eshelman, Caruna, and Schaffer (1989), both theoretically and empirically. It was found that none of them is the consistent winner, and there was not more than 20% difference in speed among the techniques.

REAL CROSSOVER

In real encoding implementations, each chromosome is encoded as a vector of real numbers with the same length as the solution vector. In recent years, several crossover schemes have been proposed for real-number encoding, and most of them belong to the following two categories: conventional crossovers (including simple and random) and arithmetic crossovers.

The conventional crossovers are made by extending the canonical crossover operators for binary representation into the real coding case. The arithmetic crossovers

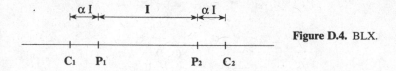

Figure D.4. BLX.

are constructed by borrowing the concept of linear combination of vectors from the area of convex sets theory. These crossover operators are described in detail below.

Simple Crossovers

Simple crossovers (Spears and Jong, 1991; Syswerda, 1989) include one-point, two-point, multipoint, and uniform crossover. These crossover operators are analogous to those of the binary implementation, except that each gene is now a real number rather than a binary bit (0 or 1).

Random Crossovers

Essentially, this kind of crossover creates offspring randomly within a hyperrectangle defined by the parent points. *Flat crossover* is a basic random crossover given by Radcliffe (1990), which produces an offspring by uniformly picking a value for each gene from the range formed by the values of two corresponding parents' genes. A generalized crossover, called *blend crossover* (denoted as BLX-α), is proposed in Eshelman and Schaffer (1991) to introduce more variance into the operator. BLX-α also uniformly picks values from a range, but this range is not formed by the values of the two parents' genes. As illustrated by Figure D.4, P_1 and P_2, the gene values (real number) of two parents, form the interval I, and α is a user-defined real parameter. BLX-α first extends interval I with αI on both sides such that the *crossover interval* is limited by points C_1 and C_2. Then a point is randomly picked from the crossover interval. By extending the interval formed by the parents to include regions beyond the parents, blend crossover introduces more exploration into the search and often increases the robustness of the algorithm. Also, because the area of the extension is proportional to the area of the interval formed by the parents, the amount of exploration introduced is actually determined by the closeness of the parents. As the evolution proceeds, chromosomes in the population become converged, and the parents become similar, and the disruptive exploration introduced by blend crossover becomes less significant. The parameter α controls the amount of disruptive exploration, and it is usually set as 0.5, such that the probability that an offspring lies outside its parents is equal to the probability that it lies between them.

Arithmetic Crossovers

Let x_1 and x_2 be two parent chromosomes (real vectors). Arithmetic crossovers are defined as the combination of two vectors as follows:

$$x_1' = \lambda_1 x_1 + \lambda_2 x_2,$$
$$x_2' = \lambda_1 x_2 + \lambda_2 x_1.$$

By restricting the coefficients λ_1 and λ_2 in different ways, the equations above produce three different kinds of arithmetic crossover (Wright, 1991; Michalewicz, 1994; Gen,

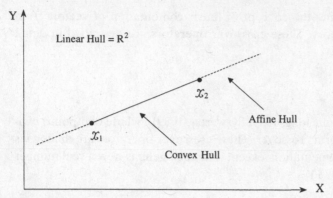

Figure D.5. Arithmetic crossover.

Ida, and Cheng, 1995):

linear crossover λ_1, λ_2 are real,
affine crossover $\lambda_1 + \lambda_2 = 1$,
convex crossover $\lambda_1 + \lambda_2 = 1$, $\lambda_1 > 0, \lambda_2 > 0$.

The names *linear*, *affine*, and *convex* are borrowed from convex set theory, in which a combination of two vectors x_1 and x_2 is called linear, affine, or convex when λ_1 and λ_2 are restricted to be real, $\lambda_1 + \lambda_2 = 1$, or $\lambda_1 + \lambda_2 = 1, \lambda_1 > 0, \lambda_2 > 0$. The convex crossover appears to be the most commonly used. When the restriction that $\lambda_1 = \lambda_2 = 0.5$ is applied, the special case of the averaging crossover is produced (Davis, 1991).

A geometric explanation of arithmetic operators for a two-dimensional case is shown in Figure D.5. Here x_1 and x_2 are two parent vectors. The offspring generated with convex crossover constitute the so-called convex hull. Similarly, the offspring generated with affine crossover constitute the affine hull and the offspring generated with linear crossover constitute the linear hull. In the two-dimensional case (Figure D.5), the solid line connecting the two parents is the convex hull, and the solid and dashed lines are the affine hull and the linear hull is the whole space.

APPENDIX : MUTATION

In a binary-coded GA, consider a particular bit position, say bit 10, which has the same value (0) for all chromosomes in the population. In such a case, crossover alone cannot explore new solutions because it does not generate a new gene value; that is, crossover cannot create a chromosome with a value of 1 for bit 10, as its value is 0 in all parents. If a value of 0 for bit 10 turns out to be suboptimal, then, without mutation, the algorithm will have no chance of finding the best solution. However, if mutation is applied, there will be a probability that value 1 will be introduced in bit 10 of some chromosomes, and if this results in an improvement in fitness, those chromosomes will be multiplied by the selection algorithm and the value 1 at bit 10 will be inherited by other offspring through crossover. Although the crossover operator is the most efficient and important genetic operator, by itself it does not guarantee the reachability of the entire search space with a finite population size. It is mutation that makes the entire search space reachable.

In real encoding, the basic mutation operator, called *uniform mutation*, simply replaces a gene (real number) with a randomly selected real number within a specified range. Other real mutation schemes have been developed. For example, *boundary mutation* replaces a gene with either the lower bound or the upper bound (Michalewicz et al., 1994). Gen, Liu, and Ida (1994) present a direction-based mutation,

$$x' = x + r \cdot d,$$

where r is a randomly generated nonnegative real number, and d is the approximate gradient direction vector of the objective function f with its kth component defined as

$$d_k = \frac{f(x_1, \ldots, x_k + \Delta x_i, \ldots, x_n) - f(x_1, \ldots, x_k, \ldots, x_n)}{\Delta x_k}$$

More details about implementations of real coded mutation operators can be found in Gen and Cheng (1996).

The mutation probability P_M is defined as the probability of mutating each gene. It controls the rate at which new gene values are introduced into the population. In general, mutation is implemented as a background operator and is applied with low probability. If the mutation probability is too high, too many random perturbations will occur, and the offspring will lose their resemblance to their parents. The ability of the algorithm to learn from the history of the search will then be lost.

REFERENCES

Andreatta, A. and Ribeiro, C. (1994). "A graph partitioning heuristic for the parallel pseudo-exhaustive logical test of VLSI combinational circuits," *Annals of Operations Research Speed Computing*, **50**:1–36.

Bäck, T. (1994). "Selective pressure in evolutionary algorithms: a characterization of selection mechanisms." In *Proceedings of the First IEEE Conference on Evolutionary*, D. Fogel (ed.), IEEE, New York, pp. 57–62.

Bäck, T., Hammel, U., and Schwefel, H.-P. (1997). "Evolutionary computation: comments on the history and current state," *IEEE Transactions on Evolutionary Computation*, **1**(1):3–15.

Bäck, T. and Hoffmeister, F. (1991). "Extended selection mechanism in genetic algorithms." In *Proceedings of the Fourth International Conference on Genetic Algorithms*, Morgan Kaufmann Publishers, San Mateo, CA, pp. 92–99.

Baker, J. E. (1985). "Adaptive selection methods for genetic algorithm." In *Proceedings of the International Conference on Genetic Algorithms and Their Applications*, J. Grefenstetle (ed.), Lawrence Erlbaum Associates, Hillsdale, NJ, pp. 101–111.

Baker, J. E. (1987). "Reducing bias and inefficiency in the selection algorithm." In *The Second International Conference on Genetic Algorithms*, Lawrence Erlbaum Associates, Hillsdale, NJ.

Balasubramanian, S. and Norrie, D. (1995). "A multi-agent intelligent design system integrating manufacturing and shop flow control." In *Proceedings of the First International Conference on Multiagent Systems*, MIT Press, Cambridge, MA.

Banzhaf, W., Nordin, P., Keller, R., and Francone, F. (1998). *Genetic Programming: An Introduction*. Morgan Kaufmann, New York.

Battiti, R. and Tecchiolli, G. (1994). "The reactive Tabu search," *ORSA Journal on Computing*, **6**(2):126–140.

Battiti, R. and Tecchiolli, G. (1995). "Training neural nets with the reactive Tabu search," *IEEE Transactions on Neural Networks*, **6**(5):1185–1200.

Bensøe, M. P. and Kikuchi, N. (1988). "Generating optimal topologies in structural design using a homogenization method," *Computer Methods in Applied Mechanics and Engineering*, **71**:197–224.

Bilbro, G. (1991). "Efficient generators in simulated annealing," Technical Report TR-91/12, North Carolina State University, Raleigh, NC.

Bonabeau, E., Dorigo, M., and Theraulaz, G. (1999). *Swarm Intelligence – From Natural to Artificial Systems*. Oxford University Press, New York.

Brady, M. (1984). "A wire-wrap design aid, written in PROLOG," *Computer-Aided Design*, **16**(5):253–263.

Brotherton, T. and Chadderdon, G. (1998). "Automated rule extraction for engine health monitoring." In *Proceedings of the 7th International Evolutionary Programming Conference*, Springer-Verlag, Heidelberg, Germany.

Burke, E., DeCausmaecker, P., and VandenBerghe, G. (1998). "A hybrid Tabu search algorithm for the nurse rostering problem." In *Proceedings of the Second Asia-Pacific Conference on Simulated Evolution and Learning*, Springer-Verlag, Heidelberg, Germany.

Cagan, J., Clark, R., Dastidar, P., Szykman, S., and Weisser, P. (1996). "HVAC CAD layout tools: a case study of university/industry collaboration." In *Proceedings of the 1996 ASME Design Engineering Technical Conferences and Computers in Engineering Conference: Design Theory and Methodology Conference*, ASME, New York, 96-DETC/DTM-1505.

Cagan, J., Degentesh, D., and Yin, S. (1998). "A simulated annealing-based algorithm using hierarchical models for general three-dimensional component layout," *Computer-Aided Design*, **30**(10):781–790.

Cai, L., Wu, Q., and Yong, Z. (2000). "A genetic algorithm with local search for solving jobshop problems." In *Proceedings of EvoWorkshops 2000*, Springer-Verlag, Heidelberg.

Campbell, M., Amon, C., and Cagan, J. (1997). "Optimal three-dimensional placement of heat generating electronic components," *Journal of Electronic Packaging*, **119**(2):106–113.

Campbell, M., Cagan, J., and Kotovsky, K. (1999a). "A-design: an agent-based approach to conceptual design in a dynamic environment," *Research in Engineering Design*, **11**(3):172–192.

Campbell, M., Cagan, J., and Kotovsky, K. (2000). "Agent-based synthesis of electromechanical design configurations," *ASME Journal of Mechanical Design*, **122**(1):61–69.

Campbell, M. I. (2000). "The A-Design invention machine: a means of automating and investigating conceptual design," Ph.D. Dissertation, Carnegie Mellon University, Pittsburgh, PA.

Caro, G. D. and Dorigo, M. (1998). "Antnet: distributed stigmergetic control for communications networks," *Journal of Artificial Intelligence Research*, **9**:317–365.

Cartwright, H. and Tuson, A. (1994). "Genetic algorithms and flowshop scheduling: towards the development of a real-time process control system," *Lecture Notes in Computer Science*, **865**:277–290. (AISB Workshop: Evolutionary Comput.).

Cartwright, H. M. and Hopkins, J. A. (1996). "Evolutionary design of synthetic routes in chemistry." In *Evolutionary Computation, AISB Workshop*. Heidelberg, T. C. Fogarty (ed.), Springer-Verlag, pp. 23–380.

Chakrpani, J. and Skorin-Kapov, J. (1995). *Mapping Tasks to Processors to Minimize Communication Time in a Multiprocessor System*, Kluwer Academic, Boston, MA.

Chapman, C. and Jakiela, M. (1996). "Genetic algorithm-based structural topology design with compliance and topology simplification considerations," *ASME Journal of Mechanical Design*, **118**(1):89–98.

Chapman, C., Saitou, K., and Jakiela, M. (1994). "Genetic algorithms as an approach to configuration and topology design," *ASME Journal of Mechanical Design*, **116**(4):1005–1012.

Chatterje, A. and Hartley, R. (1990). "A new simultaneous circuit partitioning and chip placement approach based on simulated annealing." In *Proceedings of the Design Automation Conference*, ASME, New York, pp. 36–39.

Chen, W., Wiecek, M., and Zhang, J. (1998). "Quality utility – a compromise programming approach to robust design." In *Proceedings of DETC98: 1998 ASME Design Engineering Technical Conference*, ASME, New York.

Chiang, W.-C., and Kouvelis, P. (1994). "Simulated annealing and Tabu search approaches for unidirectional flowpath design for automated guided vehicle systems," *Annals of Operations Research*, **50**.

Colorni, A., Dorigo, M., Maniezzo, V., and Trubian, M. (1994). "Ant system for jobshop scheduling," *Belgian Journal of Operations Research, Statistics and Computer Science*, **34**:39–53.

Corne, D. and Ross, P. (1997). "Practical issues and recent advances in job- and open-shop scheduling." In *Evolutionary Algorithms in Engineering Applications*, D. Dasgupta and Z. Michalewicz (eds.), Springer-Verlag, Berlin, pp. 531–546.

Corne, D., Ross, P., and Fang, H.-L. (1994). "Fast practical evolutionary timetabling," *Lecture Notes in Computer Science*, **865**:250–263 (AISB Workshop: Evolutionary Computing).

Daouas, T., Ghedira, K., and Muller, J.-P. (1995). "Distributed flow shop scheduling problem: global versus local optimization." In *Proceedings of the First International Conference on Multiagent Systems*, MIT Press, Cambridge.

Davis, L. (1991). *Handbook of Genetic Algorithms*. Van Nostrand Reinhold, New York.

De Jong, K. A. (1975). "An analysis of the behavior of a class of genetic adaptive systems," Ph.D. Dissertation, The University of Michigan, Ann Arbor.

de Souza, P. S. and Talukdar, S. (1995). "Asynchronous teams: autonomous agents organizations for solving combinatorial problems." In *Proceedings of the First International Conference on Multiagent Systems*, MIT Press, Cambridge.

DeSanctis, M., Benjamin, R., and Taylor, W. (1988). "Node placement algorithms to display communications topology to network controllers." In *Theoretical Foundations of Computer Graphics and CAD*, R. A. Earnshaw (ed.), Vol. F40 of NATO ASI, Springer-Verlag, New York, pp. 599–616.

Dighe, R. and Jakiela, M. J. (1995). "Solving pattern nesting problems with genetic algorithms employing task decomposition and contact detection," *Evolutionary Computation*, **3**(3):239–266.

Dimopoulos, C. and Zalzala, A. M. S. (2000). "Recent development in evolutionary computation for manufacturing optimization: problems, solutions, and comparisons," *IEEE Transactions on Evolutionary Computation*, **4**(2):93–113.

Domschke, D., Forst, P., and Voss, S. (1992). "Tabu search techniques for the quadratic semi-assignment problem." In *New Directions for Operations Research in Manufacturing*, G. Fandel, T. Gulledge, and A. Jones (eds.), Springer, Berlin.

Drechsler, R. (1998). *Evolutionary Algorithms for VLSI CAD*. Kluwer Academic, Boston, MA.

Dubois, D. and Prade, H. (1980). *Fuzzy Sets and Systems: Theory and Applications*. Academic, New York.

Duda, J. and Jakiela, M. (1997). "Generation and classification of structural topologies with genetic algorithm speciation," *ASME Journal of Mechanical Design*, **119**(1):127–131.

Eshelman, L. J. (1991). "The CHC adaptive search algorithm: how to have safe search when engaging in nontraditional genetic recombinition." In *Foundation of Genetic Algorithms*, Morgan Kaufmann, San Mateo, CA, pp. 265–283.

Eshelman, L. J., Caruna, R., and Schaffer, J. D. (1989). "Biases in the crossover landscape," in *The Third International Conference on Genetic Algorithms*, Lawrence Erlbaum Associates, Hillsdale, NJ, pp. 10–19.

Eshelman, L. J. and Schaffer, J. (1991). "Real-coded genetic algorithms and interval-schemata." In *Foundation of Genetic Algorithms*, Morgan Kaufmann, San Mateo, CA, pp. 187–202.

Fang, H.-L., Corne, D., and Ross, P. (1996). "A genetic algorithm for job-shop problems with various schedule quality criteria." In *Evolutionary Computation, AISB Workshop*, T. C. Fogarty (ed.), Springer-Verlag, Berlin, pp. 39–49.

Fanni, A., Giacinto, G., and Marchesi, M. (1996). "Tabu search for continous optimization of electromagnetic structures." In *International Workshop on Optimization and Inverse Problems in Electromagnetism*, Brno, Czech Republic.

Ferber, J. (1999). *Multi-Agent Systems: An Introduction to Distributed Artificial Intelligence*, Addison-Wesley, Reading, MA.

Figueira Pujol, J. C. and Poli, R. (1998). "Efficient evolution of asymmetric recurrent neural networks using a PDGP-inspired two-dimensional representation." In *Proceedings of the First European Workshop on Genetic Programming*, Springer-Verlag, Berlin.

Fischer, K., Mueller, J., and Pischel, M. (1995). "A model for cooperative transportation scheduling." In *Proceedings of the First International Conference on Multiagent Systems*, MIT Press, Cambridge.

Fogarty, T. (1994). "Co-evolving co-operative populations of rules in learning control systems," *Lecture Notes in Computer Science*, **865**:195–209 (AISB Workshop: Evolutionary Computing).

Fogel, D., Ed. (1998). *Evolutionary Computation: The Fossil Record*. IEEE, New York.

Fogel, D. and Simpson, P. (1993). "Evolving fuzzy clusters." In *Proceedings of the International Conference on Neural Networks*, IEEE, New York.

Fogel, D. B. (1994). "An introduction to simulated evolutionary optimization," *IEEE Transaction on Neural Networks*, **5**(1):3–14.

Fogel, D. B., Fogel, L. J., and Porto, V. W. (1990). "Evolving neural networks," *Biological Cybernetics*, **63**:487–493.

Fogel, L. J. (1962). "Antonomous automata," *Industrial Research*, **4**:14–19.

Fogel, L. J. (1964). "On the organization of intellect," Ph.D. Dissertation, University of California, Los Angeles.

Garcia, S., Gonzalez, F., and Sanchez, L. (1999). "Evolving fuzzy rule based classifiers with GA-P: a grammatical approach." In *Proceedings of the Second European Workshop on Genetic Programming (EuroGP'99)*, Springer-Verlag, Berlin.

Geman, S. and Geman, D. (1984). "Stochastic relaxation, Gibbs distribution, and the Bayesian restoration in images," *IEEE Transactions on Pattern Analysis and Machine Intelligence*, **6**(6):721–741.

Gen, M. and Cheng, R. (1996). *Genetic Algorithms and Engineering Design*. Wiley, New York.

Gen, M., Ida, K., Cheng, R. (1995). "Multirow machine layout problem in fuzzy environment using genetic algorithms, computers & industrial engineering," *Engineering Design and Automation*, 29:519–523, Sept. 1995.

Gen, M., Liu, B., and Ida, K. (1994). "Evolution program for constrained nonlinear optimization." In *Proceedings of the 16th International Conference on Computers and Industrial Engineering*, Springer-Verlag, Berlin.

Gendreau, M., Soriano, P., and Salvail, L. (1995). "Metaheuristics for the vehicle routing problem." In *Local Search Algorithms*, J. Lenstra and E. Aarts (eds.), Wiley, Chichester.

Gero, J. S., Kazakov, V. A., and Schnier, T. (1997). "Genetic engineering and design problems." In *Evolutionary Algorithms in Engineering Applications*, D. Dasgupta and Z. Michalewicz (eds.), Springer-Verlag, Berlin. pp. 47–68.

Glover, F. (1986). "Future paths for integer programming and links to artificial intelligence," *Computers and Operations Research*, **13**:533–549.

Glover, F. and Laguna, M. (1997). *Tabu Search*. Kluwer Academic, Boston, MA.

Goldberg, D. E. (1987). "Simple genetic algorithms and the minimal deceptive problem." In *Genetic Algorithms and Simulated Annealing*, Morgan Kaufmann, San Mateo, CA, pp. 74–88.

Goldberg, D. E. (1989). *Genetic Algorithms in Search, Optimization, and Machine Learning*. Addison-Wesley, Reading, MA.

Goldberg, D. E. (1990). "Real-coded genetic algorithms, virtual alphabets, and blocking," IlliGAL Report 90001, University of Illinios at Urbana-Champaign.

Goldberg, D. E., and Deb, K. (1991). "A comparative analysis of selection schemes used in genetic algorithms." In *Foundations of Genetic Algorithms*, Morgan Kaufmann, San Mateo, CA, pp. 69–93.

Goldberg, D. E., Korb, B., and Deb, K. (1989). "Messy genetic algorithms: motivation, analysis, and first results," *Complex Systems*, **3**(5):493–530.

Gupta, R. and Jakiela, M. (1994). "Simulation and shape synthesis of kinematic pairs via small-scale interference detection," *Research in Engineering Design*, **6**(2):103–123.

Hancock, P. J. B. (1990). "GANNET: design of a neural net for face recognition by genetic algorithm." In *Proceedings of IEEE Workshop on Genetic Algorithms, Neural Networks, and Simulated Annealing Applied to Problems in Signal and Image Processing*, IEEE, New York.

Harik, G. R. (1997). *Learning Linkage to Efficiently Solve Problems of Bounded Difficulty Using Genetic Algorithms*. Ph.D. Thesis, The University of Michigan.

Haynes, T., Wainwright, R., and Sen, S. (1995). "Evolving cooperation strategies." In *Proceedings of the First International Conference on Multiagent Systems*, MIT Press, Cambridge, MA.

Hewitt, C. (1977). "Viewing control structures as patterns of message passing," *Artificial Intelligence*, **8**(3):323–374.

Holland, J. H. (1962). "Outline for a logical theory of adaptive systems," *Journal of the Association for Computing Machinery*, **3**:297–314.

Holland, J. H. (1975). *Adaptation in Natural and Artificial Systems*. The University of Michigan Press, Ann Arbor.

Huang, M., Romeo, F., and Sangiovanni-Vincentelli, A. (1986). "An efficient general cooling schedule for simulated annealing." In *ICCAD-86: IEEE International Conference on Computer Aided Design–Digest of Technical Papers*, IEEE, New York, pp. 381–384.

Husbands, P., Jermy, G., McIlhagga, M., and Ives, R. (1996). "Two applications of genetic algorithms to component design." In *Evolutionary Computation, AISB Workshop*, T. C. Fogarty (ed.), Springer-Verlag, Berlin, pp. 50–61.

Hustin, S., and Sangiovanni-Vincentelli, A. (1987). "TIM, a new standard cell placement program based on the simulated annealing algorithm." In *IEEE Physical Design Workshop on Placement and Floorplanning*, IEEE, New York.

Iffenecker, C. and Ferber, J. (1992). "Using multi-agent architecture for designing electromechanical products." In *Proceedings of Avignon '92 Conference on Expert Systems and their Application*, Kluwer, Boston/London/Dordrecht.

Ingber, L. (1992). "Generic messoscopic neural networks based on statistical mechanics of neocortical interactions," *Physical Review A*, **45**(4):R2183–R2186.

Ingber, L. (1993). "Simulated annealing: practice versus theory," *Journal of Mathematical and Computational Modelling*, **18**(11):29–57.

Itoh, T., Watanabe, T., and Yamaguchi, T. (1995). "Self organizational approach for integration of distributed expert systems." In *Proceedings of the First International Conference on Multiagent Systems*, AAAI Press, MIT Press, Cambridge, MA.

Jakiela, M., Chapman, C., Duda, J., Adewuya, A., and Saitou, K. (2000). "Continuum structural topology design with genetic algorithms," *Computer Methods in Applied Mechanics and Engineering*, **186**(2–4):339–356.

Jang, J.-S., and Sun, C.-T. (1993). "Functional equivalence between radial basis function and fuzzy inference systems," *IEEE Transactions on Neural Networks*, **4**(1):156–159.

Janilow, C. Z., and Michalewicz, Z. (1991). "An experimental comparison of binary and floating point representations in genetic algorithms." In *The Fourth International Conference on Genetic Algorithms*, Morgan Kaufmann, San Mateo, CA.

Kaise, N. and Fujimoto, Y. (1998). "Applying the evolutionary neural networks with genetic algorithms to control of a rolling inverted pendulum." In *Proceedings of the Second Asia-Pacific Conference on Simulated Evolution and Learning*, Springer, New York.

Keeney, R. and Raiffa, H. (1976). *Decisions with Multiple Objectives: Preferences and Value Tradeoffs*. Wiley, New York.

Kim, D. and Ahn, S. (1999). "A MS-GS VQ codebook design for wireless image communication using the genetic algorithm," *IEEE Transactions on Evolutionary Computation*, **3**(1): 35–52.

Kincaid, R. and Berger, R. (1993). "The damper placement problem on space truss structures," *Location Science*, **1**(3):219–234.

Kincaid, R., Martin, A., and Hinkley, J. (1995). "Heuristic search for the polymer straightening problem," *Computational Polymer Science*, **5**:1–5.

Kirkpatrick, S., Gelatt Jr., C. D., and Vecchi, M. P. (1983). "Optimization by simulated annealing," *Science*, **220**(4598):671–679.

Klir, G. J., and Yuan, B. (1995). *Fuzzy Sets and Fuzzy Logic: Theory and Applications*. Prentice-Hall, Englewood Cliffs, NJ.

Klir, G. J., and Folger, T. A. (1988). *Fuzzy Sets, Uncertainty, and Information*. Prentice-Hall, Englewood Cliffs, NJ.

Kolli, A., Cagan, J., and Rutenbar, R. (1996). "Packing of generic, three dimensional components based on multi-resolution modeling." In *Proceedings of the 1996 ASME Design Engineering Technical Conferences and Computers in Engineering Conference: 22nd Design Automation Conference*, ASME, New York, DAC-1479.

Koza, J., Bennett F. B., Andre, D., and Keane, M. (1999). *Genetic Programming III: Darwinian Invention and Problem Solving*. Morgan Kaufmann, San Mateo, CA.

Langdon, W. B. (1995). "Scheduling planned maintenance of the national grid." In *AISB Workshop on Evolutionary Computation*, T. C. Fogarty (ed.), Springer, New York, pp. 132–153.

Lech, M. and Hua, Y. (1991). "Vector quantization of images using neural networks and simulated annealing." In *Proceedings of the IEEE Workshop on Neural Networks for Signal Processing*, B. H. Juang, S. Y. Kung, and C. A. Kamm (eds.), IEEE, New York, pp. 552–561.

Lech, M. and Hua, Y. (1992). "Image vector quantization using neural networks and simulated annealing." In *International Conference on Image Processing and its Applications*, IEEE, New York.

Lee, B. and Sheu, B. (1991). *Hardware Annealing in Analog VLSI Neurocomputing*. Kluwer Academic, Boston, MA.

Lee, C.-Y., and Antonsson, E. (2000a). "Dynamic partitional clustering using evolution strategies." In *Proceedings of the Third Asia Pacific Conference on Simulated Evolution and Learning*, IEEE, New York.

Lee, C.-Y. and Antonsson, E. K. (2000b). "Self-adapting vertices for mask synthesis." In *MSM2000, Modeling and Simulation of Microsystems, Semiconductors, Sensors and Actuators*, Applied Computational Research Society, Cambridge, MA.

Levenick, J. R. (1995). "Metabits: generic endogenous crossover control." In *Proceedings of the Sixth International Conference on Genetic Algorithms*, Morgan Kaufmann, San Mateo, CA, pp. 88–95.

Li, H. and Antonsson, E. K. (1999). "Mask-layout synthesis through an evolutionary algorithm." In *MSM'99, Modeling and Simulation of Microsystems, Semiconductors, Sensors and Actuators*, Applied Computational Research Society, Cambridge, MA.

Lienig, J. (1997a). "A parallel genetic algorithm for performance-driven VLSI routing," *IEEE Transactions on Evolutionary Computation*, **1**(1):29–39.

Lienig, J. (1997b). "Physical design of VLSI circuits and the application of genetic algorithms." In *Evolutionary Algorithms in Engineering Applications*, D. Dasgupta and Z. Michalewicz (eds.), Springer, New York, pp. 277–292.

Lipson, H. and Pollack, J. (2000). "Automatic design and manufacture of artificial lifeforms," *Nature*, **406**, 974–978.

Liu, B. Y., Haftka, R. T., Akgun, M. A., et al. (2000). "Permutation genetic algorithms for stacking sequence design of composite laminates," *Computer Methods in Applied Mechanics and Engineering*, **186**(2–4):357–372.

Lobo, F. G., Deb, K., Goldberg, D. E., Harik, G., and Wang, L. (1998). "Compressed Introns in a linkage learning genetic algorithm." In Koza, J.R., Banzhaf, W., Chellapilla, K., Deb, K., Dorigo, M., Fogel, D. B., Garzon, M. H., Goldberg, D. E., Iba, H., and Riolo, R. (eds.), *Genetic Programming 1998: Proceedings of the Third Annual Conference*, Morgan Kaufmann, San Mateo, CA, pp. 551–558.

Lunn, K. and Johnson, C. (1996). "Spatial reasoning with genetic algorithms – an application in planning of safe liquid petroleum gas sites." In *Evolutionary Computation, AISB Workshop*, T. C. Fogarty (ed.), Springer, New York, pp. 73–84.

Ma, L. and Antonsson, E. K. (2000a). "Applying genetic algorithms to MEMS synthesis." In *ASME International Mechanical Engineering Congress and Exposition*, ASME, New York.

Ma, L. and Antonsson, E. K. (2000b). "Mask-layout and process synthesis for MEMS." In *MSM2000, Modeling and Simulation of Microsystems, Semiconductors, Sensors, and Actuators*, Applied Computational Research Society, Cambridge, MA.

Marshall, J. and Bansal, V. (1992). *Financial Engineering: A Complete Guide to Financial Innovation*. New York Institute of Finance, New York.

Mazumder, P. and Rudnick, E. M. (1999). *Genetic Algorithms for VLSI Design, Layout and Test Automation*, Prentice–Hall, Englewood Cliffs, NJ.

Mazzola, J. and Schantz, R. (1995). "Single-facility resource allocation under capacity-based economies and diseconomies of scope," *Management Science*, **41**(4):669–689.

Meng, L., Wu, Q., and Yong, Z. (2000). "A faster genetic clustering algorithm." In *Proceedings of EvoWorkshops 2000*, Springer, New York.

Metropolis, N., Rosenbluth, A., Rosenbluth, M., Teller, A., and Teller, E. (1953). "Equation of state calculations by fast computing machines," *Journal of Chemical Physics*, **21**(6):1087–1092.

Michalewicz, Z. (1994). *Genetic Algorithms + Data Structures = Evolution Program*, Springer-Verlag, New York.

Michalewicz, Z., Logan, T., and Swaminathan, S. (1994). "Evolutionary operations for continuous convex parameter spaces," Anthony V. Sebald and L. J. Fogel (eds.), World Scientific, Singapore. In *Proceedings of the Third Annual Conference on Evolutionary Programming*, Springer, New York.

Mihaila, D. (1997). "Genetic algorithms with optimal structure applied to evolvable hardware." In *Proceedings of the International Conference on Computational Intelligence: Theory and Applications*, Springer, New York.

Morley, R. and Schelberg C. (1993). "An analysis of a plant-specific dynamic scheduler." In *Proceedings of the NSF Workshop on Dynamic Scheduling*, Springer, New York.

Mukherjee, B. (1996). "Some principles for designing a wide-area WDM optical network," *IEEE/ACM Transactions on Networking*, **4**(5):684–695.

Murata, T., Ishibuchi, H., Nakashima, T., and Gen, M. (1998). "Fuzzy partition and input selection by genetic algorithms for designing fuzzy rule-based classification systems." In *Proceedings of the 7th International Evolutionary Programming Conference*, Springer, New York.

Naito, T., Odagiri, R., Matsunaga, Y., Tanifuhi, M., and Murase, K. (1996). "Genetic evolution of a logic circuit which controls and autnomous mobile robot." In *Proceedings of the First International Conference on Evolvable Systems (ICES)*, Springer, New York.

Neal, R. (1993). "Probabilistic inferences using Markov Chain Monte Carlo methods," Technical Report CRG-TR-93-1, University of Toronto.

Niittylahti, J. (1992). "Hardware implementation of Boolean neural network using simulated annealing," Technical Report 8-92, Tampere University of Technology.

Nowicki, E. and Smutnicki, C. (1995). "The flow shop with parallel machines: a Tabu search approach," Technical Report ICT PRE 30/95, Technical University of Wroclaw.

Nwana, H. and Ndumu, D. (1997). "An introduction to agent technology." In *Lecture Notes in Artificial Intelligence 1198: Software Agents and Soft Computing: Towards Enhancing Machine Intelligence*, Springer, New York, pp. 3–26.

Obayashi, S., Sasaki, D., Takeguchi, Y., and Hirose, N. (2000). "Multiobjective evolutionary computation for supersonic wing-shape optimization," *IEEE Transactions on Evolutionary Computation*, **4**(2):182–187.

O'Neill, M. and Ryan, C. (1999). "Evolving multiline compilable C programs." In *Proceedings of the Second European Workshop on Genetic Programming (EuroGP'99)*, Springer, New York.

Paechter, B. (1994). "Optimising a presentation timetable using evolutionary algorithms," *Lecture Notes in Computer Science*, **865**:264–276 (AISB Workshop: Evolutionary Computing). Springer, New York.

Peterson, C. and Soderberg, B. (1992). "Artificial neural networks and combinatorial optimization problems." In *Local Search in Combinatorial Optimization*, E. Aarts and J. Lentra (eds.), Wiley, New York.

Porto, S. and Ribeiro, C. (1995a). "Parallel Tabu search message-passing synchronous strategies for task scheduling under precedence constraints," *Journal of Heuristics*, 1(2): 207–225.

Porto, S. and Ribeiro, C. (1995b). "A Tabu search approach to task scheduling on heterogeneous processors under precedence constraints," *International Journal of High Speed Computing*, 7:45–71.

Purohit, B., Clark, T., and Richards, T. (1995). "Techniques for routing and scheduling services on a transmission network," *BT Technology Journal*, 13(1):64–72.

Radcliffe, N. (1990). "Genetic neural networks on MIMD computers," Ph.D. Dissertation, University of Edinburgh, U.K.

Raittinen, H. and Kaski, K. (1990). "Image deconvolution with simulated annealing method," *Physica Scripta*, T33:126–130.

Rechenberg, I. (1965). "Cybernetic solution path of an experimental problem," Technical Report Library Translation No. 1122, Royal Aircraft Establishment, Farnborough, Hants, U.K.

Rechenberg, I. (1973). *Evolutionsstrategie: Optimierung technischer systeme nach prinzipien der biolgischen evolution*. Frommann-Holzboog Verlag, Stuttgart.

Reddy, G. and Cagan, J. (1995a). "An improved shape annealing algorithm for truss topology generation," *ASME Journal of Mechanical Design*, 117(2A):315–321.

Reddy, G. and Cagan, J. (1995b). "Optimally directed truss topology generation using shape annealing," *ASME Journal of Mechanical Design*, 117(1):206–209.

Riche, R. L. and Haftka, R. T. (1997). "Evolutionary optimization of composite structures." In *Evolutionary Algorithms in Engineering Applications*, D. Dasgupta and Z. Michalewicz (eds.), Springer, New York, pp. 87–102.

Rochat, Y. and Semet, F. (1994). "A Tabu search approach for delivering pet food and flour," *Journal of the Operational Research Society*, 45(11):1233–1246.

Rochat, Y. and Taillard, E. (1995). "Probabilistic diversification and intensification in local search for vehicle routing," *Journal of Heuristics*, 1(1):147–167.

Rosenberg, R. S. (1967). *Simulation of Genetic Populations with Biochemical Properties*. Ph.D. Thesis, The University of Michigan, Ann Arbor, MI.

Ross, P. and Corne, D. (1995). "Comparing genetic algorithms, simulated annealing, and stochastic hillclimbing on timetable problems." In *Evolutionary Computation, AISB Workshop*, T. C. Fogarty (ed.), Springer, New York, pp. 94–102.

Ross, T. J. (1995). *Fuzzy Logic with Engineering Applications*. McGraw-Hill, New York.

Saitou, K. and Jakiela, M. (1995). "Automated optimal design of mechanical conformational switches," *Artificial Life*, 2(2):129–156.

Salami, M., Murakawa, M., and Higuchi, T. (1996). "Data compression based on evolvable hardware." In *Proceedings of the First International Conference on Evolvable Systems (ICES)*, Springer, New York.

Savic, D. A. and Walters, G. A. (1995). "Genetic operators and constraint handling for pipe network optimization." In *Evolutionary Computation, AISB Workshop*, T. C. Fogarty (ed.), Springer, New York, pp. 154–165.

Schmidt, L. C. and Cagan, J. (1995). "Recursive annealing: a computational model for machine design," *Research in Engineering Design*, 7(2):102–125.

Schnecke, V. and Vorngerger, O. (1997). "Hybrid genetic algorithms for constrainted placement problems," *IEEE Transactions on Evolutionary Computation*, 1(4):266–277.

Schoonderwoerd, R., Holland, O., Bruten, J., and Rothkrantz, L. (1996). "Ant-based load balancing in telecommunications networks," *Adaptive Behavior*, 5:169–207.

Schwefel, H. P. (1965). "Kybernetische evolution als strategie der experimentellen forschung in der strmungstechnik," Ph.D. Dissertation, Technical University of Berlin.

Shea, K. and Cagan, J. (1997). "Innovative dome design: applying geodesic patterns with shape annealing," *Artificial Intelligence in Engineering Design and Manufacturing*, 11(5):379–394.

Shea, K., Cagan, J., and Fenves, S. (1997). "A shape annealing approach to optimal truss design with dynamic grouping of members," *ASME Journal of Mechanical Design*, **119**(3):388–394.

Skorin-Kapov, J. and Labourdette, J.-F. (1994). "On Tabu search for the location of interacting hub facilities," *European Journal on Operational Research*, **73**(3):501–508.

Skorin-Kapov, J. and Labourdette, J.-F. (1995). "On minimum congestion in logically rearrangeable multihop lightwave networks," *Journal of Heuristics*, **1**(1):129–146.

Skorin-Kapov, J. and Vakharia, A. (1993). "Scheduling a flow-line manufacturing cell: a Tabu search approach," *International Journal of Production Research*, **31**(7):1721–1734.

Skourikhine, A. (1998). "An evolutionary algorithm for designing feedforward neural networks." In *Proceedings of the 7th International Evolutionary Programming Conference*, Springer, New York.

Spears, W. M. and Jong, K. D. (1991). "On the virtues of parameterized uniform crossover." In *The Fourth International Conference on Genetic Algorithms*, Morgan Kaufmann, San Mateo, CA, pp. 230–236.

Syswerda, G. (1989). "Uniform crossover in genetic algorithms." In *The Fourth International Conference on Genetic Algorithms*, Morgan Kaufmann, San Mateo, CA, pp. 2–9.

Szu, H. and Hartley, R. (1987). "Fast simulated annealing," *Physical Letters A*, **122**(3–4):157–162.

Szykman, S. and Cagan, J. (1995). "A simulated annealing-based approach to three-dimensional component packing," *ASME Journal of Mechanical Design*, **117**(2A):308–314.

Szykman, S. and Cagan, J. (1996). "Synthesis of optimal non-orthogonal routes," *ASME Journal of Mechanical Design*, **118**(3):419–424.

Szykman, S. and Cagan, J. (1997). "Constrained three dimensional component layout using simulated annealing," *ASME Journal of Mechanical Design*, **119**(1):28–35.

Szykman, S., Cagan, J., and Weisser, P. (1998). "An integrated approach to optimal three dimensional layout and routing," *ASME Journal of Mechanical Design*, **120**(3): 510–512.

Tan, K., Lee, T., and Khor, E. (2000). "Automatic design of multivariable QFT control system via evolutionary computation." In *Proceedings of EvoWorkshops 2000*, Springer, New York.

Tezuka, M., Hiji, M., Miyabayashi, K., and Okumura, K. (2000). "A new genetic representation and common cluster crossover for job shop scheduling problems." In *Proceedings of EvoWorkshops 2000*, Springer, New York.

Tsujimura, Y. and Gen, M. (1996). "Genetic algorithms for solving multiprocessor scheduling problems." In *Proceedings of the First Asia-Pacific Conference on Simulated Evolution and Learning*, Springer, New York.

Urquhart, N., Chisholm, K., and Paechter, B. (2000). "Optimising an evolutionary algorithm for scheduling." In *Proceedings of EvoWorkshops 2000*, Springer, New York.

Vaessens, R., Aarts, E., and Lenstra, J. (1995). "Job shop scheduling by local search," Technical Report Memorandum COSOR 94-05, Eindhoven University of Technology, Eindhoven, Netherlands.

Veeramani, D. and Stinnes, A. (1996). "Optimal process planning for four-axis turning centers." In *Proceedings of the 1996 ASME Design Engineering Technical Conference and Computers in Engineering Conference*, ASME, New York.

Wicker, D., Rizki, M., and Tamburino, L. (1998). "A hybrid evolutionary learning system for synthesizing neural network pattern recognition systems." In *Proceedings of the 7th International Evolutionary Programming Conference*, Springer, New York.

Wiesmann, D., Hammel, U., and Back, T. (1998). "Robust design of multilayer optical coatings by means of EAs," *IEEE Transactions on Evolutionary Computation*, **2**(4):162–167.

Wong, D. F., Leong, H. W., and Liu, C. L. (1989). *Simulated Annealing for VLSI Design*. Kluwer Academic, Boston, MA.

Wong, K. and Wong, S. (1994). "Development of hybrid optimisation techniques based on genetic algorithms and simulated annealing." In *Progress in Evolutionary Computation: AI '93 and AI '94 Workshops on Evolutionary Computation*, Volume 956 of Lecture Notes in Artificial Intelligence, Springer-Verlag, New York.

Wright, A. H. (1991). "Genetic algorithms for real parameter optimization." In *Foundations of Genetic Algorithms*, Morgan Kaufmann, San Mateo, CA, pp. 205–217.

Wu, A. S. and Lindsay, R. K. (1996). "A survey of Intron research in genetics." In Ebeling, W., Rechenberg, I., Schwefel, H.-P., and Voigt, H.-M. (eds.). In *Proceedings of the 4th International Conference on Parallel Problem Solving from Nature (PPSN IV)*, Springer-Verlag, Berlin Lecture Notes in Computer Science, Volume 1141.

Wu, Q. and Sloane, T. H. (1992). "CMOS leaf-cell design using simulated annealing," Technical report, Bucknell University, Lewisburg, PA.

Xiao, J., Michalewicz, Z., Zhang, L., and Trojanowski, K. (1997). "Adaptive evolutionary planner/navigator for mobile robots," *IEEE Transactions on Evolutionary Computation*, **1**(1):18–28.

Yin, S. (2000). "A pattern search-based algorithm for automated product layout," Ph.D. Dissertation, Carnegie Mellon University, Pittsburgh, PA.

Yin, S. and Cagan, J. (2000). "An extended pattern search algorithm for three-dimensional component layout," *ASME Journal of Mechanical Design*, **122**(1):102–108.

Yin, S., Cagan, J., Hodges, P., and Li, X. (1999). "Layout of an automobile transmission using three-dimensional shapeable components." In *Proceedings of the 1999 ASME Design Engineering Technical Conferences: Design Automation Conference*, ASME, New York, DETC99/DAC-8564.

Yue, K. K. and Lilja, D. J. (1997). "Designing multiprocessor scheduling algorithms using a distributed genetic algorithm system." In *Evolutionary Algorithms in Engineering Applications*, D. Dasgupta and Z. Michalewicz (eds.), Springer, New York, pp. 207–222.

Zhang, F., Zhang, Y., and Lee, A. (1997). "Using GAs in process planning for job shop machining," *IEEE Transactions on Evolutionary Computation*, **1**(4):278–289.

Zhang, Q. and Leung, Y.-W. (1999). "An orthogonal GA for multimedia multicast routing," *IEEE Transactions on Evolutionary Computation*, **3**(1):53–62.

Zhao, Q. (1996). "A study on co-evolutionary learning of neural networks." In *Proceedings of the First Asia-Pacific Conference on Simulated Evolution and Learning*, Springer, New York.

Zimmermann, H.-J. (1985). *Fuzzy Set Theory – and Its Applications*, Management Science/Operations Research, Kluwer-Nijhoff, Boston, MA.

CHAPTER NINE

Kinematic Synthesis

J. Michael McCarthy and Leo Joskowicz

INTRODUCTION

This chapter surveys recent results in the kinematic synthesis of machines. A machine is generally considered to be a device that directs a source of power to a desired application of forces and movement. The earliest machines were simply wedges, levers, and wheels that amplified human and animal effort. Eventually wind and water power were captured to drive gearing that rotated millstones and pumps (Dimarogonas, 1993). Today chemical, nuclear, hydroelectric, and solar energy drive machines that build, manufacture, transport, and process items that affect all aspects of our lives.

Kinematic synthesis determines the configuration and size of the mechanical elements that shape this power flow in a machine. In this chapter we show how researchers in machine design and robotics use the concepts of workspace and mechanical advantage as the criteria for the kinematic synthesis of a broad range of machines.

The workspace of a machine part is the set of positions and orientations that it can reach. We will see that there are many representations of this workspace; however, the best place to start is with the kinematics equations of the chain of bodies that connect the part to the base frame. The range of values of the configuration parameters that appear in these kinematics equations defines the configuration space of the system. The velocity of the part is then obtained by computing the Jacobian of these kinematics equations in terms of the rate of change of these configuration parameters. The principle of virtual work links this Jacobian directly to the mechanical advantage of the system.

We begin by showing how virtual work relates the input–output speed ratios of a system to its mechanical advantage. This is developed for general serial and parallel chains of parts within a machine. Then we explore the current design theory for serial and parallel robots and find that workspace and mechanical advantage, clothed in various ways, are the primary considerations. We then develop in detail the current results in the kinematic synthesis of constrained movement that focuses on satisfying workspace constraints formulated as algebraic equations. Finally, we outline the configuration space analysis of machines that shows the versatility of these concepts for representing assembly, tolerances, and fixtures.

A survey of computer-aided kinematic synthesis software shows that although specialized systems exist in research laboratories, there is very little available in engineering practice. Also, current design systems leave untouched many of the machine topologies that are available for invention. Designer–inventors need a software environment that allows them to specify a workspace and an associated distribution of mechanical advantage, and then simulate and display candidate designs to evaluate their performance, including issues of tolerancing and assembly. The result would be a remarkable opportunity for the design of new devices to serve our needs.

KINEMATICS AND KINETICS

The term *kinematics* refers to geometric properties of movement such as position, velocity, and acceleration of components of a machine; *kinetics* refers to properties of forces acting on and exerted by the part. Newton's second law of motion relates kinetics and kinematics by defining the acceleration of a particle as proportional to the difference between the force applied to the particle and the force it exerts:

$$\mathbf{F}_{\text{in}} - \mathbf{F}_{\text{out}} = m\ddot{\mathbf{x}}. \tag{9.1}$$

If these forces are equal then the particle moves at a constant velocity.

Newton's law can summed over all particles in the machine and integrated as the system moves along a trajectory $\mathbf{x}(t)$ to define the change in energy of the system as the difference in the input and output work. It is a law of mechanics that this equality of work and energy remains unchanged for small variations $\delta\mathbf{x}$ of the trajectory. Thus, the variations in work and energy must cancel for all virtual displacements; that is,

$$\delta W_{\text{in}} - \delta W_{\text{out}} = \delta E. \tag{9.2}$$

A machine is designed to minimize energy losses, often caused by friction and fatigue, so $\delta E = 0$, which means the input and output variations in work must cancel. This is known as the *principle of virtual work* (Greenwood, 1977; Moon, 1998).

At a specific instant of time, we can introduce the virtual displacement $\delta\mathbf{x} = \mathbf{v}\delta t$, where $\mathbf{v}$ is the velocity and δt is a virtual time increment. This allows us to define the virtual work $\delta W_{\text{in}} = P_{\text{in}}\delta t$ and $\delta W_{\text{out}} = P_{\text{out}}\delta t$, where $P = \mathbf{F} \cdot \mathbf{v}$ is the instantaneous power. The result is that Equation (9.2) becomes

$$(P_{\text{in}} - P_{\text{out}})\delta t = 0. \tag{9.3}$$

Thus, the usual assumption is that the machine does not dissipate power and in every configuration the input and output instantaneous power are equal. Because power is force times velocity, we have that *the mechanical advantage of a machine is the inverse of its speed ratio*. In the next section we examine this in more detail.

MACHINE TOPOLOGY

A machine is constructed from a variety of elements such as gears, cams, linkages, ratchets, brakes, and clutches. Each of these elements can be reduced to a set of *links* connected together by *joints*. Perhaps the simplest joint to construct, though

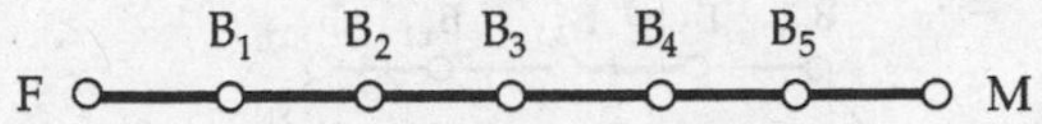

Figure 9.1. The part M is connected to the base frame F by a series of links (vertices) and joints (edges).

difficult to analyze, is the cam-and-follower joint formed by one link's pushing against a second follower link. In this case, the movement of the output link depends on the shape of the contacting surfaces. In contrast, pure rotary and sliding joints, and joints constructed from them, are considered simple joints because they are easy to analyze, though difficult to construct. These two classes of joints are often termed *higher* and *lower* pairs, respectively, (Reuleaux, 1875; Waldron and Kinzel, 1998).

Each component M of a machine is connected by a series of links and joints to the base frame F of the device, and the topology of the system is presented as a graph with the links as vertices and the joints as edges (Kota, 1993). For example, the typical robot arm has the graph shown in Figure 9.1. The mathematical relation that defines the position of each machine part M in the frame F is called its *kinematics equations*.

KINEMATICS EQUATIONS

If we model the local geometry of higher-pair joints by using rotary and sliding joints, then the position of every component of a machine can be obtained from a sequence of lines representing the axes S_j of equivalent revolute or prismatic joints. Between successive joint axes, we have the common normal lines A_{ij}, which together with the joint axes forms a serial chain, as shown in Figure 9.2. This construction allows the specification of the location of the part relative to the base of the machine by the matrix equation

$$[D] = [Z(\theta_1, \rho_1)][X(\alpha_{12}, a_{12})][Z(\theta_2, \rho_2)] \cdots [X(\alpha_{m-1,m}, a_{m-1,m})][Z(\theta_m, \rho_m)], \quad (9.4)$$

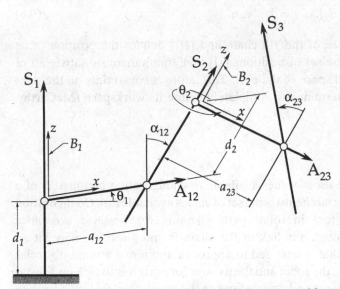

Figure 9.2. A machine part is located in space by a sequence of frames consisting of axes S_j and their common normals A_{ij}.

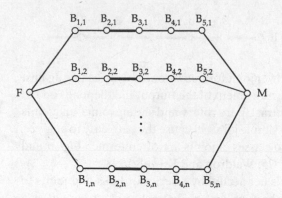

Figure 9.3. A part M is often connected to the base frame F by parallel series of links and joints.

known as the *kinematics equations* of the chain (Paul, 1981; Craig, 1989). The set of all positions $[D]$ obtained as the joint parameters vary over their range of movement defines the *workspace* of the component, also called its *configuration space* (Greenwood, 1977; Arnold, 1978).

The matrices $[Z(\theta_j, \rho_j)]$ and $[X(\alpha_{ij}, a_{ij})]$ are 4×4 matrices that define screw displacements around and along the joint axes S_j and A_{ij}, respectively (Bottema and Roth, 1979). The parameters α_{ij}, a_{ij} define the dimensions of the links in the chain. The parameter θ_j is the joint variable for revolute joints and ρ_j is the variable for prismatic joints. The trajectory $\mathbf{P}(t)$ of a point $\mathbf{p}$ in any part of a machine is obtained from the joint trajectory, $\vec{\theta}(t) = [\theta_1(t), \ldots, \theta_m(t)]^T$, so we have

$$\mathbf{P}(t) = \{D[\vec{\theta}(t)]\}\mathbf{p}. \tag{9.5}$$

A single part is often connected to the base frame by more than one serial chain, as shown in Figure 9.3. In this case we have a set of kinematics equations for each chain,

$$[D] = [G_j][D(\vec{\theta}_j)][H_j], \quad j = 1, \ldots, n, \tag{9.6}$$

where $[G_j]$ locates the base of the jth chain and $[H_j]$ defines the position of its attachment to the part. The set of positions $[D]$ that simultaneously satisfy all of these equations is the workspace of the part. This imposes constraints on the joint variables that must be determined to completely define its workspace (McCarthy, 1990; Tsai, 1999).

CONFIGURATION SPACE

Configuration space is the set of values available to configuration parameters of a mechanical system. For a serial chain it is the set of values available in $\vec{\theta}$. Configuration space is a fundamental tool in robot path planning for obstacle avoidance (Lozano-Perez, 1983). Though any link in the chain forming a robot may hit an obstacle, it is the gripper that is intended to approach and move around obstacles such as the table supporting the robot and the fixtures for parts it is to pick up. Obstacles define forbidden positions and orientations in the workspace that map back to forbidden joint angles in the configuration space of the robot. Robot path planners seek trajectories to a goal position through the free space around these joint space obstacles (Latombe, 1991).

The map of joint space obstacles provides a convenient illustration of the movement available to mechanical parts that are in close proximity (Joskowicz and Sacks, 1999). For this reason it has found applications far from robot path planning, such as the modeling of tolerances, assembly, and fixtures.

SPEED RATIOS

The speed ratio for any component of a machine relates the velocity $\dot{\mathbf{P}}$ of a point $\mathbf{P}$ to the joint rates $\dot{\vec{\theta}} = (\dot{\theta}_1, \ldots, \dot{\theta}_m)^T$. The velocity of this point is given by

$$\dot{\mathbf{P}} = \mathbf{v} + \vec{\omega} \times (\mathbf{P} - \mathbf{d}), \tag{9.7}$$

where $\mathbf{d}$ and $\mathbf{v}$ are the position and velocity of a reference point and $\vec{\omega}$ is the angular velocity of the part.

The vectors $\mathbf{v}$ and $\vec{\omega}$ depend on the joint rates $\dot{\theta}_j$ by the formula

$$\begin{Bmatrix} \mathbf{v} \\ \vec{\omega} \end{Bmatrix} = \begin{bmatrix} \dfrac{\partial \mathbf{v}}{\partial \dot{\theta}_1} & \dfrac{\partial \mathbf{v}}{\partial \dot{\theta}_2} & \cdots & \dfrac{\partial \mathbf{v}}{\partial \dot{\theta}_m} \\ \dfrac{\partial \vec{\omega}}{\partial \dot{\theta}_1} & \dfrac{\partial \vec{\omega}}{\partial \dot{\theta}_2} & \cdots & \dfrac{\partial \vec{\omega}}{\partial \dot{\theta}_m} \end{bmatrix} \begin{Bmatrix} \dot{\theta}_1 \\ \vdots \\ \dot{\theta}_m \end{Bmatrix}, \tag{9.8}$$

or

$$\mathsf{V} = [J]\dot{\vec{\theta}}. \tag{9.9}$$

The coefficient matrix $[J]$ in this equation is called the *Jacobian* and is a matrix of speed ratios relating the velocity of the part to the input joint rotation rates (Craig, 1989; Tsai, 1999).

MECHANICAL ADVANTAGE

If the machine component exerts a force $\mathbf{F}$ at the point $\mathbf{P}$, then the power output is

$$P_{\text{out}} = \mathbf{F} \cdot \dot{\mathbf{P}} = \sum_{j=1}^{m} \mathbf{F} \cdot \left[\frac{\partial \mathbf{v}}{\partial \dot{\theta}_j} + \frac{\partial \vec{\omega}}{\partial \dot{\theta}_j} \times (\mathbf{P} - \mathbf{d}) \right] \dot{\theta}_j. \tag{9.10}$$

Each term in this sum is the portion of the output power that can be associated with an actuator at the joint S_j, if one exists.

The power input at joint S_j is the product $\tau_j \dot{\theta}_j$ of the torque τ_j and joint angular velocity $\dot{\theta}_j$. Using the principle of virtual work for each joint, we can compute

$$\tau_j = \mathbf{F} \cdot \frac{\partial \mathbf{v}}{\partial \dot{\theta}_j} + (\mathbf{P} - \mathbf{d}) \times \mathbf{F} \cdot \frac{\partial \vec{\omega}}{\partial \dot{\theta}_j}, \quad j = 1, \ldots, m. \tag{9.11}$$

We have arranged this equation to introduce the six-vector $\mathsf{F} = [\mathbf{F}, (\mathbf{P} - \mathbf{d}) \times \mathbf{F}]^T$, which is the resultant force and moment at the reference point $\mathbf{d}$.

Equations (9.11) can be assembled into the matrix equation

$$\vec{\tau} = [J^T]\mathsf{F}, \tag{9.12}$$

where $[J]$ is the *Jacobian* defined above in Equation (9.8). For a chain with six joints, this equation can be solved for the output force-torque vector F:

$$F = [J^T]^{-1}\vec{\tau}. \tag{9.13}$$

Thus, we see that the matrix that defines the mechanical advantage for this system is the inverse of the matrix of speed ratios. This is a more general version of the statement that to increase mechanical advantage we must decrease the speed ratio.

SIMPLE MACHINES

Here we illustrate the basic issues of kinematic synthesis that we will discuss in more detail later. The kinematics equations, the relation between mechanical advantage and speed ratio and configuration space are easily developed for the lever, wedge, and planar RR chain (R denotes a revolute joint).

THE LEVER

A lever is a solid bar that rotates about a fixed hinge O called its *fulcrum*. This serial chain and it has the kinematics equations

$$[D] = [Z(\theta)], \tag{9.14}$$

which define a pure rotation about the fulcrum. Its configuration parameter is simply the rotation angle θ.

Let the input from a motor or applied force result in a torque $T_{in} = F_{in}a$ about the fulcrum, resulting in an output force at **B** (see Figure 9.4). If the angular velocity of the lever is $\dot{\theta}$, then the velocity of **B** is $v_{out} = b\dot{\theta}$, and the principle of virtual work yields the relationship

$$(F_{in}a\dot{\theta} - F_{out}b\dot{\theta})\delta t = 0, \tag{9.15}$$

or

$$\frac{F_{out}}{F_{in}} = \frac{a}{b}. \tag{9.16}$$

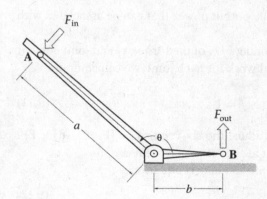

Figure 9.4. A lever is driven by a torque about its fulcrum to lift a load at **B**.

This is the well-known *law of the lever*, which defines the mechanical advantage of the lever. The speed ratio of the output to input is given by

$$\frac{v_{out}}{v_{in}} = \frac{b\dot{\theta}}{a\dot{\theta}} = \frac{b}{a}, \tag{9.17}$$

which is the inverse of its mechanical advantage. This is the simplest example of the relationship between the speed ratio and mechanical advantage.

THE WEDGE

A wedge is a right triangle with apex angle α that slides horizontally along a flat surface and lifts a load vertically by sliding it along its inclined face and against a vertical wall, as shown in Figure 9.5. This system consists of two parallel chains that support a load. One consists of the PP chain formed by the wedge itself, and the second P joint formed by load sliding against the wall (P denotes a prismatic, or sliding, joint). The kinematics equations of the system are

$$[D] = [G_1][Z(0, x)][X(\alpha, 0)][Z(0, a)][H_1],$$
$$[D] = [G_2][Z(0, y)][H_2], \tag{9.18}$$

where $[G_i]$ and $[H_i]$ locate the base and moving frames for the two chains. The horizontal slide x and vertical slide y define the configuration of the system.

The configuration parameters x and y must satisfy a constraint equation associated with the geometry of the two chains. The slide y is related to x by the slope $\tan \alpha$ of the wedge, that is,

$$y = x \tan \alpha + k, \tag{9.19}$$

where k is a constant. This equation yields the speed ratio

$$\frac{\dot{y}}{\dot{x}} = \tan \alpha. \tag{9.20}$$

The principle of virtual work now yields

$$(F_{in}\dot{x} - F_{out}\dot{y})\delta t = 0, \tag{9.21}$$

and we obtain the mechanical advantage

$$\frac{F_{out}}{F_{in}} = \frac{1}{\tan \alpha}. \tag{9.22}$$

Figure 9.5. A wedge is driven in the x direction to slide a load along the y direction.

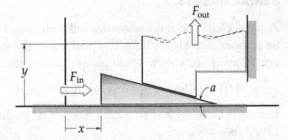

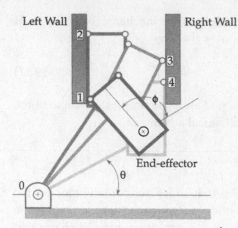

Figure 9.6. A planar RR robot moves its end effector between two walls.

The fundamental issues shown here for a lever and wedge appear in the kinematic synthesis of general serial and parallel robot systems. The primary concerns are the workspace and mechanical advantage, or speed ratio, of the system.

THE PLANAR RR CHAIN

Mechanical systems, including robots that interact with the world, often have components that make intermittent contact with other components. Thus, the system may be either an open chain or a closed chain, depending on the contact configuration.

Perhaps the simplest system that illustrates this issue is an RR chain with its rectangular end-effector moving between two walls (see Figure 9.6). The position and orientation of the rectangle is defined by the joint variables θ and ϕ; therefore its configuration space is two dimensional. The configurations excluded by the presence of the walls are said to define joint space obstacles. We show only the quadrant $0 \leq \theta \leq 180°$ and $0 \leq \phi \leq 180°$ in Figure 9.7, with the free space in white.

The boundary curves of a joint space obstacle are defined by the modes of contact of the end-effector and a wall. For obstacles and links that are polygons and circles, the robot–obstacle system forms a planar linkage that can be analyzed to determine this boundary (Ge and McCarthy, 1989). For example, when a vertex of the rectangular end effector moves along the left wall from position 1 to 2, as shown, the system forms a slider–crank linkage that is easily analyzed to determine $\phi(\theta)$. Similar calculations can be done for spatial polyhedra and spheres in contact in order to compute obstacle boundaries for spatial systems (Ge and McCarthy, 1990).

SERIAL ROBOTS

A serial chain robot is a sequence links and joints that begins at a base and ends with an gripper (see Figure 9.8). The position of the gripper is defined by the kinematics equations of the robot, which generally have the form

$$[D] = [Z(\theta_1, \rho_1)][X(\alpha_{12}, a_{12})][Z(\theta_2, \rho_2)] \cdots [X(\alpha_{56}, a_{56})][Z(\theta_6, \rho_6)], \qquad (9.23)$$

because the robot has six joints. The set of positions $[D]$ reachable by the robot is called its *workspace*.

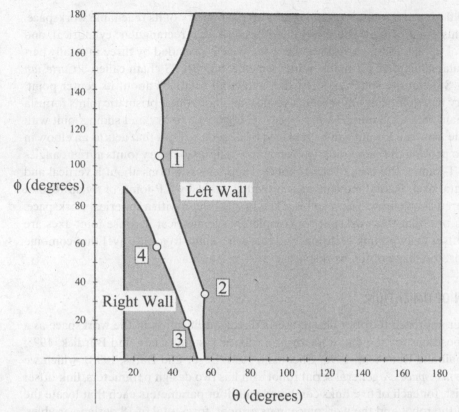

Figure 9.7. The joint angle values available to the robot are restricted by the presence of the two walls.

The links and joints of a robot are usually configured to provide separate translation and orientation structures. Usually, the first three joints are used to position a reference point in space, and the last three form the *wrist*, which orients the gripper around this point (Vijaykumar, Waldron, and Tsai, 1987; Gupta, 1987). This reference point is called the *wrist center*. The volume of space in which the wrist center can be placed is called the *reachable workspace* of the robot. The rotations available at each of these points is called the *dextrous workspace*.

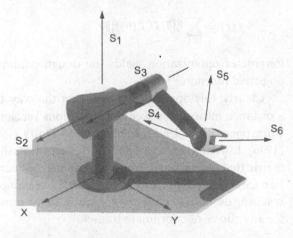

Figure 9.8. A serial robot is defined a set of joint axes S_i and the common normals A_{ij} between them.

The design of a robot is often based on the symmetry of its reachable workspace. From this point of view there are three basic shapes: rectangular, cylindrical, and spherical (Craig, 1989). A rectangular workspace is provided by three mutually perpendicular sliding, or prismatic, joints, which form a PPPS chain called a *Cartesian robot* – S denotes a spherical wrist that allows all rotations about its center point. A rotary base joint combined with a vertical and horizontal prismatic joints forms a CPS chain with a cylindrical workspace – C denotes a rotary and sliding joint with the same axis. The P-joint can be replaced by a revolute joint that acts as an elbow in order to provide the same radial movement. Finally, two rotary joints at right angles form a T-joint at the base of the robot that supports rotations about a vertical and horizontal axes. Radial movement is provided either by a P-joint, or by an R-joint configured as an elbow. The result is a TPS or TRS chain with a spherical workspace.

It is rare that the workspace is completely symmetrical because joint axes are often offset to avoid link collision, and there are limits to joint travel that combine to distort the shape of the workspace.

DESIGN OPTIMIZATION

Another approach to robot design uses a direct specification of the workspace as a set of positions for the end effector of a robotic system (Chen and Burdick, 1995; Chedmail and Ramstein, 1996; Chedmail, 1998; Leger and Bares, 1998), which we call the *taskspace*. A general serial robot arm has two design parameters, link offset and twist, for each of five links combined with four parameters each that locate the base of the robot and the workpiece in its gripper, for a total of 18 design variables. The link parameters are often specified so the chain has a spherical wrist and specific workspace shape. The design goal is usually to determine the workspace volume and locate the base and workpiece frames so that the workspace encloses the specified taskspace.

The taskspace is defined by a set of 4×4 transformations $[T_i]$, $i = 1, \ldots, k$. The problem is solved iteratively by selecting a design and using the associated kinematics equations $[D(\vec{\theta})]$ to compute the minimum relative displacements $[T_i D^{-1}(\vec{\theta}_i)]$. The invariants of each of these relative displacements are used to construct an objective function

$$f(\mathbf{r}) = \sum_{i=1}^{k} \|[T_i D^{-1}(\vec{\theta}_i)]\|. \tag{9.24}$$

Parameter optimization yields the design parameter vector $\mathbf{r}$ that minimizes this objective function.

Clearly, this optimization relies on the way the invariants are used to define a distance measure between the positions reached by the gripper and the desired workspace. Park (1995), Martinez and Duffy (1995), Zefran, Kumar, and Croke (1996), Lin and Burdick (2000), and others have shown that there is no such distance metric that is coordinate frame invariant. This means that unless this objective function can be forced to zero so the workspace completely contains the taskspace, the resulting design will not be "geometric" in the sense that the same design is obtained for any choice of coordinate frames.

If the goal is a design that best approximates the taskspace, then we cannot allow both the location for the base frame of the robot and the location of the workpiece in the gripper to be design variables. For example, in the above example, if the location of the base of the robot is given then we can find a coordinate invariant solution for the location of the workpiece in the gripper. In contrast, if the position of the workpiece is specified, the entire formulation may be inverted to allow a coordinate frame invariant solution for the location of the base frame (Park and Bobrow, 1995).

4×4 TRANSFORMS AS 4×4 ROTATIONS

Etzel and McCarthy (1996) and Ahlers and McCarthy (2000) embed 4×4 homogeneous transformations in the set of 4×4 rotation matrices in order to provide the designer control over the error associated with coordinate frame variation within a specific volume of the world that contains the taskspace.

Consider the 4×4 screw displacement about the Z axis defined by

$$[Z(\theta, \rho)] = \begin{bmatrix} \cos\theta & -\sin\theta & 0 & 0 \\ \sin\theta & \cos\theta & 0 & 0 \\ 0 & 0 & 1 & \rho \\ 0 & 0 & 0 & 1 \end{bmatrix}. \tag{9.25}$$

This defines a rigid displacement in the three-dimensional hyperplane $x_4 = 1$ of four-dimensional Euclidean space, E^4. The same displacement can be defined in parallel hyperplanes, $x_4 = R$, simply by dividing the translation component by R, that is

$$[Z(\theta, \rho)] = \begin{bmatrix} \cos\theta & -\sin\theta & 0 & 0 \\ \sin\theta & \cos\theta & 0 & 0 \\ 0 & 0 & 1 & \rho/R \\ 0 & 0 & 0 & 1 \end{bmatrix}. \tag{9.26}$$

This matrix can be viewed as derived from the 4×4 rotation:

$$[Z(\theta, \gamma)] = \begin{bmatrix} \cos\theta & -\sin\theta & 0 & 0 \\ \sin\theta & \cos\theta & 0 & 0 \\ 0 & 0 & \cos\gamma & \sin\gamma \\ 0 & 0 & -\sin\gamma & \cos\gamma \end{bmatrix}, \tag{9.27}$$

where $\tan\gamma = \rho/R$. This formalism views spatial translations as rotations of small angular values. The error ϵ associated with this approximation is less than $\sqrt{L/R}$, where L is the maximum dimension of the world space.

The result is an optimization strategy that computes both the base and workpiece frames, yielding essentially the same design for all coordinate changes within the given volume of space. This approach also provides the opportunity to tailor the number of joints in the chain and set the values of internal link parameters to fit a desired task.

SPEED RATIOS

A six-axis robot has a 6×6 Jacobian $[J]$ obtained from Equation (9.8) that is an array of speed ratios relating the components of the velocity $\mathbf{v}$ of the wrist center and the angular velocity $\vec{\omega}$ of the gripper to each of the joint velocities. The principle of virtual work yields the relationship

$$\mathsf{F} = [J^T]^{-1}\vec{\tau}, \tag{9.28}$$

which defines the force-torque F exerted at the wrist center in terms of the torque applied by each of the actuators. The link parameters of the robot can be selected to provide a Jacobian $[J]$ with specific properties.

The sum of the squares of the actuator torques of robot is often used as a measure of "effort" (Gosselin, 1998; Albro et al., 2000). From Equation (9.28) we have

$$\vec{\tau}^T\vec{\tau} = \mathsf{F}^T[J][J^T]\mathsf{F}. \tag{9.29}$$

The matrix $[J][J^T]$ is square and positive definite. Therefore, it can be viewed as defining an hyperellipsoid in six-dimensional space (Shilov, 1974). The lengths of the semidiameters of this ellipsoid are the inverse of the absolute value of the eigenvalues of the Jacobian $[J]$. These eigenvalues may be viewed as "modal" speed ratios that define the amplification associated with each joint velocity. Their reciprocals are the associated "modal" mechanical advantage, so the shape of this ellipsoid illustrates the force amplification properties of the robot.

The ratio of the largest of these eigenvalues to the smallest, called the *condition number*, gives a measure of the anisotropy or "out-of-roundness" of the ellipsoid. A six-sphere has a condition number of one and is termed *isotropic*. When the gripper of a robot is in a position with an isotropic Jacobian, there is no amplification of the speed ratios or mechanical advantage. This is considered to provide high-fidelity coupling between the input and output because errors are not amplified (Salisbury and Craig, 1982; Angeles and Lopez-Cajun, 1992). Thus, the condition number is used as a criterion in a robot design (Angeles and Chablat, 2000).

In this case, it is assumed that the basic design of the robot provides a workspace that includes the taskspace. Parameter optimization finds the internal link parameters that yield the desired properties for the Jacobian. As in minimizing the distance to a desired workspace, optimization based on the Jacobian depends on a careful formulation to avoid coordinate dependency.

PARALLEL ROBOTS

A robotic system in which two or more serial chain robots support an end-effector is called a *parallel robot*. Each of the serial chains must have six degrees of freedom; however, in general only a total of six joints in the entire system are actuated. A good example is the Stewart platform formed from six TPS robots, in which usually only the P-joint in each chain is actuated; see Figure 9.9 (Fichter, 1987; Merlet, 1999; Tsai, 1999).

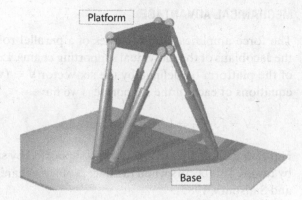

Figure 9.9. A parallel robot can have as many as six serial chains that connect a platform to the base frame.

The kinematics equations of the TPS legs are

$$[D] = [G_j][D(\vec{\theta}_j)][H_j], \quad j = 1, \ldots, 6, \tag{9.30}$$

where $[G_j]$ locates the base of the leg and $[H_j]$ defines the position of its attachment to the end effector. The set of positions $[D]$ that simultaneously satisfy all of these equations is the workspace of the parallel robot.

Often the workspace of an individual chain of a parallel robot can be defined by geometric constraints. For example, a position $[D]$ is in the workspace of the jth supporting TPS chain if it satisfies the constraint equation

$$\{[D]\mathbf{q}_j - \mathbf{P}_j\}^T\{[D]\mathbf{q}_j - \mathbf{P}_j\} = \rho_j^2. \tag{9.31}$$

This equation defines the distance between the base joint $\mathbf{P}_j$ and the point of attachment $\mathbf{Q}_j = [D]\mathbf{q}_j$ to the platform as the length ρ_j is controlled by the actuated prismatic joint. In this case the workspace is the set of positions $[D]$ that satisfy all six equations, one for each leg.

WORKSPACE

The workspace of a parallel robot is the intersection of the workspaces of the individual supporting chains. However, it is not the intersection of the reachable and dextrous workspaces separately. These workspaces are intimately combined in parallel robots. The dextrous workspace is usually largest near the center of the reachable workspace and shrinks as the reference point moves toward the edge. A focus on the symmetry of movement allowed by supporting leg designs has been an important design tool resulting in many novel parallel designs (Hervé, 1978, 1999). Simulation of the system is used to evaluate its workspace in terms of design parameters.

Another approach is to specify directly the positions and orientations that are to lie in the workspace and solve the algebraic equations that define the leg constraints to determine the design parameters (Murray et al., 1997; Murray and Hanchak, 2000). This yields parallel robots that are asymmetric but have a specified reachable and dextrous workspace.

MECHANICAL ADVANTAGE

The force amplification properties of a parallel robot are obtained by considering the Jacobians of the individual supporting chains. Let the linear and angular velocity of the platform be defined by the six-vector $V = (\mathbf{v}, \vec{\omega})^T$; then from the kinematics equations of each of the support legs we have

$$V = [J_1]\dot{\vec{\rho}}_1 = [J_2]\dot{\vec{\rho}}_2 = \cdots = [J_6]\dot{\vec{\rho}}_6. \tag{9.32}$$

Here we assume that the platform is supported by six chains, but it can be supported by fewer. This occurs when the fingers of a mechanical hand grasp an object (Mason and Salisbury, 1985).

The force on the platform applied by each chain is obtained from the principle of virtual work as

$$F_j = \left[J_j^T\right]^{-1}\vec{\tau}_j, \quad j = 1,\ldots,6. \tag{9.33}$$

There are only six actuated joints in the system, so we assemble the associated joint torques into the vector $\vec{\tau} = (\tau_1,\ldots,\tau_6)^T$. If F_i is the force-torque obtained from Equation (9.33) for $\tau_i = 1$, then the resultant force-torque W applied to the platform is

$$W = [F_1, F_2,\ldots, F_6]\vec{\tau}, \tag{9.34}$$

or

$$W = [\Gamma]\vec{\tau}. \tag{9.35}$$

The elements of the coefficient matrix $[\Gamma]$ define the mechanical advantage for each of the actuated joints. In the case of a Stewart platform, the columns of this matrix are the Plücker coordinates of the lines along each leg (Merlet, 1989).

The principle of virtual work yields the velocity of the platform in terms of the joints rates $\dot{\vec{\rho}}$ as

$$[\Gamma]^T V = \dot{\vec{\rho}}. \tag{9.36}$$

Thus, the inverse of $[\Gamma]$ defines the speed ratios between the actuated joints and the end-effector. The same equation can be obtained by computing the derivative of geometric constraint equations (9.31), and $[\Gamma]$ is the *Jacobian* of the parallel robot system (Kumar, 1992; Tsai, 1999).

The Jacobian $[\Gamma]$ is used in parameter optimization algorithms to design parallel robots (Gosselin and Angeles, 1988) with isotropic mechanical advantage. The square root of the determinant $|[\Gamma][\Gamma]^T|$ measures the six-dimensional volume spanned by the column vectors F_j. The distribution of the percentage of this volume compared to its maximum within the workspace is also used as a measure of the overall performance (Lee, Duffy, and Keler, 1996; Lee, Duffy, and Hunt, 1998). A similar performance measure normalizes this Jacobian by the maximum joint torques available and the maximum component of force and torque desired, and then it seeks an isotropic design (Salcudean and Stocco, 2000).

LINKAGE DESIGN THEORY

An assembly of links and joints, which is our general definition of a machine, can be called a *linkage*. However, this term is generally restricted to machine elements that have much less than the six degrees of freedom typical of a robotic system. Often they are one degree of freedom, single input, single output devices such as the four-bar linkage.

The kinematic synthesis theory presented above for robots is actually a generalization of an approach originally developed for linkages. Beginning with a set of task positions, Burmester (1886) obtained an exact geometric solution to the constraint equations of a planar RR chain, which he then assembled into a four-bar linkage. This has grown into a rich theory for the exact solution of the geometric constraint equations for RR and RP planar chains, RR spherical chains, and TS, CC, and RR spatial chains. See Chen and Roth (1967) and Suh and Radcliffe (1978).

THE SPATIAL RR CHAIN

The principles of kinematic synthesis of linkages can be seen in the direct solution of the constraint equations of an RR chain. Figure 9.10 shows a spatial RR chain, which can be considered the simplest robot. It has 10 design parameters, four each for the fixed axis G and the moving axis W, and the offset ρ and twist angle α.

The kinematics equations of the spatial RR chain are given by

$$[D] = [G][Z(\theta, 0)][X(\alpha, \rho)][Z(\phi, 0)][H], \tag{9.37}$$

which define its workspace. Choose a reference position $[D_1]$ and right translate the entire workspace to obtain

$$
\begin{aligned}
[D_{1k}] = [D][D_1]^{-1} &= ([G][Z(\theta, 0)][X(\alpha, \rho)][Z(\phi, 0)][H]) \\
&\times ([G][Z(\theta_1, 0)][X(\alpha, \rho)][Z(\phi_1, 0)][H])^{-1}. \tag{9.38}
\end{aligned}
$$

Figure 9.10. A spatial RR chain.

This equation can be simplified to obtain

$$[D_{1k}] = [T(\Delta\theta, \mathbf{G})][T(\Delta\phi, \mathbf{W})], \tag{9.39}$$

where

$$[T(\Delta\theta, \mathbf{G})] = [G][Z(\Delta\theta, 0)][G]^{-1},$$
$$[T(\Delta\phi, \mathbf{W})] = ([G][Z(\theta_1, 0)][X(\alpha, \rho)])[Z(\Delta\phi, 0)]$$
$$\times ([G][Z(\theta_1, 0)][X(\alpha, \rho)])^{-1}. \tag{9.40}$$

This defines the workspace of the RR chain as the composition of a rotation about the moving axis W in its reference position followed by a rotation about the fixed axis G, ($\Delta\theta = \theta - \theta_1$ and $\Delta\phi = \phi - \phi_1$ measure the rotation from the reference position).

To design a spatial RR chain we determine G, W, α, and ρ such that Equation (9.39) includes the desired taskspace. The general case requires the solution of 10 algebraic equations in 10 unknowns. We can simplify the presentation significantly by restricting attention to planar RR chains for which the axes G and W are parallel, which means the twist angle $\alpha = 0$.

THE PLANAR RR CHAIN

If G and W are parallel, then they can be located in the plane perpendicular to their common direction by the coordinates $\mathbf{P} = (x, y)^T$ and $\mathbf{Q} = (\lambda, \mu)^T$. The kinematics equations of the chain become

$$[D_{1k}] = [T(\Delta\theta, \mathbf{P})][T(\Delta\phi, \mathbf{Q})], \tag{9.41}$$

which is the composition of rotations parallel to this plane about the points Q and then P.

The workspace of this chain can also be defined as the set of displacements $[D_{1k}]$ that satisfy the algebraic equation

$$\{[D_{1k}]\mathbf{Q} - \mathbf{P}\}^T\{[D_{1k}]\mathbf{Q} - \mathbf{P}\} = \rho^2. \tag{9.42}$$

This is the geometric constraint that the displaced moving pivot $[D_{1k}]\mathbf{Q}$ must remain at a constant distance ρ from the fixed pivot P.

We use Equation (9.42) to directly determine the five design parameters $\mathbf{r} = (x, y, \lambda, \mu, \rho)$. This is done by evaluating this equation at five task positions $[T_i]$, $i = 1, \ldots, 5$, so we have

$$\{[D_{1i}]\mathbf{Q} - \mathbf{P}\}^T\{[D_{1i}]\mathbf{Q} - \mathbf{P}\} = \rho^2, \quad i = 1, \ldots, 5. \tag{9.43}$$

The result is a set of five equations in the five unknown design parameters. The distance ρ is easily eliminated by subtracting the first equation from the remaining four. This also cancels the squared terms u^2, v^2, λ^2, and μ^2. The resulting four equations are bilinear in the variables (x, y) and (λ, μ), and they can be written as

$$\begin{bmatrix} A_2(x, y) & B_2(x, y) \\ A_3(x, y) & B_3(x, y) \\ A_4(x, y) & B_4(x, y) \\ A_5(x, y) & B_5(x, y) \end{bmatrix} \begin{Bmatrix} \lambda \\ \mu \end{Bmatrix} = \begin{Bmatrix} C_2(x, y) \\ C_3(x, y) \\ C_4(x, y) \\ C_5(x, y) \end{Bmatrix}. \tag{9.44}$$

In order for this equation to have a solution, the four 3×3 minors of the augmented coefficient matrix must all be identically zero. This yields four cubic equations in the coordinates x and y of the fixed pivot. These cubic equations can be further manipulated to yield a quartic polynomial in x (McCarthy, 2000). Each real root of this polynomial defines a planar RR chain that reaches the five specified task positions.

DESIGN SOFTWARE

Kaufman (1978) was the first to transform this mathematical result into an interactive graphics program for linkage design, called KINSYN. See also Rubel and Kaufman (1977). He used a modified game controller to provide the designer the ability to input a set of task positions. An important feature of this software was the decision to allow the designer to only specify four, not five, positions. Rather than obtain a finite number of RR chains, his software determined the cubic curve of solutions known as the *center-point curve*. This curve is obtained by setting the minor obtained from the first three equations in Equation (9.44) to zero. Kaufman's software would ask the designer to select two points on this curve in order to define two RR chains that it assembled into the one degree of freedom 4R linkage, or *four-bar* linkage. Analysis routines evaluate the performance of the design and provide a simulation of its movement.

Erdman and Gustafson (1977) introduced LINCAGES, which, like KINSYN, focused on four task positions for the design of a 4R planar linkage. This software introduced a "guide map" that displayed the characteristics of every four-bar linkage that could be constructed from points on the center-point curve. This software was extended by Chase et al. (1981) to design an additional 3R chain to form a six-bar linkage.

Waldron and Song (1981) introduced the design software RECSYN, which again sought 4R closed chains that guide a body through three or four task positions. Their innovation was an analytical formulation that ensured the linkage would not "jam" as it moved between the design positions. In linkage design a jam is equivalent to hitting a singular configuration in a robot, which occurs when the determinant of the Jacobian becomes zero. An important feature of this software was the growing reliance on graphical communication of geometric information regarding the characteristics of the available set of designs.

Larochelle et al. (1993) introduced the Sphinx software for the design of spherical 4R linkages, which can be viewed as planar 4R linkages that are bent onto the surface of a sphere. A spherical RR chain is obtained when the link offset is $\rho = 0$, as shown in Figure 9.11. The fixed and moving axes of spherical RR chains are defined by the unit vectors $\mathbf{G} = (x, y, z)^T$ and $\mathbf{W} = (\lambda, \mu, \nu)^T$. The workspace of relative rotations $[A_{1k}]$ reachable by this chain is defined by the algebraic equation

$$\mathbf{G}^T[A_{1k}]\mathbf{W} = \cos \alpha. \tag{9.45}$$

This equation is evaluated at five specified task orientations to obtain equations that are essentially identical to Equation (9.44) and solved in the same way (McCarthy, 2000).

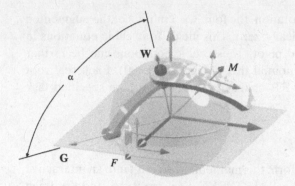

Figure 9.11. A spherical RR chain.

Following the pattern established by KINSYN and LINCAGES, Sphinx asks the designer to specify four task orientations, and then it generates the *center-axis cone*, which is the spherical equivalent of the center-point curve. The software also computes a "typemap," which classifies by means of color coding the movement of every 4R chain that can be constructed from pairs of axes on this cone. The typemap also included filters that eliminated designs with known defects. A later version of this software called SphinxPC also included planar 4R linkage design (Ruth and McCarthy, 1997). The display and typemap windows of SphinxPC are shown in Figure 9.12.

The three-dimensional nature of the interaction needed for spherical 4R linkage design presents severe visualization challenges. The designer finds that specifying a task as a set of spatial orientations is an unfamiliar experience. Furlong, Vance, and Larochelle (1998) used immersive virtual reality in their software IRIS to enhance this interaction.

Larochelle (1998) introduced the SPADES software, which provided interactive design for a truly spatial linkage system, the 4C linkage, for the first time. A CC chain is the generalized robot link that allows both rotation about and sliding along each axis. Let $G = (G, P \times G)^T$ and $W = (G, Q \times W)^T$ be the Plücker coordinates locating the fixed and moving axes in space. The workspace of this chain can be defined as the displacements $[D_{1k}]$ that satisfy the pair of geometric constraints,

$$G^T[A_{1k}]W = \cos\alpha,$$
$$(P \times G)^T[A_{1k}]W + G^T[D_{1k}](Q \times W) = -\rho\sin\alpha. \tag{9.46}$$

These equations constrain the link parameters α and ρ to be constant for every position of the moving frame. There are 10 design parameters consisting of four parameters for each of the two axes, the link offset ρ, and twist angle α.

By evaluating the two constraint equations at each of five task positions, we obtain 10 equations in 10 unknowns. Five of these are identical to those used for the synthesis of spherical RR chains and can be solved to determine the directions G and W. The remaining five equations are linear in the components of $P \times G$ and $Q \times W$ and are easily solved.

SPADES generates the *center-axis congruence*, which is the set of spatial CC chains that reach four spatial task positions. It then assembles pairs of these chains

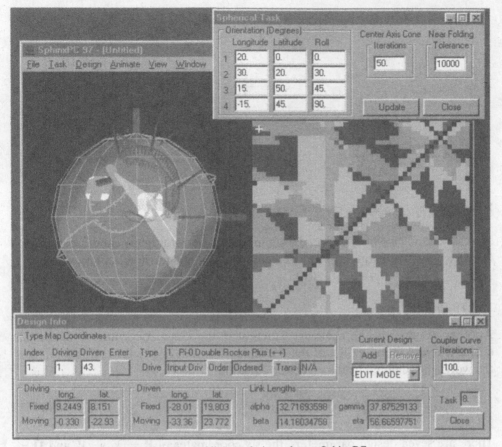

Figure 9.12. The desktop of the spherical linkage design software SphinxPC.

into two degree of freedom 4C linkages. This software demonstrates the significant visualization challenge that exists in the specification of spatial task frames and evaluation of candidate designs.

These linkage design algorithms ensure that the workspace of each linkage includes the specified taskspace, However, in each case the designer is expected to

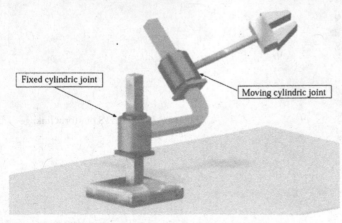

Figure 9.13. The CC open chain robot.

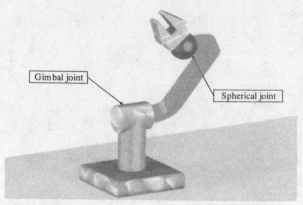

Figure 9.14. The TS open chain robot.

search performance measures and examine simulations for many candidate designs in order to verify the quality of movement between the individual task positions.

THE TS CHAIN

Another example of the challenge inherent in the kinematic synthesis of spatial chains is found in the design of the TS chain, shown in Figure 9.14. This chain has the workspace defined by the algebraic equation,

$$([D_{1k}]\mathbf{Q} - \mathbf{P})^T([D_{1k}]\mathbf{Q} - \mathbf{P}) = \rho^2. \tag{9.47}$$

There are seven design parameters, the six coordinates of $\mathbf{P} = (x, y, z)^T$ and $\mathbf{Q} = (\lambda, \mu, \nu)^T$, that define the centers of the T and S joints, respectively, and the length ρ. Therefore, we can evaluate this equation at seven spatial positions. The result, however, is that we can compute as many as 20 TS chains (Innocenti, 1995). If these are assembled into the single degree of freedom 5TS linkage, shown in Figure 9.15, then over 15,000 designs must be analyzed, which is an extreme computational burden (McCarthy and Liao, 1999). Prototype software shows that it is

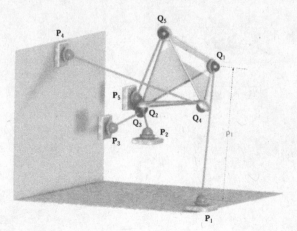

Figure 9.15. The 5TS platform linkage.

remarkably difficult for a designer to specify seven spatial positions and obtain a useful design.

Recent research has focused on developing higher-resolution tasks and providing methods for fitting a low-dimensional workspace to a general taskspace.

TASK SPECIFICATION

The workspace of a linkage is a subset of the group of all spatial displacements, denoted SE(3), reachable by a workpiece, or end effector, of the system. The taskspace is a discrete set of points in SE(3) that locate keyframes, or precision positions, that the linkage must reach. Research in motion interpolation has provided techniques to generate smooth trajectories through a set of keyframes by using Bezier-style methods.

Bezier interpolation is used in computer drawing systems to generate curves through specified points (Farin, 1997). Shoemake (1985) shows that this technique can be used to interpolate rotation keyframes specified by quaternion coordinates (Hamilton, 1860). Ge and Ravani (1994) generalized Shoemake's results by using dual quaternions to interpolate spatial displacements, and Ge and Kang (1995) refined this approach to ensure smooth transitions at each keyframe. The result is an efficient method to specify task trajectories in the manifold SE(3), as shown in Figure 9.16; see McCarthy (1990) for a discussion of quaternions and dual quaternions

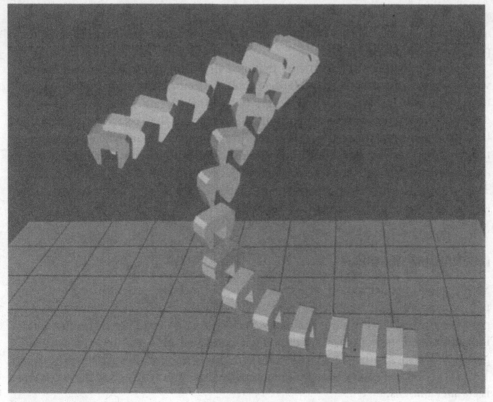

Figure 9.16. Spatial tasks can be specified by using Bezier motion interpolation.

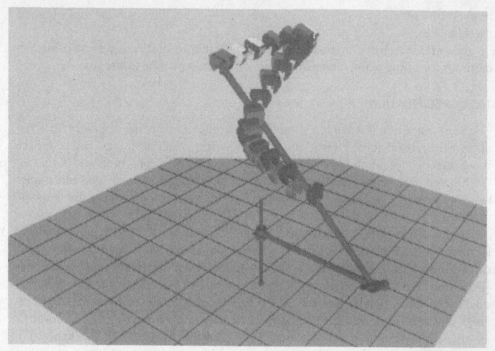

Figure 9.17. The workspace of a spatial RR robot is a manifold that can be fit to a desired trajectory by adjusting its design parameters.

as Clifford algebras, which represent the groups of rotations SO(3) and spatial displacements SE(3), respectively.

Ahlers and McCarthy (2000) show that the Clifford algebra of 4×4 rotations provides an efficient method for specifying spatial trajectories by using double quaternion equations. In McCarthy and Ahlers (2000) they design spatial CC chains for discrete subsets of a double quaternion trajectory, and they select the design that best fits the overall trajectory. This procedure combines an analytical solution for the fixed and moving axes with an optimization procedure that bounds the error associated with coordinate transformations in both the fixed and moving frames. This strategy was applied to the design of spatial RR robots in Perez and McCarthy (2000), as shown in Figure 9.17.

MECHANISM DESIGN

So far we have restricted our attention to isolated mechanisms consisting of lower-pair joints permanently attached together. In general, rigid-body systems consist of parts whose contacts are more complex and change with time. These systems include linkages moving in workspaces with obstacles, mechanisms consisting of cams, gears, geneva wheels, and other higher pairs, and general assemblies of rigid bodies (Norton, 1993; Tsai, 1993). The key issue for kinematic analysis and synthesis of these systems is *contact analysis*.

Contact analysis determines the positions and orientations at which the parts of a system touch and the ways that the touching parts interact. The interactions consist of

Figure 9.18. Ratchet mechanism: (a) pawl's advancing the ratchet; (b) pawl is fully advanced; (c) pawl is retracting. White circles indicate revolute joints.

constraints on the part motions that prevent them from overlapping. The constraints are expressed as algebraic equations that relate the part coordinates. For example, a ball rolling down a 45° slope obeys the constraint $x - y = r$, where x and y are the coordinates of the ball's center point and r is the ball radius. The constraints are a function of the shapes of the touching part features (vertices, edges, and faces); hence they change when one pair of features breaks contact and another makes contact.

To illustrate contact analysis and its role in design, consider the ratchet mechanism shown in Figure 9.18. The mechanism has four moving parts and a fixed frame. The driver, link, and ratchet are attached to the frame by revolute joints. The pawl is attached to the link by a revolute joint and is attached to a spring (not shown) that applies a counterclockwise torque around the joint. A motor rotates the driver with constant angular velocity, causing the link pin to move left and right. This causes the link to oscillate around its rotation point, which moves the pawl left and right. The leftward motion pushes a ratchet tooth, which rotates the ratchet counterclockwise. The rightward motion frees the pawl tip from the tooth, which allows the spring to rotate the pawl to engage the next tooth.

Kinematic analysis and synthesis of the ratchet mechanism requires contact analysis. We need to determine if the link oscillates far enough, if the pawl pushes the ratchet teeth far enough, if the system can jam, and so on. In the driver–link pair, the link pin interacts with the inner and outer driver profiles, creating a positive drive with small play. The ratchet–pawl pair is much harder to design because the part shapes and contact sequences are complex and because the pawl can translate horizontally, translate vertically, and rotate, whereas the other parts just rotate. We also need to validate intended interactions among all the parts, such as the indirect relation between the driver and the ratchet, and we must rule out interference, such as the pawl hitting the frame.

Contact analysis is best understood in the framework of *configuration space*. Configuration space is a geometric representation of rigid-body interaction that is

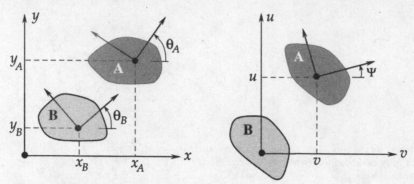

Figure 9.19. Pairwise configuration parameters coordinates: (a) absolute and (b) relative coordinates.

widely used in robot motion planning (Latombe, 1991). The configuration space of a mechanical system describes all possible part interactions. It encodes quantitative information, such as part motion paths, and qualitative information, such as system failure modes. It provides a framework within which diverse design tasks can be performed, as we explain next.

CONFIGURATION SPACE

We study contact analysis within the configuration space representation of rigid-body interaction. The configuration space of a system of rigid parts is a parameter space whose points specify the spatial configurations (positions and orientations) of the parts. The parameters usually represent part translations and rotations, but they can be arbitrary generalized coordinates. The configuration space dimension equals the number of independent part motions, called *degrees of freedom* of the system.

We begin by studying a mechanical system consisting of a pair of planar parts. We attach reference frames to the parts and define the configuration of a part to be the position and orientation of its reference frame with respect to a fixed global frame. Figure 9.19(a) shows planar parts A and B, their reference frames, and their configurations (x_a, y_a, θ_a) and (x_b, y_b, θ_b). The configuration space of the pair is the Cartesian product, $(x_a, y_a, \theta_a, x_b, y_b, \theta_b)$, of the part configurations. The configuration space coordinates represent three degrees of freedom of each part.

An alternative representation is to describe the relative position and orientation of part A with respect to B, which is fixed and whose reference frame is at the origin of the axes, as illustrated in Figure 9.19(b). In this case, three relative parameters (u, v, ψ) uniquely describe the relative position of A with respect to B. The relation between the absolute and relative coordinate systems is

$$
\begin{aligned}
u &= (x_a - x_b)\cos\theta_b + (y_a - y_b)\sin\theta_b, \\
v &= (y_a - y_b)\cos\theta_b - (x_a - x_b)\sin\theta_b, \\
\psi &= \theta_a - \theta_b.
\end{aligned}
\tag{9.48}
$$

The configuration space dimension for planar pairs is six for absolute coordinates, and three for relative coordinates. For spatial pairs, it is 12 for absolute coordinates and six for relative coordinates. Other useful parameterizations include quaternions for spatial rotations (Bottema and Roth, 1979) and Clifford algebras parameterization of planar and spatial displacements that yield algebraic surfaces that represent geometric constraints (McCarthy, 1990; Collins and McCarthy, 1998; Ge et al., 1998).

Contact analysis is simplified by considering only the varying configuration parameters. In mechanisms, parts frequently have less than six degrees of freedom. The fixed degrees of freedom correspond to constant configuration parameter values. Mathematically, this corresponds to projecting the higher-dimensional configuration space into a lower-dimensional space with identical properties. For example, in the ratchet mechanism, the ratchet and the driver are mounted on fixed axes, so only their orientation varies. A two-dimensional configuration space, showing the dependence between the two orientation parameters, fully describes the contacts between them.

Configuration space partitions into three disjoint sets that characterize part interaction: blocked space where the parts overlap, free space where they do not touch, and contact space where they touch. Contact space is the common boundary of free and blocked spaces. Free and blocked space are open sets whose dimension is identical to that of the configuration space, whereas the dimension of contact space is one lower. Intuitively, free and blocked space are open because disjoint or overlapping parts remain so under all small motions, whereas contact space is closed because touching parts separate or overlap under some small motions.

We illustrate these concepts with a simple example: a block that moves in a fixed frame. In Figure 9.20(a), the frame is fixed at the global origin, so we can ignore its coordinates from the configuration space and consider only the block coordinates (u, v, ψ) relative to the block origin. Assume first that the block translates in the displayed orientation without rotating. This yields a two-dimensional configuration space whose parameters are the horizontal and vertical parameters u and v, as shown in Figure 9.20(b). The gray region is blocked space, the white region is free space, and the black lines are contact space. The dot in free space marks the displayed position of the block. Free space divides into a central rectangle where the block is inside the frame, an outer region where it is outside, and a narrow connecting rectangle where it is partly inside. The contact constraints (lines in this case) bounding these regions represent contacts between the vertices and edges of the block and the frame. Typical configurations on each region are shown in Figure 9.20(b) on the left. A collision-free motion of the block corresponds to a continuous path in free and contact space. Changing the orientation of the block yields configuration spaces with different topologies, as shown in Figures 9.20(b)–9.20(e). Note that the free space consists now of two disconnected inside and outside regions because the block does not fit through the frame opening and thus cannot exit the inner region as before.

Consider now the same example, but with the block orientation varying. The configuration space becomes three dimensional with rotation coordinate ψ varying from $-\pi$ to π. Contact space is now two dimensional and is formed by contact patches, as shown in Figure 9.21. Typical configurations for three patches are shown on the left. To understand this space, consider it as a stack of planar slices along the rotation axis. Each slice is the configuration space of a block that translates at a fixed orientation, such as the three examples in Figures 9.20(b), 9.20(d), and 9.20(e). The full space is the union of the slices. The free space consists of an outer region,

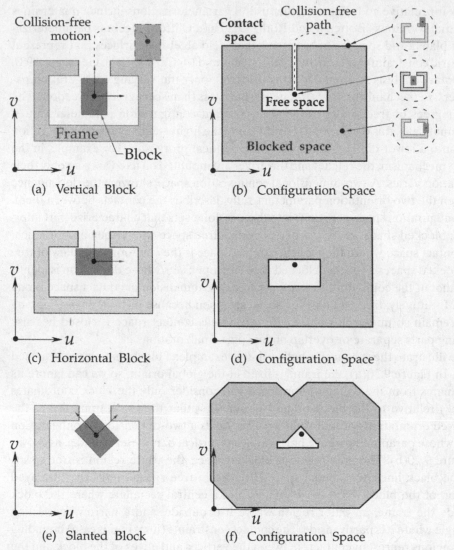

Figure 9.20. A translating block moving around a fixed frame at fixed orientations and their corresponding configuration spaces: vertical orientation (top), horizontal (middle) and slanted (bottom).

two inner regions, and two connecting channels near $\psi = \pm\pi/2$ where the block is nearly vertical. The outer envelope is the union of the outer regions, the inner envelope is the union of the inner rectangles, and the channels are the union of the connecting regions. Blocked space is the region between the envelopes and outside the channels. Contact boundaries, where contact patches intersect, correspond to simultaneous feature contacts.

Whatever its dimension, the configuration space of a pair is a complete representation of the part contacts. Any contact question is answerable by a configuration space query. For example, testing if parts overlap, do not touch, or are in contact in a given configuration corresponds to testing if the configuration point is in blocked, free, or contact space. Contacts between pairs of features correspond to contact constraints (curve segments in two dimensions and surface patches in three). The constraints' geometry encode the motion constraints. Their boundaries encode the

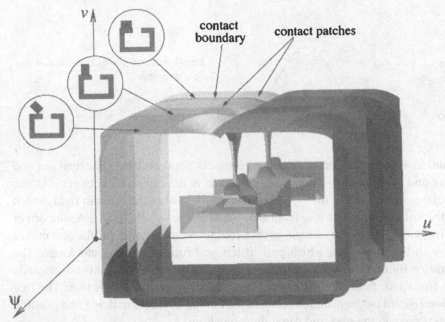

Figure 9.21. Contact space for a block with three degrees of freedom; each shade of gray denotes a contact patch.

contact change conditions. Part motions correspond to paths in configuration space. A path is legal if it lies in free and contact space, but it is illegal if it intersects blocked space. Contacts occur at configurations where the path crosses from free to contact space, break where it crosses from contact to free space, and change where it crosses between neighboring contact constraints.

The configuration space representation generalizes from pairs of parts to systems with more than two parts. A mechanical system of n parts has a $6n$-dimensional configuration space whose points specify the n part configurations. A system configuration is free when no parts touch, is blocked when two parts overlap, and is in contact when two parts touch and no parts overlap. The mechanical system configuration space can be obtained by combining the configuration spaces of its pairs (Joskowicz and Sacks, 1991), because the system is a collection of kinematic pairs (Reuleaux, 1875). System configuration spaces allow us to analyze multipart interactions but are difficult to compute.

CONFIGURATION SPACES OF THE RATCHET MECHANISM

We illustrate how pairwise configuration spaces are used in kinematic analysis and synthesis on the ratchet mechanism of Figure 9.18. There are four interacting pairs: driver–link, link–pawl, pawl–ratchet, and pawl–frame. The link–pawl pair is a revolute joint, and thus has a simple relationship: the pawl is constrained to rotate around the pin axis. The driver–link configuration space is two dimensional because both are pinned by revolute joints to the base. The pawl–frame and pawl–ratchet configuration spaces are three dimensional, because the pawl has three degrees of freedom.

Figure 9.22 shows the configuration space of the driver–link pair. The configuration space coordinates are the driver orientation θ and the link orientation ω. The

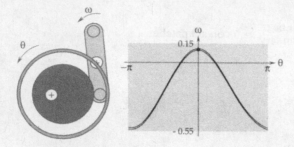

Figure 9.22. Driver–link pair and its configuration space.

upper and lower contact curves represent contacts between the cylindrical pin and the outer and inner cam profiles. The free space is the region in between. As the driver rotates from $\theta = -\pi$ to $\theta = 0$, its inner profile pushes the link pin right, which rotates the link counterclockwise from $\omega = -0.47$ rad to $\omega = 0.105$ rad. As the driver rotates from $\theta = 0$ to $\theta = \pi$, the pin breaks contact with the inner profile and makes contact with the outer one, which pulls it left and rotates the link clockwise. The configuration follows the lower contact curve from $\theta = -\pi$ to 0, travels horizontally through free space until it hits the upper contact curve, and follows it to π. The free play is determined by the distance between the curves. Changes in shape and position parameters change the play and can induce blocking.

Figure 9.23 shows the ratchet–pawl pair, a slice of its three-dimensional configuration space, and the three-dimensional contact space of the ratchet moving relative to the pawl. To best understand the figure, consider first the two-dimensional slice in Figure 9.23(b), which shows how the ratchet translates in the displayed orientation. The dot marks the displayed position of the ratchet relative to the pawl. It lies on a contact curve that represents contact between the pawl tip and the side of a ratchet tooth. The right end of the curve is the intersection point, with a second contact curve that represents contact between the left corner of the pawl and the next tooth counterclockwise. The ratchet can maintain this contact while translating right until the second contact occurs and further translation is blocked. The topology of the slice is preserved when the orientation of the pawl changes slightly, as can be seen in the contact space given in Figure 9.23(c). Changes in the orientation of the teeth and the ratchet's center of rotation can change the kinematic function of the pair.

CONFIGURATION SPACE COMPUTATION

Robotics research confirms the empirically observed difficulty of contact analysis with formal proofs that configuration space computation is worst-case exponential in the number of degrees of freedom (Latombe, 1991). Despite this result, contact analysis is manageable in practice because mechanical systems have characteristics that distinguish them from arbitrary collections of parts. Mechanism parts usually only interact with a few neighboring parts, are connected by simple joints, have few degrees of freedom, or consist of symmetric patterns of feature groups, for example, gear teeth. Typical systems have only one or two true degrees of freedom, not many. The challenge is to exploit these properties to develop specialized algorithms for important classes of mechanical systems and design tasks.

The robotics literature contains many configuration space computation algorithms (Latombe, 1991), although most are restricted to pairs of polygons and

Figure 9.23. (a) Ratchet–pawl pair; (b) configuration space slice at $\psi = 0.277$ rad; (c) contact space; the inner region is blocked space and the outer region is free space.

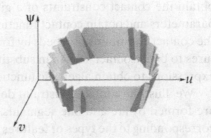

polyhedra. Special-purpose contact analysis methods for gears and cams, for which the contact sequence is known, are described in Angeles and Lopez-Cajun (1991), Gonzales and Angeles (1993), and Litvin (1994). Other higher pairs can be classified as planar or spatial and as fixed axes (one degree of freedom per part) or general. Fixed-axes planar pairs are by far the most common, followed by fixed-axes spatial pairs and by planar pairs with three degrees of freedom, according to a survey of more than 2,500 mechanisms from an encyclopedia (Joskowicz and Sacks, 1991). General spatial higher pairs with six degrees of freedom are rare. Efficient algorithms for computing two-dimensional configuration spaces of planar fixed-axes pairs whose shapes are formed by arc and line segments (Sacks and Joskowicz, 1995), and for fixed-axes spatial pairs whose shapes are formed by planar, cylindrical, and spherical patches bounded by line segments and circular arcs (Drori, Joskowicz, and Sacks 1999), are available. Sacks (1998) describes an algorithm for computing three-dimensional configuration spaces of general planar pairs, with which the figures of this section were generated. There are no algorithms computing the configuration space of general spatial pairs with six degrees of freedom, although robot motion planning research provides algorithms for a polyhedral robot moving amidst fixed polyhedral obstacles (Donald, 1987; Joskowicz and Taylor, 1996; Latombe, 1991).

Configuration space computation consists of partitioning the configuration space of a pair of interacting parts into free, contact, and blocked space. The geometric algorithms proceed in two steps: (1) compute the contact constraints for each pair of part features, and (2) compute the partition of the configuration space induced by the contact constraints. We describe each step briefly next.

Contact Constraints. The contact constraints are the configurations in which the features would touch if there were no other features to interfere. Contact constraints depend on the types of features in contact and on the part motion. For example, the contact constraints generated by translating line segments are line segments, as illustrated in Figure 9.20(b). There is one contact constraint for each combination of part features and motions. For example, there are 16 types of contact constraints for fixed-axes pairs whose parts are polygons, corresponding to all combinations of point and line segment feature pairs and all combinations of rotation and translation motions. The contact constraints are algebraic equalities specifying the contact function and inequalities specifying the contact boundary conditions. They are curves for two-dimensional configuration spaces (fixed-axes pairs) and patches for three-dimensional configuration spaces (general planar pairs).

Contact constraints can be systematically derived for each combination and stored in parametric form in a table. The parameters are the geometric characteristics of the part features, such as the line segment slope and the arc radius and origin. To obtain the contact constraints of a given feature pair, we substitute the geometric parameters and obtain contact functions and contact inequalities. The derivation of the contact constraints proceeds by formulating the geometric conditions for the features to be in contact, and then substituting the configuration parameters into these expressions to obtain algebraic functions.

We illustrate contact constraint derivation for general planar pairs whose shapes are formed by arc and line segments. There are three types of contact constraints, corresponding to the types of features in contact and their motions: moving arc–fixed line, moving line–fixed arc, and moving arc–fixed arc. Contacts involving points are identical to those for arcs of radius zero. Line–line contacts are subsumed by line–point contacts. Figure 9.24(a) shows an arc–line contact. The contact condition is that the distance between the center o of the arc and the line lm equals the arc radius r:

$$(\vec{o}_A - \vec{l}) \times (\vec{m} - \vec{l}) = dr,$$

where the multiplication sign denotes the vector cross product, d is the length of the line segment, and its interior lies to the left when traversed from l to m. Figure 9.24(b) shows an arc–arc contact. The contact condition is that the distance between the centers equals the sum of the radii:

$$(\vec{o}_A - \vec{p}) \cdot (\vec{o}_B - \vec{q}) = (r + s)^2,$$

where r and s are positive for convex arcs and negative for concave arcs.

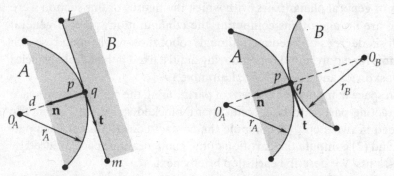

Figure 9.24. Contacts between planar features: (a) circular arc–line segment, (b) two circular arcs. Shading indicates the part interior.

We obtain the contact constraint functions from these equations by expressing the vectors in coordinate form. Let $\vec{q} = (q_x, q_y)$ be the coordinates of a point on the fixed part, and let $\vec{p} = (u, v) + R_\psi(p_x, p_y)$ be the coordinates of a moving point, where (p_x, p_y) are the part coordinates and R_ψ is a rotation matrix. After substitution, we obtain functions of the form $f(u, v, \psi) = 0$ parameterized by the part features. The moving arc–fixed line function is

$$u(m_y - l_y) - v(m_x - l_x) + (m_y - l_y)(o_x \cos \psi - o_y \sin \psi)$$
$$- (m_x - l_x)(o_x \sin \psi + o_y \cos \psi) = dr.$$

The moving line–fixed arc equation is

$$(u - o_x)(l_x \cos \psi - l_y \sin \psi) - (v - o_y)(l_x \sin \psi + l_y \cos \psi)$$
$$+ l_x(m_y - l_y) - l_y(m_x - l_x) + dr = 0.$$

The moving arc–fixed arc equation is

$$u^2 + v^2 + 2(o_x \cos \psi - o_y \sin \psi)u + 2(o_x \sin \psi + o_y \cos \psi)v$$
$$+ o_x^2 + o_y^2 + p_x^2 + p_y^2 - (r + s)^2 = 0.$$

Similar equations are obtained for contact ranges. Parametric contact constraints thus derived can be stored in a table and instantiated for particular feature pair geometry. The contact constraints of a pair are obtained by instantiating the parametric constrains for each pair of features.

Configuration Space Partition. The contact constraints partition configuration space into connected components. The component that contains the initial configuration is the reachable free space. Its boundary is a subset of the contact constraints. The other constraints lie in blocked space or form the boundaries of unreachable free space regions. Infeasible contact constraints must be eliminated. This is illustrated in Figure 9.25(a). It shows all the pairwise contact constraints for the translating block in Figure 9.20(a). Note that the two contact configurations on the left are unrealizable, and thus their contact constraints should be eliminated or restricted. Figure 9.25(b) shows the contact space (solid lines) partitioning the configuration space. Dashed lines indicate subsumed contacts.

Configuration space partition is computed by using computational geometry algorithms. The algorithms compute the intersection between contact constraints and classify free and blocked space accordingly. The algorithms are based on two- and three-dimensional geometric operations involving curve and surface intersection and sweeping algorithms (Sacks and Joskowicz, 1995; Sacks, 1998; Drori et al., 1999).

MECHANISM DESIGN WITH CONFIGURATION SPACES

Configuration spaces provide the computational basis for a wide variety of mechanism design tasks, including tolerancing, assembly, and shape and motion synthesis. The goal is to study system function under a range of operating conditions, find and correct design flaws, and optimize performance.

Tolerancing. Tolerancing consists of determining the variations in the system function that are due to manufacturing variations in its parts. Manufacturing

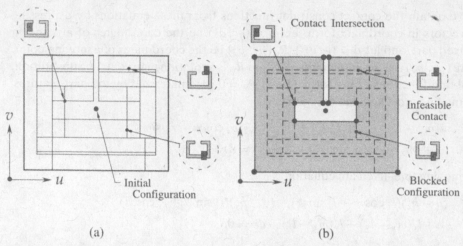

(a)　　　　　　　　　　　　　　　　　　　　　(b)

Figure 9.25. Partition of the fastener configuration space: (a) all pairwise contact constraints; (b) induced partition (solid lines are contact space; dashed lines are infeasible contacts).

variations are often expressed as tolerances, which correspond to intervals around nominal part shape and position values. For example, assume that the moving block in Figure 9.26 has nominal height $h = 20$ mm and width $w = 9.925$ mm (the frame has a fixed opening of 10 mm). A small variation corresponds to a tight tolerance interval, that is, ± 0.05 mm, whereas a larger variation corresponds to a looser interval, that is, ± 0.1 mm. The variations in the system function can be qualitative and quantitative. Qualitative variations occur when part parameter variations can cause unintended contact effects, such as jamming or interference with a frame. Quantitative variations determine the worst-case and average error in the system function. For example, the looser tolerance interval on the block width can prevent the block from exiting the frame (a qualitative change), whereas the smaller tolerance interval bounds the play between the block and the frame (a quantitative measure).

Kinematic variation can be modeled within the configuration space representation (Sacks and Joskowicz, 1998). As the parameters vary around their nominal

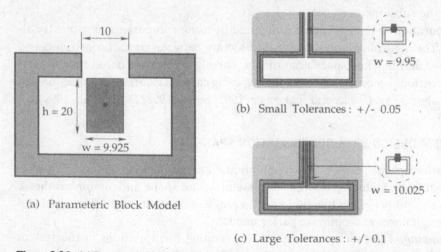

Figure 9.26. (a) Parametric block model, and details of the generalized configuration space with (b) small (± 0.05 mm) and (c) large (± 0.1 mm) tolerances.

values, the contact configurations vary around a zone around the nominal contact space. The zones define the set of all configurations for which there is one or more combinations of parameter values within the tolerance interval that generates a contact. The boundaries of the contact zone represent the worst-case variation. A standard first-order approximation of the contact zone boundaries can be obtained by summing up the partial derivatives of the parametric contact functions.

Figures 9.26(b) and 9.26(c) illustrate these concepts. They show a detail of the nominal configuration space (solid black lines and gray shading) and the contact zones (dark gray lines and dark gray shading) corresponding to the tolerance intervals. Zone boundaries (dark gray lines) correspond to worst-case variations. Configurations in the zone correspond to specific part instances in contact (inserts on the right). The distance between the inner and outer upper vertical zone boundaries quantify minimum and maximum play. For small tolerances, the topology of the space is preserved, meaning that the function of the system remains unchanged and the play interval is [0.025, 0.125] mm. However, for the larger tolerances, the zone boundaries cross, blocking the channel. This means that for some block dimensions within the tolerance interval, the block will not fit through the frame because the inner configuration space region is now disconnected from the outer one.

Assembly. Assembly consists of finding a set of part motions that brings the parts of a mechanism in their working configuration. It is a kind of motion planning problem, in which the starting configuration is the unassembled parts, the assembly sequence is a set of collision-free assembly part motions for parts, and the goal configuration is the assembled mechanism in its working configuration. Assembly planning is an integral part of mechanism design, as it influences part shapes and positions. Because it deals with part contacts, the configuration space framework described in this chapter is an appropriate framework to study the problem. Note that the assembly part motions are usually different from the motions that the parts have when functioning.

The main difficulty is that assembly motions are usually spatial, so parts can have up to six degrees of freedom, which precludes pairwise configuration space computation. Many approaches have been developed (Homem de Mello and Lee, 1991), some based on local geometric structures called *blocking graphs* (Wilson and Latombe, 1994). Latombe, Wilson, and Cazals (1997), Blind et al. (2000), and Xiao and Xuerong (2000) describe an algorithm for assembly planning with toleranced parts based on configuration space.

Shape Synthesis. One aspect of mechanism synthesis is part shape synthesis and optimization. The goal is to create and modify part shapes to satisfy design requirements. Special-purpose approaches have been developed for cam (Angeles and Lopez-Cajun, 1991; Gonzales and Angeles, 1993) pairs and for gear (Litvin, 1994) pairs. For other pairs, approaches based on configuration space have been proposed.

Shape synthesis can be seen as the inverse process of analysis: given a parametric part model and a design specification represented as a configuration space, the goal is to invert the mapping from parameter values to configuration spaces so that the values yield the desired configuration space (Joskowicz and Sacks, 1994). For nonparametric part models, part boundaries can be modified by using heuristic methods (Joskowicz and Addanki, 1988; Faltings and Sun, 1996). Similar techniques have been applied to feeding track design (Caine, 1994).

A related task is fixture synthesis, in which the goal is to design fixture shapes that eliminate the movement of parts to be machined. Eliminating movement corresponds to reducing the volume of the configuration space surrounding the machining configuration to zero (Rimon and Blake, 1996; Brost and Goldberg, 1996; Rimon and Burdick, 1998).

Motion Synthesis. Another aspect of mechanism synthesis is motion synthesis. Motion synthesis is the task of designing a mechanism topology that transforms a set of input motions into desired output motions. The mechanism topology is built as a chain of basic kinematic pair building blocks, such as joints, cams, gears, and other higher pairs. Hoover and Rinderle (1989) developed a heuristic method based on functional integration for speed ratio and geometric requirements of gear transmission mechanisms. Kota and Chiou (1992) represent kinematic pair function with matrices that are chained to obtain the desired output. Chakrabarti and Bligh (1996) successively refine motion type requirements based on kinematic pair behavior.

Configuration space provides a representation of mechanism function that has been used for mechanism retrieval (Joskowicz, 1990; Murakami and Nakajima, 1997) and motion synthesis (Subramanian and Wang, 1995; Li, Chan, and Tan, 1999). The basic building blocks are kinematic pairs, whose function is described with a qualitative representation of their configuration space. The synthesis algorithm generates design solutions by forming chains of kinematic pairs that match the desired mechanism function. The advantages of these methods are that they allow for qualitative function descriptions akin to those used in conceptual design (Stahovich, Davis, and Shrobe, 1998), and that they allow for the design of multiple-state mechanisms. To date, they are limited to simple planar mechanisms.

STATE OF THE ART

The kinematic synthesis of machines ranging from the lever to the robot has provided remarkable capabilities for redirecting power to our ends. Furthermore, it seems clear that the future holds more opportunities for mechanical systems that are tailored to enhance and augment our individual capabilities, whether at the human scale or microscale.

Research in the kinematic synthesis of mechanisms and robots has demonstrated the central importance of workspace and mechanical advantage as design criteria. The workspace, as defined by the kinematics equations, can be used for design, tolerancing, and assembly planning. The Jacobian of these equations yields the speed ratios of the system, which, by means of the principle of virtual work, define its mechanical advantage. And, optimization techniques have proven effective for fitting the workspace and mechanical advantage of serial and parallel chain systems to design specifications (Ravani and Roth, 1983; Larochelle and McCarthy, 1995).

Unfortunately, computer tools for kinematic synthesis seem to exist only as special-purpose algorithms developed by individual researchers. Commercial systems, such as LINCAGES 2000 (University of Minnesota), SyMech (www.symech.com), and Watt (heron-technologies.com), focus solely on planar four- and six-bar linkages. Kinematic synthesis in engineering applications seems to consist of iterated analysis, in which little or no attempt is made to use workspace and mechanical advantage criteria to generate designs.

A systematic procedure for the design of both serial and parallel linkages and robots that can match workspace and mechanical advantage is needed, especially one that allows a comparison of different machine topologies. Consider the following enumeration of spatial open chains of various degrees of freedom (DOF):

- Two-DOF chains: 2R, RP, C, T.
- Three-DOF chains: 3R, RRP, PPR, CR, CP, TR, RP, S.
- Four-DOF chains: 4R, 3PR, 3RP, CRR, CPR, CC, TRR, TPR, TC, TT, RS, PS.
- Five-DOF chains: 5R, 3PRR, RRS, PPS, CPC, CRC, CS, TRC, TPC, 2TR, 2TP, TS.

Only a few of these topologies have been explored for use in the design of spatial linkage systems. Clearly, computer automation of the synthesis of these chains and their assembly into parallel systems can open the door to a wealth of new devices. This is particularly true when asymmetry exists in the task, or the device must work around obstacles and people.

Configuration space analysis provides new capabilities for automation of the design and manufacture of mechanical systems. This is an extension of the concept of workspace to include all bodies, not just the end effector (Joskowicz and Sacks, 1999). In this formulation the free-space boundary for each part is obtained by analyzing the contact available with every other part. The set of positions and orientations of a body can be parameterized in many useful ways. Quaternions have been found to provide a convenient representation for spatial rotations (Bottema and Roth, 1979). Clifford algebras have been used to parameterize planar and spatial displacements in order to obtain algebraic surfaces that represent geometric constraints (McCarthy, 1990; Collins and McCarthy, 1998; Ge et al., 1998).

The generalization of Bezier curves to spatial motion provides a convenient way to specify continuous taskspaces. However, the extension of this technique to surfaces, solids, hypersolids, or hypersurfaces in $SE(3)$ does not exist. This means that we cannot directly specify the workspace of a robotic system. Currently, we rely on general symmetry requirements.

Methodologies for the design of cam systems, gear trains, and linkages have yet to be integrated into computer-aided engineering tools. Although sculptured surfaces are central to the operation of human and animal joints, they have yet to be exploited for robotics applications (Lenarčič, Stanišić, and Parenti-Castelli, 2000; Parenti-Castelli and DiGregorio, 2000).

Microelectromechanical systems provide unique opportunities for multiple small systems constructed by using layered manufacturing technologies typical of electronic devices. The small size of these systems challenges the basic assumptions of virtual work, and their design has benefited from the synthesis theory for compliant linkages (Ananthasuresh and Kota, 1995). The design of microrobotic systems requires advances in the design and construction of joints and actuators (Will, 2000).

A speculative direction for kinematic synthesis research involves the analysis of proteins and rational drug design (Wang, Lozano-Perez, and Tidor, 1998, LaValle et al., 2000). A protein can be viewed a serial chain of links, called *amino acids*, each of which are similar except for molecular radicals that extend to the side. A protein chain may contain 1000 amino acids that fold into a complex spatial configuration. This configuration can shift between two positions in order to perform a task, exactly like the machines that we have been considering in this chapter.

CONCLUSION

This chapter surveys the theoretical foundation and current implementations of kinematic synthesis for the design of machines. The presentation is organized to illustrate the importance of the concepts of workspace and mechanical advantage in the synthesis process. Although the workspace of a robot arm is a familiar entity, it is not generally recognized that concepts ranging from the geometric constraints used in linkage design through configuration spaces used to define tolerances are simply different representations of workspace. Similarly, whereas mechanical advantage is clearly important to the design of a lever and wedge, it is not as obvious that the Jacobian conditions used in the design of robotic systems are, in fact, specifications on mechanical advantage.

The result of this study are the conclusions that (i) workspace and mechanical advantage are effective specifications for the synthesis of a broad range of mechanical devices; (ii) there already exist a large number of specialized algorithms that demonstrate the effectiveness of kinematic synthesis; and (iii) there are many opportunities for the application of new devices available through kinematic synthesis. What is needed is a systematic development of computer tools for kinematic synthesis that integrates the efforts of a community of researchers from mechanical design, robotics, and computer science.

REFERENCES

Albro, J. V., Sohl, G. A., Bobrow, J. E., and Park, F. C. (2000). "On the computation of optimal high-dives." In *Proceedings of the International Conference on Robotics and Automation*, IEEE, Piscataway, NJ, pp. 3959–3964.

Ahlers, S. G. and McCarthy, J. M. (2000). "The Clifford algebra of double quaternions and the optimization of TS robot design." In *Applications of Clifford Algebras in Computer Science and Engineering*, E. Bayro and G. Sobczyk (eds.), Birkhauser, Cambridge, MA.

Ananthasuresh, G. K. and Kota, S. (1995). "Designing compliant mechanisms," *Mechanical Engineering*, 117(11):93.

Angeles, J. and Chablat, D. (2000). "On isotropic sets of points in the plane. Application to the design of robot architectures." In *Advances in Robot Kinematics*, J. Lenarčič and M.M. Stanišić (eds.), Kluwer Academic, The Netherlands, pp. 73–82.

Angeles, J. and Lopez-Cajun, C. (1991). *Optimization of Cam Mechanisms*. Kluwer Academic, Dordrecht, The Netherlands.

Angeles, J. and Lopez-Cajun, C.S. (1992). "Kinematic isotropy and the conditioning index of serial manipulators," *International Journal of Robotics Research*, 11(6):560–571.

Arnold, V. I. (1978). *Mathematical Methods of Classical Mechanics*. Translated by K. Vogtmann and A. Weinstein, Springer-Verlag, New York.

Blind, S., McCullough, C. C., Akella, S., and Ponce, J. (2000). "A reconfigurable parts feeder with an array of pins." In *Proceedings of the International Conference on Robotics and Automation*, IEEE, Piscataway, NJ, pp. 147–153.

Bodduluri, R. M. C. and McCarthy, J. M. (1992). "Finite position synthesis using the image curve of spherical four-bar motion," *ASME Journal of Mechanical Design*, 114(1):55–60.

Bottema, O. and Roth, B. (1979). *Theoretical Kinematics*, North-Holland, New York (reprinted by Dover).

Brost, R. C. (1989). "Computing metric and topological properties of configuration space obstacles." In *International Conference on Robotics and Automation*, IEEE, Piscataway, NJ, pp. 1656–1660.

Brost, R. C. and Goldberg, K. Y. (1996). "A complete algorithm for designing planar fixtures using modular components," *IEEE Transactions on Robotics and Automation*, **12**(1):31–46.

Burmester, L. (1886). *Lehrbuch der Kinematik*, Verlag Von Arthur Felix, Leipzig, Germany.

Caine, M. E. (1994). "The design of shape interactions using motion constraints." In *Proceedings of the IEEE International Conference on Robotics and Automation*, IEEE, New York, pp. 366–371.

Chakrabarti, A. and Bligh, T. P. (1996). "An approach to functional synthesis of solutions in mechanical conceptual design: parts II and III," *Research in Engineering Design*, **8**:52–62 and **2**:116–124.

Chase, T., Erdman, A. G., and Riley, D. (1981). "Synthesis of six bar linkages using and interactive package," *Proceedings of the 1981 OSU Applied Mechanisms Conference*, Oklahoma State University, Stillwater, OK.

Chedmail, P. (1998). "Optimization of multi-DOF mechanisms." In *Computational Methods in Mechanical Systems*, J. Angeles and E. Zakhariev (eds.), Springer-Verlag, Berlin, pp. 97–129.

Chedmail, P. and Ramstein, E. (1996). "Robot mechanisms synthesis and genetic algorithms." In *Proceedings of the 1996 IEEE Robotics and Automation Conference*, IEEE, New York, pp. 3466–3471.

Chen, I. and Burdick, J. (1995). "Determining task optimal modular robot assembly configurations." In *Proceedings of the 1995 IEEE Robotics and Automation Conference*, IEEE, New York, pp. 132–137.

Chen, P. and Roth, B. (1967). "Design equations for the finitely and infinitesimally separated position synthesis of binary links and combined link chains," *ASME Journal of Engineering for Industry*, **91**:209–219.

Collins, C. L., and McCarthy, J. M. (1998). "The quartic singularity surfaces of a planar platforms in the Clifford algebra of the projective plane," *Mechanism and Machine Theory*, **33**(7):931–944.

Craig, J. J. (1989). *Introduction to Robotics: Mechanics and Control.* Addison-Wesley, Reading, MA.

Dimarogonas, A. (1993). "The origins of the theory of machines and mechanisms." In *Modern Kinematics: Developments in the Last Forty Years*, A. G. Erdman (ed.), Wiley, New York.

Donald, B. R. (1987). "A search algorithm for motion planning with six degrees of freedom," *Artificial Intelligence*, **31**:295–353.

Drori, I., Joskowicz, L., and Sacks, E. (1999). "Spatial contact analysis of fixed-axes pairs using configuration spaces." In *Proceedings of the 13th IEEE International Conference on Robotics and Automation*, IEEE, New York.

Erdman, A. G. and Gustafson, J. E. (1977). "LINCAGES: linkage interactive computer analysis and graphically enhanced synthesis package," ASME Paper No. 77-DTC-5, ASME, New York.

Erdman, A. G. and Sandor, G. N. (1997). *Mechanism Design: Analysis and Synthesis*. 3rd ed., Prentice-Hall, Englewood Cliffs, NJ.

Etzel, K. and McCarthy, J. M. (1996). "A metric for spatial displacements using biquaternions on SO(4)." In *Proceedings of the 1996 International Conference on Robotics and Automation*, IEEE, Piscataway, NJ.

Faltings, B. and Sun, K. (1996). "FAMING: supporting innovative mechanism shape design," *Computer-Aided Design*, **28**(3): 207–213.

Farin, G. (1997), *Curves and Surfaces for CAGD*. Academic, Chestnut Hill, MA.

Fichter, E. F. (1987). "A Stewart platform-based manipulator: general theory and practical construction." In *Kinematics of Robot Manipulators*, J. M. McCarthy (ed.), MIT Press, Cambridge, MA, pp. 165–190.

Furlong, T. J. Vance, J. M., and Larochelle, P. M. (1998). "Spherical mechanism synthesis in virtual reality," *CD-ROM Proceedings of the ASME DETC'98*, ASME, New York, Paper No. DETC98/DAC-5584.

Ge, Q. J. and Kang, D. (1995). "Motion interpolation with G^2 composite Bezier motions," *ASME Journal of Mechanical Design*, **117**:520–525.

Ge, Q. J. and McCarthy, J. M. (1989). "Equations for the boundaries of joint obstacles for planar robots." In *Proceedings of the International Conference on Robotics and Automation*, Vol. 1, IEEE, Piscataway, NJ, pp. 164–169.

Ge, Q. J. and McCarthy, J. M. (1990). "An algebraic formulation of configuration space obstacles for spatial robots." In *Proceedings of the 1990 International Conference on Robotics and Automation*, Vol. 3, IEEE, Piscataway, NJ, pp. 1542–1547.

Ge, Q. J. and McCarthy, J. M. (1991). "Functional constraints as algebraic manifolds in a Clifford algebra," *IEEE Journal of Robotics and Automation*, 7(5):670–677.

Ge, Q. J. and Ravani, B. (1994). "Geometric construction of Bézier motions," *ASME Journal of Mechanical Design*, 116:749–755.

Ge, Q. J., Varshney, A., Menon, J. P., Chang, C.-F. (1998). "Double quaternions for motion interpolation," *CD-ROM Proceedings of the ASME DETC'98*, ASME, New York, Paper No. DETC98/DFM-5755.

Gonzales-Palacios, M. and Angeles, J. (1993). *Cam Synthesis*. Kluwer Academic, Dordrecht, The Netherlands.

Greenwood, D. T. (1977). *Classical Dynamics*. Prentice-Hall, Englewood Cliffs, NJ.

Gosselin, C. (1998). "On the design of efficient parallel mechanisms." In *Computational Methods in Mechanical Systems*, J. Angeles and E. Zakhariev (eds.), Springer-Verlag, Berlin, pp. 68–96.

Gosselin, C. and Angeles, J. (1988). "The Optimum kinematic design of a planar three degree of freedom parallel manipulator," *ASME Journal of Mechanisms, Transmissions, and Automation in Design*, 110(3):35–41.

Gupta, K. (1987). "On the nature of robot workspace," *Kinematics of Robot Manipulators*, J. M. McCarthy (ed.), MIT Press, Cambridge, MA, pp. 120–129.

Hamilton, W. R. (1860). *Elements of Quaternions*. Reprinted by Chelsea Publications, 1969, New York.

Hervé, J. M. (1978). "Analyse structurelle des méchanismes par groupe des déplacements, *Mechanism and Machine Theory*, 13(4):437–450.

Hervé, J. M. (1999). "The Lie group of rigid body displacements, a fundamental tool for mechanism design," *Mechanism and Machine Theory*, 34:719–730.

Homem de Mello, L., and Lee, S. (1991). *Computer-Aided Mechanical Assembly Planning*. Kluwer Academic, Dordrecht, The Netherlands.

Hoover, S. P. and Rinderle, J. R. (1989). "A synthesis strategy for mechanical devices," *Research in Engineering Design*, 1:87–103.

Innocenti, C. (1995). "Polynomial solution of the spatial Burmester problem," *ASME Journal of Mechanical Design*, 117:64–68.

Joskowicz, L. (1990). "Mechanism comparison and classification for design," *Research in Engineering Design*, 1(4):149–166.

Joskowicz, L. and Addanki, S. (1988). "From kinematics to shape: an approach to innovative design." In *Proceedings of the National Conference on Artificial Intelligence*, AAAI Press, Menlo Park, CA, pp. 347–352.

Joskowicz, L. and Sacks, E. (1991). "Computational kinematics," *Artificial Intelligence*, 51:381–416.

Joskowicz, L. and Sacks, E. (1994). "Configuration space computation for mechanism design," *Proceedings of the 1994 IEEE International Conference on Robotics and Automatio*, IEEE Computation Society Press, Piscataway, NJ.

Joskowicz, L. and Sacks, E. (1999). "Computer-aided mechanical design using configuration spaces," *IEEE Computers in Science and Engineering*, **Nov/Dec**:14–21.

Joskowicz, L. and Taylor, R. H. (1996). "Interference-free insertion of a solid body into a cavity: an algorithm and a medical application," *International Journal of Robotics Research*, 15(3): 211–229.

Kaufman, R. E. (1978). "Mechanism design by computer," *Machine Design*, **Oct**:94–100.

Kota, S. and Chiou, S. J. (1992). "Conceptual design of mechanisms based on computational synthesis and simulation of kinematic building blocks," *Research in Engineering Design*, 4:75–87.

Kota, S. (Ed.) (1993). "Type synthesis and creative design." In *Modern Kinematics: Developments in the Last Forty Years,* A. G. Erdman (ed.), Wiley, New York.

Kumar, V. (1992). "Instantaneous kinematics of parallel-chain robotic mechanisms," *Journal of Mechanical Design,* **114**(3):349–358.

Larochelle, P. M. (1998). "Spades: software for synthesizing spatial 4C linkages," in *CD-ROM Proceedings of the ASME DETC'98,* ASME, New York, Paper No. DETC98/Mech-5889.

Larochelle, P., Dooley, J., Murray, A., McCarthy, J. M. (1993). "SPHINX – software for synthesizing spherical 4R mechanisms." In *Proceedings of the 1993 NSF Design and Manufacturing Systems Conference,* Vol. 1, NSF, Washington, D.C., pp. 607–611.

Larochelle, P. M. and McCarthy, J. M. (1995). "Planar motion synthesis using an approximate bi-invariant metric," *ASME Journal of Mechanical Design,* **117**(4):646–651.

Latombe, J.-C. (1991). *Robot Motion Planning.* Kluwer Academic, Dordrecht, The Netherlands.

Latombe, J.-C., Wilson, R., and Cazals, F. (1997). "Assembly sequencing with toleranced parts," *Computer-Aided Design,* **29**(2):159–174.

LaValle, S. M., Finn, P. W., Kavraki, L. E., and Latombe, J. C. (2000). "A randomized kinematics-based approach to pharmacophore-constrained conformational search and database screening," *Journal of Computational Chemistry,* **21**(9):731–747.

Lee, J., Duffy, J., and Keler, M. (1996). "The optimum quality index for the stability of in-parallel planar platform devices." In *CD-ROM Proceedings of the 1996 ASME,* *Design Engineering Technical Conferences,* ASME, New York, 96-DETC/MECH-1135.

Lee, J., Duffy, J., and Hunt, K. (1998). "A practical quality index based on the octahedral manipulator," *International Journal of Robotics Research,* **17**(10):1081–1090.

Leger, C. and Bares, J. (1998). "Automated synthesis and optimization of robot configurations." In *CD-ROM Proceedings of the ASME DETC'98,* ASME, New York, Paper No. DETC98/Mech-5945.

Lenarčič, J., Staniŝić, M.M., and Parenti-Castelli, V. (2000). "A 4-DOF parallel mechanism simulating the movement of the human sternum-clavicle-Scapula complex." In *Advances in Robot Kinematics,* J. Lenarčič and M.M. Staniŝić (eds.), Kluwer Academic, Dordrecht, The Netherlands, pp. 325–331.

Li, C. L., Chan, K. W., and Tan, S. T. (1999). "A configuration space approach to the automatic design of multiple-state mechanical devices," *Computer-Aided Design,* **31**:621–653.

Lin, Q. and Burdick, J. W. (2000). "On well-defined kinematic metric functions." In *Proceedings of the International Conference on Robotics and Automation,* IEEE, Piscataway, NJ, pp. 170–177.

Litvin, F. (1994). *Gear Geometry and Applied Theory.* Prentice-Hall, Englewood Cliffs, NJ.

Lozano-Perez, T. (1983). "Spatial planning: a configuration space approach," *IEEE Transactions on Computers,* **C-32**(2):108–120.

Martinez, J. M. R., and Duffy, J. (1995). "On the metrics of rigid body displacements for infinite and finite bodies," *ASME Journal of Mechanical Design,* **117**(1):41–47.

Mason, M. T. and Salisbury, J. K. (1985). *Robot Hands and the Mechanics of Manipulation.* MIT Press, Cambridge, MA.

Merlet, J. P. (1989). "Singular configurations of parallel manipulators and Grassmann geometry," *International Journal of Robotics Research,* **8**(5):45–56, 1989.

Merlet, J. P. (1999). *Parallel Robots.* Kluwer Academic, Dordrecht, The Netherlands.

McCarthy, J. M. (1990). *Introduction to Theoretical Kinematics.* MIT Press, Cambridge, MA.

McCarthy, J. M. (2000). *Geometric Design of Linkages.* Springer-Verlag, New York.

McCarthy, J. M. and Ahlers, S. G. (2000). "Dimensional synthesis of robots using a double quaternion formulation of the workspace." In *Robotics Research: The Ninth International Symposium,* J. M. Hollerbach and D. E. Koditschek (eds.), Springer-Verlag, New York, pp. 3–8.

McCarthy, J. M. and Liao, Q. (1999). "On the seven position synthesis of a 5-SS platform linkage," *ASME Journal of Mechanical Design,* in press.

Moon, F. C. (1998). *Applied Dynamics: With Applications to Multibody and Mechatronic Systems.* Wiley, New York.

Murakami, T. and Nakajima, N. (1997). "Mechanism concept retrieval using configuration space," *Research in Engineering Design*, 9:99–111.

Murray, A. P., Pierrot, F., Dauchez, P., and McCarthy, J. M. (1997). "A planar quaternion approach to the kinematic synthesis of a parallel manipulator," *Robotica*, 15:4.

Murray, A. and Hanchak, M. (2000). "Kinematic synthesis of planar platforms with RPR, PRR, and RRR chains." In *Advances in Robot Kinematics*, J. Lenarčič and M. M. Stanišić (eds.), Kluwer Academic, Dordrecht, The Netherlands, pp. 119–126.

Norton, R.L. (Ed.) (1993). "Cams and cam followers." In *Modern Kinematics: Developments in the Last Forty Years*, A. G. Erdman (ed.), Wiley, New York.

Paul, R. P. (1981). *Robot Manipulators: Mathematics, Programming, and Control*. MIT Press, Cambridge, MA.

Park, F. C. (1995). "Distance metrics on the rigid body motions with applications to mechanism design," *ASME Journal of Mechanical Design*, 117(1):48–54.

Park, F. C. and Bobrow, J. E. (1995). "Geometric optimization algorithms for robot kinematic design," *Journal of Robotic Systems*, 12(6):453–463.

Parenti-Castelli, V. and DiGregorio, R. (2000). "Parallel mechanisms applied to the human knee passive motion simulation." In *Advances in Robot Kinematics*, J. Lenarčič and M. M. Stanišić (eds.), Kluwer Academic, Dordrecht, The Netherlands, pp. 333–344.

Perez, A. M. and McCarthy, J. M. (2000). "Dimensional synthesis of spatial RR robots." In *Advances in Robot Kinematics*, J. Lenarčič and M. M. Stanišić (eds.), Kluwer Academic, Dordrecht, The Netherlands, pp. 93–102.

Ravani, B. and Roth, B. (1983). "Motion synthesis using kinematic mapping," *ASME Journal of Mechanisms, Transmissions and Automation in Design*, 105:460–467.

Reuleaux, F. (1875). *Theoretische Kinematik*, Friedrich Vieweg und Sohn, Brunswick, Germany. Translated by A. B. W. Kennedy, *Reuleaux, Kinematics of Machinery*, Macmillan, London, 1876. Reprinted by Dover, New York, 1963.

Rimon, E. and Blake, A. (1996). "Caging 2D bodies by one-parameter two-fingered gripping systems." In *IEEE International Conference on Robotics and Automation*, IEEE, New York, pp. 1458–1464.

Rimon, E. and Burdick, J. W. (1998). "Mobility of bodies in contact," *IEEE Transactions on Robotics and Automation*, 14(5):696–717.

Rubel, A. J. and Kaufman, R. E. (1977). "KINSYN 3: a new human-engineering system for interactive computer aided design of planar linkages," *Journal of Engineering for Industry*, 99(2).

Ruth, D. A. and McCarthy, J. M. (1997). "SphinxPC: an implementation of four position synthesis for planar and spherical 4R linkages." In *CD-ROM Proceedings of the ASME DETC'97*, ASME, New York, Paper No. DETC97/DAC-3860.

Sacks, E. (1998). "Practical sliced configuration spaces for curved planar pairs," *International Journal of Robotics Research*, 17(11).

Sacks, E. and Joskowicz, L. (1995). "Computational kinematic analysis of higher pairs with multiple contacts," *Journal of Mechanical Design*, 117:269–277.

Sacks, E. and Joskowicz, L. (1998). "Parametric kinematic tolerance analysis of general planar systems," *Computer-Aided Design*, 30(9):707–714.

Salisbury, J. K. and Craig, J. J. (1982). "Articulated hands: force control and kinematic issues," *International Journal of Robotics Research*, 1(1):4–17.

Salcudean, S. E. and Stocco, L. (2000). "Isotropy and actuator optimization in haptic interface design." In *Proceedings of the International Conference on Robotics and Automation*, IEEE, Piscataway, NJ, pp. 763–769.

Shilov, G. E. (1974). *An Introduction to the Theory of Linear Spaces*, Dover. New York.

Shoemake, K. (1985). "Animating rotation with quaternion curves," *ACM SIGGRAPH*, 19(3):245–254.

Stahovich, T., Davis, R., and Shrobe, H. (1998). "Generating multiple new designs from sketches," *Artificial Intelligence*, 104:211–264.

Subramanian, D. and Wang, C. (1995). "Kinematic synthesis with configuration space," *Research in Engineering Design*, **7**(3):192–213.

Suh, C. H. and Radcliffe, C. W. (1978). *Kinematics and Mechanism Design*. Wiley, New York.

Tsai, L. W. (1993). "Gearing and Geared Systems." In *Modern Kinematics: Developments in the Last Forty Years*, A. G. Erdman (ed.), Wiley, New York.

Tsai, L.W. (1999). *Robot Analysis, The Mechanics of Serial and Parallel Manipulators*. Wiley, New York.

Vijaykumar, R., Waldron, K., and Tsai, M. J. (1987). "Geometric optimization of manipulator structures for working volume and dexterity." In *Kinematics of Robot Manipulators*, J. M. McCarthy (ed.), MIT Press, Cambridge, MA, pp. 99–111.

Waldron, K. J. and Kinzel, G. (1998). *Kinematics, Dynamics and Design of Machinery* Wiley, New York.

Waldron, K. J. and Song, S.M. (1981). "Theoretical and numerical improvements to an interactive linkage design program, RECSYN." In *Proceedings of the Seventh Applied Mechanisms Conference*, Oklahoma State University, Stillwater, OK.

Wang, E., Lozano-Perez, T., and Tidor, B. (1998). "AmbiPack: a systematic algorithm for packing of rigid macromolecular structures with ambiguous constraints," *Proteins: Structure, Function, and Genetics*, **32**:26–42.

Will, P. (2000). "MEMS and robotics: promise and problems." In *Proceedings of the International Conference on Robotics and Automation*, IEEE, Piscataway, NJ, pp. 938–946.

Wilson, R. and Latombe, J.-C. (1994). "Geometric reasoning about mechanical assembly," *Artificial Intelligence*, **71**:371–396.

Xiao, J. and Xuerong, J. (2000). "Divide-and-merge approach to automatic generation of contact states and planning contact motion." In *Proceedings of the International Conference on Robotics and Automation*, IEEE, Piscataway, NJ, pp. 750–756.

Zefran, M., Kumar, V., and Croke, C. (1996). "Choice of Riemannian metrics for rigid body kinematics." In *CD-ROM Proceedings of the ASME DETC'96*, ASME, New York, Paper No. DETC96/Mech-1148.

CHAPTER TEN

Systematic Chemical Process Synthesis

Scott D. Barnicki and Jeffrey J. Siirola

INTRODUCTION

Chemical processing involves the transformation of one composition or form of matter into another, most often via chemical reactions and purifications, but also through other transforming operations as well. Although criticized by the public and government alike, chemical processing plays a central role in the manufacture of a myriad of everyday materials, such as brick, mortar, steel, aluminum, glass, paper, plastics, rubber, concrete, asphalt, gasoline, plaster, fibers, dyes, paint, adhesives, soap, photographic film, computer chips, pharmaceuticals, vitamins, fertilizers, pesticides, foods, wine, and drinking water, as well as in such activities as food preservation, sewage treatment, power generation, smog prevention, and clothes cleaning. These activities are conducted naturally by biological organisms and intentionally by mankind in both batchwise and continuous modes. Production scales range from micrograms to gigatons, at locations as varied as the backyard pond, the kitchen, an oil refinery, a pharmaceutical plant, or a steel mill. Chemical processing primarily involves the interconversion of one material into another rather than the formation of materials into an artifact or the design of functionality into an artifact.

Process systems engineering is that branch of chemical engineering concerned with the design, operation, and control of chemical processes. Process systems engineers make much use of computer simulation and mathematical techniques to optimize potential chemical plant designs before such plants are actually constructed and operated. However, before any process design alternative may be modeled, simulated, analyzed, or optimized, it must first be invented or synthesized. The basic objective of chemical process design is the creation and specification of a process flow diagram or *flow sheet*. The flow sheet represents a means of converting a set of raw materials into desirable products utilizing physically realizable transformations, or *unit operations*, such as reactors, distillation columns, pumps, heat exchangers, and the like. These unit operations are governed by the laws of chemical and physical thermodynamics and the conservation of mass and energy. All of this must be done at the desired scale, safely, environmentally responsibly, efficiently, economically, and in a timely fashion.

Chemical process design involves a hierarchy of complex activities at both routine and innovative levels. Routine design is characterized by well-defined procedures for determining appropriate values for equipment design parameters. This activity is largely analytical and is primarily concerned with the specification and function of particular unit operations such as reactors, extractors, and distillation columns.

By contrast, *process synthesis*, the invention of overall chemical processing concepts, is a creative open-ended activity characterized by a combinatorially large number of feasible alternatives (Rudd, Powers, and Siirola, 1973). The process synthesis task lays out the major structural characteristics of the process flow sheet, the different kinds of unit operations, their specification, their placement, and their interconnection. The initial process concept will be refined and optimized on the basis of laboratory and pilot plant experimental data, detailed design calculations, application of engineering standards, expert opinion, available equipment and construction capabilities, and other input. However, it is the structure of the process flow sheet created during process synthesis that determines most of the economic potential of a proposed chemical process. The number of feasible flow sheets is potentially very large. It is very desirable that conceptual process flow sheets first synthesized prove to be better starting points for the remaining design optimization and refinement activities than other possible flow sheets. Good routine design can enhance the value of a well-synthesized flow sheet, but it cannot overcome the economic onus of a structurally inferior flow sheet. For this reason, significant effort has been directed toward systematization of the chemical process synthesis activity.

APPROACHES TO CHEMICAL PROCESS SYNTHESIS

Chemical process synthesis as an activity is characterized by a very large number of discrete decisions. In the past, conceptual process design engineers have relied heavily on a process of *evolutionary modification* to generate process flow sheets. Previously successful flow sheets for the same or a similar desired product are copied with necessary adaptive modifications to meet the objectives of the specific case at hand. This approach has also led to the cataloguing of standard flow-sheet patterns. Examples are the schemes for breaking heterogeneous low-boiling binary azeotropes, or sequences for extractive distillation (Barnicki and Siirola, 1997a). Algorithmic and heuristic approaches have been used to systematize this evolutionary modification approach. Thompson and King (1972) and also Stephanopoulos and Westerberg (1976) used thermodynamic analysis to identify portions of the design with the greatest potential for beneficial modification. The quality of process designs generated by the evolutionary modification approach depends critically on the starting flow sheet and the methods used to modify it. Domain knowledge is implicitly built into the existing starting flow sheet. Knowledge that may be critical to the evolution process may be biased already in a potentially inappropriate fashion toward the old solution. Thus by the nature of this process, the resulting designs are rarely novel, but they are generally workable. Evolutionary modification is a frequently used process synthesis technique because of the extensive existing repertoire of design heuristics, encyclopedias of complete and partial flow sheets, and because companies often are reluctant to risk millions of dollars on truly innovative designs too far removed from previous experience.

Most systematic approaches to chemical process synthesis, however, follow two alternative approaches. The first, *superstructure optimization* (Ichikawa, Nashida, and Umeda, 1969), views process synthesis as an optimization over flow-sheet structure. This approach starts with an all-encompassing super flow sheet containing many alternative and redundant designs, components, and interconnections. Superstructures have been formulated in which either the various unit operation components are specified (state equipment networks) or specific interconnections are defined (state task networks). With either formulation, mathematical optimization techniques are applied to systematically eliminate less economical parts of the superstructure or less appropriate interconnections. The remaining interconnections, component design, and operating parameters are optimized in parallel for some, usually economic, objective function. *Mixed-integer nonlinear programming* and more recently *generalized disjunctive programming* are now most often exploited in the superstructure optimization approach to chemical process synthesis (Türkay and Grossmann, 1998; Yeomans and Grossmann, 1999). The simultaneous optimization of flow-sheet structure, design, and operational parameters is quite appealing. However, "optimality" depends critically upon the completeness of the starting superstructure. This structure must come from somewhere, and generation of the initial superstructure is a rather weak link at the moment. For some simplified classes of design subproblems, such as the generation of energy-efficient heat exchanger networks, the superstructure is implicit in the mathematical formulation (Yee and Grossmann, 1990). For more general chemical process synthesis problems, however, there is not yet a formal way to generate the starting superstructures. Nevertheless, with ever-increasing computer resource availability and clever, efficient problem formulation, superstructure optimization is becoming a more practical approach for real process synthesis problems, and it offers tremendous potential for the future.

In the final chemical process synthesis approach, *systematic generation*, the flow sheet is constructed from a portfolio of basic chemical engineering unit operations such as reactors, distillation columns, heat exchangers, compressors, tanks, and the like. These operations are selected and interconnected in a directed fashion, guided by process and thermodynamic constraints such that the raw materials become progressively transformed into the desired product. Whereas superstructure optimization is essentially a brute force, computationally intensive method, systematic generation approaches rely on more and better-organized domain knowledge to make the synthesis problem tractable. Systematic generation approaches synthesize each flow sheet from scratch. Sometimes the approach leads to flow-sheet structures that are already well known, whereas at other times the approach leads to entirely novel solutions. However, in the absence of the exhaustive generation of all alternatives, systematic generation cannot guarantee design optimality. Both algorithmic and heuristic methods have been proposed for the systematic generation approach. Some mimic design expert behavior and some focus on new ways of representing the underlying sciences governing chemical behavior. Some of these methods can be implemented by hand and some have been computerized. Systematic generation has been the approach used by most formalized chemical process synthesis algorithms that have seen industrial application to the present time.

SYSTEMATIC GENERATION AND THE MEANS-ENDS ANALYSIS PARADIGM

Computer-assisted goal-directed problem-solving paradigms were formalized in the artificial intelligence literature in the 1950s and 1960s as state-space search methods, and more particularly as means-ends analysis (Simon, 1969; Nilsson, 1980). Solution methods of this sort involve the concepts of *states, properties*, and *operators*. A state is physically realizable configuration of the problem variables contained within the domain space. If the value of a particular property of an initial or intermediate state is different from the value of the corresponding property in the desired goal state, a *property difference* is detected. An operator transforms property differences of one state into another state according to certain domain-dependent rules. A solution is the result of tracking a search process. Transformation operators are applied to the initial state when property differences are detected to produce successive intermediate states in an attempt to systematically eliminate differences until the goal state is produced.

The chemical process synthesis problem can be readily adapted to such a formalism (Siirola and Rudd, 1971). Raw materials, intermediates, desired products, and other streams within a chemical process are described and characterized by a number of physical and chemical properties. These properties include the molecular identity of the species, the amount involved, the composition or purity of a material, thermodynamic phase, temperature, pressure, and possibly particle size, shape, color, and other physical characteristics. In the means-ends analysis formulation of the systematic generation approach to chemical process synthesis, an appropriate collection of raw materials is considered to be an initial state. The desired products are the goal state. If the value of a particular property of a raw material is different from the desired value of the corresponding property in the desired product, a property difference is detected. The purpose of an industrial chemical process is to apply technologies, that is, transformation operators, in a defined sequence such that these property differences in identity, amount, concentration, phase, temperature, pressure, form, and so on are systematically eliminated and the raw materials thereby become transformed into the desired products.

In the chemical processing domain there are well-known transformation operators for reducing or eliminating property differences. Familiar examples include chemical reaction to change molecular identity, mixing and splitting (and purchase) to change amount, separation to change concentration and purity, enthalpy modification to change phase, temperature, pressure, and the like. These operators are governed and constrained by domain-dependent rules including the laws of chemistry and physics. In many cases there is such a close relationship between an abstract property-changing operator and a physically realizable unit operation to accomplish the task (reactor, distillation column, heat exchanger, compressor, pump, etc.) that some designers think directly in terms of unit operations when developing conceptual process flow sheets. Property differences are translated directly into unit operation solutions. This may not always be the best strategy. Sometimes there is a distinct advantage in keeping task (transformation operator) identification quite distinct from equipment (unit operation) specification, as some of the more innovative flow sheets result from the synergistic integration of usually independent transformations. For example, when reaction and distillation operators are identified separately but later combined under the appropriate conditions into a single piece of equipment, the result can lead to novel superior chemical process flow-sheet solutions.

Some property-changing operators can only be applied to an intermediate state, say a stream in a partially synthesized flow sheet, when certain other properties of the stream are within specified values. These necessary preconditions may not be true at the desired time of application. For example, a separation method to select only crystals greater than a given size can only be applied if a stream contains a solid phase. Similarly, a separation method expected to exploit relative volatility differences (like distillation) can be applied only if enthalpy conditions permit simultaneous liquid and vapor phases. If the preconditions for the immediate application of a property-difference-reducing operator believed to be useful are not met, a new process design subproblem may be formulated whose objective is to reduce property differences between the state at hand and the conditions necessary for the application of the desired operator. This recursive strategy is an important feature of the means-ends analysis paradigm. Note that the acceptable preconditions for the application of an operator may not be specific property values, but may be a range of values for a particular property or a complex combination of values for several properties.

With the means-ends analysis paradigm, alternative solutions are generated when more than one operator is identified that can reduce or eliminate a property difference. The decision of which operator to choose might be made on the basis of some evaluation at the time the operators are being examined. Alternatively, each may be chosen separately, then the consequences followed separately (leading to alternative design solutions) and each final flow-sheet solution evaluated separately.

Yet another possibility is that all feasible alternative operators are selected and applied in parallel, leading to a superstructure flow sheet. When all property differences have been resolved, the entire superstructure is evaluated and optimized, with the less economical redundant portions being eliminated. This is one version of the superstructure optimization approach to chemical process synthesis in which the superstructure itself is created by a systematic generation procedure.

The heart of the systematic generation paradigm is the formalization of a control strategy for the detection of property differences and selection and application of the most appropriate difference-reduction phenomena and operators at each point in the process synthesis sequence.

OPPORTUNISTIC STRATEGIES

The most straightforward control method is to select operators in a forward-chaining or *opportunistic* direction. Development of the process flow sheet proceeds in the direction of material flow. One starts with the initial state, the description of the available raw materials, and successively applies transformation operators to produce intermediate states with fewer and fewer differences compared with the goal state until the product is reached. In simplistic terms, the control strategy can be stated as, "Given the current situation and the tools at hand, where can I go from here?" Such an approach is easier to execute automatically because the effect of operators is generally better understood from a simulation or rating point of view. Given an input state and values for various operating parameters, the consequence of the transforming unit operation often can be calculated.[1] Commercial chemical process simulation software is quite capable of performing these transformation calculations.

BALTAZAR (Mahalec and Motard, 1977) was an example of an early computerized systematic synthesis system that generally followed the means-ends analysis approach with an opportunistic control strategy. Both the starting materials initial state and the desired product goal state were given. The conceptual process design was developed opportunistically in the same direction as material flow, by applying property-changing technologies to existing streams to produce new streams closer in properties to the desired product. Material and energy-conserving mass and thermal recycles were considered at the end of the procedure. Any point in the partially completed flow sheet was a feasible consequence of the raw materials and the technologies so far specified. Alternative solutions were generated if property differences were eliminated in different orders or if more than one operator were identified that could accomplish a property difference elimination task.

The synthesis of process flow sheets often involves the subproblem of the generation of separation sequences for streams whose various components have to be sent to different destinations, which arises during raw material preparation, product isolation and finishing, and waste treatment. An example might be the effluent from a reaction operation that could contain the one or more desired products, some undesired byproducts or coproducts, incompletely reacted reactants, solvents, and so on, each of which may have to be separated from the others and purified before being sent on. Separations are accomplished by exploiting phenomena and conditions under which the various components behave somehow differently, generally leading to two or more outlet streams in which the relative concentrations of at least some of the components are different. For example, if some of the components are solids under conditions where other components are liquids, then a filtration operation might effect a separation. If all of the components are liquids but with different boiling points, then exploitation of relative volatility (distillation) might be selected. Because of the large number of phenomena that might be exploited to effect separations and the different orders in which the different phenomena might be applied, the synthesis of separation sequences is a complex problem in its own right.

Even if the flow sheet is restricted to the exploitation of only few phenomena such as distillation for separations, the number of separation sequences may be large.

---

[1] One notable exception to this approach direction is the field of synthetic organic chemistry, in which the chemical reaction pathway leading to the desired product is typically identified. Because of the extremely large number of feasible operators (the entire domain of chemistry) and the fact that the goal state but not the initial state may have been specified, chemists more generally work backward in a retrosynthetic direction from the desired target compund. They search for reactions and candidate precursor molecules at each stage until readily available starting materials are identified.

Early investigators utilized list processing and branch-and-bound search techniques for simple ideal distillation sequencing where the order of volatility is based only on the boiling points of each of the individual components (Siirola and Rudd, 1971; Hendry and Hughes, 1972; Mahalec and Motard, 1977). In common among these studies, the space search is pruned and directed by cost functions and heuristics that estimated the economic penalty for each proposed operation. Because composition differences for each species may be resolved in different orders, multiple sequences exist to accomplish the desired separations for multicomponent separation problems. Superior solutions can be found by straightforward generation and evaluation of all possible sequences, or by the use of well-developed rules or *heuristics* for the generation of only the most promising sequences (Seader and Westerberg, 1977). In some cases, the heuristics themselves are often examined and selected in a hierarchical manner (Barnicki and Fair, 1990, 1992). Design knowledge can be organized into a structured query system in which an operation at each stage is selected by a series of heuristics ordered based on pure component properties and on process characteristics.

Such flow sheets are generated in an opportunistic manner, starting with the original mixture and applying separation operators one at a time until all of the desired product compositions are met. At any point in the flow-sheet synthesis procedure, the partial design generated is a feasible consequence of the initial feed composition and the separation methods so far specified. The resolution of composition differences between the intermediate streams and the remaining goals is addressed by specifying additional separation, stream splitting, or mixing operations. If distillation is feasible for all separations, this strategy will not lead to intermediate streams from which it is impossible to resolve the remaining composition differences, so no attempt is made to look ahead to anticipate or accommodate potential difficulties. This approach has been used to effectively synthesize flow sheets for nonazeotropic systems using a variety of separation operations for both liquid and gaseous systems, thus demonstrating that an opportunistic control strategy can be successful for at least a limited class of industrially relevant problems. One such application using a heuristic-based sequencing method is illustrated below for the separation of coconut fats.

EXAMPLE   Fractionation of Coconut Fatty Acids

A mixture of coconut fatty acids consisting of lauric, myristic, palmitic, and oleic acid (in order of increasing boiling point) is to be separated into its four pure components. Assuming only sharp separations between adjacent key components and limiting the separation operators to single-feed distillation, one could construct five different column sequences by exhaustive enumeration (Figure 10.1). Each of these column sequences could be simulated in detail and the lowest cost option selected. Alternatively, any of a number heuristic methods could be used to construct an optimal or near-optimal sequence. The method outlined here is presented in more detail in Barnicki and Siirola (1997b). The feed composition (with components listed in order of increasing boiling point), adjacent relative volatilities, and a separation co-efficient (the ratio of distillate and bottoms fractions multiplied by one minus the relative volatility) are given as follows.

	Component	Mole %	Boiling Point at 8.5 KPa (°C)	Adjacent Relative Volatility	Separation Coefficient
Product A	Lauric acid	45	214.6		
				2.32	1.08
Product B	Myristic acid	7	236.3		
				2.23	0.75
Product C	Palmitic acid	11	257.1		
				1.68	0.25
Product D	Oleic acid	27	270.3		

The heuristics indicate that the easiest separation that directly gives a product and contains a large fraction of the feed should be completed first. This translates into removal of the lauric acid product as the distillate of the first column. The mole fractions and separation coefficients are recalculated for the bottoms product.

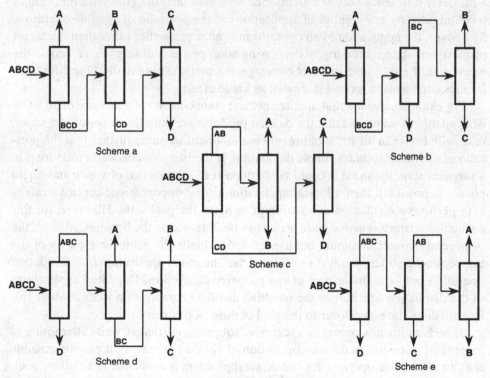

Figure 10.1. Standard distillation sequences for four-component mixtures.

	Component	Mole %	Boiling Point at 8.5 KPa (°C)	Adjacent Relative Volatility	Separation Coefficient
Product B	Myristic acid	16	236.3		
				2.23	0.23
Product C	Palmitic acid	26	257.1		
				1.68	0.49
Product D	Oleic acid	58	270.3		

The most volatile product (myristic acid) is a small fraction of the feed, but it has a higher relative volatility (as is thus easier). The oleic acid fraction is the majority of the feed, but has a lower relative volatility. In such a case the heuristics suggest that the split with the highest separation coefficient be selected next. The last column separates products B and C. The finished sequence is pictured in Figure 10.1(b). Detailed simulations of the five alternative flow sheets confirm that the capital cost of this design is ~10% less than the next cheapest sequence.

One drawback of the purely opportunistic control strategy quickly becomes apparent when more complicated problems are tackled. The application of a particular operator may result in the partial elimination of a property difference or may lead to a property difference between an intermediate state and the goal state that cannot be eliminated by any number of applications of the portfolio of possible operators. Moreover, the application of an operator may alter properties other than the target property, creating, increasing, or decreasing other property differences as well. In the worst case, the algorithm may not converge, but oscillate repeatedly. One difference is eliminated while a second is created as a side effect.

For example, assume that an intermediate state consists of a mixture of 10 wt.% isopropanol in water and that the desired products are pure isopropanol and water, each with less than 0.1 wt.% of the other component. Assume further that the portfolio of operators includes simple distillation. If the isopropanol–water mixture is in a physical state such that it could be distilled (i.e., coexistence of vapor and liquid phases is possible), then a logical application of the opportunistic control strategy is to perform a distillation in an attempt to reach the goal state. However, for this particular mixture, whereas pure water can be obtained as the bottoms product, the isopropanol–water minimum boiling azeotrope limits the achievable purity of the isopropanol distillate product to ~70 mol.%. The initial application of a distillation operation only eliminates part of the property differences. Repeated applications of the distillation operator to the resulting distillate composition will not move this intermediate state any closer to the goal of pure isopropanol.

If such a difficulty occurs in a recursive subproblem (that is, while attempting to meet the preconditions for the application of another operator), it may be possible that an alternative operator for the original problem is available. If no other available operators are applicable at the top level, then the partial solution has reached a dead end. Often the only recourse is to backtrack to an earlier intermediate state, discarding part or all of the current flow sheet. Generation of dead-end partial solutions is not in itself a fatal flaw, but this can increase greatly the computational burden of an algorithm.

NONMONOTONIC PROCESS SYNTHESIS REASONING

Generally when the means-ends analysis paradigm is applied, if a particular difference does not exist between the existing state and the goal state, there is no incentive to create one. However, consider the example of a pure component stream in the current state of a subcooled liquid and whose goal state is a superheated vapor at the same pressure. Because the initial state and the goal state are at the same pressure, a straightforward opportunistic controlled means-ends analysis paradigm might specify methods to increase the temperature of the initial subcooled state until the stream is saturated, then apply a phase change operator to produce a vapor phase, and then finally apply a vapor phase temperature increase method until the final goal state is achieved, as shown in Figure 10.2(a). However, such a synthesis approach could not generate the alternative in which the initial state was first increased in pressure (corresponding to the saturation temperature equal to the final goal temperature), then increased in temperature to saturation, phase changed, then decreased in pressure to the final goal state, as shown in Figure 10.2(b). The difference in designs is that vapor phase temperature change, which turns out to be expensive to implement because of the particularities of vapor phase heat transfer, is avoided. Characteristic of the alternative design is that a property difference was intentionally created where

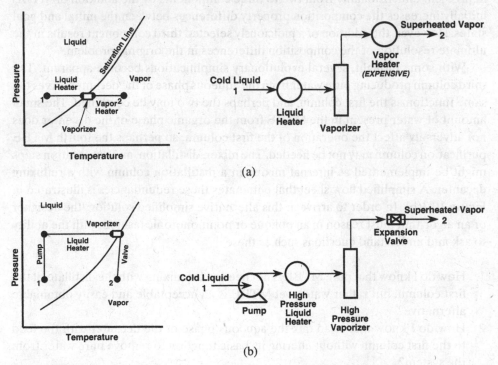

Figure 10.2. Flow sheets (a) for superheated vapor at constant pressure and (b) with pressure excursion for superheated vapor, avoiding vapor heating.

one did not exist. This alternative requires some sort of problem anticipation and some modification to the opportunistic control of the means-ends analysis paradigm.

Consider again the example of the isopropanol–water mixture. One way to get around a low-boiling azeotrope known from the literature is to add to the system a new third component with additional azeotrope-forming characteristics. An example of such a component is methyl tertiary-butyl-ether (MTBE). The MTBE–water azeotrope has a lower boiling point than the isopropanol–water azeotrope and in addition forms two liquid phases upon settling. This characteristic can be exploited to solve the original separations problem. For example, the initial mixture could be distilled to give a pure water bottoms product and a distillate mixture close in composition to the isopropanol–water azeotrope. MTBE is mixed with the distillate of the first column and distilled a second time to produce pure isopropanol as the bottoms product and the low-boiling MTBE–water azeotrope as the distillate. The MTBE–water azeotrope can be separated into two liquid phases by decantation. The aqueous phase can be distilled a third time to give a pure water bottoms product and again the MTBE–water azeotrope as distillate, while the organic phase from the decanter may likewise be distilled to give a pure MTBE bottoms product and a third time the MTBE–water azeotrope as distillate. The azeotropic distillates from both of these last two distillation columns may be recycled to the decanter. The MTBE is mixed with the feed to the second column. The resulting flow sheet is given in Figure 10.3(a). The characteristic feature of this isopropanol–water separation is the intentional addition of a new component, called a *mass-separating agent*, and the creation of associated composition property differences where they did not previously exist. The change in problem dimensionality from two to three components by the addition of MTBE initially increases the composition property differences between the initial and goal states. However, the addition of a judiciously selected third component results in the ultimate resolution of the composition differences in the original problem.

With some thought, several evolutionary simplifications become apparent. The third column producing pure water from the aqueous phase of the decanter serves the same function as the first column, and perhaps the two may be combined. The small amount of water present in the MTBE from the organic phase of the decanter does not adversely affect the operation of the first column, so perhaps the fourth MTBE purification column may not be needed. The mixer, distillation, and decantation steps might be implemented as internal mixing in a distillation column with a refluxing decanter. A simplified flow sheet that eliminates these redundancies is illustrated in Figure 10.3(b). In order to arrive at this alternative simplified solution, the designer or an algorithm must reason in an oblique or nonmonotonic fashion, with the ability to ask and understand questions such as these:

1. How do I know that pure MTBE is not needed for mixing with the distillate of the first column, but rather water-wet MTBE is an acceptable and easily obtainable alternative?
2. How do I know that I can mix the aqueous phase of the decanter with the feed to the first column without altering its basic function to remove pure water from the system?
3. How do I know that the three separate unit operations of mixing, distillation,

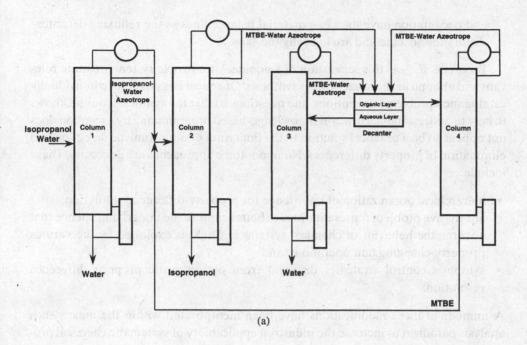

(a)

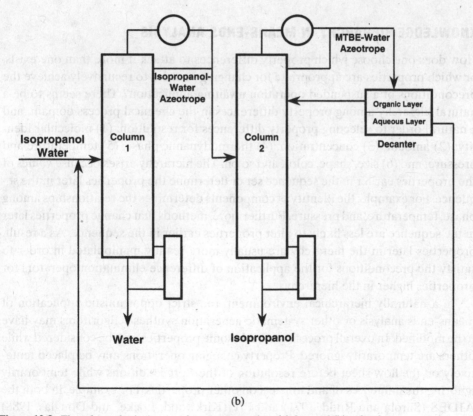

(b)

Figure 10.3. Flow sheets for isopropanol–water separation: (a) first and (b) refined.

and decantation have the same material balance lines as the refluxing decanter–distillation scheme and are logically the same?

Problems such as this separation of a nonideal azeotropic system are quite relevant and abound in chemical process synthesis. Of course it is possible to individually catalog such problems, exceptions, and questions so that the synthesis routine knows to how to address them. However, knowledge-based programming by exception does not appear to be a practical solution to the limitation of opportunistic detection and elimination of property differences. Nonmonotonic approaches are necessary. These include:

- hierarchical organization of knowledge for property-difference resolution;
- alternative problem representations to better capture the underlying science that governs the behavior of chemical systems and that is exploited by the various property-changing unit operations; and
- synthesis control strategies different from opportunistic property-difference resolution.

A number of these modifications have been incorporated within the means-ends analysis paradigm to increase the industrial applicability of systematic chemical process synthesis methods.

KNOWLEDGE HIERARCHY IN MEANS-ENDS ANALYSIS

How does one choose which property differences to attack if more than one exists, or which properties are appropriate for changing in order to recursively achieve the preconditions of a suspended operation awaiting application? There seems to be a natural hierarchy among property differences in the chemical process domain, and a natural order to selecting property differences for resolution: (1) molecular identity; (2) amount; (3) concentration; (4) thermodynamic phase; (5) temperature and pressure; and (6) size, shape, color, and so on. The hierarchy arises because values of the properties earlier in the sequence set or determine the properties later in the sequence. For example, the identity of components determines the relationships among phase, temperature, and pressure. Furthermore, methods that change properties later in the sequence are less likely to alter properties earlier in the sequence. As a result, properties later in the hierarchy are usually more readily manipulated in order to satisfy the preconditions for the application of difference elimination operators for properties higher in the hierarchy.

In a naturally hierarchical environment, the strict opportunistic application of means-ends analysis or other systematic generation synthesis algorithms may have to be modified. In overall process synthesis, some properties may be considered while others are temporarily ignored. Property changing operations may be placed tentatively on the flow sheet before resolution of their preconditions while temporarily ignoring the influences of and impacts on other properties. For example, in both the AIDES (Siirola and Rudd, 1971) and PIP (Kirkwood, Locke, and Douglas, 1988) hierarchical chemical process synthesis procedures, a *reaction path* or *plant network* is first constructed that considers only the resolution of molecular identity property differences. A chain of identity-changing reactors or plants are specified to convert raw materials into desired products while temporarily ignoring all other properties and property differences. Reactors are placed strategically like islands on the flow sheet, without first satisfying preconditions, as it is speculated from the molecular identity differences that they will be required, although it is not known in detail exactly how they will be connected to the raw materials, the products, or to each other.

At the next hierarchical stage, known as *species allocation* (in AIDES) or *plant connection and recycle* (in PIP), reactor interconnections are tentatively determined considering only the role of each species (reactant, byproduct, catalyst, etc.) and the amount property. At this point, target flows among the raw material sources, reactions, products, wastes, and so on are determined, including gross recycle of catalysts and incompletely converted reactants. AIDES attempts to "look ahead" and judge the severity of the separations that may be required to implement the proposed species allocation among the different sources and destinations. PIP has a somewhat more fixed allocation of reaction products to the next plant and catalysts, incompletely converted reactants, and unneeded coproducts back to the same plant or to waste treatment. In both systems, the allocation is based only on the identity, role, and amount properties while all other properties are temporarily ignored. The reaction operators may not be applied in a strict sense because their feed streams are not completely specified in terms of all properties, and therefore neither can the reaction effluent streams be completely specified with certainty.

In the next hierarchical levels, PIP resolves composition differences for each re-actor effluent by a rule-based decision procedure specifying, as appropriate, a *phase separation system* followed by separate *vapor, liquid*, and *solid recovery systems* consisting of various separation methods (Douglas, 1988, 1995). PIP follows this with a flow-sheet-wide *energy integration*. In AIDES, the remaining property differences are resolved in the means-ends manner in a series of subproblems, using each raw material and reactor effluent stream as an initial state and the required conditions at whatever destinations the contained species were allocated to as the corresponding goal states. In many cases, the differences in concentration, phase, temperature, pressure, and form can be hierarchically detected and resolved, recursively if necessary, without disturbing the prior species allocation (amount property) decisions. As an example of an additional hierarchy in focus, AIDES specifically keeps distinctly separate the identification of methods to resolve property differences, *task identification*, the association of one or more of these methods with specific pieces of processing equipment, *task integration*, and the detailed specification of such units, *equipment design*, sometimes with dramatic results (Siirola, 1995). Figure 10.4 shows the progression of a process flow sheet being generated by AIDES, whereas Figure 10.5 shows the generation of a flow sheet using PIP.

When means-ends analysis approaches are applied in a hierarchical environment, partial results do not represent necessarily a feasible, if partial, flow sheet, as in purely opportunistic systematic generation approaches such as BALTAZAR. The final design must be carefully reevaluated to ensure that assumptions implicitly made about properties not considered in the decision-making at any point, either because they

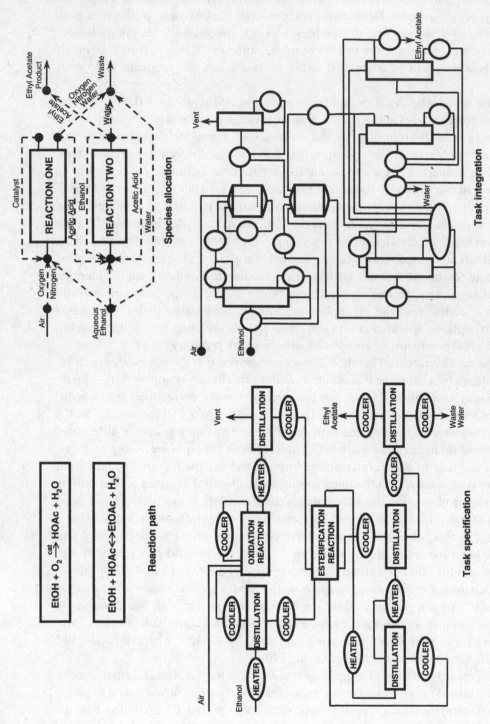

Figure 10.4. Flow sheet development, using the AIDES process synthesis procedure.

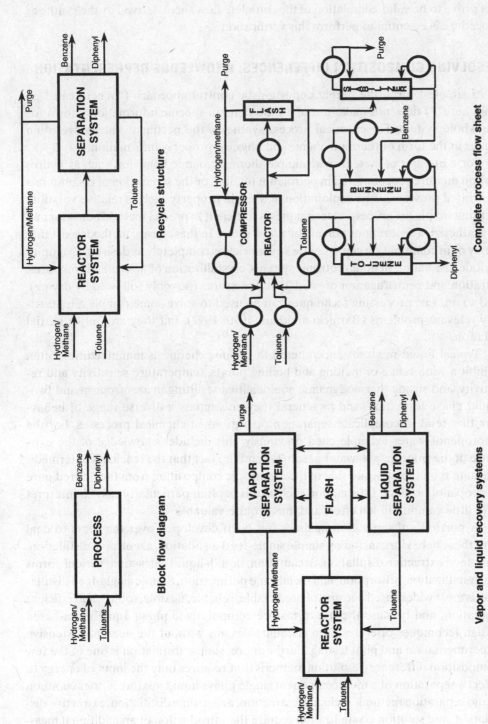

Figure 10.5. Flow sheet development, using the PIP process synthesis procedure.

were yet to be considered, or because they were thought to be previously resolved, in fact prove to be valid. Simulation of the complete flow sheet derived by the synthesis procedure is essential to perform this verification.

RESOLVING COMPOSITION DIFFERENCES: KNOWLEDGE REPRESENTATION

In its simplest form, the purely opportunistic control approach ("Where can I go from here?") does not make use of all the domain-specific information known for the whole system. For chemical process synthesis, the pertinent knowledge often comes in the form of thermodynamic and physical property information.

For a number of systems with simple thermodynamic behavior, such as hydrocarbon mixtures encountered in petroleum refining or the separation of coconut fats illustrated previously, the exploitation of a single property such as relative volatility by using simple list processing techniques is sufficient to predict method behavior and to synthesize a separations portion of a flow sheet. In these cases, methods exist that singly or in combination can separate key species as completely as desired. Examples include distillation of nonazeotropic systems, crystallization of noneutectic mixtures, filtration and centrifugation of solid–liquid mixtures (possibly followed by drying), and so on. List processing techniques can be used to solve some complex, industrially relevant problems (Barnicki and Fair, 1990, 1992), but they are only a partial solution.

Typical liquid mixtures encountered in organic chemicals manufacturing often exhibit a wide range of melting and boiling points, temperature sensitivity and reactivity, and strong thermodynamic nonidealities resulting in azeotropism and two-liquid phase formation, and in general they encompass a diverse range of behaviors that tend to complicate separation operations in chemical processes. For the isopropanol–water example cited previously, this includes knowledge of the existence of the isopropanol–water azeotrope and the fact that the reachable intermediate state is on the opposite side of the azeotropic composition from the desired pure isopropanol product. Information such as this is often particularly easy to interpret with little computational effort and thus is quite valuable.

A portfolio of separation methods has been developed over the years to deal with these behaviors, including simple single-feed distillation, azeotropic distillation, dual-feed extractive distillation, decantation, liquid–liquid extraction, various forms of crystallization, adsorption, and membrane permeation. Simple single-feed distillation is most widely used because of predictable, reliable, flexible, robust, and efficient operation, and because the use of mature computerized phase equilibrium-based design techniques often result in adequate designs without the need for extensive experimentation and pilot testing. Furthermore, simple distillation is one of the few composition difference-resolution methods that requires only the input of energy to effect a separation of a multicomponent single phase liquid mixture. Other common liquid separation methods including extraction, azeotropic distillation, extractive distillation, and solution crystallization require the introduction of an additional mass-separating agent (MSA) – MTBE for the case illustrated in Figure 10.3. The MSA must not only be selected but must also be recovered and recycled for economical operation, adding further complexity to a process flow sheet.

Useful graphical representations that encapsulate much of the underlying thermodynamics of separations phenomena and aid in understanding solution behavior include *residue curve maps* for vapor-liquid equilibria (VLE), *miscibility diagrams* for liquid–liquid equilibria (LLE), and *solubility diagrams* for solid–liquid equilibria (SLE) (Barnicki and Siirola, 1997a; Wibowo and Ng, 2000). Residue curves trace the multicomponent liquid phase composition of a particular experimental distillation device, the simple single-stage batch stillpot, as a function of time. Residue curves also approximate the liquid compositions profiles in continuous staged or packed distillation columns operating at infinite reflux, and they are also indicative of many aspects of the behavior of continuous columns operating at practical reflux ratios. Residue curves start at the initial composition charged to the stillpot and terminate at the composition of the last drop left as the stillpot runs dry. Families of residue curves are generated experimentally (or mathematically extrapolated both forward and backward in time, if an accurate model of the relevant solution thermodynamics is known), starting with different initial stillpot compositions. The composition of the last drop may or may not always be the same, depending on the system and the initial composition. Families of all residue curves that originate at one composition and terminate at another composition define a residue curve *region*. All systems with no azeotropes and even some systems with azeotropes have only one region, the entire composition space. All residue curves in such systems originate at the lowest-boiling composition of the system and terminate at the highest-boiling composition. Other systems, however, in which not all residue curves originate or terminate at the same compositions have more than one region. The demarcation between regions in which adjacent residue curves originate from different compositions or terminate at different compositions is called a *separatrix*. Separatrices related to the existence of azeotropes. In the composition space for a binary system, the separatrix is a point (the azeotropic composition). With three components, the separatrix becomes a (generally curved) line, with four components a surface, and so on.

All pure components and azeotropes in a system lie on residue curve region boundaries. Within each region, the most volatile composition (either a pure component or minimum-boiling azeotrope and the origin of all residue curves) is the *low-boiling node*. The least volatile composition (either a pure component or a maximum-boiling azeotrope and the terminus of all residue curves) is the *high-boiling node*. All other pure components and azeotropes are called *intermediate-boiling saddles* because no residue curves originate, terminate, or quite pass through these compositions. Adjacent regions may share nodes and saddles. Pure components and azeotropes are labeled as nodes and saddles as a result of the boiling points of all of the components and azeotropes in the system. If one component is removed from the mixture, the labeling of all remaining pure components and azeotropes, specifically those that were saddles, may change. Region-defining separatrices always originate or terminate at saddle azeotropes, but never at saddle pure components. An example residue curve map for the ternary isopropanol–water–MTBE system is given in Figure 10.6(a).

As a result of the physics of distillation, to a first approximation, the distillate and bottoms of a single-feed distillation column lie on the same residue curve (and hence are in the same region). Also in composition space, by mass balance the distillate,

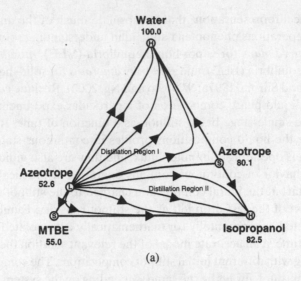

(a)

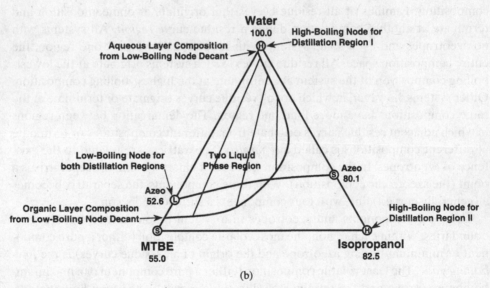

(b)

Figure 10.6. Isopropanol–water–MTBE system: (a) residue curve map and distillation regions, and (b) compositions of interest.

bottoms, and net feed are collinear. For a given multicomponent mixture, a single-feed distillation column can be designed with sufficient stages, reflux, and material balance control to produce a variety of different separations ranging from the *direct* mode of operation (pure low-boiling node taken as distillate) to the *indirect* mode (pure high-boiling node taken as bottoms). This range of operability results in an approximately bow-tie-shaped set of *reachable compositions* that may be opportunistically achieved by single-feed distillation. Because residue curves are deflected by saddles, it is generally not possible to obtain a saddle product (pure component or azeotrope) from a simple single-feed column. For preliminary design purposes, the lowest- and highest-boiling nodes and the compositions on the residue curve region boundary directly opposite the feed composition from these two nodes are opportunistically easy to achieve.

There are few points and regions in the composition space that are of particular significance in separations process synthesis. These *compositions of interest* include the feed and desired product compositions, as well as pure components, azeotropes, eutectics, and selected points on liquid–liquid binodal curves. The choice of composition for a MSA, if required, is critical and usually a composition that can be conveniently regenerated within the process. Binary and ternary azeotropes that are also high- or low-boiling nodes (as well as the corresponding two liquid phase compositions, if heterogeneous) are of particular interest in this regard, as these points are easily reachable compositions by distillation and decantation. Pure components and azeotropes that are saddles are, by contrast, poor choices for MSA recycle, as they are not included in the set of reachable compositions (unless they become nodes when another component in the system is removed). Potential compositions of interest for the isopropanol–water–MTBE system are labeled in Figure 10.6(b).

ANTICIPATING DIFFICULTIES: STRATEGIC PROCESS SYNTHESIS

An alternative approach to purely opportunisitc control for flow-sheet generation is to exploit structural knowledge available through effective representation methods. Such knowledge can be used to look ahead to potential difficulties and develop contingencies early in the synthesis process to deal with them before running into dead ends. This is the basis of what we will call *strategic* control methods.

It is clear that although distillation is a favored separation method, quite a number of situations prevent or interfere with the use of simple direct or indirect mode distillation schemes specified in an opportunistic manner. In general, a designer may be faced with avoiding, overcoming, or exploiting a limited set of *critical features* in order to accomplish the overall objective of the separation system. These critical features include the following.

1. Distillation boundaries: If a goal (or MSA) composition is in a different distillation region from the source, the product cannot be obtained directly by simple single-feed distillation,
2. Saddle products: If a goal (or MSA) is a saddle in a particular distillation region, that product cannot be obtained at high purity directly by simple single-feed distillation.
3. Pinched or close boiling regions: If a source and goal composition are separated by a region of low relative volatility, simple single-feed distillation is not precluded, but tends to require a large number of stages and high reflux ratio.
4. Overlapping melting–boiling points: Some solutions may contain components with melting points that are higher than the boiling points of other components. The distillation of such mixtures may result in solidification within the column.
5. Temperature-sensitive components: Some mixtures may contain components that degrade, decompose, polymerize, or otherwise react in an undesirable manner at a temperature conditions of distillation. Some other separation method must be used.

One might consider an opportunistic approach to composition difference resolution for nonideal mixtures with critical features. Guided by general sequencing heuristics, one may pick any separation method that is feasible and applicable to the current state, and proceed to build a flow-sheet solution toward the desired goal compositions. However, the existence of critical features can preclude the successful or complete application of simple distillation. In systems exhibiting critical features, it is quite possible to reach an intermediate composition from which it is impossible to reduce remaining composition differences by distillation or any other separation method. This is where a nonmonotonic approach of anticipating difficulties comes into play.

Once a critical feature has been identified, it is useful to examine *strategic* methods for crossing, breaking, bypassing, reaching, or exploiting the critical feature. The strategies and resulting separation methods for handling a given critical feature may differ considerably, and a given strategy can often be implemented in several different ways. Examples of strategies for dealing with some critical features include the following.

For overcoming distillation boundaries, do as follows.

1. Exploit a LLE (decantation or extraction) or SLE (crystallization or adsorption) tie line or kinetic phenomena (membrane) to cross the distillation boundary.
2. Add a MSA or other composition to reach a new region (mixing).
3. Expand dimensionality (add a MSA) such that the boundary does not extend into the new residue curve map (extractive distillation).
4. Change the system pressure to shift the position of the boundary (pressure swing distillation).
5. Exploit an extreme boundary curvature (feed in one region and both distillate and bottoms in another region).
6. Relax problem specification so that distillation boundary need not be crossed.

For reaching saddle products, do as follows.

1. Exploit two-feed distillation not restricted to nodal distillates (extractive distillation).
2. Reduce dimensionality (remove a different component) such that saddle composition becomes a node.
3. Exploit LLE, SLE, or kinetic phenomena not limited by VLE nodes and saddles.

For pinched or close boiling regions, do as follows.

1. Use multiple separators and multiple separation paths to bypass pinch.
2. Exploit LLE, SLE, or kinetic phenomena not related to VLE pinches.
3. Change system pressure to lessen or eliminate pinch.
4. Distill through pinch anyway.

For overlapping melting–boiling points, do as follows.

1. Build a band of intermediate solvent with a sidedraw–return column.
2. Add an intermediate solvent.
3. Alter the system pressure.

For temperature-sensitive components, do as follows.

1. Exploit LLE, SLE, or kinetic phenomena operable at a lower temperature.
2. Reduce system pressure.

A more complete list of strategies and implementation schemes for overcoming or exploiting critical features for separation scheme synthesis is given in Barnicki and Siirola (1997b).

It may not be known exactly where in the flow sheet a strategic operation will end up, only that it will be required some place in some form in order to overcome or exploit a particular critical feature. Thus, strategic operations, similar to the reactors in the AIDES hierarchical synthesis procedure, are often placed as "islands" on the flow sheet without well-defined feeds or products. The region where the feed must be located may be known as well as a general idea of the types of products expected, but no definite compositions for either. For example, for a strategic decant operation, all that may be known initially is that the feed must be somewhere in the two-phase liquid region and the products will be on the binodal curve at opposite ends of a tie line through the feed, possibly specified to be in two different distillation regions. Even the choice of an MSA itself is a strategic decision that is at first assumed to exist before a process is designed to regenerate it.

Along with strategic operations, opportunistic separations may also be possible. When critical features are present, opportunistic separations can be thought of as links among sources, strategic operations, and goals. An opportunistic operation often is used to reach a composition where a strategic separation is applicable (e.g., opportunistically distill into a two-phase liquid region and then strategically decant the mixture to cross a distillation region boundary). Sometimes an opportunistic separation will be equivalent to a strategic separation. Alternatively, if no critical features are present, the entire separations flow sheet may be synthesized heuristically by a series of opportunistic separations alone, as in the separation of coconut fats illustrated previously.

The recycling of a stream to a point upstream in the process is often a powerful alternative to performing additional processing. In particular, regeneration and recycle of an MSA composition is essential to the economic operation of such separation methods as extraction, extractive distillation, and azeotropic distillation. However, recycling cannot be done indiscriminately and material balances must be carefully considered. Unit operations that worked before closing the recycle may no longer function in the same manner, or they may become infeasible. For example, the recycle may cause an upstream composition to move into another residue curve region, making a previously specified distillation now impossible. Alternatively, the recycle may result in the infinite buildup of a particular component if there is no outlet for that component from the recycle loop.

It is often beneficial to reexamine a completed flow sheet and look for opportunities for simplification and consolidation of unit operations. A complicated series of unit operations can sometimes be replaced by a simpler structure with equivalent material balances. When two or more sections of the flow sheet perform similar functions, one section often can be eliminated by recycling its feed stream to the remaining section. An impure MSA that is also an easily regenerable composition of interest will often function as effectively as a pure component MSA, without the need for additional purification operations.

PUTTING IT ALL TOGETHER

Any practical algorithm for the automatic generation of flow sheets must emphasize the use of the appropriate knowledge and the identification of critical features and strategies to deal with them, and provide a means for selecting and sequencing both opportunistic and strategic separations. One implementation of the procedure begins by the construction of lists of sources (initial states) and destinations (goal states) for the separations process. A source may be the original feed mixture or a stream created by a strategic or opportunistic separation. A source list is maintained as the algorithm progresses. It contains streams that have not been identified as destinations, have not been recycled, and have not been fed into another unit operation. Destinations are the final (or sometimes intermediate) goals of the separation flow sheet. They may be products, by-products, MSA compositions that must be regenerated and recycled, or the semidefined feed to a strategic operation. The destination list also changes as the design proceeds.

The thermodynamics and physical properties of the mixture to be separated are examined. VLE nodes and saddles, LLE binodal curves, and the like are labeled. Critical features and compositions of interest are identified. A stream is selected from the source list and is identified either as meeting all the composition objectives of some destination, or else as in need of further processing. If a critical feature must be overcome, a strategic operation may be proposed and either applied to the stream or placed on the flow sheet as an island. Otherwise an opportunistic separation is selected and applied to the stream. In either event, any new sources (output streams generated from these operations) and destinations (other input streams required by the selected operation) are added to the respective lists. The process is repeated until the source list is empty and all the destination specifications including mass balances have been satisfied, or unless the algorithm fails because no strategic or opportunistic operation could be found for a stream remaining in the source list. Next the flow sheet is examined for redundancies, equivalencies, and recycle opportunities. These are implemented within material balances constraints. Finally, the flow sheet must be simulated by rigorous heat and material balances to ensure that all physical, thermodynamic, and economic constraints are met.

EXAMPLE Production of Diethoxymethane

As an example of the separation system synthesis method, consider a process for the production of diethoxymethane. Diethoxymethane (DEM), the ethanol acetal of formaldehyde, is a useful chemical intermediate. The raw materials are paraformaldehyde and ethanol, and the reaction also results in water as a byproduct. Volatile DEM, ethanol, and water are readily removed from the reaction mass. The process species allocation is such that DEM is the desired product, the water is waste, and any ethanol is to be recycled back to the reactor. The conceptual separations synthesis problem then is to produce from an effluent stream evaporated from the reactor DEM fit for commercial use, water pure enough to be discarded, and ethanol pure enough to be recycled.

The problem here is that the ternary DEM–ethanol–water azeotrope is homogeneous and cannot be further purified by simple distillation. Using traditional patterns for breaking homogeneous azeotropes, one might consider adding a heavy component to the system to extract away one of the components, or a light immiscible component to form some heterogeneous (two liquid phase) ternary or possibly quaternary azeotrope that can be separated by decantation. Is an alternative separation system not requiring the use of an external MSA possible?

The DEM–ethanol–water system has three binary minimum-boiling azeotropes, and one ternary minimum-boiling azeotrope. The system vapor-liquid thermodynamics are represented by a fairly common residue curve map, shown in Figure 10.7(a). Distillation boundaries connect the ternary azeotrope

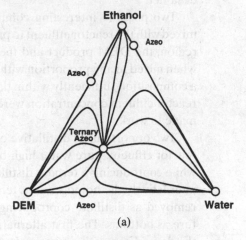

(a)

Figure 10.7. Diethoxymethane–ethanol–water system: (a) residue curve map, and (b) distillation and liquid immiscibility regions.

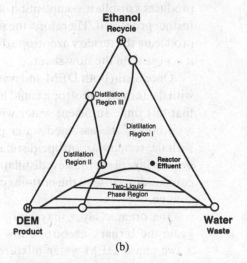

(b)

with each of the three binary azeotropes, producing three distinct distillation regions. The desired separation system products DEM, ethanol, and water are each high-boiling nodes in their respective distillation regions. The ternary azeotrope is the low-boiling node common to all distillation regions. The system exhibits a small region of liquid–liquid immiscibility, shown in Figure 10.7(b), which, however, does not include the reactor effluent concentration (the feed to this separation system), nor the ternary azeotrope.

The reactor effluent is in the same distillation region as water. Because DEM is in a different region, a critical feature is the distillation boundary between these two regions, which somehow must be crossed. When this strategic problem is attacked first, it is noticed that some liquid–liquid tie lines span the distillation boundary. If the composition from the reactor effluent can be moved into the lower part of the two-phase region, then liquid–liquid decantation exploiting the density difference between organic and aqueous liquid phases in equilibrium may be used to cross the distillation boundary. Therefore, a strategic operation that may be first placed on the flow sheet is a decanter.

Two possible interesting compositions may be assumed that might be mixed with the reactor effluent to produce a composition in the heterogeneous region: the DEM product and the water product. However, neither choice when mixed in any proportion with the reactor effluent composition results in a composition sufficiently within the two-liquid phase region. However, if the reactor effluent concentration were altered somewhat, such a scheme may be made to work.

Two opportunistic distillative operations might be applied directly to the reactor effluent: pure water high-boiling node could be removed as bottoms while coproducing a ternary distillate on the distillation region boundary, or alternatively the homogeneous ternary azeotrope low-boiling node could be removed as distillate coproducing a binary DEM-free ethanol–water mixture as bottoms. The first alternative seems plausible, as it produces one of the desired separation system products, water. However that alternative also produces a distillate composition on an interior boundary that cannot be easily further processed. Therefore, the second opportunistic alternative is selected, producing the ternary azeotrope distillate and an ethanol–water bottoms, and it is placed on the flow sheet.

Once again both DEM and water products are examined to see if mixture with the ternary azeotrope could lead to a two-phase mixture. It is easily seen that this time, sufficient water will result in a two-liquid phase mixture, so a portion of the assumed water product is added to the flow sheet, mixed with the ternary azeotrope distillate, and sent to the decanter. The resulting organic layer is in the new distillation region, whereas the aqueous layer has a composition close to the bottoms composition of the opportunistic distillation column.

The organic layer may be opportunistically distilled to produce either again the ternary azeotrope low-boiling node distillate while coproducing a two-phase DEM-water mixture as bottoms, or alternatively pure DEM high-boiling node as bottoms coproducing a ternary distillate on the distillation boundary close to the ternary azeotrope. The second alternative is selected because it achieves one of the separation system goals, and it is added to the flow sheet.

The distillate from this second opportunistic distillation column is next considered. It lies on a distillation boundary and ordinarily would be difficult to process. However, it also lies near the ternary azeotrope, which, however, is a composition that has already been processed. This distillate stream is therefore a candidate for recycle. Such a recycle must be carefully checked to see if premixing the two similar streams alters or invalidates any of the decisions previously made. It turns out that mixing the two distillate streams does change the amount of water necessary to bring the resulting composition into the two-liquid phase region, but other than affecting the size of the associated equipment, does not change the validity of the previous partially synthesized flow sheet. Therefore the proposed recycle is accepted and added to the design.

The aqueous layer from the decanter and the bottoms of the first column remain to be processed, and the goals of producing pure water and pure ethanol remain to be achieved. These two streams could be processed separately, but as they have similar compositions they could also be combined. The combined mixture could be opportunistically distilled to produce either yet again the ternary azeotrope low-boiling node distillate while coproducing another DEM-water mixture (to very little advantage), or alternatively pure water high-boiling node as bottoms coproducing a ternary distillate on the distillation boundary highly enriched in ethanol. The second alternative is selected, producing as bottoms both the desired water product from the separation system and the water assumed for mixing in the decanter, and it is placed on the flow sheet.

Two of the three goals of the separation problem, pure DEM and pure water, have now been met, although neither component has been perfectly recovered. The last remaining stream, with a composition lying on a distillation boundary, contains all of the ethanol and small amounts of both DEM and water. Getting pure ethanol will require another boundary crossing. With the use of an approach similar to that already employed, some pure DEM product could be returned to produce a mixture in the third distillation region from which pure ethanol could ultimately produced as a high-boiling node distillation bottoms. However, because the ethanol is only going to be recycled to the reactor, the required purity specification might be questioned. What would be the effect of small amounts of DEM and water in the recycled ethanol? Acetal formation is equilibrium controlled, so there is some deleterious effect of including reaction products with reactants. However, additional analysis proved the effect to be small, so the ethanol purity specification was relaxed and the distillate from the third column was used directly as produced. Figure 10.8 shows on an overlaid distillation region and liquid phase region diagram how compositions are mixed and separated and how the reactor effluent composition is eventually transformed into the three goal compositions. Figure 10.9

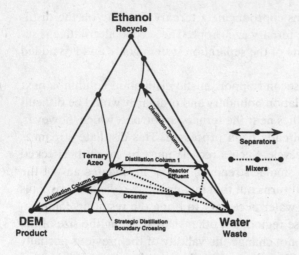

Figure 10.8. Synthesizing the diethoxy-methane recovery process.

shows the resulting process flow sheet synthesized. It is interesting to note that the key feature of the flow sheet is the strategic distillation-boundary-crossing decanter. It is perhaps even more remarkable that a decanter is the key unit operation in a flow sheet in which none of the streams has two liquid phases!

Separation process synthesis is an open-ended design activity capable of generating a large number of feasible alternatives. Often very different flow sheets result if the choice of MSA is changed, a different separation method is applied to a particular stream, or if the order in which streams on the source list are examined is changed. Each decision point can be revisited and another alternative developed.

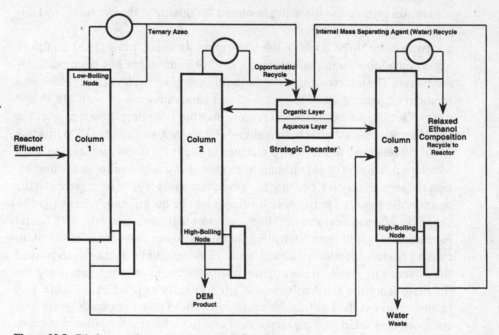

Figure 10.9. Diethoxymethane recovery process flow sheet.

Unfortunately, at present there are no heuristics for automatically synthesizing the best separation method for nonideal mixtures without systematically generating and evaluating multiple alternatives.

CONCLUSIONS

Process synthesis, the invention of process concepts represented as flow sheets, is one of the more creative and challenging endeavors encountered in the chemical processing industries. A structurally innovative flow sheet may be one of the critical determiners of economic viability. Considerable effort has been devoted to the development of systematic approaches to chemical process synthesis. These have begun to have measurable industrial impact. With these techniques an increased number of higher value, lower energy, lower environmental impact, and sometimes even novel design alternatives have been synthesized and actually implemented. Energy reductions of 50% and net present cost reductions of 35% typically are achievable.

Certain features of systematic process synthesis approaches appear to have special merit. These include hierarchical means-ends analysis architectures as a recursive problem-solving paradigm, combinations of opportunistic and strategic control methods, thinking specifically in terms of tasks to be accomplished before equipment to be employed, thinking of nonconventional properties to be exploited to accomplish identified tasks, and recognizing the importance of knowledge representations to encapsulate analysis within synthesis activities.

Although this discussion has focused on systematic generation approaches, advances in problem formulation and in computational hardware and software capabilities offer the promise of a new generation of practical chemical process synthesis techniques based directly on structural optimization. Soon the goal of synthesizing provably unbeatable conceptual process flow sheets may be at hand.

REFERENCES

Barnicki, S. D. and Fair, J. R. (1990). "Separation system synthesis: A knowledge-based approach. 1. Liquid mixture systems," *Industrial and Engineering Chemistry Research*, **29**: 421–432.

Barnicki, S. D. and Fair, J. R. (1992). "Separation system synthesis: A knowledge-based approach. 2. Gas/vapor mixtures," *Industrial and Engineering Chemistry Research*, **31**:1679–1694.

Barnicki, S. D. and Siirola, J. J. (1997a). "Enhanced distillation." In *Chemical Engineers' Handbook*, 7th ed., McGraw-Hill, New York.

Barnicki, S. D. and Siirola, J. J. (1997b). "Separations process synthesis." In *Encyclopedia of Chemical Technology*, 4th ed., Wiley, New York.

Douglas, J. M. (1988). *Conceptual Design of Chemical Processes*. McGraw-Hill, New York.

Douglas, J. M. (1995). "Synthesis of separation system flowsheets," *American Institute of Chemical Engineers Journal*, **41**:2522–2536.

Ichikawa, A., Nashida, N., and Umeda, T. (1969). "An approach to the optimal synthesis problem," presented at the 34th Annual Meeting, Society of Chemical Engineers, Japan.

Hendry, J. E. and Hughes, R. R. (1972). "Generating separation process flowsheets," *Chemical Engineering Progress*, **68**:71–76.

Kirkwood, R. L., Locke, M. H., and Douglas, J. M. (1988). "A prototype expert system for synthesizing chemical process flowsheets," *Computers and Chemical Engineering*, **12**:329–343.

Mahalec, V. and Motard, R. L. (1977). "Procedures for the initial design of chemical processing system," *Computers and Chemical Engineering*, **1**:57–68.

Nilsson, N. J. (1980). *Principles of Artificial Intelligence*. Morgan Kauffman, Los Altos, CA.

Rudd, D. F., Powers G. J., and Siirola, J. J. (1973). *Process Synthesis*, Prentice-Hall, Englewood Cliffs, NJ.

Seader, J. D. and Westerberg, A. W. (1977). "A combined heuristic and evolutionary strategy for synthesis of simple separation sequences," *American Institute of Chemical Engineers Journal*, **23**:951–954.

Siirola, J. J. and Rudd, D. F. (1971). "Computer-aided synthesis of chemical process design," *Industrial and Engineering Chemistry Fundamentals*, **10**:353–362.

Siirola, J. J. (1995). "An industrial perspective on process synthesis." In *Foundations of Computer-Aided Process Design*, American Institute of Chemical Engineers Symposium Series, **91**(304):222–233.

Simon, H. A. (1969). *Science of the Artificial*. MIT Press, Cambridge, MA.

Stephanopoulos, G. and Westerberg, A. W. (1976). "Studies in process synthesis: Part II. Evolutionary synthesis of optimal process flowsheets," *Chemical Engineering Science*, **31**:195–204.

Thompson, R. W. and King, C. J. (1972). "Systematic synthesis of separation schemes," *American Institute of Chemical Engineers Journal*, **18**:941–948.

Türkay, M. and Grossmann, I. E. (1998). "Structural flowsheet optimization with complex investment cost functions," *Computers and Chemical Engineering*, **22**:673–686.

Yee, T. F. and Grossmann, I. E. (1990). "Simultaneous optimization models for heat integration-II. Heat exchanger network synthesis," *Computers and Chemical Engineering*, **14**:1165–1184.

Yeomans, H. and Grossmann, I. E. (1999). "A systematic modeling framework of superstructure optimization in process synthesis," *Computers and Chemical Engineering*, **23**:709–731.

Wibowo, C. and Ng, K. M. (2000). "Unified approach for synthesizing crystallization-based separation processes," *American Institute of Chemical Engineers Journal*, **46**:1400–1421.

CHAPTER ELEVEN

Synthesis of Analog and Mixed-Signal Integrated Electronic Circuits

Georges G. E. Gielen and Rob A. Rutenbar

INTRODUCTION

In today's microelectronics, and in particular the markets for so-called *application-specific integrated circuits* (ASICs) and high-volume custom ICs, demand is being driven by the increasing level of circuit integration. Today, we can build chips with 100 million transistors, yielding ten million useful logic gates. As a result, complete systems that previously occupied one or more printed circuit boards are being integrated on a few chips or even one single chip. Examples of these so-called *system-on-chip* (SoC) designs include single-chip digital television and single-chip cameras (ISSCC, 1996). New generations of telecommunication systems include analog, digital, and even radio-frequency (RF) sections on one chip. The technology of choice for these systems remains industry-standard *complementary metal oxide semiconductors* (CMOS), because of their unsurpassed *scaling* abilities; that is, when geometric feature sizes shrink with each new generation of CMOS technology, the resulting circuits and devices are not only smaller but also faster, and they consume less power. However, technologies that allow both CMOS and higher-speed bipolar transistors (e.g., BiCMOS, and more recently, embedded SiGe) are also used when needed for high-performance analog or RF circuits. Although most functions in these systems are implemented with digital circuitry, the analog circuits needed at the interface between the electronic system and the "real" world are also being integrated on the same chip for reasons of cost and performance. A next-generation SoC design is illustrated in Figure 11.1 and contains an embedded processor, embedded memories, reconfigurable logic, and analog interface circuits to communicate with the continuous-valued external world.

Despite the continuing trend to replace analog circuit functions with digital computations (e.g., digital signal processing has supplanted analog filtering in many applications), there are important typical functions that will *always* remain analog.

1. External world inputs: signals from a sensor, microphone, antenna, wireline network, and the like must be sensed or received and then amplified and filtered up to a level that allows digitization with a sufficient signal-to-noise-and-distortion ratio. Typical analog circuits used here are low-noise amplifiers, variable-gain amplifiers, filters, oscillators, and mixers (in case of downconversion). Applications are, for instance, instrumentation (e.g., data and biomedical), sensor interfaces (e.g.,

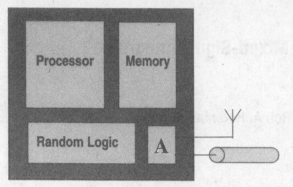

Figure 11.1. Future System on a chip.

airbag accelerometers), process control loops, telecommunication receivers (e.g., telephone or cable modems, wireless phones, set-top boxes, etc.), recording (e.g., speech recognition, cameras), and smart cards.

2. External world outputs: the signal is reconverted from digital to analog form and then "strengthened" so that it can drive the outside load (e.g., actuator, antenna, loudspeaker, wireline network) without too much distortion. Typical analog circuits used here are drivers and buffers, filters, oscillators, and mixers (in case of upconversion). Applications are, for instance, process control loops (e.g., voltage regulators for engines), telecommunication transmitters, audio and video (e.g., CD, DVD, loudspeakers, TV, PC monitors, etc.), and biomedical actuation (e.g., hearing aids).

3. Mixed-signal interfaces: these are the circuits that interface the above analog circuits with the digital part of the system. This "digital part" is today very often a *digital signal processor* (DSP) unit running its own software program to manipulate these signals in digitized form. Typical analog circuits used here are sample-and-hold circuits for signal sampling, analog-to-digital converters for amplitude discretization, digital-to-analog converters for signal reconstruction, and phase-locked loops and frequency synthesizers to generate a timing reference or perform timing synchronization.

4. Absolute references: in addition, the above circuits require stable absolute references for their operation, which are generated by voltage and current reference circuits, crystal oscillators, and the like. These provide a highly precise electrical value (e.g., a voltage, a current, a capacitance) that is stable in the face of both manufacturing variations and environmental variations such as temperature.

5. Maximum-performance digital: finally, we note that the largest analog circuits today are high-performance (high-speed, low-power) *digital* circuits. Typical examples are state-of-the-art microprocessors, which are largely custom designed like analog circuits, to push performance limits.

Analog circuits are indispensable in applications that interface with the outside world, and they will become even be more prevalent in our lives as we move toward intelligent homes, mobile road–air offices, and the wireless workplaces of the future.

When both analog and digital circuits are needed in a system, it becomes obvious to integrate them together to optimize cost and performance–if the technology allows us to do so. Ten years ago, such a system would not only require multiple chips, but also multiple semiconductor technologies: CMOS devices for small, fast digital logic, and bipolar devices for high-performance analog circuits. The continued shrinking of CMOS device sizes (SIA, 1994) not only enables larger and more complex designs, but also offers analog MOS transistor performance that approaches the performance of bipolar transistor designs (long the preferred technology for analog), but with higher density and lower cost. This explains why CMOS is the technology of choice today, and why other technologies, for example, those that mix CMOS and various bipolar devices on the same chip, such as BiCMOS or SiGe, are only used when more aggressive device characteristics are really needed. Today, RF designs that require gigahertz operating frequencies are the ones mostly likely to use such technologies. Designs from telecommunications, consumer, computing, and automotive applications, among others, are taking advantage of these recent trends in circuit integration (Liang, 1998).

The ultimate goal is the ability to design even high-performance analog circuits in "plain" digital-style CMOS technology. However, this sort of highly integrated design introduces a whole new set of problems that have to be accounted for.

1. Complexity: these integrated systems are not just large in terms of the number of transistors; they are complex in terms of the design hierarchies they implement. Multiple digital and analog subsystems, processors and software, continuous and discrete signal processing, and the like, must all be designed concurrently.
2. Design time: many mixed-signal chips are characterized by very short product life cycles and tight time-to-market constraints. The time-to-market factor is really critical for chips that eventually end up in consumer, telecom, or computer products: if one misses the initial market window relative to the competition, prices and therefore profit can be seriously eroded.
3. Portability: Digital designs are inherently more portable than analog designs, in the sense that the digital abstraction insulates many aspects of the design from the underlying circuit details. As a result, one can *migrate* a digital design from one semiconductor manufacturer (a *foundry*) to another by using existing digital synthesis tools. This is not true for the analog components on these chips, which exploit rather than abstract the underlying physics of the fabrication process to make precision circuits.

As a result, we have in the past few years the problem of increasing demand (and underlying semiconductor capability) for high-performance analog circuits, but decreasing design time, and in particular, no time to handcraft every circuit as has been traditionally done with the analog circuit.

Automation is essential for creating these chips quickly, optimally, and successfully. However, to date, it is really the digital part of these designs that has been highly automated. In the digital IC domain, synthesis tools are well developed and commercially available today. Unlike analog circuits, a digital system can naturally be represented in terms of Boolean representation and programming language constructs, and its functionality can easily be represented in algorithmic form, thus paving the way for automation of many aspects of design. Today, many aspects of the digital design process are fully automated. The hardware is described in a *hardware description language* (HDL) such as VHDL or VERILOG, either at the behavioral level or most often at the structural level. *High-level synthesis* tools attempt to synthesize the behavioral HDL description into a structural representation. *Logic synthesis* tools then translate the structural HDL specification into a gate-level netlist. *Semicustom layout* tools (place and route) map this netlist into a correct-by-construction mask-level layout based on a cell library specific for the semiconductor technology. Research interest is now moving in the direction of system synthesis in which a system-level specification is translated into a hardware–software coarchitecture with high-level specifications for the hardware, the software, and the interfaces. *Reuse methodologies* are being developed to further reduce the design effort for complex systems. One example is the design reuse methodology of assembling a system by reusing so-called *macrocells* or *virtual components* that represent fully or partially designed, parameterizable subblocks that can easily be mixed or matched in system design if they comply with certain interface standards; one example is the virtual socket interface (VSI Alliance, 1997). *Platform-based design* is another emerging system-level design methodology (Chang et al., 1999). Of course, the level of automation is far from push-button here, but the developments are keeping up reasonably well with the chip complexity offered by the technology.

Unfortunately, the story is quite different on the analog side. Analog synthesis tools are only today appearing. Existing analog design tools are mainly circuit simulators, in most cases some flavor of the ubiquitous SPICE simulator (Nagel, 1975), and mask layout editing environments and their accompanying tools (e.g., some limited optimization capabilities around the simulator, or layout verification tools). One reason for this lack of automation is that analog design is perceived as less systematic, more heuristic, and more knowledge-intensive than digital design; it has not yet been possible for analog designers to establish rigorous and accepted abstractions that hide all the device-level and process-level details from higher-level design. In addition, analog circuits are far more sensitive to nonidealities, second-order effects, and parasitic disturbances (cross talk, substrate noise, supply noise, etc.) than are digital circuits. These differences from digital design also explain why analog CAD tools cannot simply adapt digital synthesis algorithms, and why analog solutions have to be developed that are specific to the analog design paradigm and complexity. Analog circuits today are still primarily handcrafted; the design cycle for analog and mixed-signal ICs remains long and error prone. As a result, although analog circuits typically occupy only a small fraction of the total area of mixed-signal ICs, they are often the bottleneck in mixed-signal chip design.

So, our survey in this chapter really begins from these summary observations. First, digital synthesis is common. Synthesis tools for digital logic and digital layout have revolutionized the design of digital chips and have made possible our ability to design very large, very complex systems quickly and with good reliability. Second, mixed-signal integration is inevitable. Semiconductor technologies have advanced to the point where analog systems can and will be integrated onto the same chip as the digital systems. These *mixed-signal* designs are increasingly the norm, as the current generation of electronics has to communicate with the external physical world. Third, manual analog design is inefficient. Pressures on design time, design complexity, and design correctness have increased dramatically for the analog sections of these chips; these sections have traditionally been designed by hand. Fourth, analog synthesis is needed. Analog circuit and physical synthesis tools are the proper solution to these problems, and two decades of research work in this area is finally today coming into practical use.

The goals for today's emerging analog synthesis tools are, perhaps, fairly obvious: Reduce the design time and cost for analog circuits, from specification to successful silicon; reduce the risk for design errors that prevent first-pass functional chips; and increase the performance of complex mixed-signal chip designs.

The ability to automate the design of an individual analog circuit – say, a basic block with 50–100 devices, called a *cell* – offers several advantages. Given the time, expense, and risk associated with each circuit design, automation can allow more higher-level exploration and optimization of the overall system architecture. Decisions at these higher levels of analog–digital architecture have a very large impact on overall chip parameters such as power and area. Even in small analog circuits, designers find difficulty in considering multiple conflicting trade-offs at the same time – computers don't. Typical examples are fine-tuning an initial handcrafted design and improving robustness with respect to environmental variations or manufacturing tolerances. The need to redesign a circuit for a new semiconductor technology (either the next process generation, or even just a change from one foundry to another) is a large burden on analog designers. Synthesis tools can take over a large part of these so-called *technology retargeting* efforts, and they can make analog design easier or migrate to new technologies. Finally, design reuse methodologies require executable models and other information for the analog blocks used in these systems. The modeling techniques that underlie most synthesis strategies should help to make this possible.

This need for analog tools beyond circuit simulation has also clearly been identified in recent projections; for example, the Semiconductor Industry Association 1997 technology roadmap in Figure 11.2 predicted analog synthesis as an essential technology somewhere beyond the year 2000 (SIA, 1994). Interestingly enough, analog circuit synthesis tools are indeed just now appearing commercially. This is mainly due to academic and industrial research over the past 20 years (Carley et al., 1996). Simulation tools here are already well developed; most recently, mixed-signal hardware description languages such as VHDL-AMS (IEEE, 1997) and VERILOG-A/MS

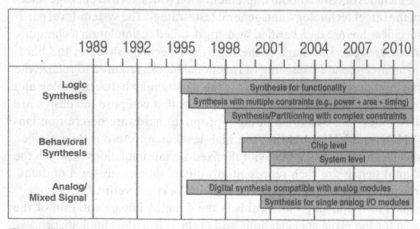

Figure 11.2. SIA synthesis potential solutions roadmap (SIA; 1994).

(Open Verilog International, 1996) provide a unifying trend that will link the various analog design automation tasks in a coherent framework that supports a more structured analog design methodology.

In this chapter, we survey the design context, history, and state of the art for analog and mixed-signal synthesis; additional details may be found in the more circuits-oriented survey offered in Gielen and Rutenbar (2000). The chapter is organized as follows. The second section reviews the mixed-signal chip design process, as well as a hierarchical design strategy for the analog blocks. The third section then describes progress on analog circuit and layout synthesis. This will be illustrated with several examples. Conclusions are then provided in the fourth section, as well as a discussion of the problems that remain. An extensive list of references completes the chapter.

THE ANALOG AND MIXED-SIGNAL DESIGN PROCESS

We will now first describe the design flow for mixed-signal integrated systems from concept to chip, followed by the description of a hierarchical design methodology for the analog blocks that can be adopted by analog CAD systems.

THE MIXED-SIGNAL IC DESIGN FLOW

Figure 11.3 illustrates a possible scenario for the design flow of a complex analog or mixed-signal IC. The various stages that are traversed in the design process are as follows.

Conceptual Design. This is the product concept stage, in which the specifications for a design are gathered and the overall product concept is developed. Careful checking of the specifications is crucial for the later success of the product. Mathematical tools such as Matlab/Simulink are often used at this stage. This stage also includes setting high-level management goals such as final cost and time to market.

System Design. This is the first stage of the actual design, in which the overall architecture of the system is designed and partitioned. Hardware and software parts are defined, and both are specified in appropriate languages. The hardware components are described at the behavioral level, and in addition the interfaces have to be specified. This stage includes decisions about implementation issues, such as package selection, choice of the target technology, and general test strategy. The system-level partitioning and specifications are then verified by using detailed cosimulation techniques.

Architectural Design. This stage is the high-level decomposition of the hardware part into an architecture consisting of functional blocks required to realize the specified behavioral description. This includes the partitioning between analog and digital blocks. The specifications of the various blocks that compose the design are defined, and all blocks are described in an appropriate hardware description language (e.g., VHDL and VHDL-AMS). The high-level architecture is then verified against the specifications by using behavioral mixed-signal simulations in which the analog and digital signals are each represented with an appropriate level of detail, and the interactions between the analog and digital blocks are verified.

Cell Design. For the analog blocks, this is the detailed implementation of the different blocks for the given specifications and in the selected technology process,

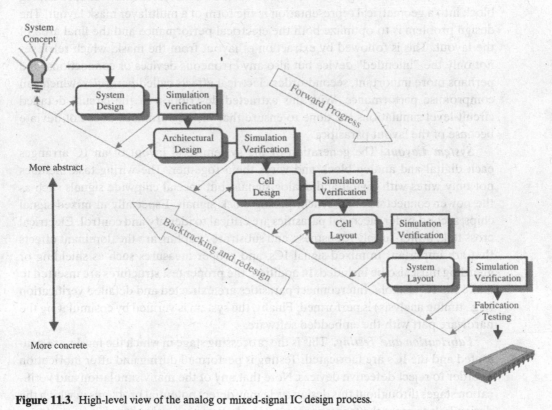

Figure 11.3. High-level view of the analog or mixed-signal IC design process.

resulting in a fully designed device-level circuit schematic. The stage encompasses both a selection of the proper circuit topology and the (usually continuous valued) circuit parameters. In analog design, this parameter design problem is called *sizing*, in reference to the fact that these parameters often determine layout properties that affect the physical (mask) size of each implemented transistor. Throughout this process, more complex analog blocks will be further decomposed into a set of subblocks. This whole process will be described in more detail in the second part of this section. Manufacturability considerations (tolerances and mismatches) are taken into account to guarantee a high yield, robustness, or both. The resulting circuit design is then verified against the specifications by using detailed circuit simulations.

Cell Layout. This stage is the translation of the electrical schematic of each analog block into a geometrical representation in the form of a multilayer mask layout. The design problem is to optimize both the electrical performance and the final area of the layout. This is followed by extraction of layout from the mask, which retrieves not only the "intended" device but also any erroneous devices or connections, and perhaps more important, second-order electrical effects called *parasitics*, which can compromise performance. With this extracted description of the circuit, detailed circuit-level simulations are done to ensure that the performance does not deviate because of the layout parasitics.

System Layout. The generation of the system-level layout of an IC arranges each digital and analog block and wires them together. The wiring task includes not only wires with digital and analog signals, but special chipwide signals such as the power connections and synchronizing clock signals. Especially in mixed-signal chips, second-order electrical parasitics are critical to identify and control. Electrical cross talk, power-supply robustness, and substrate coupling are the dominant effects that are important in mixed-signal ICs, and proper measures such as shielding or guarding must also be included. In addition, the proper test structures are inserted to make the IC testable. Interconnect parasitics are extracted and detailed verification (e.g., timing analysis) is performed. Finally, the system is verified by cosimulating the hardware part with the embedded software.

Fabrication and Testing. This is the processing stage in which the masks are generated and the ICs are fabricated. Testing is performed during and after fabrication in order to reject defective devices. Note that any of the many simulation and verification stages throughout this design cycle may detect potential problems in which the design fails to meet the target requirements. In these cases, *backtracking or redesign* will be needed, as indicated by the upward arrow on the left in Figure 11.3.

HIERARCHICAL ANALOG DESIGN METHODOLOGY

This section focuses on the design methodology adopted for the design of analog integrated circuits. These analog circuits could be part of a larger mixed-signal IC. Although at the present time there is no dominant, unique, general design methodology for analog circuits, we outline here the hierarchical methodology prevalent in many of the emerging experimental analog CAD systems (Harjani, Rutenbar, and Carley, 1989; Chang et al., 1992, 1997; Gielen, Swings, and Sansen, 1993; Malavasi et al., 1993; Donnay et al., 1996).

For the design of a complex analog block such as a phase-locked loop or an analog-to-digital converter, the analog block is typically decomposed into smaller subblocks (e.g., a comparator or a filter). The specifications of these subblocks are then derived from the initial specifications of the original block, after which each of the subblocks can be designed on its own, possibly by further decomposition into even smaller subblocks. In this way constraints are passed down the hierarchy in order to make sure that the top-level block in the end meets the specifications. This whole process is repeated down the decomposition hierarchy until a level is reached that allows a physical implementation (either the transistor level or a higher level in the case in which existing analog cells are used). The top-down synthesis process is then followed by a bottom-up layout implementation and design verification process. The need for detailed design verification is essential, because manufacturing an IC is expensive, and a design has to be ensured to be fully functional and meet all the design requirements within a window of manufacturing tolerances, before the actual fabrication is started. When the design fails to meet the specifications at some point in the design flow, redesign iterations are needed.

Most experimental analog CAD systems today use a *performance-driven design strategy within such an analog design hierarchy*. This strategy consists of the alternation of the following steps in between any two levels i and $i + 1$ of the design hierarchy (see Figure 11.4): for the top-down path, there is topology selection, specification translation (or circuit sizing), and design verification; for the bottom-up path, there is layout generation and detailed design verification (after extraction).

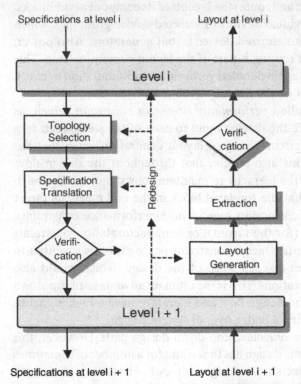

Figure 11.4. Hierarchical design strategy for analog circuits.

Topology selection is the step of selecting the most appropriate interconnection (the *topology*) of basic devices (e.g., transistors, diodes, resistors, capacitors, inductors, etc.) that can best meet the given specifications. It is quite common to select this from a set of already known circuit topologies; an alternative is that the designer develops his or her own new topology. A topology can be defined hierarchically in terms of lower-level subblocks. For an analog-to-digital converter, for instance, topology selection could be selecting between a flash converter, a successive approximation converter, or an architecture that employs fewer analog devices and replaces them with digital filtering, such as the so-called delta–sigma ($\Delta-\Sigma$) converter.

Specification translation is the step of mapping the specifications for the block under design at a given level (e.g., a converter) into individual specifications for each of the subblocks within the selected block topology, so that the complete block meets its specifications, while possibly optimizing the design toward some more global objectives (e.g., minimal power consumption). The translated specifications are then verified by means of (behavioral or circuit) simulations before one proceeds down the hierarchy. Behavioral simulations are needed at higher levels in the design hierarchy because no device-level implementation is yet available; circuit simulations are used at the lowest level in the design hierarchy. At this lowest level, the subblocks are single devices and specification translation reduces to *circuit sizing* (also called *circuit dimensioning*), which determines all device sizes, element values, and bias parameters in the circuit.

Layout generation is the step of generating the geometrical layout of each block under design, by assembling (place and route) the layouts of its component subblocks. At the lowest level these are individual devices or selected device groupings, which themselves are generated by parameterized device layout generators. Also power, ground, and substrate connection routing is part of the layout generation step. This step is followed by extraction and again detailed verification and simulation to check the impact of the layout parasitics on the overall circuit performance.

The above methodology is called *performance driven or constraint driven*, as the performance specifications are the driving input to each of the steps: each step tries to perform its action (e.g., circuit sizing or layout generation) such that the input constraints are satisfied. This also implies that throughout the design flow, constraints are propagated down the hierachy to maintain consistency in the design as it evolves, and to make sure that the top-level block in the end meets its target specifications. These propagated constraints may include performance constraints, but also geometrical constraints (for the layout), or manufacturability constraints (for yield), or even test constraints. Design constraint propagation is essential to ensure that specifications are met at each stage of the design, which would also reduce the number of redesign iterations. This is the ultimate advantage of top-down design: catch problems early in the design flow and therefore have a higher chance of first-time success, while obtaining a better overall system design.

Ideally, we would like to have *one* clean top-down design path. However, this rarely occurs in practice, as a realistic design has to account for a number of sometimes hard-to-quantify second-order effects as the design evolves. For instance, a choice of a particular topology for a function block may fail to achieve the required specifications.

Or, performance specifications may be too tight to achieve, in which case a redesign step is necessary to alter the block topology or loosen the design specifications. In the above top-down–bottom-up design flow, *redesign by means of backtracking iterations* may therefore be needed at any point where a design step fails to meet its input specifications. In that case, one or more of the previously executed steps will have to be redone; for example, another circuit topology can be selected instead of the failing one, or another partitioning of subblock specifications can be performed. One of the big differences between analog or mixed-signal designs and the more "straight" top-down digital designs is exactly the larger number of design iterations needed to come to a good design solution. The adoption of a top-down design methodology is precisely intended to reduce this disadvantage.

A question that can be posed here is, Why must analog circuits be redesigned or customized for every new application? The use of a library of carefully selected analog cells can be advantageous for certain applications, but is in general inefficient and insufficient. Because of the large variety and range of circuit specifications for different applications, any library will only have a partial coverage for each application, or will result in an excess power or area consumption that may not be acceptable for given applications. Many high-performance applications require an *optimal* design solution for *each* constituent analog circuit, in terms of power, area, and overall performance. A library-based approach would require an uneconomically large collection of infrequently used cells. Instead, analog circuits are most commonly customized toward each specific application; tools should be available to support this. In addition, the porting of the library cells whenever the process changes is a serious effort that would also require a set of tools to automate.

The following section will describe the progress and the current state of the art in analog circuit and layout synthesis.

ANALOG CIRCUIT AND LAYOUT SYNTHESIS

ANALOG CIRCUIT SYNTHESIS AND OPTIMIZATION

The first step in the analog design flow of Figure 11.4 is analog circuit synthesis, which consists of two tasks: *topology selection* and *specification translation*. Synthesis is a critical step, because most analog designs require a custom design and the number of (often conflicting) performance requirements to be taken into account is large. Analog circuit synthesis is the "inverse" operation of circuit analysis. During analysis, the circuit topology and the subblock parameters (such as device sizes and bias values) are given and the resulting performance of the overall block is calculated, as is done in any circuit simulator. During synthesis, in contrast, the block performance is specified and an appropriate topology to implement this block has to be decided first. This step is called *topology selection*. Subsequently, values for the sub block parameters have to be determined, so that the final block meets the specified performance. This step is called *specification translation* at higher levels in the design hierarchy, where performance specifications for subblocks are determined. At the lowest level of the analog hierarchy, this is called *circuit sizing*, where sizes and biasing of all devices are determined. See Figure 11.5 for an illustration of this flow for low-level cells.

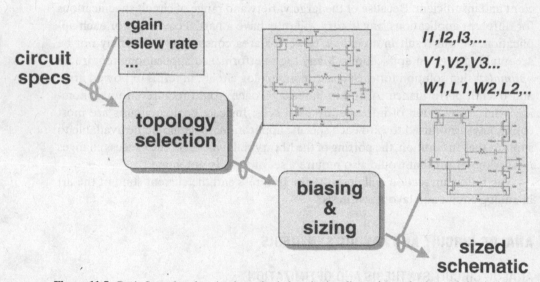

Figure 11.5. Basic flow of analog circuit synthesis for a basic cell: topology selection and circuit sizing.

The inversion process inherent to synthesis, however, is not a one-to-one mapping, but typically is an underconstrained problem with many degrees of freedom. The different analog circuit synthesis systems that have been explored up until now can be classified based on how they perform topology selection and how they eliminate the degrees of freedom during specification translation or circuit sizing. In many cases, the initial sizing produces a near-optimal design that is further fine-tuned with a circuit optimization tool. The performance of the resulting design is then verified by using detailed circuit simulations with a simulator, and, when needed, the synthesis process is iterated to arrive at a close-fit design. We now discuss the two basic steps in more detail.

Topology Selection

Given a set of performance specifications and a technology process, a designer or a synthesis tool must first select a circuit schematic that is most suitable to meet the specifications at minimal implementation cost (power, chip area). This problem can be solved by selecting a schematic from among a known set of alternative topologies such as stored in a library, or by generating a new schematic, for example by modifying an existing schematic. Although the earliest synthesis approaches considered topology selection and sizing together, the task of topology selection has received less attention in recent years, in which the focus has been primarily on the circuit sizing. Finding the optimal circuit topology for a given set of performance specifications is rather heuristic in nature and brings to bear the real expert knowledge of a designer. Thus it was only natural that the first topology selection approaches such as those in OASYS (Harjani et al., 1989), BLADES (El-Turky and Perry, 1989), or OPASYN (Koh, Séquin, and Gray, 1990) were rather heuristic in nature in that they used rules in one format or another to select a proper topology (possibly hierarchically) out of a predefined set of alternatives stored in the tool's library.

Later approaches worked in a more quantitative way in that they calculated the feasible performance space of each topology that fits the structural requirements, and then compared that feasible space to the actual input specifications during synthesis to decide on the appropriateness and the ordering of each topology. This can, for instance, be done by using interval analysis techniques (Veselinovic et al., 1995) or by using interpolation techniques in combination with adaptive sampling (Harjani and Shao, 1996). In all these programs, however, topology selection is a separate step. There are also a number of optimization-based approaches that integrate topology selection with circuit sizing as part of one overall optimization loop. This was done by using a mixed integer-nonlinear programming formulation with Boolean variables representing topological choices (Maulik, Carley, and Rutenbar, 1995), or by using a nested simulated evolution–annealing loop where the evolution algorithm looks for the best topology and the annealing algorithm for the corresponding optimum device sizes (Ning et al., 1991). Another approach that uses a genetic algorithm to find the best topology choice was presented in DARWIN (Kruiskamp and Leenaerts, 1995). Of these methods, the quantitative and optimization-based approaches are the more promising developments that address the topology selection task in a deterministic fashion, as compared with the rather ad hoc heuristic methods.

Analog Circuit Sizing

Once an appropriate topology has been selected, the next step is specification translation, in which the performance parameters of the subblocks in the selected topology are determined based on the specifications of the overall block. At the lowest level in the design hierarchy, this reduces to circuit *sizing*, where the sizes and biasing of all devices are designed. This mapping from performance specifications into proper – preferably optimal – device sizes and biasing for a selected analog circuit topology involves solving the set of physical equations that relate the device sizes to the electrical performance parameters. However, solving these equations explicitly is in general not possible, and analog circuit sizing typically results in an underconstrained problem with many degrees of freedom. The two basic ways to solve for these degrees of freedom in the analog sizing process are either by exploiting analog design knowledge and heuristics, or by using optimization techniques. These two basic methods, which are schematically depicted in Figure 11.6, correspond to the two broad classes of approaches adopted toward analog circuit synthesis, that is, the *knowledge-based* approaches and the *optimization-based* approaches (Gielen and Sansen, 1991; Carley et al., 1996).

Knowledge-Based Analog Sizing Approaches. The first generation of analog circuit synthesis systems presented in the middle to late 1980s were knowledge based. Specific heuristic design knowledge about the circuit topology under design (including the design equations but also the design strategy) was acquired and encoded explicitly in some computer-executable form, which was then executed during the synthesis run for a given set of input specifications to directly obtain the design solution. This approach is schematically illustrated in Figure 11.6(a). The knowledge was encoded in different ways in different systems.

The IDAC tool (Degrauwe et al., 1987) used manually derived and prearranged design plans or design scripts to carry out the circuit sizing. The design equations specific for a particular circuit topology had to be derived and the degrees of freedom in the design had to be solved explicitly during the development of the design plan by using simplifications and design heuristics. Once the topology was chosen by the designer, the design plan was loaded from the library and executed to produce a first-cut design that could further be fine-tuned through local optimization. The big advantage of using design plans is their fast execution speed, which allows for fast performance space explorations. The approach also attempts to take advantage of the knowledge of analog designers. IDAC's schematic library was also quite extensive and it included various analog circuits such as voltage references, comparators, and

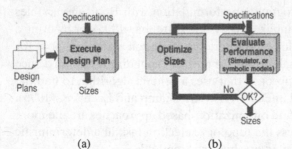

Figure 11.6. The two basic approaches toward analog circuit synthesis: (a) the knowledge-based approach using procedural design plans, and (b) the optimization-based approach.

(a)   (b)

the like, besides operational amplifiers. The big disadvantages of the approach are the lack of flexibility in the hardcoded design plans and the large time needed to acquire the design equations and to develop a design plan for each topology and design target, as analog design heuristics are very difficult to formalize in a general and context-independent way. It has been reported (Beenker et al., 1993) that the creation of a design script or plan typically took four times more effort than is needed to actually design the circuit once. A given topology must therefore at least be used in four different designs before it is profitable to develop the corresponding design plan. Considering the large number of circuit schematics in use in industrial practice, this large setup time essentially restricted the commercial usability of the IDAC tool and limited its capabilities to the initial set of schematics delivered by the tool developer. Also, the integration of the tool in a spreadsheet environment under the name PlanFrame (Henderson et al., 1993) did not fundamentally change this. Note that as a result of its short execution times, IDAC was intended as an interactive tool: users had to choose the topology themselves, and they also had to specify values for the remaining degrees of freedom left open in the design plan.

OASYS (Harjani et al., 1989) adopted a similar design-plan-based sizing approach in which every (sub)block in the library had its own handcrafted design plan, but the tool explicitly introduced hierarchy by representing topologies as an interconnection of subblocks. For example, a circuit such as an opamp was decomposed into subcircuits like a differential pair, current mirrors, and the like and was not represented as one atomic device-level schematic as in IDAC. OASYS also added a heuristic approach toward topology selection, as well as a backtracking mechanism to recover from design failures. As shown in Figure 11.7, the complete flow of the tool was then alternated topology selection and specification translation (the latter by executing the design plan associated with the topology) down the hierarchy until the device level was reached. If the design did not match the desired performance characteristics at any stage in this process, OASYS backtracked up the hierarchy, trying alternate configurations for the subblocks. The explicit use of hierarchy allowed the reuse of design plans of lower-level cells while building up higher-level-cell design plans, and therefore also leveraged the number of device-level schematics covered by one top-level topology template. Although the tool was used successfully for some

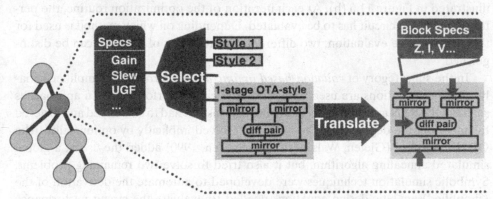

Figure 11.7. Hierarchical alternation of topology selection and specification translation down the design hierarchy (UGF, unity gain frequency; OTA, operational transconductance amplifier).

classes of opamps, comparators, and even a data converter, collecting and ordering all the design knowledge in the design plans still remained a huge, manual, and time-consuming job, restricting the practical usefulness of the tool. The approach was later adopted in the commercial MIDAS system (Beenker, Conway, and Schrooten, 1993), which was used successfully in house for certain types of data converters. Also AZTECA (Horta, Franca, and Leme, 1991) and CATALYST (Vital and Franca, 1992) use the design-plan approach to the high-level design of successive-approximation and high-speed CMOS data converters, respectively. Inspired by artificial intelligence research, other ways to encode the knowledge have also been explored, such as in BLADES (El-Turky and Perry, 1989), which is a rule-based system to size analog circuits, in ISAID (Toumazou and Makris, 1995, Makris and Toumazou, 1995) in STAIC (Harvey, Elmasry, and Leung, 1992), or in (Sheu, Fung, and Lai, 1988).

In all these methods, the heuristic design knowledge of an analog designer turned out to be difficult to acquire and to formalize explicitly, and the manual acquisition process was very time consuming. In addition to analytic equation-based design knowledge, procedural design knowledge is also required to generate design plans, as well as specialized knowledge to support tasks such as failure handling and backtracking. The overhead to generate all this was too large compared to a direct design of the circuit, restricting the tools basically to those circuits that were delivered by the tool developers. Their coverage range was found to be too small for the real-life industrial practice, and therefore these first approaches failed in the commercial marketplace.

Optimization-Based Analog Sizing Approaches. In order to make analog synthesis systems more flexible and extendible for new circuit schematics, an alternative approach was followed starting in the late 1980s. This research resulted in a second generation of methods, the *optimization-based approaches*. These use numerical optimization techniques to implicitly solve for the degrees of freedom in analog design while optimizing the performance of the circuit under the given specification constraints. These strategies also strive to automate the generation of the required design knowledge as much as possible, for example, by using symbolic analysis techniques to automatically derive many of the design equations and the sizing plans, or to minimize the explicitly required design knowledge by adopting a more equation-free simulation-oriented approach. This optimization-based approach is schematically illustrated in Figure 11.6 (b). At each iteration of the optimization routine, the performance of the circuit has to be evaluated. Depending on which method is used for this performance evaluation, two different subcategories of methods can be distinguished.

In the subcategory of *equation-based optimization approaches*, (simplified) analytic design equations are used to describe the circuit performance. In approaches such as OPASYN and STAIC the design equations still had to be derived and ordered by hand, but the degrees of freedom were resolved implicitly by optimization. The OPTIMAN tool (Gielen, Walscharts, and Sansen, 1990) added the use of a global simulated annealing algorithm, but it also tried to solve two remaining problems. Symbolic simulation techniques were developed to automate the derivation of the (simplified) analytic design equations needed to evaluate the circuit performance at every iteration of the optimization (Gielen and Sansen, 1991). Today the AC behavior (both linear and weakly nonlinear) of relatively large circuits can already be generated automatically. The second problem is then the subsequent ordering of the design equations into an application-specific design or evaluation plan. Also this step was automated by using constraint programming techniques in the DONALD tool (Swings and Sansen, 1991). Together with a separate topology-selection tool based on boundary checking and interval analysis (Veselinovic et al., 1995) and a performance-driven layout generation tool (Lampaert, Gielen, and Sansen, 1999), all these tools are now integrated into the AMGIE analog circuit synthesis system (Gielen et al., 1995), which covers the complete design flow from specifications over topology selection and circuit sizing down to layout generation and automatic verification. An example of a circuit that has been synthesized with this AMGIE system is the particle–radiation detector front end of Figure 11.8, which consists of a charge-sensitive amplifier (CSA) followed by an n-stage pulse-shaping amplifier (PSA). All opamps are complete circuit-level schematics in the actual design, as indicated in the figure. A comparison between the specifications and the performances obtained by an earlier manual design of an expert designer and by the fully computer-synthesized circuit is given in Table 11.1. In the experiment, a reduction of the power consumption with a factor of 6 (from 40 to 7 mW) was obtained by the synthesis system compared to the manual solution. In addition, the final area is slightly smaller. The layout generated for this example is shown in Figure 11.9.

The technique of equation-based optimization has also been applied to the high-level synthesis of Δ–Σ modulators in the SD-OPT tool (Medeiro et al., 1995). The converter architecture is described by means of symbolic equations, which are then used in a simulated-annealing-like optimization loop to derive the optimal subblock specifications from the specifications of the converter. Recently a first attempt was presented toward the full behavioral synthesis of analog systems from an (annotated) VHDL-AMS behavioral description. The VASE tool follows a hierarchical two-layered optimization-based design-space exploration approach to produce sized subblocks from behavioral specifications (Doboli et al., 1999). A branch-and-bound algorithm with efficient solution-space pruning first generates alternative system topologies by mapping the specifications by means of a signal-flow-graph representation onto library elements. For each resulting topology, a genetic-algorithm-based heuristic method is then executed for constraint transformation and subblock synthesis, which concurrently transforms system-level constraints into subblock design

TABLE 11.1. EXAMPLE OF ANALOG CIRCUIT SYNTHESIS EXPERIMENT WITH THE AMGIE SYSTEM

Performance	Specification	Manual Design	Automated Synthesis
Peaking time (ms)	<1.5	1.1	1.1
Counting rate (kHz)	>200	200	294
Noise (rms e-)	<1000	750	905
Gain (V/fc)	20	20	21
Output range (V)	>−1–1	−1–1	−1.5–1.5
Power (mW)	minimal	40	7
Area (mm^2)	minimal	0.7	0.6

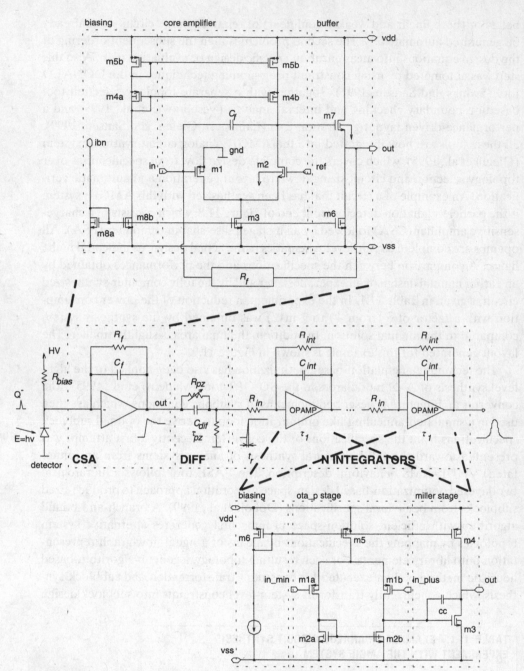

Figure 11.8. Particle–radiation detector front end as an example for analog circuit synthesis. (The opamp and filter stage symbols represent full circuit schematics as indicated).

parameters (e.g., a bias current) and fixes the topologies and transistor sizes for all subblocks. For reasons of efficiency, the performances at all levels in the considered hierarchy are estimated by using analytic equations relating design parameters to performance characteristics. The genetic algorithms operating at the different levels are speeded up by switching from traditional genetic operators to directed-interval-based

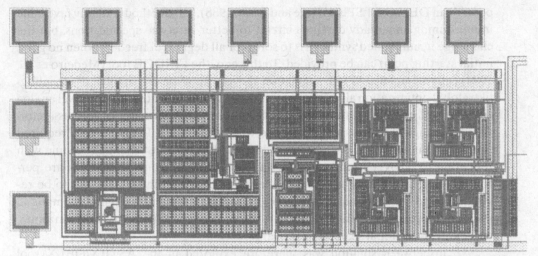

Figure 11.9. Layout of the particle–radiation detector front end generated with the AMGIE analog synthesis system.

operators that rely on characterization tables with qualitative-sensitivity information to help focus the search process in promising local regions.

In general, the big advantages of these analytic approaches are their fast evaluation time and their flexibility in manipulation possibilities. The latter is reflected in the freedom to choose the independent input variables, which has a large impact on the overall evaluation efficiency (Leyn, Gielen, and Sansen, 1998); as well as the possibility to perform more symbolic manipulations. Recently, it has been shown that the design of CMOS opamps can be formulated (more precisely, it can be fairly well approximated) as a posynomial convex optimization problem that can then be solved by using geometric programming techniques, producing a close-by first-cut design in an extremely efficient way (Hershenson et al., 1998). The initial optimization time literally reduces to seconds. The same approach has been applied to some other circuits as well (Hershenson et al., 1999). The big drawback of the analytic methods, however, is that the design equations still have to be derived, and, despite the progress in symbolic circuit analysis, not all design characteristics (such as transient or large-signal responses) are easy to capture in analytic equations with sufficient accuracy. For such characteristics either rough approximations have to be used, which undermines the sense of the whole approach, or one has to fall back on numerical simulations. This problem has sparked research efforts to try to develop equation-free approaches.

Therefore, in recent years and with improving computer power, a second subcategory of *simulation-based optimization approaches* toward analog circuit synthesis has been developed. These methods perform some form of full numerical simulation to evaluate the circuit's performance in the inner loop of the optimization; see Figure 11.6(b). Although the idea of optimization-based design for analog circuits dates back at least 30 years (Director and Rohrer, 1969), it is only recently that the computer power and numerical algorithms have advanced far enough to make this really practical. For a limited set of parameters, circuit optimization was already possible in DELIGHT.SPICE (Nye and Riley, 1988). This method is clearly favorable in fine-tuning an already designed circuit to better meet the specifications, but the challenge in automated synthesis is to solve for all degrees of freedom when no good initial starting point can be provided. To this end, the FRIDGE tool (Medeiro et al., 1994) calls a simplified SPICE simulation at every iteration of a simulated-annealing-like global optimization algorithm. In this way it is able to synthesize low-level analog circuits (e.g., opamps) with full SPICE accuracy. Performance specifications are divided in design objectives and strong and weak constraints. The number of required simulations in reduced as much as possible by adopting a fast cooling schedule with reheating to recover from local minima. Nevertheless, many simulations are performed, and the number of optimization parameters and their range has to be restricted in advance by the designer. The introduction of a new circuit schematic in such an approach is relatively easy, but the drawback remains the long run times, especially if the initial search space is large.

An in-between solution was therefore explored in the ASTRX/OBLX tool (Ochotta, Rutenbar, and Carley, 1996), where the simulation itself is speeded up by analyzing the linear (small-signal) characteristics more efficiently than in SPICE by using Asymptotic waveform evaluation (Pillage and Rohrer, 1990). For all other characteristics, equations still have to be provided by the designer. Thus this is essentially a mixed equation-simulation approach. The ASTRX subtool complies the initial synthesis specification into an executable cost function. The OBLX subtool then numerically searches for the minimum of this cost function by means of simulated annealing, hence determining the optimal circuit sizing. To achieve accuracy in the solution, encapsulated industry-standard models are used for the MOS transistors. For efficiency, the tool also uses a dc-free biasing formulation of the analog design problem, where the dc constraints (i.e., Kirchhoff current law, or KCL, at every node) are not imposed by construction at each optimization iteration, but are solved by relaxation throughout the optimization run by adding the KCL violations as penalty terms to the cost function. At the final optimal solution, all the penalty terms are driven to zero, thus resulting in a KCL-correct and thus electrically consistent circuit in the end. ASTRX/OBLX has been applied successfully to a wide variety of cell-level designs, such as a $2\times$ gain stage (Ochotta et al., 1998), but the CPU times remain large. The tool is also most suited only when the circuit behavior is relatively linear, because the other characteristics still require equations to be derived.

In the quest for industry-grade quality, most recent approaches therefore use complete SPICE simulations for all characteristics. As a way to cut down on the large synthesis time, more efficient optimization algorithms are used or the simulations are executed as much as possible in parallel on a pool of workstations. In the work of Schwencker et al. (2000), the generalized boundary curve is used to determine the step length within an iterative trust-region optimization algorithm. Using the full nonlinear cost function based on the linearized objectives significantly reduces the total number of iterations in the optimization. The ANACONDA tool, in contrast, uses a global optimization algorithm based on stochastic pattern search that inherently contains parallelism and therefore can easily be distributed over a pool of workstations, to try out and simulate 50,000–100,000 circuit candidates in a few hours (Phelps et al., 1999). MAELSTROM is the framework that provides the

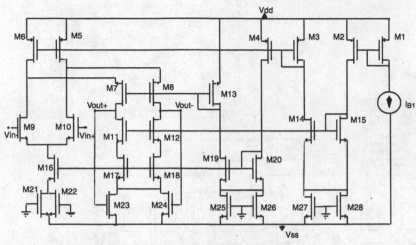

Figure 11.10. Example circuit for analog circuit synthesis.

simulator encapsulation and the environment to distribute both the search tasks of the optimization algorithm as well as the circuit evaluations at every iteration of the optimization over parallel workstations in a network (Krasnicki et al., 1999). It uses another parallel global algorithm, a combined annealing-genetic algorithm, to produce fairly good designs in a few hours. These approaches require very little advance modeling to prepare for any new circuit topology and have the same accuracy as SPICE. Figure 11.10 shows an example of an opamp circuit that has been synthesized with FRIDGE and MAELSTROM. The results are summarized in Table 11.2. In the work of Phelps et al. (2000), ANACONDA, in combination with macromodeling techniques to bridge the hierarchical levels, was applied to an industrial-scale analog system (the equalizer–filter front end for an ADSL CODEC). The experiments demonstrated that the synthesis results are comparable with or sometimes better than manual design, see Figures 11.11 and 11.12. In this particular case, a design process that took roughly 7 weeks by hand was reduced to an overnight synthesis run on a pool of CPUs. Although appealing, these methods still have to be used with care by designers because the optimizer may produce improper designs if the right design constraints are not added to the optimization problem. Reducing the CPU time remains a challenging area for further research.

TABLE 11.2. EXAMPLE OF ANALOG CIRCUIT SYNTHESIS RESULTS WITH FRIDGE AND MAELSTROM

Performance	Specification	FRIDGE	MAELSTROM
No. of variables	—	10	31
CPU time (min)	—	45	184
DC gain (dB)	≥ 70	79	76
UGF (MHz)	≥ 30	35	31
Phase margin (deg)	≥ 60	66	61
Output swing (V)	≥ 3.0	3.2	2.9
Settling time (ns)	minimal	90	62

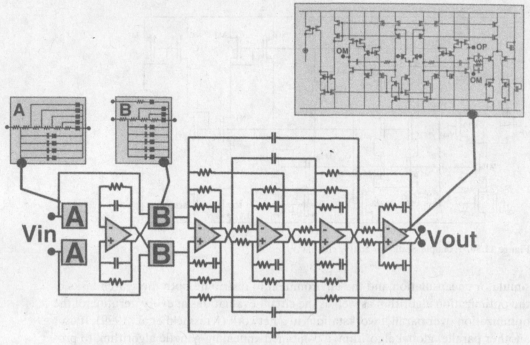

Figure 11.11. System-level schematic for an industrial equalizer filter front end of an ADSL CODEC (top) with sizing synthesized by the ANACONDA tool.

Other simulation-based approaches can be found in tools such as OAC (Onodera, Kanbara, and Tamaru, 1990), which is a specific nonlinear optimization tool for operational amplifiers and which is based on redesign starting from a previous design solution stored in the system's database. It also performs physical floorplanning during the optimization, which accounts for physical layout effects during circuit synthesis. A recent application of the simulation-based optimization approach to the high-level optimization of analog RF receiver front ends was-presented by Crols et al. (1995). A dedicated RF front-end simulator was developed and used to calculate the

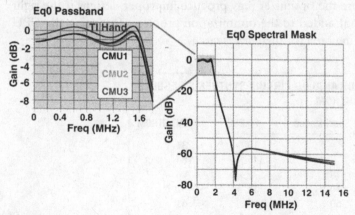

Figure 11.12. Portions of spectral mask (frequency response) for the original manual CODEC design in Figure 11.10 (from Texas Instruments; Phelps et al., 2000) and three ANACONDA designs (labeled CMU1, CMU2, CMU3). Note that the designs are essentially identical.

ratio of the wanted signal to the total power of all unwanted signals (noise, distortion, aliasing, phase noise, etc.) in the frequency band of interest. An optimization loop then determines the optimal specifications for the receiver subblocks such that the desired signal quality for the given application is obtained at the lowest possible power consumption for the overall front-end topology. Behavioral models and power estimators are used to evaluate the different front-end subblocks at this high architectural level.

In summary, the initial design systems such as IDAC were too closed and restricted to their initial capabilities and therefore failed in the marketplace. The current trend is toward open analog design systems that allow the designer to easily extend or modify the design capabilities of the system without too much software overhead. Compared with the initial knowledge-based approaches, the big advantages of the more recent optimization-based approaches are their high flexibility and extendibility, both in terms of design objectives (by altering the cost function) and in terms of the ease to add new circuit schematics. Although some additional research is still needed, especially to reduce the CPU times, it can be concluded that a lot of progress has been achieved in the field of analog circuit synthesis over the past 10 years. This has resulted in the development of several experimental analog synthesis systems, with which several designs have successfully been synthesized, fabricated, and measured. This has been accomplished not only for opamps but also for filters (Assael, Senn, and Tawfik, 1988) and data converters (Gielen and Franca, 1996). Based on these recent methods, several commercial tools are currently being introduced in the marketplace.

Finally, it has to be added that industrial design practice not only calls for fully optimized nominal design solutions, but also expects high robustness and yield in the light of varying operating conditions (supply voltage or temperature variations) and statistical process tolerances and mismatches (Director, Maly, and Strojwas, 1990). Techniques to analyze the impact of this on the yield or the capability index Cpk of the circuit (Zhang and Styblinski, 1995) after the nominal design has been completed will be discussed in detail in a later subsection. Here we briefly describe the efforts to integrate these considerations in the synthesis process itself. Yield and robustness precautions were already hardcoded in the design plans of IDAC, but they are more difficult to incorporate in optimization-based approaches. Nevertheless, first attempts in this direction have already been presented. The ASTRX/OBLX tool has been extended with manufacturability considerations and uses a nonlinear infinite programming formulation to search for the worst-case "corners" at which the evolving circuit should be evaluated for correct performance (Mukherjee, Carley, and Rutenbar, 1995). The approach has been successful in several test cases but does increase the required CPU time even further (roughly by 4–10×). In addition, the OPTIMAN program has been extended by fully exploiting the availability of the analytic design equations to generate closed-form expressions for the sensitivities of. the performances to the process parameters (Debyser and Gielen, 1998). The impact of tolerances and mismatches on yield or Cpk can then easily be calculated at each optimization iteration, which then allows the circuits to be synthesized simultaneously for performance and for manufacturability (yield or Cpk). The accuracy of the statistical predictions still has to be improved. The approach in Schwencker et al. (2000) uses parameter distances as robustness objectives to obtain a nominal design that satisfies all specifications with as large a safety margin as possible for process variations. The resulting formulation is the same as for design centering and can be solved efficiently by using the generalized boundary curve. Design centering, however, still remains a second step after the nominal design. Therefore, more research in this direction is still needed.

ANALOG AND MIXED-SIGNAL LAYOUT SYNTHESIS

The next important step in the analog design flow of Figure 11.4 after the circuit synthesis is the generation of the layout. The field of analog layout synthesis is more mature than circuit synthesis, in large part because it has been able to leverage ideas from the mature field of digital layout. However, real commercial solutions are just now appearing. Below we distinguish analog circuit-level layout synthesis, which has to transform a sized transistor-level schematic into a mask layout, and system-level layout assembly, in which the basic functional blocks are already laid out and the goal is to make a floorplan, place, and route them, as well as to distribute the power and ground connections.

Analog Circuit-Level Layout Synthesis

The earliest approaches to analog cell layout synthesis relied on *procedural module generation*, such as that explained by Kuhn (1987), in which the layout of the entire circuit was precoded in a software tool that generates the complete layout at run time for the actual parameter values entered. Today this approach is frequently used during interactive manual layout for the single-keystroke generation of the entire layout of a single device or a special group of (e.g., matched) devices by means of parameterized procedural device generators. For circuits, however, the approach is not flexible enough, and large changes in the circuit parameters (e.g., device sizes) may result in inefficient area usage. In addition, a module generator has to be written and maintained for each individual circuit.

A related set of methods is called *template driven*. For each circuit, a geometric template, for example, a sample layout (Beenker et al., 1993) or a slicing tree (Koh et al., 1990), is stored that fixes the relative position and interconnection of the devices. The layout is then completed by correctly generating the devices and the wires for the actual values of the design according to this fixed geometric template, thereby trying to use the area as efficiently as possible. These approaches work best when the changes in circuit parameters result in little need for global alterations in the general circuit layout structure, which is the case, for instance, during technology migration or porting of existing layouts, but which is not the case in general.

In practice, changes in the circuit's device sizes often require large changes in the layout structure in order to get the best performance and the best area occupation. The performance of an analog circuit is indeed affected by the layout. Parasitics introduced by the layout, such as the parasitic wire capacitance and resistance or the cross talk capacitance between two neighboring or crossing wires, can have a negative impact on the performance of analog circuits. It is therefore of utmost importance to generate analog circuit layouts such that (1) the resulting circuit still satisfies all performance specifications, and (2) the resulting layout is as compact as possible. This requires full custom layout generation, which can be handled with a *macrocell-style layout strategy*. The terminology is borrowed from digital floorplanning algorithms, which manipulate flexible layout blocks (called "macros"), arrange them topologically, and then route them. For analog circuits, the entities to be handled are structural groups of one single or a special grouping of devices (e.g., a matching pair of transistors). These device-level macros are to be folded, oriented, placed, and interconnected to make up a good overall layout. Note that many analog devices and special device groupings, even for the same set of parameters, can be generated in different geometrical variants; for example, two matching devices can be laid out in interdigitated form, or stacked, or in a quad-symmetric fashion, and so on. For each variant of each marcocell structure used, procedural module generators have to be developed to generate the actual layouts of the cells for a given set of parameter values. A drawback is that these generators have to be maintained and updated whenever the technology process changes, which creates some pressure to limit the number of different generators. Whatever the macrocells considered in a custom analog circuit layout synthesis tool, a placement routine optimally arranges the cells, while also selecting the most appropriate geometrical variant for each; a router interconnects them, and sometimes a compactor compacts the resulting layout, all while taking care of the many constraints such as symmetry and matching typical for analog circuits, and also attending to the numerous parasitics and couplings to which analog circuits (unfortunately) are sensitive. This general analog circuit layout synthesis flow is shown in Figure 11.13.

The need to custom optimize analog layouts led to the *optimization-based macrocell-place-and-route layout generation approaches*, in which the layout solution is not predefined by some template, but is determined by an optimization program according to some cost function. This cost function typically contains minimum area and net length and adherence to a given aspect ratio, but also other terms could be added, and the user normally can control the weighting coefficients of the different

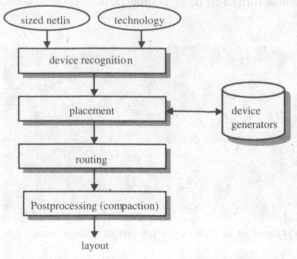

Figure 11.13. General flow of an analog circuit layout synthesis tool.

contributions. The advantage of the optimization-based approaches is that they are generally applicable and not specific to a certain circuit, and they are flexible in terms of performance and area as they find the most optimum solution at run time. The penalty to pay is their larger CPU times, and the dependence of the layout quality on the setup of the cost function. Examples of such tools are ILAC (Rijmenants et al., 1989) and the different versions of KOAN/ANAGRAM (Cohn et al., 1994). ILAC borrowed heavily from the best ideas from digital layout: efficient slicing tree floorplanning with flexible blocks, global routing by means of maze routing, detailed routing by means of channel routing, and area optimization by means of compaction (Rijmenants et al., 1989). The problem with the approach was that it was difficult to extend these primarily digital algorithms to handle all the low-level geometric optimizations that characterize expert manual design. Instead, ILAC relied on a large, very sophisticated library of device generators.

ANAGRAM and its successor, KOAN/ANAGRAM II, kept the macrocell style, but it reinvented the necessary algorithms from the bottom up, incorporating many manual design optimizations (Garrod, Rutenbar, and Carley, 1988; Cohn et al., 1991, 1994). The device placer KOAN relied on a very small library of device generators and migrated important layout optimizations into the placer itself. KOAN, which was based on an efficient simulated annealing algorithm, could dynamically fold, merge, and abut MOS devices and thus discover desirable optimizations to minimize parasitic capacitance on the fly during optimization. Its companion, ANAGRAM II, was a maze-style detailed area router capable of supporting several forms of symmetric differential routing, mechanisms for tagging compatible and incompatible classes of wires (e.g., noisy and sensitive wires), parasitic cross-talk avoidance, and over-the-device routing. Other device placers and routers operating in the macrocell-style have also appeared, such as LADIES (Mogaki et al., 1989) and ALSYN (Meyer zu Bexten et al., 1993). Results from these tools can be quite impressive. For example, Figures 11.14 and 11.15 show two automated industrial layouts from Neolinear, Inc., 2000. The larger (roughly 100 devices) layout of Figure 11.14 required approximately 1 day; the smaller (roughly 50 devices) layout was created automatically in approximately 1 hour, and we include it to show more geometric detail. These both

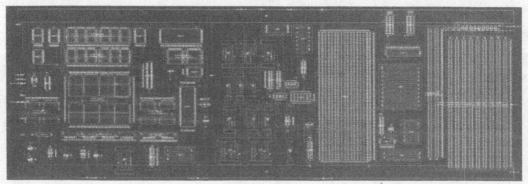

Figure 11.14. Automatic 0.25-μm CMOS layout of an ~100 device full-custom industrial analog cell (courtesy of Neolinear, Inc.).

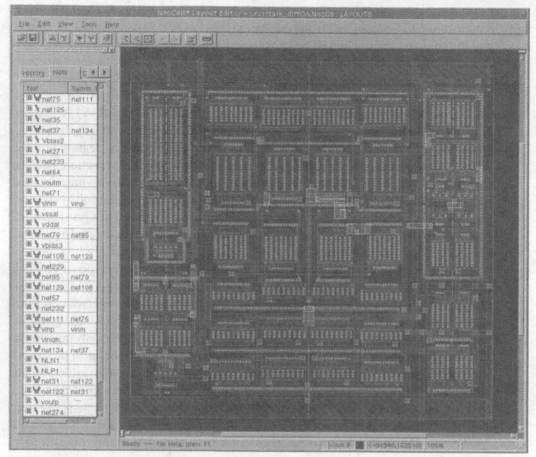

Figure 11.15. Automatic CMOS layout of an ~50 device full-custom analog industrial cell (courtesy of Neolinear, Inc.).

compare favorably in density and performance to manual layout, which would take roughly 10 times longer to accomplish.

An important improvement in the next generation of optimization-based layout tools was the shift from a rather qualitative consideration of analog constraints to an explicit quantitative optimization of the performance goals, resulting in the *performance-driven* or *constraint-driven* approaches. For example, KOAN maximized MOS drain-source merging during layout and ANAGRAM II minimized cross talk, but without any specific, quantitative performance targets. The performance-driven approaches, in contrast, explicitly quantify the degradation of the performance caused by layout parasitics, and the layout tools are driven such that this extra layout-induced performance degradation is within the margins allowed by the designer's performance specifications (Malavasi et al., 1996). In this way more optimum solutions can be found, as the importance of each layout parasitic is weighed according to its impact on the circuit performance, and the tools can much better guarantee by construction that the circuit will meet the performance specifications also after the layout phase (if possible).

Tools that adopt this approach include the area router ROAD (Malavasi and Sangiovanni-Vincentelli, 1993), the placement tool PUPPY-A (Charbon et al., 1992); and the compaction tool SPARCS-A (Malavasi et al., 1995). The routers ROAD and ANAGRAM III (Basaran, Rutenbar, and Carley, 1993) have a cost function that drives them such that they minimize the deviation from acceptable bounds on wire parasitics. These bounds are provided by designers or derived from the margins on the performance specifications by means of sensitivities. The LAYLA system (Lampaert et al., 1995, 1996, 1999) consists of performance-driven analog placement and routing tools that penalize excess layout-induced performance degradation by adding the excess degradation directly as an extra term to the cost function. Effects considered include, for instance, the impact of device merging, device mismatches, parasitic capacitance and resistance of each wire, parasitic coupling caused by specific proximities, thermal gradients, and so on. The router can manage not just parasitic wire sensitivities, but also yield and testability concerns. A layout generated by means of LAYLA is shown in Figure 11.9.

In all the above tools, sensitivity analysis is used to quantify the impact on final circuit performance of low-level layout decisions, and this has emerged as the critical glue that links the various approaches being taken for circuit-level layout and for system assembly. Choudhary and Sangiovanni-Vincentelli (1993) presented an influential early formulation of the sensitivity analysis problem, which not only quantified layout impacts on circuit performance but also showed how to use nonlinear programming techniques to map these sensitivities into maximum bounds on parasitics, which serve as constraints for various portions of the layout task. Later approaches (e.g., that of Lampaert et al., 1995), however, showed that this intermediate mapping step may not be needed. Other work (Charbon, Malavasi, and Sangiovanni-Vincentelli, 1993) showed how to extract critical constraints on symmetry and matching directly from a device schematic.

A recent innovation in CMOS analog circuit layout generation tools is the idea of separating the device placement into two distinct tasks: device stacking followed by stack placement. By rendering the circuit as an appropriate graph of connected drains and sources, one can identify natural clusters of MOS devices that ought to be merged, called *stacks*, to minimize parasitic capacitance, instead of discovering these randomly over the different iterations of the placement optimization. The work by Malavasi and Pandini (1995) presented an exact algorithm to extract all the optimal stacks and dynamically choose the right stacking and the right placement of each stack throughout the placement optimization. Because the underlying algorithm has exponential time complexity, enumerating all stacks can be very time consuming. The work by Basaran and Rutenbar (1996) offered a variant that extracts one optimal set of stacks very fast. The idea is to use this in the inner loop of a placer to evaluate fast trial merges on sets of nearby devices.

One final problem in the macrocell place-then-route style is the separation of the placement and routing steps. The problem is to estimate how much wiring space to leave around each device for the subsequent routing. Estimates that are too large result in open space; those that are too small may block the router and require changes to the placement. One solution is to get better estimates by carrying out simultaneous placement and global routing, which has been implemented for slicing-style structures (Prieto et al., 1997). An alternative is to use dynamic wire space estimation, in which space is created during routing when needed. Another strategy is analog compaction, in which extra space is left during placement, which after routing is then removed by compaction. Analog compactors that maintain the analog constraints introduced by the previous layout steps were, for instance, presented by Malavasi et al. (1995) and Okuda et al. (1989). A more radical alternative is to perform simultaneous device place and route. An experimental version of KOAN (Cohn et al., 1991) supported this alternative by iteratively perturbing both the wire and the devices, but the method still has to be improved for large practical circuits.

Performance-driven macrocell-style custom analog circuit-level layout schemes are maturing nowadays, and the first commercial versions have already started to be offered. Of course, there are still problems to solve. The wire space problem is one. Another open problem is "closing the loop" from circuit synthesis to circuit layout, so that layouts that do not meet the specifications can, if necessary, cause actual circuit design changes (by means of circuit resynthesis). Even if a performance-driven approach is used, which should generate layouts correct by construction, circuit synthesis requires accurate estimates of circuit wiring loads to obtain good sizing results, and circuit synthesis has to leave sufficient performance margins for the layout-induced performance degradation later on. How to control this loop, and how to reflect layout concerns in synthesis and synthesis concerns in layout, remains difficult.

Mixed-Signal System Layout Assembly

A mixed-signal system is a set of analog and digital functional blocks. System-level layout assembly means floorplanning, placement, and global and detailed routing (including the power grid) of the entire system, where the layouts of the individual blocks are generated by lower-level tools such as discussed in the previous section for the analog circuits. In addition to sensitivities to wire parasitics, an important new problem in mixed-signal systems is the coupling between digital switching noise and sensitive analog circuits (e.g., capacitive cross talk or substrate noise couplings).

As at the circuit level, procedural layout generation remains a practicable alternative for well-understood designs with substantial regularity (e.g., switched-capacitor filters, as given by Yaghutiel, Sangiovanni-Vincentelli, and Gray, 1986). More generally though, work has focused on custom placement and routing at the block level. For row-based layout, an early elegant solution to the coupling problem was the segregated-channels idea of Kimble et al. (1985) to alternate noisy digital and sensitive analog wiring channels in a row-based cell layout. The strategy constrains digital and analog signals never to be in the same channel, and it remains a practical solution when the size of the layout is not too large. For large designs, analog channel routers were developed. In the work of Gyurcsik and Jeen (1989), it was observed that a well-known digital channel routing algorithm, based on a gridless constraint graph formulation, could easily be extended to handle critical analog problems that involve varying wire widths and wire separations needed to isolate noisy and sensitive signals. The channel router ART extended this strategy to handle complex analog symmetries in the channel, and the insertion of shields between incompatible signals (Choudhury and Sangiovanni-Vincentelli, 1993).

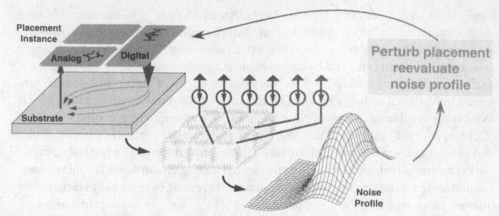

Figure 11.16. Principal flow of the WRIGHT mixed-signal floorplanner that incorporates a fast substrate noise coupling evaluator.

The WREN (Mitra et al., 1992) and WRIGHT (Mitra et al., 1995) tools generalized these ideas to the case of arbitrary layouts of mixed functional blocks. WREN comprises both a mixed-signal global router and channel router. The tool introduced the notion of SNR (signal-to-noise ratio)-style constraints for incompatible signals, and both the global and detailed routers strive to comply with designer-specified noise rejection limits on critical signals. WREN incorporates a constraint mapper that transforms input noise rejection constraints from the across-the-whole-chip form used by the global router into the per channel, per segment from necessary for the channel router. WRIGHT, in contrast, uses simulated annealing to make a floorplan of the blocks, but with an integrated fast substrate noise coupling evaluator so that a simplified view of substrate noise influences the floorplan. Figure 11.16 shows the flow of this tool. Accurate methods to analyze substrate couplings have been developed as well, but in the frame of layout synthesis tools, however, the CPU times of these techniques are prohibitive and there is a need for fast yet accurate substrate noise coupling evaluators to explore alternative layout solutions.

Another important task in mixed-signal system layout is power grid design. Digital power grid layout schemes usually focus on connectivity, pad-to-pin ohmic drop, and electromigration effects. However, these are only a small subset of the problems in high-performance mixed-signal chips, which feature fast-switching digital systems next to sensitive analog parts. The need to mitigate unwanted substrate interactions, the need to handle arbitrary (nontree) grid topologies, and the need to design for transient effects such as current spikes are serious problems in mixed-signal power grids. The RAIL system (Stanisic et al., 1994, 1996) addresses these concerns by casting mixed-signal power grid synthesis as a routing problem that uses fast asymptotic-waveform-evaluation-based (Pillage and Rohrer; 1990) linear system evaluation to electrically model the entire power grid, package, and substrate in the inner loop of grid optimization. Figure 11.17 shows an example RAIL redesign of a data channel chip in which a demanding set of dc, ac, and transient performance constraints were met automatically.

Most of these mixed-signal system-level layout tools are of recent vintage, but because they often rely on mature core algorithms from similar digital layout problems,

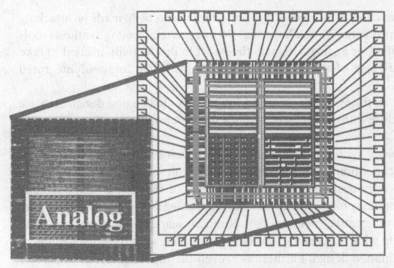

Figure 11.17. RAIL redesign of power grid for an IBM industrial magnetic data channel mixed-signal IC.

many have been prototyped both successfully and quickly. Although there is still much work to be done to enhance existing constraint mapping strategies and constraint-based layout tools to handle the full range of industrial concerns, the progress obtained opens possibilities for commercialization activities in the near future to make these tools available to all practicing mixed-signal designers.

CONCLUSIONS AND REMAINING PROBLEMS

The increasing level of integration available in today's semiconductor technologies is bringing mixed analog-digital systems together on a single chip. Despite enormous and fundamental progress concerning synthesis tools for digital circuits, the analog circuits on these chips are most often designed the old-fashioned way: by hand, one transistor at a time. The demand for electronics that connects us to the continuous external world (networks, wireless links, audio–video applications) has put significant pressure on today's inefficient and unsystematic analog design methodologies. Twenty years of mainly academic work on analog synthesis has rather rapidly come under active scrutiny by designers in search of practical solutions to these difficult circuit problems.

In this chapter we reviewed the industrial context and design flow for these mixed-signal chips, and we surveyed the historical progress and the state of the art in analog synthesis for these designs. Cast mostly in the form of numerical and combinatorial optimization tasks, linked by various forms of analysis and constraint mapping, leveraged by ever faster workstations, some of these tools show promise of practical application, and commercial start-ups have begun to bring these to industrial practice.

One conclusion is clear: to meet the requirements (time to market, cost, and quality) of future semiconductor products, analog design will have to be carried out in a much more systematic and structured way, supported by rigorous methodologies and synthesis tools. This chapter has shown that real progress has been made on attacking this very difficult problem. We can hope that the commercial analog synthesis tools now appearing will offer analog designers the boost in productivity needed to take them from the dark ages of analog "black magic" to the bright future of integrated systems on a chip.

Of course, not all the synthesis problems in this very challenging design area are solved today. We close with a short list of significant problems that remain in this area.

1. Practical topology synthesis: sizing is fairly well automated by using a mix of global numerical optimization and appropriate circuit modeling and simulation. However, topology design – in particular, the act of creating a fundamentally *novel* circuit topology for a difficult *new* design problem – is not well understood. This is clearly a "creative" act of design. Random search techniques have been able to come up with "interesting" circuits, but nothing with the compactness or elegance of a human design. Furthermore, even the topology selection process itself, given a library of alternatives, remains most often a simple and exhaustive enumeration. We need more powerful selection–creation algorithms here.

2. System-level synthesis: even assuming we can select subblock topologies from a finite list, system-level mixed-signal synthesis efforts to date have been highly specific to a *particular* class or family of circuits. Are there more general *abstractions* here that might liberate synthesis to attack a somewhat broader array of circuit targets? Will the recently developed standards for mixed-signal simulation languages (Verilog-AMS, VHDL-AMS) play the lead role here?

3. Practical circuit block modeling: synthesis at the higher levels of the design hierarchy relies on simplified behavioral models of the lower-level blocks. A good, abstract, efficient block model is an enormous enabler for higher-level synthesis. However, to date, essentially all these block models have been created by hand, by experts familiar with what these lower-level blocks ought to do. We need techniques that can create these models *automatically* from lower-level descriptions, reliably predict which ranges of performance are achievable, and smoothly trade off computing time and predictive accuracy.

4. Chip-level geometric assembly: we have seen a variety of promising individual techniques for the layout problem at chip level – floorplanning, signal routing, substrate analysis, and optimization – but not a coherent top-to-bottom layout flow. Noise problems from large, high-speed digital logic on the same chip as the sensitive analog are only going to increase. Too many of today's layout strategies are defensive, qualitative, and ad hoc.

5. High-frequency synthesis: synthesis to date has mainly targeted medium-performance analog applications, for example, those that are megahertz, not gigahertz. However, the pervasiveness of wireless technologies creates design opportunities for very high-frequency analog circuits on many new chips. These radio frequency circuits are enormously sensitive to second-order electrical effects, and they defy the common topology-select–size–place–route flow we would prefer. Radio frequency designers commonly manipulate sizing and placement and routing *concurrently* in standard practice, such is the level of intimate coupling among the available degrees of freedom in these circuits. We need new techniques that are

efficient enough, and complete enough, to consider circuit and physical design issues concurrently for this important class of circuits.

REFERENCES

Assael, J., Senn, P., and Tawfik, M. (1998). "A switched-capacitor filter silicon compiler," *IEEE Journal of Solid-State Circuits*, **23**(1):166–174.

Basaran, B. and Rutenbar, R. (1996). "An O(n) algorithm for transistor stacking with performance constraints." In *Proceedings of the ACM/IEEE Design Automation Conference*, IEEE, New York.

Basaran, B., Rutenbar, R., and Carley, L. R. (1993). "Latchup-aware placement and parasitic-bounded routing of custom analog cells." In *Proceedings of the ACM/IEEE International Conference on Computer-Aided Design*, IEEE, New York.

Beenker, G., Conway, J., Schrooten, G., and Slenter, A. (1993). "Analog CAD for consumer ICs." In *Analog Circuit Design* J. Huijsing, R. van der Plassche, and W. Sansen (eds.), Kluwer Academic, Boston, MA, Chap. 15, pp. 347–367.

Carley, L. R., Gielen, G., Rutenbar, R., and Sansen, W. (1996). "Synthesis tools for mixed-signal ICs: progress on frontend and backend strategies." In *Proceedings of the ACM/IEEE Design Automation Conference* IEEE, New York, pp. 298–303.

Chang, H., et al. (1992). "A top-down, constraint-driven design methodology for analog integrated circuits." In *Proceedings of the IEEE Custom Integrated Circuits Conference*, IEEE, New York, pp. 8.4.1–8.4.6.

Chang, H., et al. (1997). *A Top-Down, Constraint-Driven Design Methodology for Analog Integrated Circuits*, Kluwer Academic, Boston, MA.

Chang, H., et al. (1999). *Surviving the SOC Revolution – A Guide to Platform-Based-Design*. Kluwer Academic Boston, MA.

Charbon, E., Malavasi, E., Choudhury, U., Casotto, A., and Sangiovanni-Vincentelli, A. (1992). "A constraint-driven placement methodology for analog integrated circuits." In *Proceedings of the IEEE Custom Integrated Circuits Conference*, IEEE, New York, pp. 28.2.1–28.2.4.

Charbon, E., Malavasi, E., Sangiovanni-Vincentelli, A. (1993). "Generalized constraint generation for for analog circuit design." In *Proceedings of the ACM/IEEE International Conference on Computer-Aided Design*, IEEE, New York, pp. 408–414.

Choudhury, U. and Sangiovanni-Vincentelli, A. (1993). "Automatic generation of parasitic constraints for performance-constrained physical design of analog circuits," *IEEE Transactions on Computer-Aided Design*, **12**(2):208–224.

Choudhury, U. and Sangiovanni-Vincentelli, A. (1993). "Constraint-based channel routing for analog and mixed analog/digital circuits," *IEEE Transactions on Computer-Aided Design*, **12**(4):497–510.

Cohn, J., Garrod, D., Rutenbar, R., and Carley, L. R. (1991). "KOAN/ANAGRAM II: new tools for device-level analog placement and routing," *IEEE Journal of Solid-State Circuits*, **26**(3):330–342.

Cohn, J., Garrod, D., Rutenbar, R., and Carley, L. R. (1991). "Techniques for simultaneous placement and routing of custom ananog cells in KOAN/ANAGRAM II." In *Proceedings of the ACM/IEEE International Conference on Computer-Aided Design*, IEEE, New York, pp. 394–397.

Cohn, J., Garrod, D., Rutenbar, R., and Carley, L. R. (1994). *Analog Device-Level Layout Generation*, Kluwer Academic, Boston, MA.

Crols, J., Donnay, S., Steyaert, M., and Gielen, G. (1995). "A high-level design and optimization tool for analog RF receiver front-ends." In *Proceedings of the ACM/IEEE International Conference on Computer-Aided Design*, IEEE, New York, pp. 550–553.

Debyser, G. and Gielen, G. (1998). "Efficient analog circuit synthesis with simultaneous yield and robustness optimization." In *Proceedings of the ACM/IEEE International Conference on Computer-Aided Design*, IEEE, New York, pp. 308–311.

Degrauwe, M., et al. (1987). "IDAC: an interactive design tool for analog CMOS circuits," *IEEE Journal of Solid-State Circuits*, **22**(6):1106–1115.

Director, S. and Rohrer, R. (1969). "Automated network design – The frequency domain case," *IEEE Transactions on Circuit Theory*, **16**(5):330–337.

Director, S., Maly, W., and Strojwas, A. (1990). *VLSI Design for Manufacturing: Yield Enhancement*. Kluwer Academic, Boston, MA.

Doboli, A., Nunez-Aldana, A., Dhanwada, N, Ganesan, S., and Vemuri, R. (1999). "Behavioral synthesis of analog systems using two-layered design space exploration." In *Proceedings of the ACM/IEEE Design Automation Conference*, IEEE, New York, pp. 951–957.

Donnay, S., et al. (1996). "Using top-down CAD tools for mixed analog/digital ASICs: a practical design case," *Kluwer International Journal on Analog Integrated Circuits and Signal Processing*, Special Issue, **10**(1/2):101–117.

El-Turky, F. and Perry, E. (1989). "BLADES: an artificial intelligence approach to analog circuit design," *IEEE Transactions on Computer-Aided Design*, **8**(6):680–691.

Garrod, D., Rutenbar, R., and Carley, L. R. (1988). "Automatic layout of custom analog cells in ANAGRAM." In *Proceedings of the ACM/IEEE International Conference on Computer-Aided Design*, IEEE, New York, pp. 544–547.

Gielen, G. and Franca, J. (1996). "CAD tools for data converter design: an overview," *IEEE Transactions on Circuits and Systems, Part II – Analog and Digital Signal Processing*, **43**(2):77–89.

Gielen, G. and Rutenbar, R. A. (2000). "Computer aided design of analog and mixed-signal integrated circuits," *Proceedings of the IEEE*, **88**(12):1825–1849.

Gielen, G. and Sansen, W. (1991). *Symbolic Analysis for Automated Design of Analog Integrated Circuits*. Kluwer Academic, Boston, MA.

Gielen, G., Swings, K., and Sansen, W. (1993). "Open analog synthesis system based on declarative models," in *Analog Circuit Design*, J. Huijsing, R. van der Plassche, and W. Sansen (eds.), Kluwer Academic, Boston, MA, Chap. 18, pp. 421–445.

Gielen, G., Walscharts, H., and Sansen, W. (1990). "Analog circuit design optimization based on symbolic simulation and simulated annealing," *IEEE Journal of Solid-State Circuits*, **25**(3):707–713.

Gielen, G., et al. (1995). "An analog module generator for mixed analog/digital ASIC design," *John Wiley International Journal of Circuit Theory and Applications*, **23**:269–283.

Gyurcski, R. and Jeen, J. (1989). "A generalized approach to routing mixed analog and digital signal nets in a channel," *IEEE Journal of Solid-State Circuits*, **24**(2):436–442.

Harjani, R. and Shao, J. (1996). "Feasibility and performance region modeling of analog and digital circuits," *Kluwer International Journal on Analog Integrated Circuits and Signal Processing*, **10**(1):23–43.

Harjani, R., Rutenbar, R., and Carley, L.R. (1989). "OASYS: a framework for analog circuit synthesis," *IEEE Transactions on Computer-Aided Design*, **8**(12):1247–1265.

Harvey, J., Elmasry, M., and Leung, B. (1992). "STAIC: an interactive framework for synthesizing CMOS and BiCMOS analog circuits," *IEEE Transactions on Computer-Aided Design*, **11**(11):1402–1416.

Henderson, R., et al. (1993). "A spreadsheet interface for analog design knowledge capture and reuse." In *Proceedings of the IEEE Custom Integrated Circuits Conference*, IEEE, New York, pp. 13.3.1–13.3.4.

Hershenson, M., Boyd, S., and Lee, T. (1998). "GPCAD: a tool for CMOS op-amp synthesis." In *Proceedings of the IEEE/ACM International Conference on Computer-Aided Design*, IEEE, New York, pp. 296–303.

Hershenson, M., Mohan, S., Boyd, S., and Lee, T. (1999). "Optimization of inductor circuits via geometric programming." In *Proceedings of the IEEE/ACM Design Automation Conference*, IEEE, New York, pp. 994–998.

Horta, N., Franca, J., and Leme, C. (1991). "Automated high level synthesis of data conversion systems." In *Analogue-Digital ASICs – Circuit Techniques, Design Tools and Applications*, Soin, Maloberti, and Franca (eds.), Peregrinus, London.

IEEE (1996). "Systems-on-a-chip." (General theme of the IEEE International Solid-State Circuits Conference.)

IEEE (1997). "IEEE Standard VHDL 1076.1 Language Reference Manual – analog and mixed-signal extensions to VHDL 1076," IEEE 1076.1 Working Group, IEEE, New York.

Kimble, C., et al. (1985). "Analog autorouted VLSI." In *Proceedings of the IEEE Custom Integrated Circuits Conference*, IEEE, New York.

Koh, H., Séquin, C., and Gray, P. (1990). "OPASYN: a compiler for CMOS operational amplifiers," *IEEE Transactions on Computer-Aided Design*, 9(2):113–125.

Krasnicki, M., Phelps, R., Rutenbar, R., and Carley, L. R. (1999). "MAELSTROM: efficient simulation-based synthesis for custom analog cells." In *Proceedings of the ACM/IEEE Design Automation Conference*, IEEE, New York, pp. 945–950.

Kruiskamp, W. and Leenaerts, D. (1995). "DARWIN: CMOS opamp synthesis by means of a genetic algorithm." In *Proceedings of the ACM/IEEE Design Automation Conference*, IEEE, New York, pp. 550–553.

Kuhn, J. (1987). "Analog module generators for silicon compilation," *VLSI System Design*, **May**.

Lampaert, K., Gielen, G., and Sansen, W. (1995). "A performance-driven placement tool for analog integrated circuits," *IEEE Journal of Solid-State Circuits*, 30(7):773–780.

Lampaert, K., Gielen, G., and Sansen, W. (1996). "Analog routing for performance and manufacturability." In *Proceedings of the IEEE Custom Integrated Circuits Conference*, IEEE, New York, pp. 175–178.

Lampaert, K., Gielen, G., and Sansen, W. (1999). *Analog Layout Generation for Performance and Manufacturability*, Kluwer Academic, Boston, MA.

Leyn, F., Gielen, G., and Sansen, W. (1998). "An efficient DC root solving algorithm with guaranteed convergence for analog integrated CMOS circuits." In *Proceedings of the IEEE/ACM International Conference on Computer-Aided Design*, IEEE, New York, pp. 304–307.

Liang, J. (1998). "Mixed-signal IC market to surpass $10 billion in 1997 and $22 billion by 2001." (Report from Dataquest, San Jose, CA.)

Makris, C. and Toumazou, C. (1995). "Analog IC design automation: II. Automated circuit correction by qualitative reasoning," *IEEE Transactions on Computer-Aided Design*, 14(2):239–254.

Malavasi, E., Charbon, E., Felt, E., and Sangiovanni-Vincentelli, A. (1996). "Automation of IC layout with analog constraints," *IEEE Transactions on Computer-Aided Design*, 15(8):923–942.

Malavasi, E., Felt, E., Charbon, E., and Sangiovanni-Vincentelli, A. (1995). "Symbolic compaction with analog constraints," *Wiley International Journal on Circuit Theory and Applications*, 23(4):433–452.

Malavasi, E. and Pandini, D. (1995). "Optimum CMOS stack generation with analog constraints," *IEEE Transactions on Computer-Aided Design*, 14(1):107–122.

Malavasi, E. and Sangiovanni-Vincentelli, A. (1993). "Area routing for analog layout," *IEEE Transactions on Computer-Aided Design*, 12(8):1186–1197.

Malavasi, E., et al. (1993). "A top-down, constraint-driven design methodology for analog integrated circuits." In *Analog Circuit Design*, J. Huijsing, R. van der Plassche, and W. Sansen (eds.), Kluwer Academic, Boston, MA, Chap. 13, pp. 285–324.

Maulik, P., Carley, L. R., and Rutenbar, R. (1995). "Simultaneous topology selection and sizing of cell-level analog circuits," *IEEE Transactions on Computer-Aided Design*, 14(4):401–412.

Medeiro, F., Pérez-Verdú, B., Rodríguez-Vázquez, A., and Huertas, J. (1995). "A vertically-integrated tool for automated design of $\Sigma\Delta$ modulators," *IEEE Journal of Solid-State Circuits*, 30(7):762–772.

Medeiro, F., et al. (1994). "A statistical optimization-based approach for automated sizing of analog cells." In *Proceedings of the ACM/IEEE International Conference on Computer-Aided Design*, IEEE, New York, pp. 594–597.

Meyer zu Bexten, V., Moraga, C., Klinke, R., Brockherde, W., and Hess, K. (1993). "ALSYN: flexible rule-based layout synthesis for analog ICs," *IEEE Journal of Solid-State Circuits*, **28**(3): 261–268.

Mitra, S., Nag, S., Rutenbar, R., and Carley, L. R. (1992). "System-level routing of mixed-signal ASICs in WREN." In *ACM/IEEE International Conference on Computer-Aided Design*, IEEE, New York.

Mitra, S., Rutenbar, R., Carley, L. R., and Allstot, D. (1995). "Substrate-aware mixed-signal macrocell placement in WRIGHT," *IEEE Journal of Solid-State Circuits*, **30**(3):269–278.

Mogaki, M., et al. (1989). "LADIES: an automated layout system for analog LSIs." In *Proceedings of the ACM/IEEE International Conference on Computer-Aided Design*, IEEE, New York, pp. 450–453.

Mukherjee, T., Carley, L. R., and Rutenbar, R. (1995). "Synthesis of manufacturable analog circuits," *Proceedings of the ACM/IEEE International Conference on Computer-Aided Design*, IEEE, New York, pp. 586–593.

Nagel, L. (1975). "SPICE2: a computer program to simulate semiconductor circuits," Memo UCB/ERL M520, University of California, Berkeley.

Neolinear, Inc. (2000). NeoCell synthesis tool; http://www.neolinear.com.

Ning, Z., et al. (1991). "SEAS: a simulated evolution approach for analog circuit synthesis." In *Proceedings of the IEEE Custom Integrated Circuits Conference*, IEEE, New York, pp. 5.2.1–5.2.4.

Nye, W., Riley, D., Sangiovanni-Vincentelli, A., and Tits, A., (1998). "DELIGHT.SPICE: an optimization-based system for the design of integrated circuits," *IEEE Transactions on Computer-Aided Design*, **7**(4):501–518.

Ochotta, E., Mukherjee, T., Rutenbar, R., and Carley, L. R. (1998). *Practical Synthesis of High-Performance Analog Circuits*. Kluwer Academic, Boston, MA.

Ochotta, E., Rutenbar, R., and Carley, L. R. (1996). "Synthesis of high-performance analog circuits in ASTRX/OBLX," *IEEE Transactions on Computer-Aided Design*, **15**(3): 273–294.

Okuda, R., Sato, T., Onodera, H., and Tamuru, K. (1989). "An efficient algorithm for layout compaction problem with symmetry constraints." In *Proceedings of the ACM/IEEE International Conference on Computer-Aided Design*, IEEE, New York, pp. 148–151.

Onodera, H., Kanbara, H., and Tamaru, K. (1990). "Operational-amplifier compilation with performance optimization," *IEEE Journal of Solid-State Circuits*, **25**(2):466–473.

Open Verilog International (1996). "Verilog-A: Language Reference Manual: analog extensions to Verilog HDL," Version 0.1, Los Gatos, CA.

Phelps, R., Krasnicki, M., Rutenbar, R., Carley, L. R., and Hellums, J. (1999). "ANACONDA: robust synthesis of analog circuits via stochastic pattern search." In *Proceedings of the IEEE Custom Integrated Circuits Conference*, New York, pp. 567–570.

Phelps, R., Krasnicki, M., Rutenbar, R., Carley, L. R., and Hellums, J. (2000). "A case study of synthesis for industrial-scale analog IP: redesign of the equalizer/filter frontend for an ADSL CODEC." In *Proceedings of the ACM/IEEE Design Automation Conference*, IEEE, New York, pp. 1–6.

Pillage, L. and Rohrer, R. (1990). "Asymptotic waveform evaluation for timing analysis," *IEEE Transactions on Computer-Aided Design*, **9**(4):352–366.

Prieto, J., Rueda, A., Quintana, J., and Huertas, J. (1997). "A performance-driven placement algorithm with simultaneous place and route optimization for analog ICs." In *Proceedings of the IEEE European Design and Test Conference*, IEEE, New York, pp. 389–394.

Rijmenants, J., et al. (1989). "ILAC: an automated layout tool for analog CMOS circuits," *IEEE Journal of Solid-State Circuits*, **24**(4):417–425.

Schwencker, R., Schenkel, F., Graeb, H., and Antreich, K. (2000) "The generalized boundary curve – a common method for automatic nominal design and design centering of analog circuits." In *Proceedings of the IEEE Design Automation and Test in Europe Conference*, IEEE, New York, pp. 42–47.

Semiconductor Industry Association (1994). "The national technology roadmap for semiconductors," SIA; San Jose, CA.

Sheu, B., Fung, A., and Lai, Y.-N. (1998). "A knowledge-based approach to analog IC design," *IEEE Transactions on Circuits And Systems*, **35**(2):256–258.

Stanisic, B., Rutenbar, R., and Carley, L. R. (1996). *Synthesis of Power Distribution to Manage Signal Integrity in Mixed-Signal ICs*. Kluwer Academic, Boston, MA.

Stanisic, B., Verghese, N., Rutenbar, R., Carley, L. R., and Allstot, D. (1994). "Addressing substrate coupling in mixed-mode ICs: simulation and power distribution synthesis," *IEEE Journal of Solid-State Circuits*, **29**(3).

Swings K. and Sansen, W. (1991). "DONALD: a workbench for interactive design space exploration and sizing of analog circuits." In *Proceedings of the IEEE European Design Automation Conference*, IEEE, New York, pp. 475–479.

Toumazou, C. and Makris, C. (1995). "Analog IC design automation: I. Automated circuit generation: new concepts and methods," *IEEE Transactions on Computer-Aided Design*, **14**(2):218–238.

Veselinovic, P., et al. (1995). "A flexible topology selection program as part of an analog synthesis system." In *Proceedings of the IEEE European Design & Test Conference*, IEEE, New York, pp. 119–123.

Vital, J. and Franca, J. (1992). "Synthesis of high-speed A/D converter architectures with flexible functional simulation capabilities." In *Proceedings of the IEEE International Symposium on Circuits And Systems*, IEEE, New York, pp. 2156–2159.

VSI Alliance (1997). "Virtual Socket Interface Architecture Document," Version 1.0, Los Gatos, CA.

Yaghutiel, H., Sangiovanni-Vincentelli, A., and Gray, P. (1986). "A methodology for automated layout of switched-capacitor filters." In *Proceedings of the ACM/IEEE International Conference on Computer-Aided Design*, IEEE, New York, pp. 444–447.

Zhang, J. and Styblinski, M. (1995). *Yield and Variability optimization of Integrated Circuits*. Kluwer Academic, Boston, MA.

CHAPTER TWELVE

Mechanical Design Compilers

Allen C. Ward

Compilers – programs that accept a high-level description of desired systems and produce implementable versions – dominate design automation in software and integrated digital circuits. Neither word processors nor the 3 million transistor microprocessors they run on could be constructed without compilers.

In 1989, I published work describing a mechanical design compiler of a sort. (It accepted schematics and utility functions for, e.g., hydraulic systems, and it returned catalog numbers for optimal implementations. However, it represented only the crudest of geometric information, such as the heights of connecting shafts above a base plate: a better *mechanical* compiler would deal with complex geometry.) Yet if imitation is the sincerest form of flattery, my work has been little flattered. For example, the journal *Artificial Intelligence In Engineering Design, Analysis, and Manufacturing* has since published three special issues on machine learning in design (Spring 1994, April 1996, April 1998), but no papers other than mine on compilers.

This chapter will address two questions: Why is it that compilers are so central in software and integrated circuit design and so peripheral in mechanical design? What does the reason tell us about useful directions for further research? These questions, as we will see, have to be answered in abstract form, because no single company or researcher will be able to create a satisfactory design compiler. We need general guidelines to allow convergence in multiple avenues of development.

AN OUTLINE OF THE ARGUMENT

By a mechanical design compiler, I mean software that does the following.

1. It provides a high-level and reasonably flexible language in which users can accurately define what sort of mechanical system they want.
2. It is able to accept any syntactically correct and semantically meaningful input in that language.
3. It automatically transforms that description into one acceptable to a manufacturing system, which would then produce a high-quality design. (The manufacturing system might include human beings. I require only that the design be complete in the sense that those accepting it can think fairly strictly about manufacturing rather than design.)

Succinctly, compilers have to take the transformation "all the way down" to something buildable – unlike, for example, linkage synthesis tools, which stop at deciding the length of the links and omit their actual construction. Nor can they specialize so narrowly as to avoid the difficulty of dealing with new combinations of components and functions. Mechanical design programs with these characteristics would have similar impact to software and integrated digital circuit (VLSI) compilers, revolutionizing mechanical engineering.

Compilers ought to represent a central challenge for research in mechanical design automation for two reasons. First, commercial utility: Compilers by definition are useful across a wide range of products, and they handle the "boring details" that humans do badly, freeing humans for more creative thought. Second, as we shall see, systems that do not go all the way down, or that address only a narrow range of products, make it too easy to "cheat" by defining the problem to be solved in such a way as to avoid the real difficulties of mechanical design.

Moreover, I believe that mechanical design compilers are interesting in a way that software and VLSI compilers are not. In particular, to achieve the three defining characteristics of compilers for mechanical systems, we will have to give up a fourth characteristic of those simpler compilers: that of operating by fairly strict successive top-down parsing and transformation operations. In fact, I will argue (in the third section, drawing on my own work and some observations from first principles), successful mechanical design compilers will have seven more characteristics.

4. The elements of the high-level language will stand for sets of possible implementing "components" organized in an abstraction lattice (some of these components may be features of a single physical object).
5. Component modules will be written by the companies that design the components: the software will have to be sufficiently robust that system designers can assemble modules defined by multiple companies.
6. Other modules will represent various kinds of analysis codes, such as heat transfer modeling systems. These will receive appropriate abstractions of component set behavior, and impose constraints on the set of possible solutions.
7. Modules will communicate directly with each other in an "agent-based" way, so that the design emerges in a fashion more akin to the growth of a living creature than to a top-down grammatical transformation or application of production rules.
8. "Ports" or generalized interfaces will be key elements of these modules.
9. The communication will involve a new formal language for describing sets of design possibilities, based on the predicate calculus, as well as "local" marginal approximate price functions.
10. In order to handle complex continuous objects, such as many molded parts and the molds that make them, compilers will allow the parts to "divide," rather like cells in a growing tissue.

Before more thoroughly describing and making the case for characteristics 4–10, I will review some relevant literature and commercial experience to answer my first question: Why have mechanical design compilers been peripheral whereas VLSI and software compilers are so central?

WHY IS MECHANICAL DESIGN COMPILATION HARD?

I will begin by summarizing and adding to Whitney (1996), arguing that the primary problem is one of interaction among the components. I will touch on an earlier commentary on (Whitney, Antonsson, 1997), and I will discuss several academic and commercial experiments to illustrate the point.

Simplified, Whitney's argument is that because mechanical systems transmit significant power, their components interact in much more complex ways than those of digital VLSI systems. In particular (and in my words, rather than Whitney's):

a. Nonfunctional power-flow interactions such as heat flow and vibration must be taken into account.
b. Devices in a power chain must be impedance matched, so that downstream devices affect upstream devices.
c. Many components must normally be designed specifically for the system, rather than reused from previous designs: a component-to-system interaction.
d. The conflict among weight, packaging, and power handling is so severe that each component must serve multiple functions, which interact in different ways, precluding analysis or synthesis by a single-domain computer package.
e. These severe conflicts require mechanical components to be geometrically complex (with many, continuously variable geometric degrees of freedom), whereas VLSI components are simple rectangles in layers: this induces much more complex geometric interactions among components. (I would add that in VLSI design, "functional blocks" are defined as subsystems of strong interaction and are therefore geometrically compact. Many functional blocks in mechanical design, such as wiring harnesses, must be geometrically dispersed, increasing the complexity of their interactions with other components.)
f. The manufacturing processes for mechanical components are highly specialized to the component, whereas all VLSI systems are built from metallic traces and transistors by essentially similar processes. This increases the complexity of interactions between design and manufacture.
g. Mechanical components show more complex interactions among phenomena at different scales. For example, fatigue life on the macrolevel is strongly affected by events at the atomic level. Conversely, the function of a well-designed digital circuit is shielded from microphenomena.
h. Conflicts among packaging, power handling, and weight make "component design" much more challenging for mechanical systems than for VLSI systems. Components are constantly evolving. There is no level of a few common, simple building blocks for mechanical systems, to correspond with the rectangular transistors and metal traces from which all VLSI systems are constructed. In addition, point c here says that many components must be designed uniquely to each system. We are therefore unlikely to construct a software package that "knows about" all these components in the way that VLSI synthesis packages can carry a design from VHDL logic through netlists to mask layout. (However, as Whitney forecast in his article, VLSI design is in this respect becoming more like mechanical design, because the capacitive and thermal interactions become

more complex as line sizes decrease. Rather than automating the design from general logic through to mask, designers are increasingly building their designs around previously debugged components – "cores" – and using the logic-to-mask compilers only for the "glue" functions connecting the cores. I'm reminded of a fine designer, Erik Vaaler, telling me "design brackets; buy mechanisms.")

Antonsson's (1997) reply seems both too optimistic and too pessimistic. Antonsson concedes what Whitney did not say: that compilation is impossible for highly integrated mechanical systems. I think Whitney's remarks imply only that automation will be different. Antonsson goes on to outline an interesting research agenda focusing on finding less integrated mechanical systems; he mentions, for example, bicycles and kitchen appliances. But bicycles may be so little integrated that we will not need compilers for them: users will continue to select their components based on the desired feel and cost of the individual components. And (I have consulted a little for Viking Range), kitchen appliances may be too integrated for top-down transformational techniques. The flow of heat among range components, for example, seems far too complex to address through design rules.

I believe that Whitney has made a powerful case that mechanical design compilers will have to operate by extensive communication among the component representations as opposed to top-down transformations. But does the history of mechanical design automation suggest that failure to understand Whitney's point accounts for failure to develop compilers? Yes: I will discuss some relevant papers from AIEDAM (selected because their titles seemed to promise something like a design compiler), then an important commercial application.

For example, consider the work of Esterline and Kota (1992) – "A General Paradigm For Routine Design – Theory And Implementation." The "general paradigm" (stripped of mathematical detail) comprises (1) dividing the set of possible solutions to a design problem into clustered subsets; (2) establishing typical parameter values for each subset; (3) using constraints and utilities on parameter values to select the best matching subset; (4) replacing this shallow model of the subset with a deep model, "which contains sufficient information for the modeled device, process, or system to be realized"; and (5) modifying the deep model to improve the fit.

The title of the Esterline and Kota research certainly suggests something like a compiler. The mechanism is much in the spirit of the top-down transformations of software and VLSI compilers. Kota has assured me (personal communication) that he constructed a compiler based on this paradigm, for hydraulic systems.

Yet, this paradigm cannot produce high quality, buildable hydraulic system designs, because it includes no means for addressing component-to-component interactions. For example, the size of the electric or mechanical prime mover depends in part on efficiency of the hydraulic pump, and the appropriate choice of pump efficiency depends on the relative costs of electricity and prime-mover power compared with pump efficiency. Nor can we select hoses without knowing the type of connection built into the pump and hydraulic cylinders; indeed, major choices for all three components may be driven by the need for compatibility. Further, the choices may be absolutely dictated by geometric constraints from the rest of the system.

In other words, Esterline and Kota have not constructed a compiler, nor a shell for building compilers: they have constructed a system for looking up components to match specifications. Whitney was right: the VLSI paradigm has misled the researchers not only into a blind alley, but into seeing results they could not have achieved by the means they describe.

Shin and Lee (1998) describe an expert system that "enables the user the optimal design of pneumatic system [sic]." What they actually describe is another component look-up system. For example, given a stroke and load requirement for a cylinder (and an air pressure), the system will return a catalog number for a cylinder with sufficient piston area and stroke. This may be useful for low-performance pneumatic systems, where, for example, flow rates are not a major consideration. If they are, selection must again account for interactions; we may, for example, need an oversized cylinder to account for pressure drop elsewhere in the system.

Seddon and Brereton (1994), in "Component Selection Using Temporal Truth Maintenance," correctly label their system as a component picker, for orifice plates. However, they describe it in the context of system design without mention of interaction effects.

All of these papers suggest complex software performing a straightforward task: matching user specifications against a component specification database ("the crucial task" according to Esterline and Kota).

Balkany, Birmingham, and Tommelein (1993) describe and analyze several "configuration design" systems; Air-Cyl, a system for designing air cylinders; VT, a system for designing elevators; PRIDE, a system for designing paper-handling systems for photocopiers; and M1, a system for designing small computer systems. These programs are clearly intended to be system designers, not just component pickers. Except perhaps for M1, they are too specialized to be considered compilers, nor is it clear whether they take the design all the way down. However, the review offers an interesting insight: each program works by top-down decomposition. The descriptions do not make it clear how or whether they addressed interactions, *nor did the authors consider this point important enough to emphasize in these descriptions*.

Enough – Whitney is right in believing that researchers in mechanical design automation have grossly underestimated the significance of interaction effects.

What about commercial efforts? I will discuss only one, ICAD (from Knowledge Technologies International), of which I have some knowledge as a consultant to the companies that have developed it over the past decade. ICAD integrates LISP and a geometric representation to provide a language for programming the top-down definition of particular mechanical objects; it does not claim to be a compiler (though I wish I had had it when I was constructing compilers: it would have provided an excellent foundation). As far as I know, it is by far the best commercial tool for top-down mechanical synthesis. It has achieved some remarkable results: for example, more than half the framing of the Boeing 777 is said to have been designed by the ICAD code, reducing the time required from weeks to hours. However, ICAD as conventionally used handles interactions among components only through code built into a higher-level assembly; one cannot assemble predefined components that look as though they should mate and expect them to handle their interactions.

The commercial history of the tool suggests an interesting consequence. The company then called ICAD was founded at about the same time as the Parametric Technologies Corporation. Its tools for synthesis are far more powerful than those of PTC's Pro-Engineer. Yet Pro-E has expanded orders of magnitude more rapidly, and been widely imitated, whereas ICAD remains unusual. This might be explained by the difficulty of training engineers to use a LISP-based top-down synthesis tool, except that a number of companies have developed apparently successful applications, then abandoned them. The reason seems to be that the applications are far too brittle. Without the ability to readily connect existing components into new systems, and have them behave well, it is simply too much trouble to maintain many programs as technology changes. Pro-Engineer, conversely, is much weaker at synthesis, but makes it easier for the human operator to manage interaction effects by direct intervention.

So, the conclusion: Whitney is right, and mechanical design automation progress has been slow primarily because of inattention to interaction effects. But what does this imply for the future?

NECESSARY CHARACTERISTICS OF SUCCESSFUL MECHANICAL DESIGN COMPILERS

This section discusses in detail the characteristics that successful mechanical design compilers will need to have. Where relevant, it references my own work. Ironically, that work did in fact have many of the needed characteristics, but missed one so fundamental that I have not listed it: its support for describing complex geometry was inadequate.

Succinctly, that compiler allowed the user to assemble a mechanical schematic: say, motors connected to transmissions or pumps, connected to ... connected to loads. Each symbol in the schematic represented a set of possible implementations: a catalog. Users also supplied a function to be optimized, and any constraints they wished to impose: "The load will vary between zero and 100 ft lb," or "use only ac motors." These constraints were expressed in a language based on the predicate calculus; the internal constraints describing the behavior of the "part agents" were also expressed in this language and in algebraic equations.

The part agents exchanged constraints, and they eliminated from their catalogs implementations that provably could not work in the design. These elimination steps were interleaved with a search process guaranteed to find an optimal solution, which is described further below. I will draw on experience with this compiler in arguing for the following conclusions.

LINGUISTIC TERMS WILL REPRESENT SETS OF COMPONENTS

Section 1 asserted that

4. The elements of the high-level language will stand for sets of possible implementing components organized in an abstraction lattice (some of these components may be features of a single physical object).

I mean literally that the terms of our language will be things like "electric motor," representing an abstraction of dc motor and ac motor, each of which represents an abstraction of more specific motor classes, down to actual cataloged motors, or pieces of code that know how to design motors. This abstraction system is a lattice rather than a tree because things may be abstracted in different ways for different purposes: motors may be ac or dc, or constant speed, series wound, or servo, and these classifications intersect in complex ways. The second classification also illustrates another problem: it is customary and useful rather than logical and complete. For example, servo motors may be series wound, and "constant speed" may involve speed variations ranging from one part in 100,000 to 1 in 10.

This focus on abstracted component sets is in contrast with high-level languages for VLSI and software, which focus on describing function. Why? For three reasons. First, we need finer control over the way functions are implemented than is provided by purely functional statements. For example, designers will rarely want to work with the function "provide rotary motion." This function can be provided by electric and hydraulic motors of many kinds, internal and external combustion engines, and a variety of hand tools, all operating through transmissions of many kinds, or not, with very different characteristics. It will usually simply be easier to place a hydraulic motor in a schematic than to apply enough constraint to eliminate "rotory motion" solutions that don't makes sense.

Second, it will generally not be possible to reduce mechanical function components down to some base level, like the transistors in VLSI design, or the small number of machine-code instructions in software. Instead, we will be forced to work at a base level with a very large number of components. It therefore makes more sense to think in terms of abstracting these components to form higher-level classes than to have functions searching, say by keyword, for all the possible instantiations, most of which will make no sense. Catalogs are organized by abstraction hierarchy, for good reason.

Third, and most important, much of our representational detail will be specific to the components. Torque, for example, is a concept common to rotary motion from wrenches and turbines, but speed applies only to turbines. The sets of components are going to have to tell us what specifications we can apply, and some specifications will have to be added as we move down the abstraction lattice.

All of this was implemented in my compiler. For example, the compiler included tools for scanning a catalog page – say, of motors – and formulating relevant abstract specifications, such as that all of these motors operated at speeds between 1700 and 1750 rpm, or that both ac and dc motors were represented. What was not implemented, and will no doubt prove more difficult, was the use of modules that represent a part of a geometrically complex component. For example, consider stamped parts, such as those in an autobody. Will it turn out to be desirable to define common features from which body parts can be constructed, such as sharp corners, ruled surfaces, doubly stretched surfaces, holes, and so on? Or will we be forced to go no lower than "door outer panel"? And what information will we embed in the features, and what in the components?

COMPILERS WILL BE CONSTRUCTED BY MANY COMPANIES

I claimed in Section 1 that

5. Component modules will be written by the companies that design the components: the software will have to be sufficiently robust that system designers can assemble modules defined by multiple companies.

This is simply a consequence of the complexity of the modules: a module such as "heavy-duty ac high-slip 440-V electric motors" is a whole world in itself, and only the designers of such systems are competent to write the code defining the motors. Note that this is already happening in VLSI design with the use of proprietary cores. However, the VLSI designers have the enormous advantage that the cores are described in one of a few well established languages; mechanical design description languages are much more varied, and none is adequate to represent all the required information.

My own compiler supported this requirement fairly well: modules could be developed independently.

ANALYSIS MODULES WILL BE INTEGRATED INTO COMPILERS

6. Other modules will represent various kinds of analysis codes, such as heat transfer modeling systems. These will receive appropriate abstractions of component set behavior, and they will impose constraints on the set of possible solutions.

This is a consequence of the complex interactions on components. VLSI designers can design to rule, then perform a relatively small amount of analysis and optimization, because the design rules take care of most interactions. But good mechanical designers interleave choice and analysis; it does not make sense to fill in all the details of a design before performing some analysis based on generalized models. And some of this analysis extends beyond components to the system as a whole (e.g., vibration analysis), and must be conducted by software working with appropriate specialized abstractions of the components.

Much of the abstraction information will comprise intervals on appropriate generalized impedances: for example, in a vibration model, the set of ratios between inertia and stiffness possible for a component. In return, the model may impose constraints on stiffness, inertia, and their ratio, in order to make the system behave properly.

This set-based approach is so different from the way most software now works that I need to explicitly address the naïve idea that complex systems can be designed by specifying a complete design, running analysis codes on the result, and hill-climbing the output. According to a personal communication from Bob Phillips, General Electric tried this for turbine engines, probably the best modeled mechanical systems. The result was designs that did not work at all. Mechanical design is just too complex to make all the decisions first, then do the analysis. Instead, we will have to pick hydraulic motors based on high-level analysis before we pick hoses, and we will need information about the set of possible hoses while we pick the hydraulic motors.

My compiler conducted analysis only at the level of algebraic and predicate logic constraints that were embedded in the component descriptions (still a major improvement over conventional design systems.) Future systems will need to include complex analysis and simulation codes; the hard part will be learning how to apply them to abstract designs, rather than single, fully detailed designs.

MODULES WILL BE AGENTS

7. Modules will communicate directly with each other in an agent-based way, so that the design emerges in a fashion more akin to the growth of a living creature than to a top-down grammatical transformation or application of production rules.

Specifically, modules will exhibit "intentionality" in the sense of Dennett (1995). That is, first, we will be able to understand a great deal about the behavior of the agent module by assuming it has goals, such as "fit into the design, providing all required power, while minimizing weight and cost." Second, agents will rely on such intentionality in other agents. For example, agents will have to be able to inform each other about costs that they may impose on each other – "if you (the motor module) go to 220-V motors, I (the power supply module) will be able to save $320." They (more precisely, their creators) need to know that the connecting module will do something appropriate, making a decision likely to lead to low cost for the whole design. They do not need to know *how* the decisions will be made; we can appropriately interface with the rest of the system by understanding the "intentions" of the other modules, without understanding their mechanisms. (In the same way, we routinely assume that plants will grow toward the light, without needing to know the mechanisms by which they do so.)

In other words, we will have to give up the idea of centralized control. There will be no "main" program that calls on subprograms when needed because it contains the knowledge of when their expertise is relevant. Instead, each module will decide for itself when its knowledge is relevant, based on its "intentions". It will intervene to make this knowledge effective based on its own continually updated assessment of the situation around it.

This is the way that living tissues grow. The chromosomes do not carry a blueprint for a body. Rather, they carry instructions for a set of possible cells, telling each tissue type how to respond to its chemical environment. As the cells divide, they differentiate into tissue types, narrowing their possibilities to a subset of the original set. Different cell types respond to their environment in different ways: nerve cells grow, divide, and die in response to different signals than muscle cells or kidney cells.

That this strategy works seems astounding. (This may be because humans naturally think in terms of top-down design rather than bottom-up evolution. Consider with what difficulty and for how few people Darwin's simple, powerful, and well-supported ideas have displaced the top-down assumptions of creationism.) But it is astonishingly robust. Human and chimpanzee DNA are said to be only approximately 1% different; this means that almost all the knowledge used to design a chimpanzee is reused to design a human being.

It is also astonishingly compact. In bytes, the human genome is approximately the size of the Encyclopedia Britannica. Yet, it implicitly specifies not only the Britannica but every other work written, painted, sung, or otherwise constructed by human beings – and not least, human beings ourselves. Or, more accurately, it specifies the basic design that interacts with the environment to produce human beings and all their works.

We can see evidence of this strategy at the Toyota Motor Co. (Sobek, Ward, and Liker, 1999). Toyota's engineers and suppliers routinely carry multiple alternatives for many subsystems far into the design process. They leave ambiguous the exact location of interfaces. They do not employ a packaging or advanced design group to lay out the geometry and assign space. Instead, the design emerges from an exchange of information among the experts representing each subsystem, including an expert who represents the design as a whole. But this system-level expert, the chief engineer, does not have absolute power over the system design, because the system design must emerge from interaction with the subsystem designs. Thus, when we asked one chief engineer, "What makes a great car?", he replied, "lots of conflict." (Sobek et al., 1999.) This conflict is *not* resolved by referring it to a manager, but rather through negotiation among near equals.

We can expect, then, five kinds of module. Parts modules will represent a set of possible instantiations of an abstract part. Their goal will be to instantiate themselves in the way that best fits into their environment: the system design and the design of connected parts. System and subsystem modules will represent configurations of abstract parts. They also will seek to instantiate themselves in the way that best fits into their environment: the user's needs, the environment of operation, and the technical and economic characteristics of their components. Environment agents will not be parts of the design, but will represent users, regulatory agencies, and physical features of the environment of use. Manufacturing agents will connect to parts and system agents, representing sets of ways to make the design. They to will seek to fit themselves in. Of course, because the manufacturing system is itself a designed object, we could equally accurately see the manufacturing design modules as including component agents and system agents, and the product agents as representing parts of the environment of the manufacturing system. Finally, analysis agents will represent issues that cut across components and subsystems, such as vibration. They will attempt to ensure that the design performs on some set of criteria.

The primary reason to use this agent-oriented approach is simple. We cannot expect any "central authority" to know when to call on the expertise of the modules involved. The modules will have been written by many different suppliers and will represent a wide range of issues. Programmers of one module type will have to assume appropriate intentions for the other modules, ignore the ways in which they accomplish those intentions, and allow those modules to eliminate alternatives and provide information as they see fit.

My compiler handled constraints in this decentralized way (though it included only part agents, not system or environment agents). However, it handled the search for optimal solutions in a centralized fashion. Given a function to be optimized, each part agent provided a value, and the overall value was assumed to be the sum of the part values. Once as many alternatives as possible had been eliminated, the design was split into two alternative design sets by dividing one of the catalogs (selected by the user). The possible values for the value function were calculated for each alternative set, and the lower cost of the sets pursued recursively. My work with Parunak (described below) provides a completely distributed method for handling value issues.

PORTS

8. Ports or generalized interfaces will allow mating parts to mate.

Now for the tricky bit. If these modules, written in many different companies, are to interact, they need a common interaction language. When, for example, we connect a motor to a transmission, we need these modules to pass appropriate information about torque and speed to each other. When the time is right, we may also need this connection to generate a new parts module – a coupling – or we may prefer a direct spline coupling. Other connections are far more complex, such as the geometrically complex connection between a ship hull and any of its frames.

We will therefore need ports – elements that know that the output of a motor is identical to the input of the transmission, and the outer surface of a frame is minutely smaller then the inner surface of a hull. The ports will provide the context establishing the meaning of terms in which part modules will communicate.

PREDICATE CONSTRAINTS AND PRICE

9. The communication will involve a new formal language for describing sets of design possibilities, based on the predicate calculus, as well as local marginal approximate price functions.

Now comes the question of what kind of communication these modules or agents must exchange. Certainly they need to influence each other's choices, communicating about the effects those choices will have on them. This is the local approximate marginal price function: for example, a motor module may need to inform the transmission connected to it that horsepower supplied at 3600 rpm is cheaper than the same power at 1150 rpm. This information is local because it accounts only for what the motor module knows. It is marginal because it refers only to the change in utility associated with change in speed; it is a partial derivative of the utility function. And it is approximate because it is local, and because the motor module represents a large set of motors, which have differing marginal costs. In making "bids" the motor module may have to take into account the probability of selecting various types of motors with these different marginal costs (Chang, 1996).

But why do I refer to price rather than utility? Because the communication must be two way. The motor must supply information about the cost of providing power at different speeds: the price it intends to charge. However, the transmission must supply information about the value of power at different speeds: the price it is willing to pay. The design team can thus be thought of as an economy, with the players haggling to maximize the difference between what they receive and what they pay. Parunak et al. (1999) assumed that each member was a monopoly and developed bidding algorithms for this situation. Future research will deal with situations in which each abstract component represents multiple independent instantiations, which must bid to be selected.

We will also need to represent constraints, much more complex than those commonly addressed in mechanical design automation. Consider, for example, a constraint on the output speed of an electric motor. Suppose we specify that the motor speed must be between 1700 and 1800 rpm. Do we mean that we can accept any speed in this interval, with the speed changing as the load changes? Or do we mean that we can accept any speed in this interval as long as the speed is constant with load? Or do we mean that we want to be able to adjust the speed over this interval?

We could try to solve this problem by specifying the kind of motor: a constant speed motor, or an adjustable speed motor. However, we will soon be hopelessly confused by motor terminology; some constant speed motors are truly constant speed, whereas others are not. We are really trying to say something about the speed we want, not about the motors. We should be able to do so directly. These problems will only get worse as we try to represent more complex constraints, about changes in geometry, for example.

Of course, we had little difficulty explaining what we wanted in English. The problem is to find a mathematical formalism we can program into computers. I made an initial attempt in the original compiler work (see Ward and Seering, 1993). My students and I have continued to pursue such a formalism for the past decade (see, e.g., Chen and Ward, 1997; Finch and Ward, 1997; Habib and Ward, 1997; and Finch, 1998). We have achieved some interesting results, but not enough to exactly specify the best approach. I can, however, draw some conclusions with reasonable confidence.

First, the formalism will have to include the quantification features of the predicate calculus. For example, we will need to say things like "for every torque in the interval from -3 to 30 ft lb, the speed will fall in the interval from 1700 to 1800 rpm." This can be compactly expressed as

$$\forall \text{torque} \in [-3, 30], \quad \text{speed} \quad \in [1700, 1800].$$

Second, the formalism will need to unite the predicate calculus with the mathematical representations of engineering physics in a novel fashion. In addition to simple interval constraints such as those of the previous paragraph, engineering quantities are subject to constraints expressible through algebra, differential equations, and the like. Computing, and representing, the sets satisfying these constraints is often difficult; some combination of the predicate calculus with the theory of ideals may be needed. (An ideal is the set of points in an n-dimensional space that is represented by a set of algebraic equations or inequalities.)

Third, the formalism will have to represent the causal relationships among design decisions, manufacturing variation, decisions by operators, environmental variations, and feedback effects. For example, if we want constant speed, do we mean that the speed will be fixed by the motor design? Or is it acceptable that the speed be controlled by a servo loop? This remains the most difficult representation issue of all.

CELL DIVISION

10. In order to handle complex continuous objects, such as many molded parts and the molds that make them, compilers will allow the parts to divide, rather like cells in a growing tissue.

Finally, let me speculate on the most difficult problem of all: complex continuous shapes, such as molded parts and the molds that make them. My mold-making friends assure me that a great deal of money can be saved if we design the mold and the parts at the same time. Yet the part features – the "components" – that are relevant to the use of the parts are irrelevant to the flow of plastic in the mold.

How then, shall we generate these complex and smoothly blended shapes while dividing them into comprehensible pieces for which we can specify constraints and utilities? I believe that we may do so by imagining that initially we begin with three rectangular solids: an undifferentiated part, and an upper and lower mold. As we impose constraints – for example, by defining surfaces – these undifferentiated parts will begin to divide, forming regions that the system can manipulate with some independence. The shape grammar work discussed in Chapters 2 and 3 of this book will be critically important in this division process.

CONCLUSIONS

I have summarized Dan Whitney's explanation for the lack of mechanical design compilers as a claim that the interactions are too complex to address by using design rules; I have suggested that these interactions can probably be accounted for by having the system components communicate among themselves about the sets of possibilities they represent; and I have outlined a research program for making this possible. The problem is that the research program is hard – very hard. But I do not think we will need to see the entire research program completed before we begin to see commercial results. Indeed, I think that the capabilities of such software, like the designs the software will create, will emerge out of many trial efforts, rather than being designed from the top down. Already, for example, the Parametric Technology Corporation uses the term "sets of designs" in its advertising. Like the Internet substrate that will serve such systems, and like the designs they create, these capabilities will result from ferocious competition and desperate cooperation. What fun!

REFERENCES

Antonsson, E. (1997). "The potential for mechanical design compilation," *Research in Engineering Design*, 9:191–194.

Balkany, A., Birmingham, W., and Tommelein, I. (1993). "An analysis of several configuration design systems," *Artificial Intelligence in Engineering Design, Analysis, and Manufacturing*, 7(1):1–17.

Chang, T. (1996). "Conceptual robustness in distributed decision making," Ph.D. Thesis, University of Michigan, Ann Arbor.

Chen, R., and Ward, A. (1997). "Generalizing interval matrix operations for design," *ASME Journal of Mechanical Design*, **March**:21–67.

Dennet, D. C. (1995). *Darwin's Dangerous Idea*. Simon and Schuster, New York.

Esterline, A. and Kota, S. (1992). "A General paradigm for routine design – theory and implementation," *Artificial Intelligence in Engineering Design, Analysis, and Manufacturing*, 6(2):73–93.

Finch, W. (1998). "An overview of inference mechanisms for quantified relations." In *Proceedings of DETC98, 1998 ASME Design Engineering Technical Conference*, ASME, New York.

Finch, W. and Ward, A. (1997). "A system for achieving consistency in networks of constraints among sets of manufacturing, operating, and other variations." In *Proceedings of the 1997 ASME Design Engineering Technical Conferences and Computers in Engineering Conference*, ASME, New York.

Habib, W. and Ward, A. (1997). "Causality in constraint propagation," *Artificial Intelligence in Engineering Design, Analysis, and Manufacturing*, 11:419–433.

Parunak, H. V. D., Ward, A. C., Fleischer, M., and Sauter, J. A. (1999). "The RAPPID Project: symbiosis between industrial requirements and MAS research," *Autonomous Agents and Multi-Agent Systems*, 2:2 (June):111–140.

Parunak, H. V. D., Ward, A. C., and Sauter, J. A. (1999). "The MarCon algorithm: a systematic market approach to distributed constraint problems," *Artificial Intelligence for Engineering Design, Analysis and Manufacturing*, 13(3):217–234.

Seddon, A. and Brereton, P. (1994). "Component selection using temporal truth maintenance," *Artificial Intelligence in Engineering Design, Analysis, and Manufacturing*, 8:13–25.

Shin, H. and Lee, J. (1998). "An expert system for pneumatic design," *Artificial Intelligence in Engineering Design, Analysis, and Manufacturing*, 12:3–11.

Sobek, D., Ward, A., and Liker, J. (1999). "Principles of Toyota's set-based concurrent engineering process," *Sloan Management Review*, Winter:31–40.

Ward, A. and Seering, W. (1993). "Quantitative inference in a mechanical design compiler," *ASME Journal of Mechanical Design*, 115(1):29–35.

Whitney, D. (1996), "Why mechanical design cannot be like VLSI design," *Research in Engineering Design*, 8:124–138.

CHAPTER THIRTEEN

Scientific Discovery and Inventive Engineering Design

Cognitive and Computational Similarities

Jonathan Cagan,[1] Kenneth Kotovsky, and Herbert A. Simon[1]

INTRODUCTION

Michael Faraday was a scientist and an inventor. It was the same piece of work, the discovery that a magnetic field could generate electricity, and the resultant invention of the electric generator, that entitled him to both of these appellations. The goal of this chapter is to explain how a single piece of work could qualify him for both titles, and more importantly how these seemingly disparate activities are surprising similar. To do this we will attempt to delineate the domains of invention and discovery, showing how they are structured and how the underlying cognitive processes that generate both activities are broadly equivalent, even though their goals may differ. It is our contention that the activities that are commonly included in design creation and scientific discovery are often heavily intertwined, have similarly structured domains, are accomplished by very similar cognitive activities, can be simulated or emulated by similar computational models, and, many times, even yield similar artifacts, processes, or knowledge. In these latter cases, not only do they have similarly structured domains and equivalent mechanisms, but their outputs can be viewed both as invention and as discovery. With regard to Faraday, it was the invention of the electric generator (motor) and discovery of magnetism – electricity equivalency that entitles him to his dual status as inventor and scientist.

This book focuses on formal methods for design synthesis, with "formal" implying approaches that are driven by specified goals and computationally realized. In the scientific domain, the attempt to understand how people have made discoveries has similarly led to formal models of discovery. These formal models have been instantiated as computer programs that have proven able to rediscover many fundamental scientific findings. If many activities used in the discovery process are equivalent to many activities in the design process, then many techniques found in automated discovery will be useful in automated design and vice versa. We show in this chapter that at the process level this is quite accurate, that approaches used in automated discovery programs are quite similar to approaches used in automated design programs. Note that we include invention in design; just as paradigm-shifting breakthroughs

---

[1] The authors for this chapter are listed in alphabetical order.

are at the creative end of the continuum of scientific discovery, so too is invention at the creative end of the design continuum.

This chapter compares historical, cognitive, and algorithmic aspects of discovery and design. We will survey what is known about the theory of invention, by which we mean innovative engineering design, and the theory of discovery, by which we mean innovative science, with a particular interest in seeing in what ways they are the same, and in what ways different. We begin with a brief look at three historically important examples in order to demonstrate the intertwining of science and design in both domains: first, Michael Faraday's research on the magnetic induction of electricity, a piece of scientific discovery that also produced an invention – the electric generator; second, Hans Krebs' research that rested on a new experimental technique (the tissue-slice method invented by Warburg for studying metabolic processes in vivo) and led to Krebs' discovery of the reaction path for the synthesis of urea in mammals; third, the Wright Brothers' invention of the airplane, which, in addition to being a major invention, led them to discover significant aerodynamic principles. We next explore cognitive models of discovery and design. We follow with computational models in the discovery programs BACON and KEKADA and the design program A-Design. Throughout, we look across these human and computer efforts to understand cognitive and algorithmic similarities and differences across the two domains.

CASE STUDIES OF HUMAN INVENTION AND DISCOVERY

MICHAEL FARADAY

Because his complete laboratory notebooks have been published, Michael Faraday's work provides us with excellent insights into the processes of scientific discovery: not only the finished work, but the experimental and conceptual steps along the way. His discovery, in the autumn of 1831, of the magnetic induction of electrical currents illustrates what we can learn from this kind of step-by-step account of experimentation. It is especially interesting because Faraday's discovery, a basic scientific advance that provided one of the two components in Maxwell's equations, also constituted the essential step in the invention of the electric generator.

From the time of H. C. Oersted's discovery in 1821 that an electric current generates a surrounding magnetic field, a number of leading scientists, the young Faraday among them, entertained the idea that, by symmetry, perhaps a magnet could generate an electric current in a nearby wire. During the succeeding decade, Faraday conducted a number of experiments that produced significant electrical phenomena, but not the desired magnet-generated electricity. After a lapse of several years, he resumed his search in 1831. Stimulated by news of the powerful electromagnets that Joseph Henry was designing in America, Faraday had an iron ring constructed, and he began investigating whether, by winding a battery-connected wire around one side of it to magnetize it and a circuit containing an ammeter around the other side, closing the battery connection would create a current in the latter circuit.

To Faraday's surprise, he obtained no steady current, but he observed a momentary transient electrical pulse just when the battery connection was made, and another, in the reverse direction, when the connection was broken again. He then undertook dozens of experiments to see how the transient would be strengthened or weakened (and implicitly, whether it could be converted into a continuous current) if he changed his apparatus in any of a large number of ways (the number of windings, the material of the wires, the distances between parts of the apparatus, and so on). He made a number of attempts to design an apparatus that would yield the desired results, experimenting with electromagnets activated by a Voltaic pile, natural iron magnets, and the Earth's magnetism. He was testing no specific hypothesis, but looking for electrical phenomena.

After several months of effort and over 400 experiments, in which the only effects were stronger or weaker electrical transients, and noticing that the transient was of a little longer duration when a natural magnet was pushed rapidly into a circuit in the form of a hollow coil of wire, he was reminded of an experiment of Arago, in which the rotation of a magnet above a copper disk caused rotation of the disk, an effect that the existing theory of magnetism could not explain (copper is not magnetic). Faraday began to wonder if, just as plunging a magnet into a coil produced a transient electric pulse, the rotation of the magnet in Arago's experiment might have caused electrical effects in the copper disk. He was quickly led to the design of analogous experiments (inverting the roles of magnet and disk) rotating a disk between the closely spaced poles of a powerful magnet, and was rewarded, when the rotation was continuous, with the production of a continuous current. He also recognized and commented that this arrangement could be the basis for a generator, leading to our conclusion that his was an act of invention as well as an act of discovery.

The records that Faraday left behind in his notebooks demonstrate several important characteristics of his methods of discovery that were quite different from those we find in standard discussions of scientific method (e.g., Popper, 1959). Such discussions typically assume that an experiment begins with a clearly formulated hypothesis, which is to be validated or rejected by the experimental data. They ignore the question of where hypotheses come from, and consequently do not focus on discovery at all, but only upon validation of what has already been discovered.

Faraday began his research with only a very vague hypothesis: that, by symmetry with Oersted's Phenomenon, it should be possible to use magnetism to generate an electrical current. Bringing circuits into the proximity of magnetic forces should produce the desired effect or some other phenomenon of interest. It produced the latter, a surprising effect: an electrical transient. Faraday reacted to the surprise as other innovative scientists have done to their surprises. He asked how, by changing the experimental arrangement, he could magnify the effect, define its boundaries, and if possible, convert it into the quite different effect he had hoped for: a steady current.

Faraday's experimental strategy (not only in this case, but throughout his career) was not to take given hypotheses and test them, but, guided by general and often rather vague concepts, to construct conceivably relevant situations in which to look for interesting phenomena, that, if found, could be modified and enhanced. He behaved like a miner searching for an ore body rather than the same miner extending a mine in a previously discovered lode. He had a goal, in the form of a desired kind of effect, but no strong theory providing a plan for reaching that goal. In all of these respects, he resembled an inventor who has in mind a function to be performed, and searches through a large space of possible devices to contrive one that will perform it.

Faraday was not antagonistic to theory. For example, when he observed the unexpected transient in his first experiment, he spent some time formulating a hypothetical mechanism (which he called the "electrotonic state") to explain the phenomenon as produced by a change of state in the electrical circuit. In best textbook fashion, he carried out, unsuccessfully, a number of experiments to obtain empirical evidence of the wire's change in state.

Much later, after he had found ways to produce steady current, and in order to clarify some ambiguities about the direction of the current, he used iron filings (a technique that had been known for several centuries) to examine the shape of the magnetic field in which the phenomena were being produced. He observed (again without prehypothesizing it) that currents occurred in a closed circuit whenever the wire moved across the lines of magnetic force marked by the filings, regardless of whether it was the electrical wire or the magnet that moved. Now he formulated a new (and essentially correct) theory: that current was produced when lines of force were cut, the theory that ultimately made its way into Maxwell's Equations. Finally, he explained the transient that had appeared in his first experiments by the further assumption that when the battery circuit was closed, the magnetic lines around the magnet ring expanded; when the circuit was opened, they collapsed, thus cutting the ammeter circuit in both cases, but in opposite directions.

Having satisfied himself that the theory accorded with the phenomena he had observed (verification), he put aside the earlier speculative and unverified notion of the "electrotonic state." In this research episode, the principal task was not to test given theory but to produce phenomena that were interesting or desired. A very general goal led to the observation of the initial electrical transients; this led to further exploratory experimentation that evoked a memory of Arago's experiment and discovery of an experimental procedure that produced steady current. Confusion about the directions of the currents under different conditions led to experiments using iron filings to visualize the magnetic field, calling attention to a circuit's cutting magnetic lines as the condition for the induction of current and the determinant of its direction, and immediately to a theory of electromagnet induction that later became a component of Maxwell's Equations and the progenitor of the electric generator. Important aspects of design are evident in a number of ways in this work. These include the actual design of various magnetic artifacts with which to produce electricity, the iterative process (cycling between synthesis of attempted designs and analysis of obtained results), and the final results consisting of both an important discovery (the interactive equivalence of electricity and magnetism) and an important invention (the electric generator).

HANS KREBS

The lab notebooks of Hans Krebs, who discovered the Krebs Cycle of Organismic Metabolism, and earlier, the reaction path for the synthesis of urea in the liver, are also available to us. In the case of urea synthesis, they reveal events very similar to those in the previous example. This time, as in Faraday's case, the experiments took place over a period of a few months. The process begins with a problem: to discover the reaction path that synthesizes urea in the liver. Biochemists already knew that the likely sources of the nitrogen in the urea were ammonia (from the decomposition of amino acids) or amino acids (from the deaminization of proteins) as well as purines from DNA and RNA.

Experimenters had already searched unsuccessfully for the reaction path, but they used cumbersome methods involving entire organs. Krebs took up the task because he had acquired his new "secret weapon" as a postdoc in Otto Warburg's lab – a method had been designed that used thin tissue slices, which enabled him to carry out experiments many times faster than could be done by using perfused whole organs.

Krebs had a straightforward strategy: using this newly designed tissue-slice testbed, he applied ammonia and various amino acids and purines to the tissue slices and measured the yield of urea. Obtaining no interesting results, he began (for no reason that has been clearly established) to use combinations of ammonia and an amino acid. One day, when testing ammonia and ornithine (an amino acid that appeared mainly in the liver, and at the time of Krebs' experiment had no known function), he obtained a large yield of urea. With the aid of the chemical literature, he was fairly soon able to find a plausible reaction path that combined the ammonia and ornithine; then, after several further steps, used the nitrogen derived from the ammonia to form urea, recreating and releasing the ornithine for repeated use. Thus, although the experiment was probably run to test whether ornithine might be a source of urea nitrogen, the nitrogen was found to come, in fact, entirely from the ammonia. The ornithine, retaining its nitrogen, served as a powerful catalyst to speed up the reaction by a large factor.

Here we see another example of the generation of an important phenomenon "for the wrong reason." Recognized with surprise, the phenomenon leads to the desired outcome and provides evidence that leads rather directly to the correct theoretical explanation. The catalytic action of ornithine was discovered by experiment, not verified by testing after it had been hypothesized on other grounds. As in the case of Faraday, we see the scientist seeking "interesting" phenomena related to some broad goal (e.g., exploiting the supposed symmetry of electricity and magnetism), finding a law that describes the observed phenomena, and then searching for theoretical explanations of the law. At a general level of description, this rapid convergence on a new solution (explanation or theory) once a surprising discovery is made can be viewed as a sudden move to a new area of the problem space, or even to searching a new or different problem space. In design, as we will see below, a similar phenomenon arises when a genuinely new design is generated: it rapidly leads to the exploration of a new portion of the design space or even a new space in cases in which, for example, the search shifts from seeking a component that performs a certain function to searching for an entirely new design approach, producing a new set of solutions to the design problem.

THE WRIGHT BROTHERS

In turning to an examination of design, we will consider the Wright Brothers' invention of the airplane as a classic case of inventive design. We base our analysis of the Wright Brothers' contributions as inventors–scientists on the extensive treatment afforded their work in Voland's *Engineering by Design* (Voland, 1999), where he presents their work as his example of "methodical design," and in *The Wright Flyer, an Engineering Perspective*, edited by Howard S. Wolko (1987). The Wright Brothers' historic feat provides a useful view of the design process.

Contrary to the popular view that the Wright Brothers' accomplishment was primarily an experimental, empirical enterprise that relied on trial-and-error attempts to fly in windswept Kitty Hawk, North Carolina, in reality their efforts were strongly rooted in the science of fluid mechanics and in addition, made substantive contributions to that science. As Voland puts it,

> In contrast, the Wright Brothers more closely followed the structured problem-solving procedure that we now call the engineering design process to develop and refine their glider and airplane concepts. As a result, Wilbur and Orville Wright behaved more like today's engineers than yesteryear's inventors: They were not only creative but truly methodical and analytical in their work, adding the steps of synthesis and analysis that were missing in the earlier efforts of others (p. 23).
>
> G. Voland, *Engineering by Design* (1999)

With regard to their scientific contributions, the following quotations are typical evaluative comments: "The Wrights' understanding of the true aerodynamic function of a propeller, and their subsequent development of a propeller theory are important firsts in the history of aeronautical engineering" (Anderson, 1983, p. 16)[2], and "probably the best-known scientific work by the Wrights is their program to obtain data for airfoils and wings" (Culick and Jex, 1987, p. 20). In addition to their contributions to many aspects of aircraft design, one lasting contribution they made to the science of aeronautics was their deduction of a correct value for Smeaton's coefficient for drag (discussed below), which had been overestimated by a third for the previous 150 years (Culick and Jex, 1987).

The basic relations between drag and lift and its implication for power had been worked out prior to the time of the Wright Brothers' achievement, and they formed the basis for the design of their airplane. In particular, lift (the upward component of force) $L = kV^2 A C_1$, where V is the velocity, k is a constant, A is the area of the lifting surface (the wing), and C_1 is a lift coefficient that depends on both the wing camber (height to cross-sectional length ratio) and angle of attack (angle with respect to the horizontal line of travel). For drag (the force resisting forward travel) there are two components, induced drag (D_i) and parasitic drag (D_p). The relations are $D_i = kV^2 A C_d$ and $D_p = kV^2 A_f C_d$, where k is a constant, C_d is the coefficient of drag, A is the wing area, and A_f is the frontal surface area. (They achieved a reduction in frontal surface area by having the exposed pilot lie flat rather than being seated.) They calculated the power necessary to propel their vehicle in a very straightforward manner as $P = D \times V$, where D is the total drag or force needed to overcome drag and V is the desired velocity. As discussed below, the coefficient k above, known as Smeaton's coefficient, formed the basis for one important scientific discovery of the Wright Brothers. According to Voland, they designed with these relations in mind and iterated between experimentation and redesign. As he puts it,

> With these rather rudimentary mathematical models of flight behavior, the Wrights were able to develop their aircraft designs. Whenever failure struck (as it did

---

[2] Others (Lanchester, Drzweiecke, and Prandtl in Germany) had started to evolve a propeller theory independently of the Wrights, but according to Lippincott, " – the Wright Brothers evolved this theory independently without knowledge of the work of the earlier scientists" (Lippincott, 1987, p. 79).

repeatedly), they returned to careful experimentation and observation in order to correctly adjust the values of constants and variables appearing in the above equations. This methodical and iterative transition between scientific theory and its practical application – through which both the theory itself and its application are successively developed and refined – is the essence of good engineering practice (p. 32).

G. Voland (1999)

It is also, as we have seen above, the core of much scientific practice!

As they moved to the issue of control, there was a similar alternation between theory and design, especially in regard to the issue of sideslipping in the control of yaw. Over subsequent summers from 1900 to 1902, they tested a number of gliders, working out control systems for each of the three major axes of turning. They adopted a horizontal elevator in the front of the glider for pitch control and twistable wings for roll axis control. In 1901, they embarked on a sizable research project to find ways of combating some problems arising from inadequate lift. They constructed a wind tunnel to experiment with the variables that determine lift and, through testing of over 200 wing designs, found that the values in the equations for lift (reported earlier by Lilienthal) were generally accurate, with the quite surprising exception of the Smeaton's coefficient, k, which they had to significantly revise downward by a little more than a third (from 0.0049 to within a few percent of the correct value of 0.003; see Culick and Jex, 1987; Jakab, 1990).

With the use of this revised value, they modified their wing to yield more lift. They then confronted the issue of sideslipping that occurred when the airplane banked in a turn. The raised wing (as a result of changes in the attack angle) had increased drag, resulting in a turning around the yaw axis, and a decrease in lift, leading to downward sideslipping, a dangerous condition in low-flying airplanes or gliders. In 1902, they at first experimented with a pair of fixed vertical rudders to prevent sideslipping, but they soon replaced them by a single vertical rudder that could be controlled so as to both prevent sideslipping and allow control around the yaw axis. This solved a major problem in their design and allowed them to go on to successful flight. The final stage in the design of the initial airplane was the propulsion unit. Orville Wright determined specifications for the engine from an analysis of the power required and the allowable weight. This was based on a number of variables, including all of those that contribute to lift, drag, weight of the airplane and pilot, and desired velocity. His calculations yielded a value of 8.4 hp with a maximum weight of 200 lb, and his mechanic, Charles Taylor, delivered an engine that could generate 12 hp and weighed 180 lb, thus enabling the successful manned flight in December of 1903.

Through the progression of the design process by the Wright Brothers, we repeatedly observe goal-directed problem solving, both in the main goal of obtaining flight, but also in subgoals toward the main goal such as solution of the sideslip problem. We also observe an iterative process between configuration synthesis and evaluation (both analytical and experimental). Finally, although the focus is on the design of a complex device, the experimental observations led to fundamental scientific findings in fluid dynamics (Smeaton's coefficient), propeller design, and sideslip and its control. Note that although the Wright Brothers may not have made any major discoveries, they clearly made scientific advances as well as used scientific results in their invention. Their scientific impact was from the cumulative effect of their both using science and advancing our understanding of the science of aerodynamics. However, as we have noted, in both design and science there is a continuum from the routine to the creative. The important lesson is that, in order to move to the creative end of one continuum, aspects of the other continuum frequently must be addressed; that is, some level of science is usually needed to effectively invent things, and some level of design is usually needed to support scientific exploration.

SUMMARY

Krebs was considered a scientist who made a great discovery. To do so he invented new methods to perform his experiments. The Wright Brothers are known as inventors or designers who invented an important machine. However, it is clear that along the way they discovered new scientific principles. Faraday is considered both an inventor and discoverer, having needed to be both to accomplish his goals. In many ways the classification of these people comes from knowledge of the domain in which they operated. However, Krebs could be considered an inventor and the Wright brothers could be considered scientists. At the process level, in all three cases there is an iteration between a synthesis phase (of an artifact or scientific experiment, explanation, or theory) and an evaluation phase, until a desired goal is achieved, the sought-after invention or discovery. We now move on to a comparison of cognitive models of discovery and design.

COGNITIVE MODELS

THE PROCESSES OF DISCOVERY

We turn now from these historic examples to a more general discussion of what is involved in scientific discovery. There is a large literature of research on scientific discovery, produced by historians and philosophers of science, cognitive psychologists, and scientists (some of it autobiographical), and a gradual convergence is taking place toward a common picture of the discovery process. Some recent examples of, and guides to, this literature are Finke, Ward, and Smith (1992), Giere (1988), Holland et al. (1986), Holmes (1991), Langley et al. (1987), Nickles (1978), Weisberg (1993), and Dunbar (1993).

As noted above, scientific discovery can range from the breakthrough major discovery, as in the examples cited above, to the more mundane or routine small extensions of previous work that are often referred to as *paradigmatic science* (such as determining the dose-response curve for some newly discovered medication or whether an experimental result obtained with one species also holds in another). By examining a large range of examples of major discoveries, we learn that "discovery" is not limited to a single kind of scientific activity. Scientists do many things, although individual scientists may specialize in just one or two of them (Langley et al., 1987). Among the important kinds of discoveries that result from scientific activity are the following.

1. Discovery of ways of describing and thinking about some domain of phenomena; that is to say, forming representations of the domain: the problem spaces.
2. Discovery of an interesting puzzle or problem in the domain (e.g., the problem of describing and explaining the orbits of the planets around the Sun).
3. Discovery (or design) of instruments and experimental strategies for attacking empirical or theoretical problems (e.g., the thermometer, recombinant DNA, the calculus).
4. Discovery of new, often surprising and puzzling, phenomena. (We have already provided several examples.)
5. Discovery of laws or creation of models that explain the described phenomena (e.g., Newton's Law of Gravitation, which explains Kepler's descriptive Laws of Planetary Motion; Bohr's model of the hydrogen atom, which explains Balmer's Law, a description of the hydrogen spectrum).
6. Discovery of laws that describe bodies of phenomena (e.g., Hooke's Law, Ohm's Law, Mendel's Laws).

The list is undoubtedly incomplete, but it covers the main categories of discovery. Historical research has revealed a great deal about how discoveries of each of these kinds are achieved, and our understanding of the processes has been augmented and sharpened by a number of computer-simulation models that constitute testable and tested theories of these processes: see Grasshoff and May (1995), Langley et al. (1987), Shrager and Langley (1990), Shen (1993), and Valdes-Perez (1995). In the fourth section, we will briefly describe two examples of such models: one, called BACON, works solely from data to construct descriptive laws that fit them (Category 4, above); another, called KEKADA, plans sequences of experiments in order to produce observations that can be used to formulate descriptive and explanatory theories of problematic phenomena (Categories 3–6, above, especially 4 and 5).

From historical analyses and computer simulations of the kinds we illustrate below, an empirically based and rather comprehensive theory of discovery has emerged. Perhaps the most important feature of the theory is that it is built around the same two processes that have proved central to the general theory of expert human problem solving: the process of recognizing familiar patterns in the situations that are presented, and the process of selective (heuristic) search of a problem space. Basic to these are the processes of formulating a problem space and of representing available information about the problem, as well as new information acquired through observation and experimentation, in that problem space.

Put in simplest terms, human experts solve difficult problems (including problems of discovery) by selective searches through problem spaces that are sometimes very large, but they greatly abridge their searches by recognizing familiar patterns, associated in memory with information about how to proceed when a particular pattern is present. To do this, they have to represent the problem in a form that will reveal the familiar patterns and access the memory already stored for dealing with them (Simon and Kotovsky, 1963; Newell and Simon, 1972).

Efficient problem solving may require both general knowledge and a large body of knowledge that is peculiar to particular problem domains. This is true both of knowledge required for selective search (the heuristics that guide selection) and knowledge required for recognizing patterns and exploiting that recognition. Hence, expertise, including expertise in scientific discovery, is transferable from one domain to another only to a very limited degree. However, knowledge about the calculus and differential equations, and knowledge about computer programming, are examples of knowledge that has potential for broad transfer.

In summary, everything that has been learned about making discoveries suggests that discovery is just a special case of expert problem solving. The faculties we call "intuition," or "insight," or "creativity" are simply applications of capabilities for pattern recognition or skillful search to problems of discovery, albeit in very large and ill-defined problem spaces.

THE PROCESSES OF DESIGN

Design, as illustrated by the Wright Brothers, is an activity whose goal is the creation of a new or improved artifact or process. It is characterized by deliberate planned activity directed toward a goal or set of goals and is subject to constraints and often-conflicting evaluative criteria. In addition to deliberate stepwise progress, design can and often does involve flashes of insight, evidence of incubation (off-line work) and what has broadly been termed "intuitive" behavior. As argued above, such behavior, although sometimes seemingly mysterious, is amenable to a cognitive analysis. The domain of activities that constitute design can range from the fairly mundane (selection of gear ratios in a transmission or substitution of different materials to make a lighter-weight knife handle) to the highly creative (design of the first cellular device for communicating or, in the case of the airplane, creation of the magic of flight).

The latter design activities are often labeled *invention*, a term that designates one subset (the creative or "new artifact" extreme) of the design continuum. One of our claims is that invention is not essentially different from other types of design activities, but rather represents this more creative or innovative end of the design continuum. In addition, as the Wright Brothers' example demonstrates, the activity of invention or design often involves operating in a number of different modes. Artifact synthesis by means of design and construction, scientific analysis and evaluation, and empirical experimentation, all conducted in an iterative manner that (when successful) converges on a good design, are three of the major activity modes that comprise the overall activity of design.

When humans engage in a creative endeavor such as designing a new artifact, they are assumed to operate under the guidance of a goal or purpose, that is, a conception of what is to be created (Ullman, Dietterich, and Stauffer, 1988). The process has been likened to that of human problem solving, whereby people search a "problem space" that consists of possible configurations of elements, applying move operators that take them from state to state until they find a solution (Newell and Simon, 1972; Anderson, 1983). Attempts have been made to analyze the cognitive problem-solving processes (primarily search and analogizing – see Gentner, 1983; Falkenhainer, Forbus, and Gentner, 1989; Forbus, Gentner, and Law, 1994) that operate within the design process (Adelson et al., 1988; Ullman et al., 1988; Adelson, 1989). One finding is that search often goes on in multiple problem spaces. For example, in science, people often alternate between searching a space of experiments with searching a space of hypotheses (Klahr and Dunbar, 1988; Dunbar, 1993). Similarly, in design, the search may move between a space of components to instantiate or realize a given design idea (e.g., the search for an appropriate airfoil shape for a fixed wing) and a space of design conceptions or new approaches to a design problem (whether to have flapping, birdlike wings or a fixed airfoil, for example). The general cognitive architecture, Soar (Laird, Newell, and Rosenbloom, 1987), uses this multiple problem space search as a basic, almost universal, method for search that is practicable in a wide variety of problem domains.

Representational issues along with problem space size are also important in problem solving in diverse problem domains. Cagan and Kotovsky (1997) have modeled the problem-solving process based on data from human problem-solving performance through a learning algorithm in conjunction with a simulated annealing search strategy. Although the problem space might be vast, consisting of all possible legal configurations of the elements that make up the problem, nonetheless, the designer does not always know the whole space or have to search it in its entirety, given adequate heuristics for selecting propitious portions of the space to search. In fact, the size of the problem space has been shown to be but one of a number of factors that determine problem difficulty. Representational issues have also been shown to be a major determinant of problem difficulty in humans (Hayes and Simon, 1977; Kotovsky, Hayes, and Simon, 1985; Kotovsky and Simon, 1990). As we have seen, choice of an appropriate representation of the problem is a central issue in scientific discovery as well.

The processes underlying design include a broad array of cognitive activities, including search through a space or spaces of possibilities (often conducted in a stepwise or piecemeal fashion as the design is progressively instantiated with a set of components), the use of analogical reasoning to import useful ideas or substructures from previous designs, other ways of accessing memory for relevant general or domain-specific knowledge, and an evaluative phase or set of tests for whether a satisfactory end state has been reached. They include both analysis and synthesis, often (as we have seen) operating iteratively in the creation of a new design. Once a successful design is created, it is often progressively generalized or otherwise extended to create a range of possible options or extensions. In our example of a truly creative design, there is a loop between analysis and synthesis as a design concept evolves. It is our claim that this is not only true of design, but that the processes that yield new artifacts are similar to those found in a related enterprise – scientific discovery.

COMPARISON

Similarities

We have, to this point, seen similarities between discovery and invention or design in:

- the intertwining of the two (invention depending on and yielding scientific discoveries and scientific discoveries depending on and yielding inventions),
- the iterative synthesis–analysis cycle,

- the underlying cognitive activities based on problem solving, pattern recognition, analogical reasoning, and other cognitive knowledge retrieval mechanisms; and
- heuristically guided search in large ill-defined problem spaces.

In both design and scientific discovery, the respective domain of activities can range from the fairly routine or mundane to the highly creative, in which major scientific advances or designed inventions create new paradigms or change our way of interacting with or understanding the world. In both design and scientific discovery, each is generally an intrinsic part of the other process (design is needed to make discoveries, and science is needed to succeed in invention). However, the real similarity between the activities that fall under the rubric of design invention and those that are labeled scientific discovery is not only that the process of doing science is comingled with that of design, but that the underlying cognitive processes are essentially the same.

Design is a planned activity, directed toward a goal or set of goals, and subject to an evolving set of constraints. The processes underlying design include a broad array of cognitive activities, including search through a space or spaces of possibilities or components, reliance on general and domain-specific knowledge including the use of analogical reasoning to import useful ideas or substructures recognized as familiar and relevant from previous designs, and an evaluative phase or set of tests for whether a satisfactory end state has been reached. It operates iteratively between analysis and synthesis in creating new designs. Design is thus accomplished through knowledge-based pattern recognition and problem-solving search processes. Such pattern recognition activities are an important part of human cognition and are fairly central to scientific reasoning as well (Simon and Kotovsky, 1963).

The processes underlying scientific discovery also include a broad range of cognitive activities including search through a space of hypotheses and activities designed to test those hypotheses (often conducted in a stepwise fashion as the space is progressively searched.) This is often accomplished by use of analogical reasoning (from similar scientific findings) and testing of possible "solutions" (scientific results or conclusions) to determine whether an adequate understanding or conclusion about a hypothesis or theory has been attained. There is frequently an additional attempt to broaden the finding to an extended range of situations or circumstances. Finally, the process is iterative, moving between experiment generation and theory refinement. As with design, scientific discovery is characterized by deliberate planned activity directed toward a goal or set of goals and is also subject to constraints and multiple evaluative criteria (e.g., accuracy of prediction vs. parsimony of free parameters). It too involves stepwise search through a space (or multiple spaces) of possible hypotheses, experiments, and theories, as well as flashes of insight, incubation, and intuition, all of which can be described as pattern recognition, memory retrieval, and problem-solving activities operating to generate and search within an appropriate representation or problem space.

This broad-stroked description of the cognition of scientific discovery matches in surprising detail the description of the design process; they are both problem-solving activities that to date have occurred primarily in the human mind, and thus reflect its operating characteristics. Thus, when considered from a fairly broad perspective, the structure of the domains and the underlying cognitive activities exhibit a high degree of unity.

Differences

Of course there are obvious differences between discovery and invention. These differences focus on the purpose or goal of the endeavor, the knowledge used to generate the problem space being searched, and the training of engineers and scientists in today's educational environment. The fundamental difference that differentiates the two domains is the *goal* of the process: scientific explanation versus creation of a new artifact. Scientists wish to discover new knowledge about how the existing world works, whereas designers invent new devices that function within that world. Design starts with a desired function and tries to synthesize a device that produces that function. Science starts with an existing function and tries to synthesize a mechanism that can plausibly accomplish or account for that function; the mechanism already exists but it is unknown, so the scientist must still create or synthesize a model or theory that replicates or explains it.

The second significant difference between scientific discovery and invention is the knowledge base, that is, the knowledge of facts and bounds on the physical world, and the experience one has in an area. Note that knowledge is a part of cognition in that it forms the knowledge base that cognitive processes operate on, but it should not be confused with the cognitive processes themselves that are our focus. The knowledge base for engineering designers includes an understanding of the behavior of components, materials or fluids, an understanding of the applicable science of the domain, and an understanding of heuristics about the bounds on the feasible space. Designers also have developed intuition through experience and stored that knowledge in memory. Similarly, scientific discovery requires knowledge about what is known in say chemistry, biology, or physics, and also heuristics about the bounds on the feasible space. Scientists also have intuition and experience in a domain. But we might also point out that differences in the knowledge base occur not only between the two areas (invention and discovery) but within each as well; a chemist and a physicist or geologist have very different knowledge bases even though they are all scientists, and an automobile designer and the designer of chemical processing plants similarly have quite different knowledge bases.

This knowledge defines the problem space in which designers and scientists search. We differentiate between two spaces: First is the potential problem space that includes all combinations of parts or articulations of, say, geometries, whether feasible or infeasible, efficient or inefficient. The other space is the heuristic space generated by an efficient or heuristic search process. The generation of the space is dependent on the domain knowledge base; however, once the space is generated, or the method of generating the heuristic space articulated, then the cognitive process of search of that space is the same.

The third difference between scientific discovery and engineering invention is only one of perception. Scientists are perceived as those that discover and engineers as those that design. It is surprising, however, that beyond the differences in their respective knowledge base, their education is not that different! Although engineers are perceived as designers, in reality very little of their training focuses on design or synthesis methods. Most courses in an engineering curriculum focus on understanding analytical models of how the world behaves, that is, science, whereas few courses focus on synthesis methods. We recognize, and have argued strongly above, that scientific knowledge is critical to engineering design; however, we also argue that synthesis is what enables engineers to practice the design process. We argue that engineering curricula should be broadened for more inclusion of synthesis methodology and the means to apply science in the synthesis process.

In a similar vein, much training of science students, at least at the undergraduate level, focuses on teaching the findings or knowledge base of science, that is, models of how the world behaves in various scientific domains. The inclusion of more teaching of the methods of science, the tools for discovering new knowledge or synthesizing scientific experiments and theories, is often relegated to advanced graduate training. In this fashion, the training of engineering and of science students exhibits a somewhat nonpropitious parallel. Our focus on the interplay of analysis and synthesis raises the issue of the optimal pedagogical approach to allow students to acquire and use both types of tools or approaches. Perhaps individuals with the breadth and perspective of Faraday, the Wright Brothers, and Krebs would be less rare.

So again, although education and training between scientific discoverers and design inventors is quite different as a result of the knowledge taught, their cognitive models of search, once the problems are defined, are the same. Under this view, an engineer given a knowledge base of biology and the experience and insight of an experienced biologist could discover new theories, and a biologist given the knowledge base, experience, and insight of a mechanical design engineer could use his or her problem-solving processes to invent new artifacts.

COMPUTER MODELS

We turn now to observation of computer models of discovery and design. The discovery programs were created to capture cognitive models of human discovery directly. The design program was developed with the intention of automating the process of invention by using the generation and search capabilities of the computer along with insights gained from human design activities where beneficial.

DISCOVERY

BACON

BACON is a computer program that simulates (by actually doing it) the process of discovering laws that describe the data provided to it (Langley et al., 1987). Basically, BACON works by generating mathematical functions, then fitting them, by adjustment of parameters, to the data. It requires no knowledge of the meanings of the data points, but operates wholly by searching for invariant patterns of data (e.g., a pattern such as $y = ax^2$, for fixed a, that holds for all y's and their corresponding x's in the data set). It does not generate functions at random, but determines what function to generate next by examining the ways in which those previously generated have failed to fit the data.

For example, if y/x decreases as x increases, and y/x^2 increases as only x not y increases, BACON considers their product: y^2/x^3, finding, in this case, that it is constant (within a given allowance for measurement error). BACON then concludes that $y = ax^{3/2}$, where a is a constant. When y is identified as the period of revolution of a planet around the Sun and x as the distance of the planet from the Sun, this can be recognized as Kepler's Third Law.

The core of BACON is a small set of productions (if $\rightarrow$ then rules) that generate a sequence of hypotheses $z = f(x)$, where x is the observed independent variable, and z is a prediction of the observed dependent variable, y. As we have seen in the example, if the hypothesis doesn't fit the data (as it usually doesn't), BACON then examines y and one of the new variables already computed, say y/x, to see if the former increases or decreases monotonically with the latter. In the first case it introduces their ratio as a new variable and tests for its constancy; in the second case it introduces their product and tests for constancy. If the test is satisfied in either case, the new variable, say w, is a law, for $y = w = f(x)$ Thus, in the example, having found that y/x^2 and y/x are not constant but vary with x in opposite directions, it tests their product, finding the constant law.

Because BACON requires no knowledge about the meanings of the variables, it is a completely general data-driven discovery engine, which is, of course, not guaranteed to find a pattern in any given application. Nevertheless, without any changes in the heuristics or other features of the program, or any prior knowledge of the nature of the variables, it has (re)discovered Kepler's Third Law, Coulomb's Law of Static Electricity, Ohm's law of electrical resistance, Archimede's Law of displacement, Snell's Law of Light Refraction, Black's Law of Specific Heat, the law of conservation of momentum, the law of gravitation, and formulas for simple chemical compounds. In all of these cases it has to generate only a small number of functions (seldom as many as a score) in order to find one that fits the data.

Not only does BACON (re)discover these and other important scientific laws, but it also introduces new theoretical concepts into its functions, assigning them as properties to various parameters. For example, in the Snell experiment, it introduces the coefficient of refraction as a property of each of the substances that the light traverses; in the Black experiment, it introduces and determines the value of a specific heat for each substance; in the experiments in mechanics, it introduces gravitational and inertial mass (and distinguishes them); in the chemical experiments, it introduces atomic and molecular weights (and distinguishes them). Hence BACON discovers theoretical terms as well as laws.

BACON introduces new concepts when it discovers that a given law holds only when a discrete experimental variable (say, whether the liquid being tested is water or alcohol) is held constant, but, with a change in parameter, also holds for other values of the variable. It then assigns that parameter as the value of a new property of the variable. Thus, in Black's Experiment it finds that it can assign a specific parameter value (which we know as "specific heat") to each distinct liquid and thereby generalize the law to apply to mixtures of liquids as well as to a single liquid.

BACON's searches can often be substantially shortened if it need not work in total "blindness," but is provided with some prior theoretical constraints: for example, conservation of mass and heat in Black's experiments on temperature equilibrium; or conservation of atoms in the chemical experiments. So, although designed as a data-driven discovery system, it can also operate in a combined data-and-theory-driven mode.

KEKEDA

In BACON, empirical data are the required inputs. KEKADA (Kulkarni and Simon, 1990) is a computational model that simulates experimental strategies for acquiring and interpreting data. It has to be provided with substantive knowledge about some domain of science, with knowledge of feasible experimental manipulations and with a scientific problem to be solved. Its task is to plan successive experiments aimed at solving the problem. After it plans an experiment, it is provided with knowledge of the data obtained by carrying it out (provided to the program by the user), and it then uses this new information, together with the knowledge it already had, to plan a new experiment. Both knowledge and experimental strategies are represented as productions, that is, if → then rules.

KEKADA's knowledge takes the form, then, of a system of productions, but, in contrast to BACON, there are a large number of these, for they must represent not only general heuristics of search and experiment construction, but also domain knowledge. As an example of a general heuristic, from the outcomes of experiments it forms estimates of the yields of certain substances that are to be expected when the next experiment is performed. If a quite different yield is observed, KEKADA reacts with "surprise." It then designs new experiments to test the scope of the surprising phenomenon, and subsequently experiments to search for causal factors.

Here is an example of domain-specific knowledge: in the search for the mechanism of urea synthesis in the mammalian liver, Krebs' problem discussed earlier, KEKADA is provided, at the outset, with the relevant chemical knowledge and knowledge of experimental methods that Krebs possessed. It plans a series of experiments using the tissue-slice method of Warburg with ammonia and/or amino acids (already known, with near certainty, to be the sources of the nitrogen in urea). It eventually comes to try an experiment with ammonia and ornithine, as Krebs did, and it then goes on to find the reaction path and the role of ornithine as a catalyst. It is interesting that the reason why KEKADA (and Krebs) experimented with ornithine (because it was a likely source of nitrogen) was unrelated to its actual function (as a catalyst of the reaction that incorporated ammonia in urea). The discovery, in both cases, was an "accident" that happened, in the words of Pasteur, "to a prepared mind."

KEKADA makes important use of a surprise heuristic in focusing its attention on ornithine after obtaining the first evidence of that amino acid's large effect on urea production. To be surprised, one must have expectations, which KEKADA forms on the basis of the results of experiments. Having formed an expectation of a small yield of urea in similar experiments, it is surprised when ammonia with ornithine gives a very much larger yield. The surprise heuristic then causes it to plan experiments that will seek to amplify the effect and to determine the range of conditions under which it will be observed: for example, what substances similar to ornithine will produce the effect (in this case, none); how will it (volume of urea produced) vary with the inputs of ammonia and ornithine (experimentally derived answer: directly with the ammonia, very little with the ornithine); and so on.

KEKADA has also been exercised, although with somewhat less attention to detail, on Faraday's experiments seeking to obtain magnetic induction of electrical currents. As in the case of the simulation of Krebs' work, the surprise heuristic played a large role in the simulation of Faraday. The initial experiment was motivated by the goal of obtaining electricity from magnetism; the observed transient was treated as a surprising phenomenon, leading to a series of experiments, similar to Faraday's, aimed at exploring the range of the phenomenon and intensifying it. Knowledge of Arago's experiment could be used to motivate experiments with disks. The simulation was in good agreement with the actual history. Of greater importance in the case of Faraday than in the case of Krebs was the former's great skill in designing and building laboratory equipment and using it to devise new experimental arrangements.

There is a growing number of other computational models of discovery systems, a number of which have been cited above.

DESIGN

We now turn to a brief discussion of a computational model of design that incorporates the same kinds of cognitive mechanisms as those seen to be central to scientific discovery; specifically, the knowledge-based recognition of patterns, the formation of a propitious representation of the design problem space, within which the design problem is formulated, and heuristically based search through that constructed problem space for an adequate solution. Lessons learned from work in cognition forms a basis for the development of the agent-based structure of A-Design and, in particular, the actions of the management agents in guiding what we refer to as top-down search.

A-Design

A-Design (Campbell, Cagan, and Kotovsky, 1999 and 2000) is a program that automates the invention of electromechanical artifacts. The theoretical basis for the program is derived from a model of the design process, namely a mapping from conceptualization to instantiation, the bridge between synthesis and analysis, and iterative refinement. These aspects of the program are true for all good design. Other aspects may or may not connect to human design but nevertheless assist the computer in invention. The approach taken in A-design is to use design agents to act on a representation of the problem domain to create design configurations, instantiations, and modifications. The program also compresses the design process by exploring multiple solutions in parallel and uses computer iteration to evolve new designs.

A-Design's search for optimal designs is accomplished through intelligent modification of alternatives by a collection of agents. The agents interact with designs and other agents based on their perception of the design problem, the relative preferences of the design goals presented by the user, and an agent's individual preferences. Agents affect their environment by adding or subtracting elements to designs, or by altering other agents. At a high level the A-Design program is a model of (human) group activity.

The program creates designs by working on different levels of abstraction. The agents are able to place components into a configuration based on their knowledge of how components behave. This knowledge is modeled in a functional representation for describing components and designs. The representation developed for this work is based strongly on qualitative physics (Forbus, 1988), bond graphs (Paynter, 1961; and Ulrich and Seering, 1989), and function block diagrams (Pahl and Beitz, 1988), and more specifically on work done by Welch and Dixon (1994), and Schmidt and Cagan (1995, 1998).

The fundamental issue in creating a representation for describing a design configuration is developing a formal method to infer how individual components behave and, further, how the connected sum of components behaves. In this representation, components are described by their ports, or points of connectivity with other components, called *functional parameters*. Information about how components are constrained at their ports, how energy and signals are transformed between ports, and how energy variables within the system relate to others throughout the design is found in the component descriptions. These descriptions, known as *embodiments*, also contain information about the kinds of parameters that describe the component such as length, spring constant, resistance, and cost or efficiency. The use of an appropriate representation defines the problem space and is thus a major contributor to A-Design's being able to solve the problem of creating successful designs. See Chapter 6 for further discussion of this representation.

Once a configuration is created, it is instantiated. Here actual components are selected from a catalog of components indexed by their functionality and ability to instantiate an embodiment. It is here that real values of the variables in the design's embodiments are set.

The need to link between synthesis and analysis is critical in all levels of A-Design. At the embodiment level, the program uses analytical models of components to set up physical constraints on the system. Though not necessarily at the level of the Wright brothers' wind tunnel experiments, the rough analysis still uses science to help direct the design process. The program also calls on analysis simulators to evaluate complete designs after instantiation, enabling a robust representation of physics in the process.

Iteration is an important part of the process. The agents initially create functionally feasible but inefficient designs. In a genetic algorithm-like fashion the process iterates by pruning out ineffective designs and mutating others to create a new population of designs. The process evolves designs toward the optimal arrangement of components that best meet user preferences across multiple design goals. Just as in scientific discovery where there is an iteration between experiment and theory or hypothesis refinement, so too in A-Design there is an iteration between the creation of new abstract designs and the instantiated realization of those designs. This iterative search in dual or multiple domains or problem spaces is a characteristic of human scientific and inventive activity (Klahr and Dunbar, 1988; Dunbar, 1993).

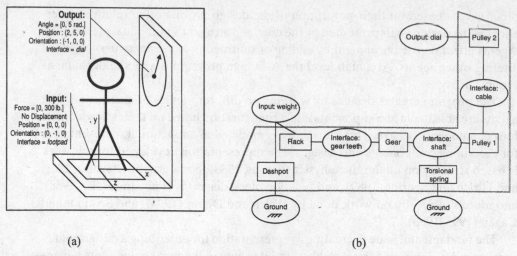

(a)   (b)

Figure 13.1. Weighing Machine: (a) user-defined functional specifications, (b) possible configuration of embodiments that meet the weighing machine specifications.

There have been several applications using A-Design. One application found in Campbell et al. (1999) shows the design of weighing machines; in another (Campbell, 2000), MEMS accelerometers were created. The weighing machine problem is outlined in Figure 13.1, where the problem description consists of an input–output specification. The input is a person's weight at a certain location, and the output is a dial reading that reflects the input weight. In this example, four objectives were chosen for optimization: minimize cost, minimize mass, minimize dial error, and minimize input displacement. Note that whereas the first two objectives are calculated by summing data provided within the catalog on each of the components used in the design, the latter two are results of the interaction of the components in the completed design and depend on the values of the components used as well as the behavior predicted by the behavioral equations (a.k.a., physics). The catalog consists of 32 embodiments and their instantiations, including motors, pistons, potentiometers, worm gears, and levers. For each of the embodiments, there exist actual components drawn from several catalogs totaling just over 300 components available for constructing designs. In Figure 13.1 is an example configuration of embodiments created by the program. Figure 13.2 shows two weighing machines created by the process for different user utilities or preference weightings for the objectives with rendering of the compound.

The A-Design program is one approach to automating the design process. Other programs with similar goals follow similar strategies, namely iteration toward improved solution, looping between synthesis and evaluation, and at least a reference to cognitive models. Included in this group are the FFREADA/GGREADA programs by Schmidt and Cagan (1995, 1998), which apply simulated annealing to a top-down functional hierarchy. Also included is the structural shape annealing program of Shea and Cagan (1997, 1999), which models structural topologies by means of shape grammars and uses simulated annealing to search the space generated by the grammar; the program has created a wide variety of structural trusses and domes, addressing a variety of design goals. The structural shape annealing work

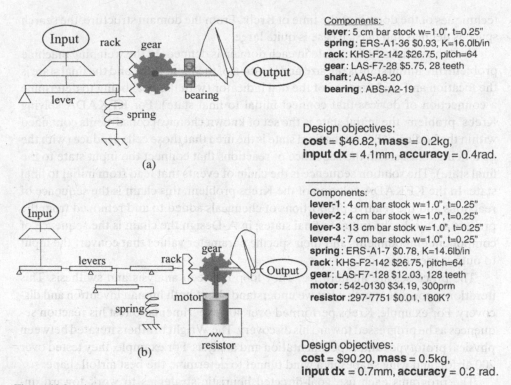

Components:
lever: 5 cm bar stock w=1.0", t=0.25"
spring: ERS-A1-36 $0.93, K=16.0lb/in
rack: KHS-F2-142 $26.75, pitch=64
gear: LAS-F7-28 $5.75, 28 teeth
shaft: AAS-A8-20
bearing: ABS-A2-19

Design objectives:
cost = $46.82, **mass** = 0.2kg,
input dx = 4.1mm, **accuracy** = 0.4rad.

Components:
lever-1: 4 cm bar stock w=1.0", t=0.25"
lever-2: 4 cm bar stock w=1.0", t=0.25"
lever-3: 13 cm bar stock w=1.0", t=0.25"
lever-4: 7 cm bar stock w=1.0", t=0.25"
spring: ERS-A1-7 $0.78, K=14.6lb/in
rack: KHS-F2-142 $26.75, pitch=64
gear: LAS-F7-128 $12.03, 128 teeth
motor: 542-0130 $34.19, 300prm
resistor: 297-7751 $0.01, 180K?

Design objectives:
cost = $90.20, **mass** = 0.5kg,
input dx = 0.7mm, **accuracy** = 0.2 rad.

Figure 13.2. Two different alternatives created by the A-Design process. Design (a) is found by an equal preference for four design objectives: minimize cost, minimize mass, minimize dial error, and minimize input displacement; design (b) is found by placing more importance on minimizing the input displacement.

calls a finite-element analysis code in each iteration to evaluate the effectiveness of the evolving design solutions (See Chapter 4 for a further discussion of this technique). In these works, simulated annealing is used as the search strategy based on a high-level analogy to human creative design, where random directions are chosen and pursued if seemingly beneficial or else reversed if seemingly inferior; Cagan and Kotovsky (1997) demonstrated the effectiveness of simulated annealing in models of human problem solving as applied to tavern problems that had been empirically explored by Kotovsky and Simon (1990).

MAPPING ACROSS PROGRAMS

As discussed throughout the presentation of this work, the five similarities across programs of design and discovery are (1) the definition of the problem through input and output states, (2) the iterative synthesis–analysis cycle, (3) the progressive evolution of solutions through iterations, (4) the large problem spaces upon which the programs search, and (5) the direct or indirect mapping of underlying cognitive activities based on problem solving and pattern recognition within the search strategies. In each of these programs the problem representation defines the domain structure. In A-Design this representation is the functional abstract description and specific embodiment and instantiation details; in the KEKADA solution of Krebs' problem the representation is the existing knowledge of chemistry and experimental techniques of the domain at the time of Krebs. From the domain structure, the search space is defined and in each case is quite large.

Next the initial and final state in each domain is defined. In the weighing machine problem, the initial state is the size and location of the input force and the final state is the location and performance of the dial indicator (with the goal being to determine a connection of devices that connect initial to final state). For KEKADA solving Krebs' problem, the initial state is the set of known chemical constituents contained within the bodies' cells and the final state is the urea that those cells produce (with the goal being to ascertain the sequence of reactions that connect the input state to the final state). The solution sequence is the chain of events that lead from initial to final state. In the KEKADA solution of the Krebs problem, this chain is the sequence of reactions and the relative proportions of chemicals added to and removed from the process as it links the initial to final states; in A-Design the chain is the sequence of component embodiments and their specific parameter values that convert the input to output specification.

The processes are all iterative. They loop between analysis and synthesis. This iteration is consistent with what we understand about both human invention and discovery. For example, Krebs performed over 400 experiments to test his reaction sequences as he progressed toward his discovery. The Wright brothers iterated between physical prototypes and experimentation and analysis. For example, they tested over 200 airfoil and wing shapes in a wind tunnel to determine the best airfoil shape.

The programs each use goal-directed heuristic strategies to work toward improved solutions. KEKADA uses heuristics to focus on improved solution concepts. A-Design uses heuristic strategies through agents and keeps alive several disparate alternatives based on a Pareto analysis (Campbell et al., 1999). Along the way the programs each "recognize" surprises in the solutions, altering their search patterns as a result; in KEKADA the recognition of ornithine's catalytic effect on urea production led to a new line of search, whereas in A-Design the use of a motor and resistor in parallel similarly led to a decidedly new solution to the subproblem of providing damping to the system; see Figure 13.2(b).

In addition to the use of KEKADA to discover the solution of Krebs' problem, it was also shown to recreate Faraday's invention of the generator. In this high-level application there appear many similarities to the approach taken by A-Design. Here experimental setups were configured to eventually discover the reason and means for magnet-generated electricity. A catalog of apparatus components was given to the program. The program chose the configuration of the experiments and instantiated parameter values, basing each decision on the outcome of previous experiments, until the right reaction sequence (or configuration) was found.

BACON, too, demonstrated the evolution of solutions through iterations, the iterative synthesis–analysis cycle, and the direct mapping of underlying cognitive activities based on problem solving and pattern recognition. Here the initial state is the data, for example Kepler's data, and the final state is the goal of determining a function that matches that data. The configuration sequence adds new multiplicative or divisor components to the function resulting in a final function that models the data; BACON's catalog consists of functional terms.

In examining the basics of the design synthesis (A-Design) versus scientific discovery (KEKADA and BACON) programs, we find that, although they were developed for very different purposes, they have strong similarities at the process level. Both types of programs search a large space in a goal-directed fashion, both use an initial and final state to define the overall characteristics of the problem, both use an appropriate representation of the problem to define the problem space, both define a catalog of components to be assembled into a chain representing the solution state, and both use a synthesize–evaluate loop to verify performance characteristics of the solution state and direct future decisions on changes to the current configuration.

CONCLUSIONS

Four conclusions emerge about the processes of discovery and design: The first is that both science and design are often intermingled with each other, and both require analysis and synthesis. So, invention often involves science and scientific discovery often involves design. This is true of both the processes (where in science experiments and their apparati have to be designed, and in design scientific principles have to be developed to support designs or design choices).

The second is that both consist of a broad array of activities that range from routine to the creative and revolutionary. Thus, science can range from routine paradigmatic science ("Does the speed of mental rotation decrease with age similarly to the decrease in the speed of memory search?") to the revolutionary discovery ("Does the constancy of the speed of light have implications for length, weight, and time as objects increase their velocity?"). Similarly, design can range from the routine ("Given the usefulness or popularity of the "GoodGrips" potato peeler and can opener, can a similar bottle opener be designed?") to the creative and revolutionary invention ("Can a germ-free environment be created in which to store food by heating and then sealing containers?")

The third and major conclusion we reach is that at a deep level, the cognitive and computational processes that accomplish both activities are virtually identical, namely pattern learning and recognition, processes of constructing adequate representations of some part of the world, and processes of intelligently searching through a vast and often ill-defined problem space that incorporates that representation.

The fourth conclusion is that the processes of design and discovery are structurally similar but differ in their goals and their knowledge bases. Both start from function and move to device or mechanism, but in design the goal is the creation of a device that accomplishes a desired function whereas in science the goal is the creation of a model of a mechanism that accomplishes an existing function. The implication of this and the previous conclusion is that computational models of the two domains operate with similar search and recognition strategies but act on different problem representations, thus allowing for fortuitous cross-fertilization of the algorithms.

Many ideas, methods, and findings from the field of scientific discovery have implications for or are involved in design, design automation, and design education, and vice versa. The similarities between the processes underlying these two superficially different areas of human (and more recently, machine) endeavor may exist for a very simple reason. They are both the product of human minds' trying to solve challenging and complex problems: The problem of discovering the nature of the world we inhabit and the problem of designing processes and artifacts that "improve" that world and, when operating correctly, make it even more inhabitable. Faraday was both a scientist and an inventor, and our conclusion is that what Faraday really was was a brilliant problem solver.

REFERENCES

Adelson, B. (1989). "Cognitive research: uncovering how designers design; cognitive modelling: explaining and predicting how designers design," *Research in Engineering Design*, 1(1):35–42.

Adelson, B., Gentner, D., Thagard, P., Holyoak, K., Burstein M., and Hammond, K. (1988). "The role of analogy in a theory of problem-solving." In *Proceedings of the Eleventh Annual Meeting of the Cognitive Science Society*, Erlbaum.

Anderson, J. R. (1983). *The Architecture of Cognition*, Harvard University Press, Cambridge.

Cagan, J. and Kotovsky, K. (1997). "Simulated annealing and the generation of the objective function: a model of learning during problem solving," *Computational Intelligence*, 13(4):534–581.

Campbell, M. (2000). "*The A-Design invention machine: a means of automating and investigating conceptual design*," Ph.D. Dissertation, Carnegie Mellon University, Pittsburgh, PA.

Campbell, M., Cagan, J., and Kotovsky, K. (1999). "A-Design: an agent-based approach to conceptual design in a dynamic environment," *Research in Engineering Design*, 11: 172–192.

Campbell, M., Cagan, J., and Kotovsky, K. (2000). "Agent-based synthesis of electromechanical design configurations," *ASME Journal of Mechanical Design*, 122(1): 61–69.

Culick, F. E. C., and Jex, H. R. (1987). "Aerodynamics, stability, and control of the 1903 Wright Flyer." In *The Wright Flyer, An Engineering Perspective*, H. S. Wolko (ed.), Smithsonian Institution, Press, Washington, DC.

Dunbar, K. (1993). "Concept discovery in a scientific domain," *Cognitive Science*, 17:397–434.

Falkenhainer, B., Forbus, K., and Gentner, D. (1989). "The structure-mapping engine: algorithm and examples," *Artificial Intelligence*, 41:1–63.

Finke, R. A., Ward, T. B., and Smith, S. M. (1992). *Creative Cognition*. MIT Press, Cambridge, MA.

Forbus, K. D. (1988). "Qualitative physics: past, present, and future." In *Exploring Artificial Intelligence*, H. Shrobe (ed.), Morgan Kaufmann, San Mateo, CA, pp. 239–296.

Forbus, K. D., Gentner, D., and Law, K. (1994). "Mac/Fac: a model of similarity-based retrieval," *Cognitive Science*, 19:141–205.

Gentner, D. (1983). "Structure mapping: a theoretical framework for analogy," *Cognitive Science*, 7:155–170.

Giere, R. N. (1988). *Explaining Science: A Cognitive Approach*. University of Chicago Press, Chicago.

Grasshoff, G. and May, M. (1995). "Methodische analyse wissenschaftlichen entdeckens," *Kognitionswissenschaft*, 5:51–67.

Hayes J. R. and Simon, H. A. (1977). "Psychological differences among problem isomorphs." In *Cognitive Theory*, W. J. Castellan, N. B. Pisoni, and G. R. Potts (eds.), Erlbaum, Englewood Cliffs, N. J., Vol. 2, pp. 21–41.

Holland, J. H., Holyoak, K. J., Nisbett, R. E., and Thagard, P. R. (1986). *Induction: Processes of Inference, Learning, and Discovery*. MIT Press, Cambridge, MA.

Holmes, F. L. (1991). *Hans Krebs: The formation of a scientific life, 1900–1933*. Oxford University Press, New York.

Jakab, P. L. (1990). *Visions of a Flying Machine, The Wright Brothers and the Process of Invention*. Smithsonian Institution Press, Washington, DC.

Klahr, D. and Dunbar, K. (1988). "Dual space search during scientific reasoning," *Cognitive Sceince*, 12:1–55.

Kotovsky, K., Hayes, J. R., and Simon, H. A. (1985). "Why are some problems hard: evidence from Tower of Hanoi," *Cognitive Psychology*, 17:248–294.

Kotovsky, K. and Simon, H. A. (1990). "What makes some problems really hard: explorations in the problem space of difficulty," *Cognitive Psychology*, **22**:143–183. Reprinted in Simon, H.A. (1989). *Models of Thought, Volume Two.* Yale University Press, New Haven, CT.

Kulkarni, D. and Simon, H. A. (1990). "Experimentation in machine discovery." In *Computational Models of Scientific Discovery and Theory Formation*, J. Shrager and P. Langley (eds.), Morgan Kaufmann, San Mateo, CA.

Laird, J. E., Newell, A., and Rosenbloom, P. S. (1987). "Soar: an architecture for general intelligence," *Artificial Intelligence*, **33**:1–64.

Langley, P., Simon, H. A., Bradshaw, G. L., and Zytkow, J. M. (1987). *Scientific Discovery*, MIT Press, Cambridge, MA.

Lippincott, H. H. (1987). "Propulsion systems of the Wright Brothers." In *The Wright Flyer, An Engineering Perspective*, H. S. Wolko (ed.), Smithsonian Institution Press, Washington, DC.

Newell, A. and Simon, H. A. (1972). *Human Problem Solving.* Prentice-Hall, Englewood Cliffs, NJ.

Nickles, T., Ed. (1978). *Scientific Discovery, Logic and Rationality.* Reidel, Boston, MA.

Pahl, G. and Beitz, W. (1988). *Engineering Design – A Systematic Approach*, Springer-Verlag, New York.

Paynter, H. M. (1961). *Analysis and Design of Engineering Systems.* MIT Press, Cambridge, MA.

Popper, K. R. (1959). *The Logic of Scientific Discovery.* Kritchmion, London.

Schmidt, L. C. and Cagan, J. (1995). "Recursive annealing: a computational model for machine design," *Research in Engineering Design*, **7**:102–125.

Schmidt, L. C. and Cagan, J. (1998). "Optimal configuration design: an integrated approach using grammars," *Journal of Mechanical Design*, **120**:2–9.

Shea, K. and Cagan, J. (1997). "Innovative dome design: applying geodesic patterns with shape annealing," *Artificial Intelligence in Engineering Design, Analysis, and Manufacturing*, **11**:379–394.

Shea, K. and Cagan, J. (1999). "The design of novel roof trusses with shape annealing: assessing the ability of a computational method in aiding structural designers with varying design intent," *Design Studies*, **20**:3–23.

Shen, W. M. (1993). "Discovery as autonomous learning from the environment," *Machine Learning*, **12**:143–165.

Shrager, J. and Langley, P., Eds. (1990). *Computational Models of Scientific Discovery and Theory Formation.* Morgan Kaufmann, San Mateo, CA.

Simon, H. A. (1993). "A very early expert system," *Annals of the History of Computing*, **15**(3):63–68.

Simon, H. A. and Kotovsky, K. (1963). "Human acquisition of concepts for sequential patterns," *Psychological Review*, **70**:534–546; reprinted in Simon, H. A. (1979). *Models of Thought.* Yale University Press, New Haven.

Ullman, D. G., Dietterich, T. G., and Stauffer, L. A. (1988). "A model of the mechanical design process based on empirical data," *Artificial Intelligence in Engineering Design, Analysis, and Manufacturing*, **2**(1):33–52.

Ulrich, K. and Seering, W. (1989). "Synthesis of schematic descriptions in mechanical design," *Research in Engineering Design*, **1**:3–18.

Valdes-Perez, R. E. (1995). "Some recent human/computer discoveries in science and what accounts for them," *Artificial Intelligence*, **16**(3):37–44.

Voland, G. (1999). *Engineering by Design.* Addison-Wesley, Reading, MA.

Weisberg, R. W. (1993). *Creativity: Beyond the Myth of Genius.* Freeman, New York.

Welch, R. V. and Dixon, J. (1994). "Guiding conceptual design through behavioral reasoning," *Research in Engineering Design*, **6**:169–188.

Wolko, H. S., Ed. (1987). *The Wright Flyer, An Engineering Perspective.* Smithsonian Institution Press, Washington, DC.

Index